Modern Astrophysics

Modern Astrophysics

Edited by **Audria Baldwin**

Callisto Reference
New York

Published by Callisto Reference,
106 Park Avenue, Suite 200,
New York, NY 10016, USA
www.callistoreference.com

Modern Astrophysics
Edited by Audria Baldwin

International Standard Book Number: 978-1-63239-736-2 (Hardback)

Printed in the United States of America.

Contents

Preface

Astrophysics is the discipline that studies all the important aspects related to heavenly bodies, galaxies and universe. These properties are examined by applying the combination of physics and chemistry. Broadly defined characteristics that are mainly studied under this area of study include chemical composition, luminosity, density and temperature. It also covers properties related to dark matter and black holes. Observational astrophysics is concerned with recording data related to quantitative implications whereas theoretical astrophysics measures data related to physical models. The aim of this book is to present researches that have transformed this discipline and aided its advancement. The case studies included herein will serve as an excellent guide to develop a comprehensive understanding of astrophysics. It will provide comprehensive knowledge to the researchers, students and experts associated with modern astrophysics.

Significant researches are present in this book. Intensive efforts have been employed by authors to make this book an outstanding discourse. This book contains the enlightening chapters which have been written on the basis of significant researches done by the experts.

Finally, I would also like to thank all the members involved in this book for being a team and meeting all the deadlines for the submission of their respective works. I would also like to thank my friends and family for being supportive in my efforts.

Editor

1

Studies on Longer Wavelength Type II Radio Bursts Associated with Flares and CMEs during the Rise and Decay Phase of 23rd Solar Cycle

V. Vasanth and S. Umapathy

School of Physics, Madurai Kamaraj University, Madurai 625 021, India

Correspondence should be addressed to V. Vasanth; vasanth_velu2007@yahoo.co.in

Academic Editor: Luciano Nicastro

A statistical study on the properties of CMEs and flares associated with DH-type II bursts in the 23rd solar cycle during the period 1997–2008 is carried out. A sample of 229 events from our recent work is used for the present study (Vasanth and Umapathy, 2013). The collected events are divided into two groups as (i) solar cycle rise phase events and (ii) solar cycle decay phase events. The properties of CMEs in the two groups were compared and the results are presented. It is noted that there is no difference in the properties of type II burst like start frequency and end frequency between the solar cycle rise phase events and decay phase events. The mean CME speed of solar cycle decay phase events (1373 km s^{-1}) is slightly higher than the solar cycle rise phase events (1058 km s^{-1}). The mean CME acceleration of solar cycle decay phase events (−15.18 m s^{-2}) is found to be higher than that of the solar cycle rise phase events (−1.32 m s^{-2}). There exists good correlation between (i) CME speed and width and (ii) CME speed and acceleration for solar cycle decay phase events ($R = 0.79$, $R = -0.80$) compared to solar cycle rise phase events ($R = 0.60$, $R = -0.57$). These results indicate that the type II bursts parameters do not depend upon the time of appearance in the solar cycle.

1. Introduction

Coronal mass ejections (CMEs) are the large eruption of magnetized plasma from sun into heliosphere and are important cause for geomagnetic storms if they are directed towards earth. The shocks driven by the CMEs accelerate electrons and produce type II bursts in the corona and interplanetary medium (IP) [1]. Type II bursts are the longest known signatures of shock wave [2]. In the dynamic spectrum they are observed as slowly drifting (from high to low frequency) emission bands. The first identification of type II bursts was made by Payne-Scott et al. [3] and later Wild and McCready [2] classified them as type II bursts to differentiate them from fast drifting type III bursts. The first observation of type II bursts in the IP medium was made by IMP-6 [4] and Voyager [5] spacecraft missions. Many IP type II bursts are observed by ISEE-3 spacecraft in the frequency range of 2 MHz to 30 kHz [6, 7]. The data on DH-type II bursts are provided by radio and plasma wave (WAVES) experiment on board wind spacecraft launched in 1994 [8]. The wind spacecraft observes the type II/IV radio bursts in the frequency range between 14 MHz and 20 kHz by RAD1 and RAD2 due to ionospheric cut-off frequency [9–11]. The frequency range in the decameter hectometric (DH) wavelength domain corresponds to the heights of 2–10 $R_{\odot}$ and bridges the gap between the metric type II bursts observed by ground based radio observatories and the kilometric type II bursts observed by space-borne radio instruments.

The CMEs associated with the DH-type II bursts are found to be fast and wide, and they show strong deceleration in the LASCO FOV [12]. It is now well accepted that the type II bursts in DH and Km range are driven by CMEs, while the origin of coronal shocks is still under debate [11, 13–15]. Lara et al. [16] studied the CMEs associated with type II bursts (m- and DH-type II bursts) wavelength range and found that the CMEs associated with metric type II bursts are more energetic than regular CMEs but less energetic than the CMEs associated with DH-type II bursts. The characteristics

TABLE 1: Properties of DH-type II bursts, flares, and CMEs in solar cycle rise phase and solar cycle decay phase events.

Serial number	Properties	Solar cycle rise phase events	Solar cycle decay phase events	t-test P–value (%)
		DH-type II bursts		
1	Start frequency (MHz)	10.65 **(0.38)**	10.12 **(0.47)**	37.71
2	End frequency (MHz)	2.48 **(0.26)**	1.78 **(0.25)**	5.55
3	Duration (min)	441 **(67.92)**	369 **(61.09)**	43.04
4	Type II speed (km s^{-1})	1135 **(71.50)**	1260 **(79.71)**	25.51
		Flares		
5	Flare duration (min)	53 **(4.27)**	59 **(4.35)**	33.64
6	Flare rise time (min)	26 **(1.89)**	31 **(2.47)**	8.84
7	Flare decay time (min)	27 **(2.75)**	28 **(2.67)**	84.99
		CMEs		
8	CME speed (km s^{-1})	1058 **(47.53)**	1377 **(58.02)**	$\ll 1$
9	Width (deg)	260 **(10.42)**	276 **(10.59)**	26.59
10	Acceleration (m s^{-2})	−1.32 **(2.29)**	−15.18 **(4.45)**	2.01

*The mean **(standard error)** values of the properties are given in this table.

of CMEs associated with DH-type II bursts are studied by several authors [11, 12, 17–19]. The DH-type II bursts are found to be associated with strongly energetic and wide CMEs, and these CMEs are good indicators of geomagnetic storms at earth if they appear on the front side of the solar disk [12].

In the present work, we aimed to study the properties of CMEs and flares associated with DH-type II bursts in the rise and decay phase of 23rd solar cycle during the period 1997–2008. The method of data analysis is described in Section 2, the results are presented in Section 3, and the conclusions are discussed in Section 4.

2. Data Analysis

A set of 344 DH-type II bursts associated with flares and CMEs listed in the Wind/WAVES type II catalog (http://cdaw.gsfc.nasa.gov/CME_list/radio/waves_type2.html) in the 23rd solar cycle during the period 1997–2008 are used in the present analysis. The characteristics of associated CME data such as speed, width, and acceleration are obtained from the LASCO CME catalog (http://cdaw.gsfc.nasa.gov/CME_list) maintained by the CDAW data center [20, 21]. CME can be directly observed from 2 to 32 $R_\odot$ from the sun using the SOHO/LASCO C2 and C3 coronagraphs [22]. The properties of CMEs listed in the LASCO CME catalog are used for the analysis without correcting the projection effects. We have used the following selection criteria to select the events for our further analysis. They are as follows.

(i) The source location of the events should be clear.

(ii) The flare importance should be clear.

(iii) Backside events should be excluded.

Using the above selection criteria, 232 events are selected out of 344 events in the type II catalog. Further out of 232 events, 3 complex events are removed from the present study therefore the total number of events reduces to 229 events. Each event contains DH-type II bursts and its associated flares and CMEs. The 23rd solar cycle covering the period 1997–2008 has double peaked solar maximum. The first peak occurred during April, 2000 and the second peak during April, 2002. Therefore, we classified our events into groups based on the behavior of solar cycle as

(i) solar cycle rise phase events (DH-type II bursts associated with flares and CMEs that are observed during the period April 1997–April 2002);

(ii) solar cycle decay phase events (DH-type II bursts associated flares and CMEs that are observed during the period May 2002–April 2008).

Out of 229 events, 127 events occurred during the solar cycle rise phase and the remaining 102 events occurred in the solar cycle decay phase of the 23rd solar cycle during the period 1997–2008. The properties of type II bursts, flares, and CMEs in the solar cycle rise phase and in solar cycle decay phase were compared and the results are discussed in the next section.

3. Results and Discussions

The statistical properties of DH-type II bursts, flares, and CMEs in the solar cycle rise and decay phase are investigated and the results are summarized in Table 1. In this table, column 1 specifies the characteristic properties; column 2 defines the statistical properties. The statistical properties of solar cycle rise phase events and those for solar cycle decay phase events are given in column 3 and column 4, respectively. The t-test P value specifies the statistical significance of the mean values between the two groups and they are given in column 5.

3.1. Properties of Type II Bursts. The distributions of start frequency of the DH-type II bursts in the solar cycle rise

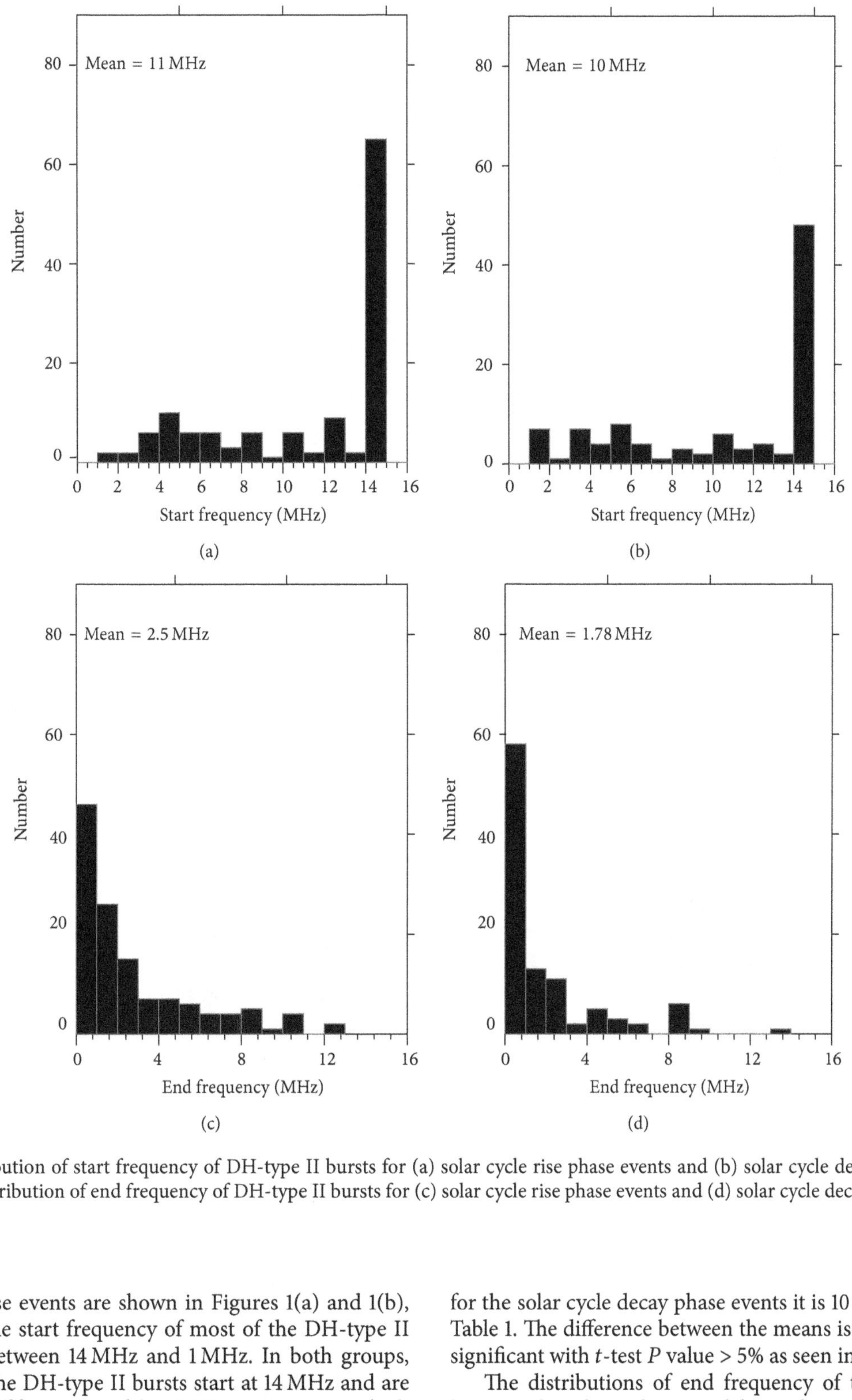

FIGURE 1: Distribution of start frequency of DH-type II bursts for (a) solar cycle rise phase events and (b) solar cycle decay phase events. Similarly the distribution of end frequency of DH-type II bursts for (c) solar cycle rise phase events and (d) solar cycle decay phase events.

and decay phase events are shown in Figures 1(a) and 1(b), respectively. The start frequency of most of the DH-type II bursts varies between 14 MHz and 1 MHz. In both groups, about 50% of the DH-type II bursts start at 14 MHz and are the upper cut-off frequency of the WAVES instrument, which indicates that there is a chance for some of the events to be continuation of m-type II bursts. DH-type II bursts are radio signatures of CME driven shocks [12]. The start frequency indicates the height at which the radio emission from the shocks becomes visible [23]. The mean start frequency of type II bursts in solar cycle rise phase events is 11 MHz, whereas for the solar cycle decay phase events it is 10 MHz as seen in Table 1. The difference between the means is not statistically significant with t-test P value > 5% as seen in Table 1.

The distributions of end frequency of the DH-type II bursts in the solar cycle rise and decay phase events are shown in Figures 1(c) and 1(d), respectively. The end frequency varies between 12 MHz and 20 kHz. The end frequency is indicative of CME energies [23, 24]. The shocks produced by fast and wide CMEs travel far into the interplanetary medium. In solar cycle rise phase 50% of the type II events and in solar cycle decay phase 61% of the type II events end

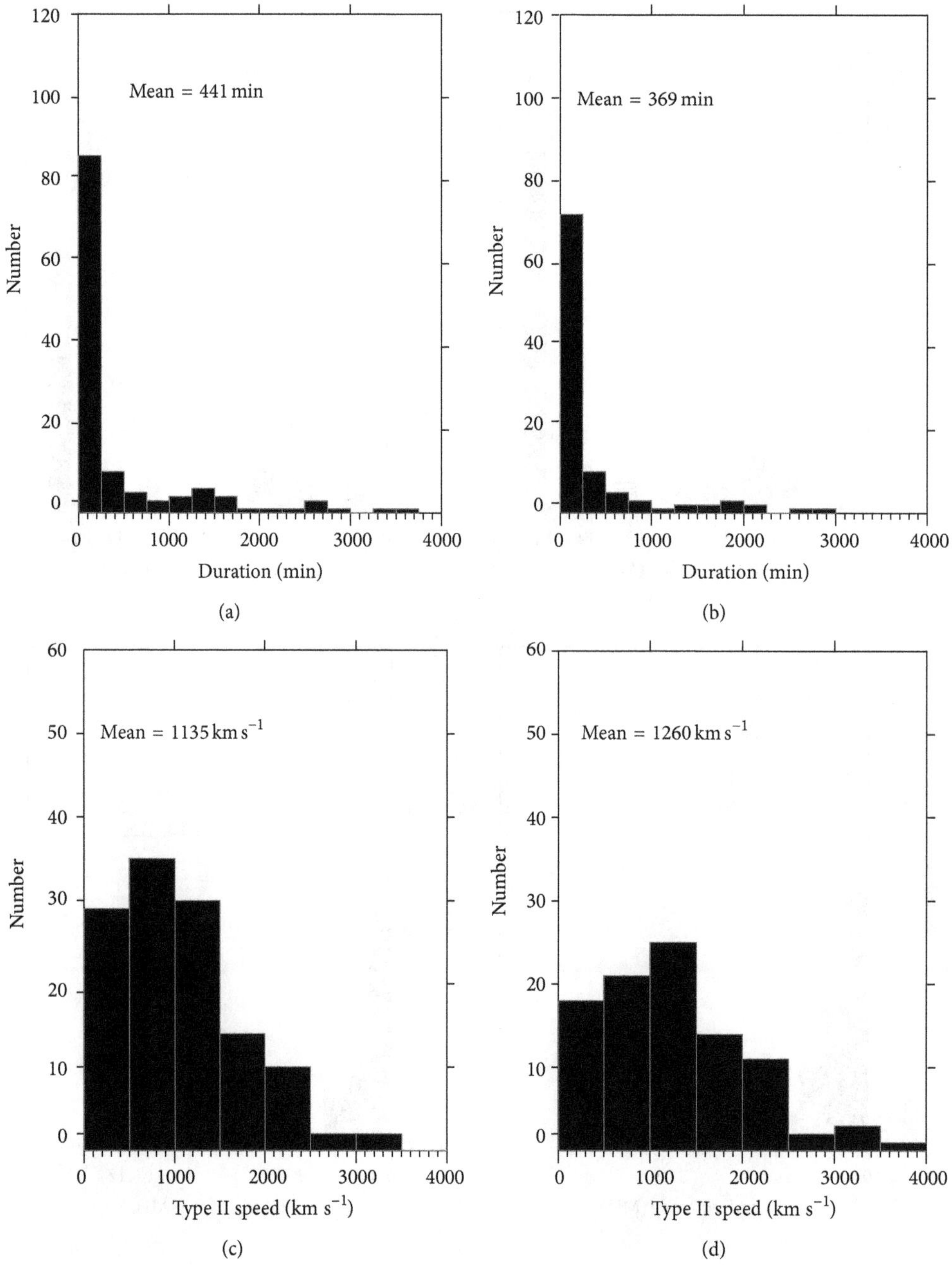

FIGURE 2: Distributions of duration of DH-type II bursts for (a) solar cycle rise phase events and (b) solar cycle decay phase events and shock speed of DH-type II bursts for (c) solar cycle rise phase events and (d) solar cycle decay phase events.

below 1 MHz. Majority of the type II events in both groups have end frequency around/below 1 MHz which indicates that they continue into km range (shown by first bin of Figures 1(c) and 1(d), resp.). The mean end frequency of solar cycle rise phase events is 3 MHz and that of the solar cycle decay phase events is 2 MHz as seen in Table 1. The difference between the means is less statistically significant with t-test P value = 6% as seen in Table 1.

Figures 2(a) and 2(b) show the distributions of duration of the DH-type II bursts in solar cycle rise and in decay phase events. The duration of DH-type II bursts varies between few minutes to long time as 4000 minutes. In solar cycle rise phase 68% and in solar cycle decay phase 70% of the DH-type II bursts duration lie within 200 min. The mean duration of DH-type II bursts in solar cycle rise phase events is 441 min and that of the solar cycle decay phase events is 369 min as seen in Table 1. The difference between the means is not statistically significant with t-test P value > 5% as seen in Table 1.

The shock speed of the DH-type II bursts was estimated by applying Leblanc electron density model [18, 25]. The model is valid for $R > 1.2\,R_{\odot}$ and not for very low heights.

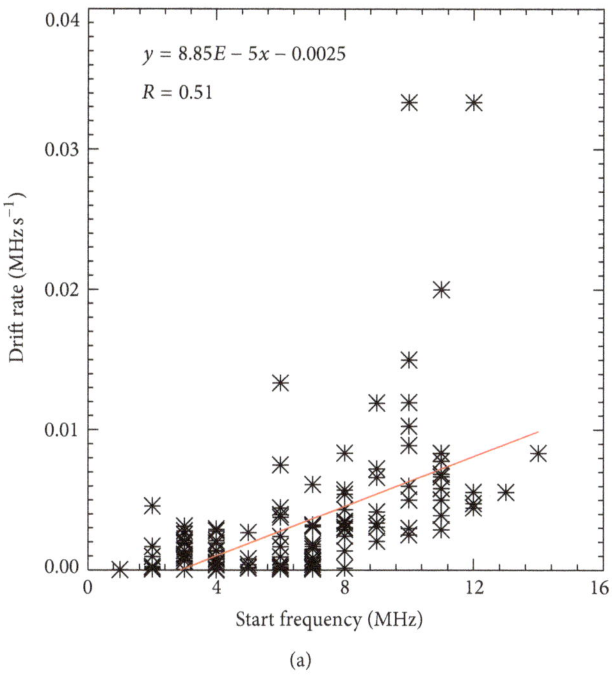

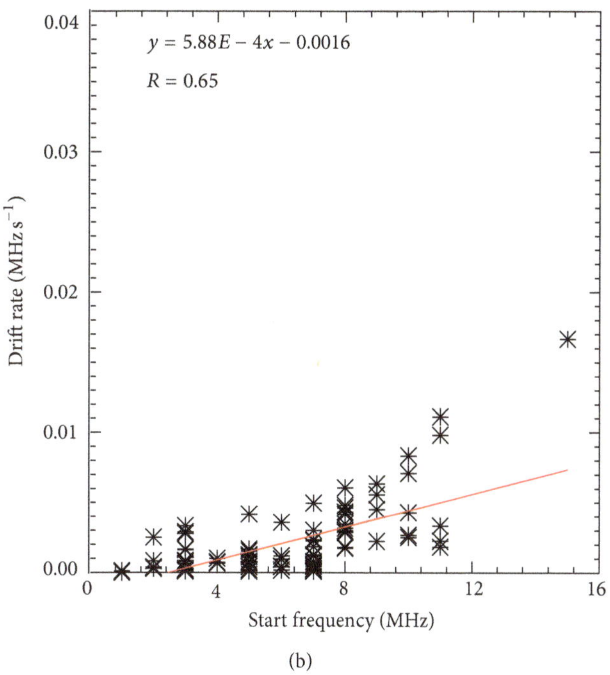

FIGURE 3: Correlation between start frequency and drift rate for (a) solar cycle rise phase events and (b) solar cycle decay phase events.

For each measured frequency, their corresponding electron density number and radial distance are obtained using the Leblanc electron density model from which the shock speed of the DH-type II bursts is estimated. The distributions of shock speeds inferred from the DH-type II burst drift rates for the solar cycle rise and decay phase events are presented in Figures 2(c) and 2(d), respectively. The shock speed varies between few hundred km s^{-1} and 4000 km s^{-1}. The mean shock speed of DH-type II bursts in solar cycle rise phase events is 1135 km s^{-1} and that of the solar cycle decay phase events is 1260 km s^{-1} as seen in Table 1. The difference between the means is not statistically significant with t-test P value > 5% as seen in Table 1.

The correlations between mean starting frequency of DH-type II bursts and drift rate for solar cycle rise and decay phase events are shown in Figures 3(a) and 3(b), respectively. Figure 3 shows that there might be dependence between the mean starting frequency and drift rate for both groups of events. The correlation is better for solar cycle decay phase events as compared to solar cycle rise phase events ($R = 0.65$ and $R = 0.51$, resp.). It is clear from the figure that the drift rate increases with start frequency [26].

3.2. Properties of Associated Flares. The distributions of flare rise time associated with DH-type II bursts in the solar cycle rise and decay phase events are shown in Figures 4(a) and 4(b), respectively. The flare rise time (the time difference between flare start and flare peak) helps in establishing the relationship between flares and type II bursts [11, 27], since type II burst starting between flare start and flare peak has a chance to be driven by flare shocks. Among solar cycle rise phase events 54% (69/127) of the flares rise within 20 min and the remaining 46% (58/127) of the events have longer flare rise time, while among solar cycle decay phase events only 39% (40/102) of the flares rise within 20 min and the remaining 61% (62/102) have longer flare rise time. The mean flare rise time in solar cycle rise phase events is 26 min and that of the solar cycle decay phase events is 31 min as seen in Table 1. The difference between the means is not statistically significant with t-test P value > 5% as seen in Table 1.

The distributions of flare decay time associated with DH-type II bursts in the solar cycle rise and decay phase events are shown in Figures 4(c) and 4(d), respectively. The flare decay time (the time difference between flare peak and flare end) varies between few minutes and 200 min. Among solar cycle rise phase events 55% (70/127) and among solar cycle decay phase events 56% (57/102) of the flares decay within 20 min. The mean flare decay time in solar cycle rise phase events is 27 min and that of the solar cycle decay phase events is 28 min. The difference between the means is less statistically significant with t-test P value = 9% as seen in Table 1.

Figures 5(a) and 5(b) show the distribution of flare duration associated with DH-type II bursts in the solar cycle rise and decay phase events. Among solar cycle rise phase events, 61% (78/127) have duration within 50 min, while among solar cycle decay phase events, only 55% (56/102) of the events have duration within 50 min. The mean flare duration in solar cycle rise phase events is 53 min and that of the solar cycle decay phase events is 59 min. The longer flare duration indicates the enhanced action of Lorentz force and the associated CMEs may have higher speed and acceleration

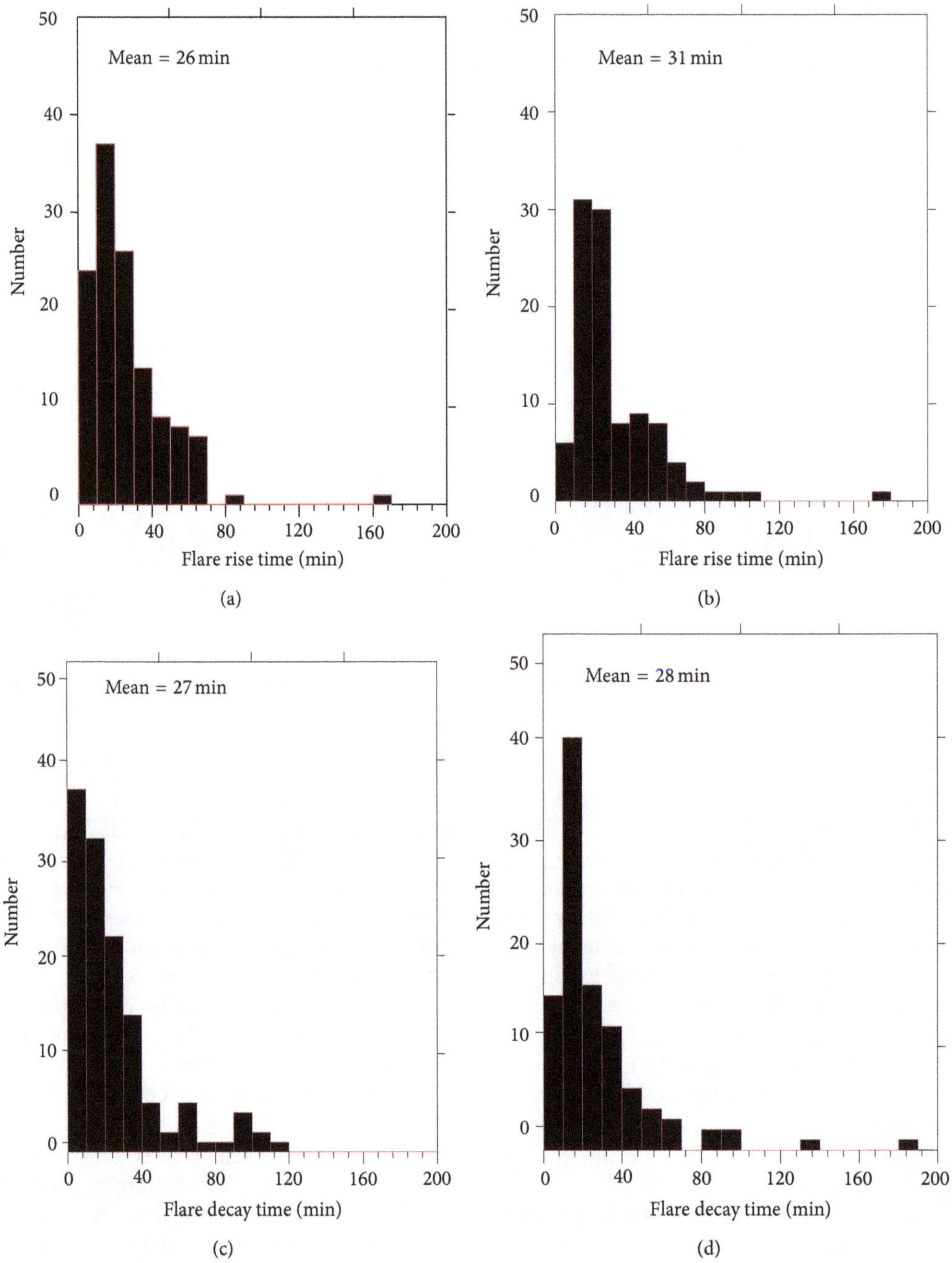

FIGURE 4: Distributions of flare rise time for (a) solar cycle rise phase events and (b) solar cycle decay phase events and flare decay time for (c) solar cycle rise phase events and (d) solar cycle decay phase events.

[28]. The difference between the means is not statistically significant with t-test P value > 5% as seen in Table 1.

The distributions of flare classes associated with CMEs and DH-type II bursts for the solar cycle rise and decay phase events are shown in Figures 5(c) and 5(d), respectively. The X-ray flare class is classified as A, B, C, M, and X depending on their intensities, where ($A = 10^{-8}$, $B = 10^{-7}$, $C = 10^{-6}$, $M = 10^{-5}$, and $X = 10^{-4}\ Wm^{-2}$) the value after the flare class is the multiplication factor. For example, M3.4 has a flux of $3.4 \times 10^{-5}\ Wm^{-2}$. In solar cycle rise phase events, 25% (32/127) are X-class flares, 52% (66/127) are M-class flares, 19% (24/127) are C-class flares, and 4% (5/127) are B-class flares, while in solar cycle decay phase events 30% (31/102) are X-class flares, 47% (48/102) are M-class flares, 22% (22/102) are C-class flares, and 1% (1/102) are B-class flares. The solar cycle decay phase events are associated with more energetic X-class flares than solar cycle rise phase events. We expect that the stronger flares will be associated with fast CMEs [29, 30].

The distributions of source location of flares associated with DH-type II bursts for the solar cycle rise phase and decay

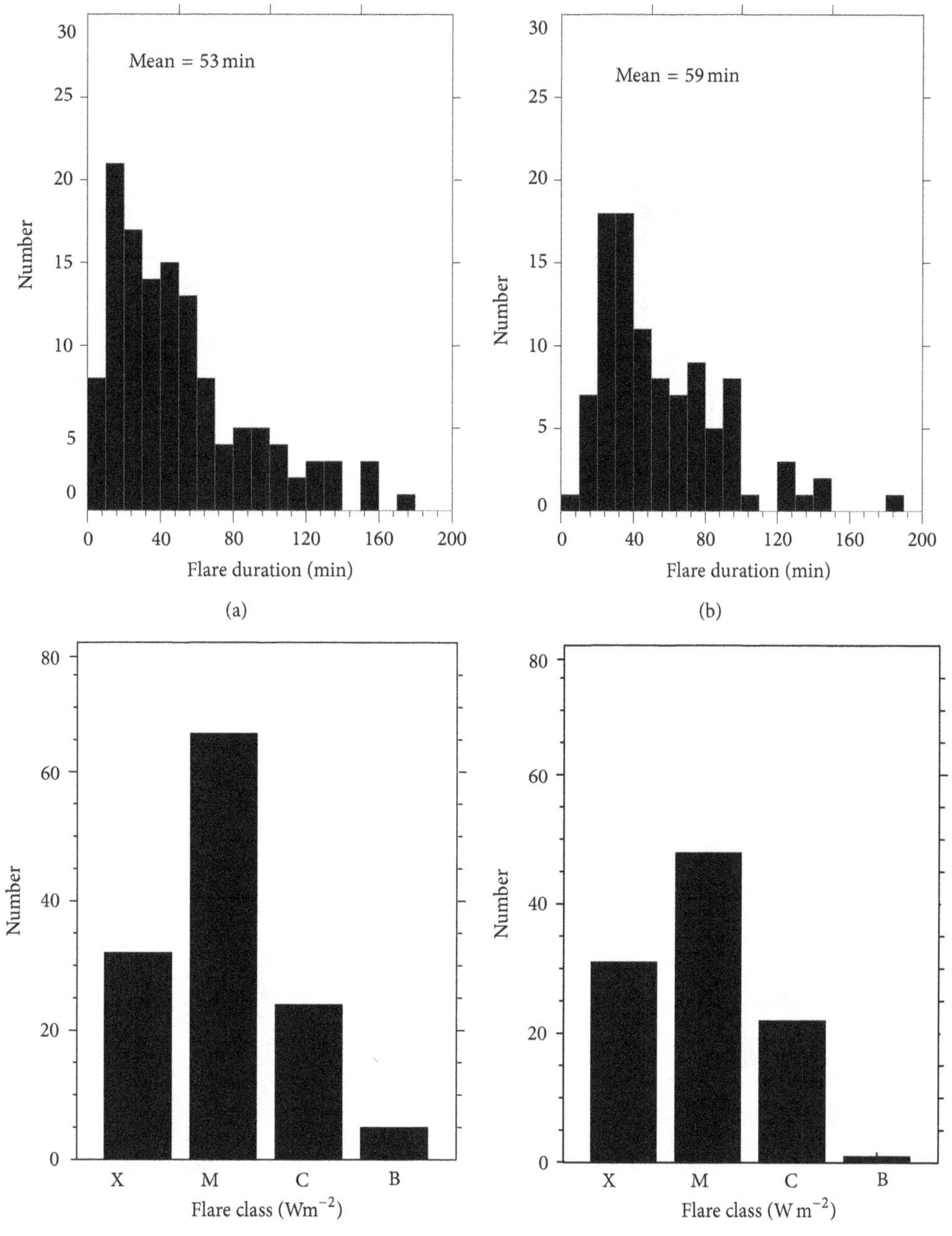

FIGURE 5: Distributions of flare duration for (a) rise phase events and (b) decay phase events and flare classes for (c) rise phase events and (d) decay phase events.

phase events are shown in Figures 6(a) and 6(b), respectively. Among solar cycle rise phase events, 64% (81/127) originate in western hemisphere and the remaining 36% (46/127) originate in eastern hemisphere. Similar result was found for solar cycle decay phase events; 60% (61/102) originate in western hemisphere and the remaining 40% (41/102) originate in eastern hemisphere. It is clear that majority of the events 62% (142/229) in both cases originate in the western hemisphere.

3.3. Properties of Associated CMEs. The properties of CMEs (like speed, width, acceleration, etc.) listed in the LASCO CME catalog are used in the present study. All the parameters measured in the plane of the sky will suffer from projection effects and no attempt has been made to correct the projection effects. The properties of CMEs in solar cycle rise and decay phase events are compared and the results are discussed.

The distributions of CME speed associated with flares and DH-type II bursts for the solar cycle rise and decay phase events are shown in Figures 6(c) and 6(d), respectively. The CME speeds are measured from the height-time measurements projected in the plane of the sky. The speed of CMEs varies between few tens of km s^{-1} and 3000 km s^{-1}. We found

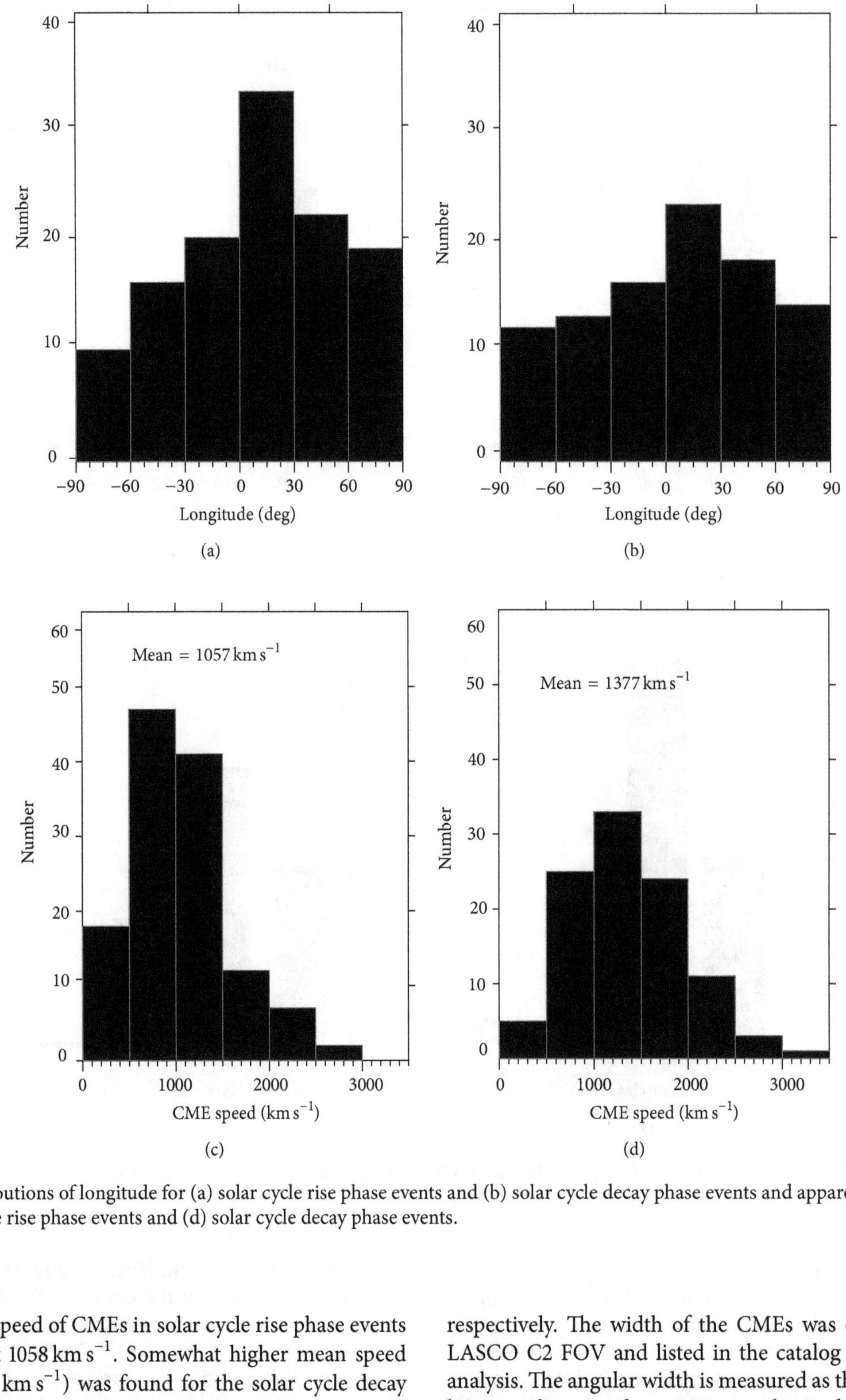

FIGURE 6: Distributions of longitude for (a) solar cycle rise phase events and (b) solar cycle decay phase events and apparent speed of CMEs for (c) solar cycle rise phase events and (d) solar cycle decay phase events.

that the mean speed of CMEs in solar cycle rise phase events amounts about 1058 km s^{-1}. Somewhat higher mean speed of CMEs (1373 km s^{-1}) was found for the solar cycle decay phase events. Therefore, the solar cycle decay phase events are found to have higher CME speed compared to solar cycle rise phase events. The difference between the means is statistically significant with t-test P value < 1% as seen in Table 1.

The distributions of projected width of CMEs associated with flares and DH-type II bursts for the solar cycle rise and decay phase events are shown in Figures 7(a) and 7(b), respectively. The width of the CMEs was estimated in the LASCO C2 FOV and listed in the catalog are used for the analysis. The angular width is measured as the angular extent between the two edge position angles in the sky plane. The width of CMEs varies between 30° and 360°. We did not find significant difference in the mean width of CMEs in solar cycle rise phase events (260°) and in mean width of CMEs in solar cycle decay phase events (275°). In solar cycle rise phase events, 54% (70/130) are halo CMEs and in solar cycle decay phase events 58% (59/102) are halo CMEs. The difference

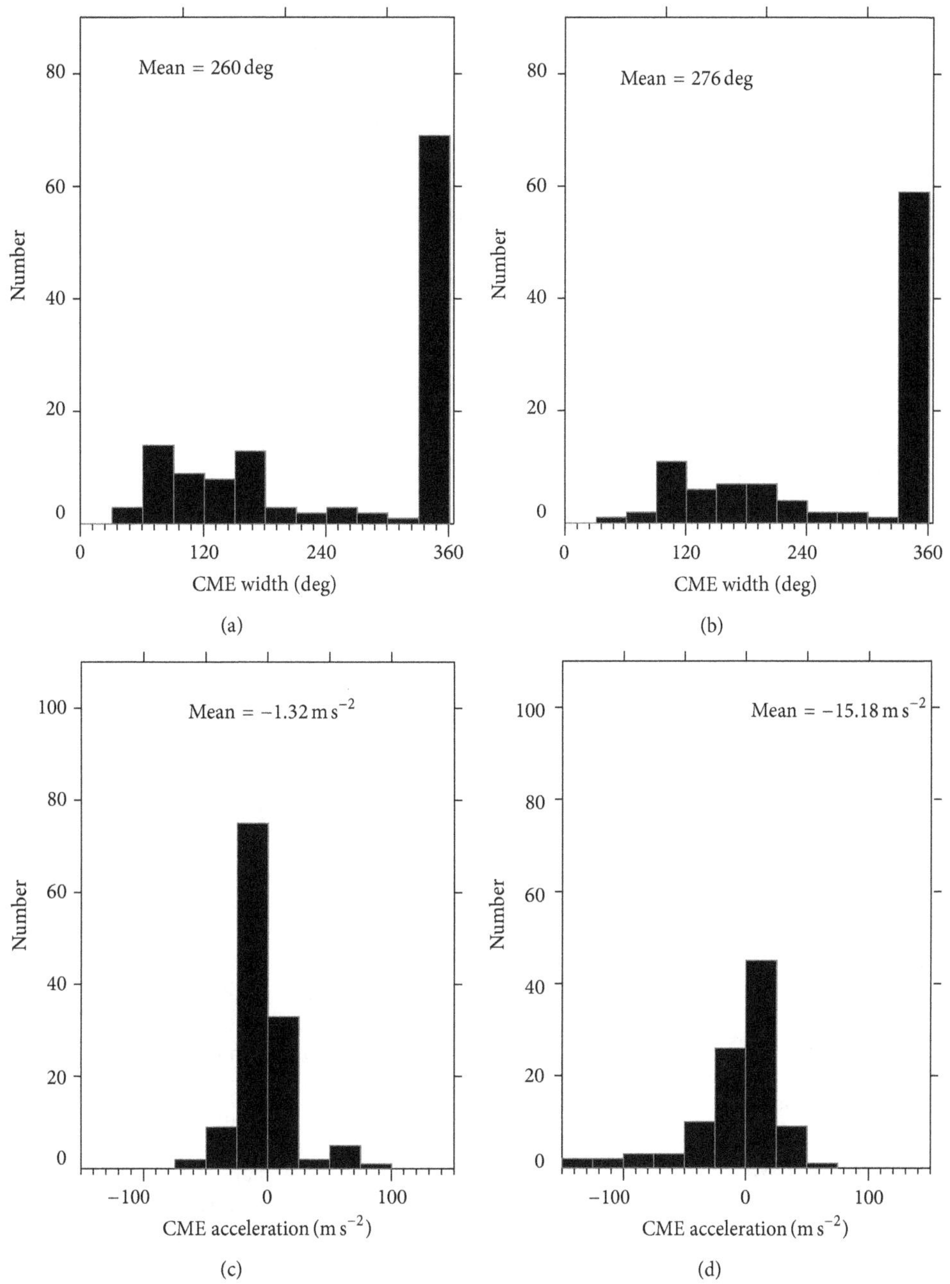

FIGURE 7: Distributions of apparent width of CMEs for (a) solar cycle rise phase events and (b) solar cycle decay phase events and acceleration of CMEs for (c) solar cycle rise phase events and (d) solar cycle decay phase events.

between the means is not statistically significant with t-test P value > 5% as seen in Table 1.

The distributions of acceleration of CMEs associated with flares and DH-type II bursts in the solar cycle rise and decay phase events are shown in Figures 7(c) and 7(d), respectively. The CMEs with data point ≥4 are used for the further analysis, so that the acceleration is reliable. Therefore, the acceleration of CMEs is estimated in the full LASCO FOV. The CMEs will accelerate/decelerate depending on their interaction with the solar wind in the LASCO FOV [19]. The fast CMEs decelerate, while the slow CMEs accelerate in the LASCO FOV [21]. The acceleration and deceleration of CMEs depend on the Lorentz force, drag force, and gravity force. The Lorentz force acts longer in accelerating CMEs and shorter in decelerating CMEs; thereby the drag force dominates the dynamics of CMEs with decrease in Lorentz and gravity force and the CME decelerates in the entire LASCO FOV [18, 19]. In Figure 7, the acceleration of CMEs in the rise phase events ranges between $-80\,\mathrm{m\,s^{-2}}$ and $100\,\mathrm{m\,s^{-2}}$, while in the solar cycle decay phase events it ranges between $-180\,\mathrm{m\,s^{-2}}$ and $60\,\mathrm{m\,s^{-2}}$. This shows that the faster CMEs in the solar cycle decay phase decelerate more than those in the solar cycle

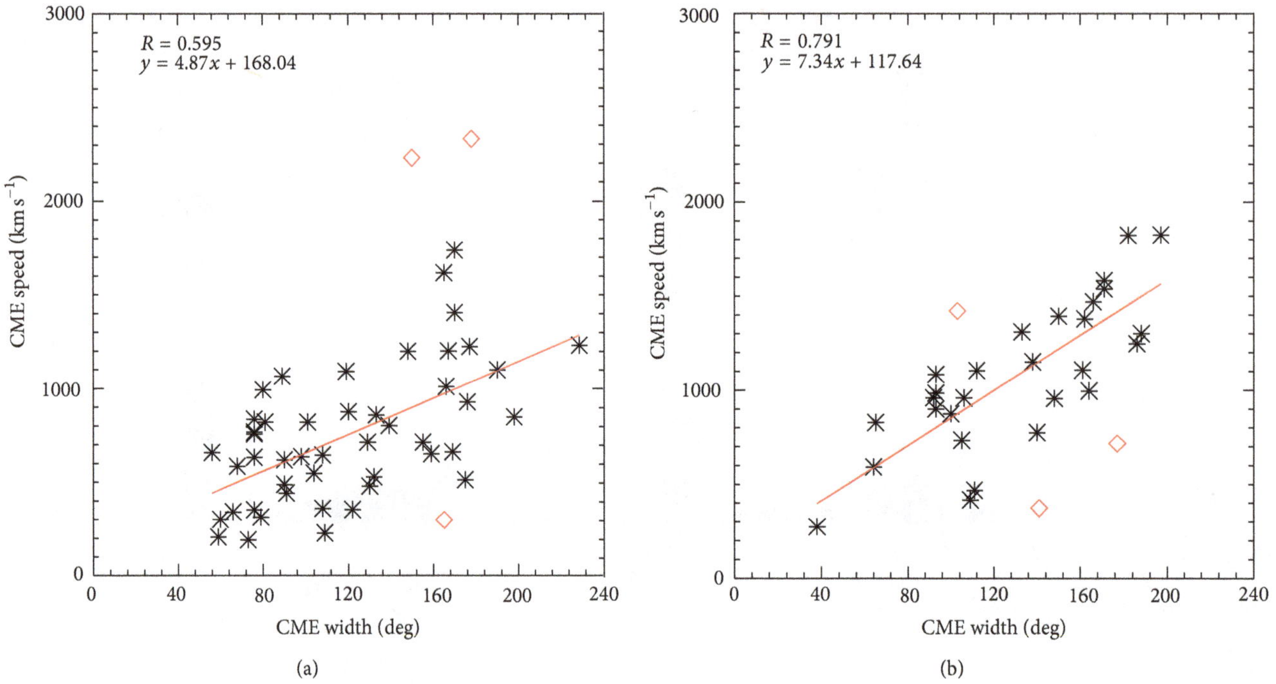

FIGURE 8: Correlation between projected speed and width of CMEs for (a) rise phase events and (b) decay phase events (red diamonds are outlier points).

rise phase events. The mean acceleration of 99 solar cycle rise phase events is $-1.32\ \mathrm{m\,s^{-2}}$ and that of the 72 solar cycle decay phase events is $-15.18\ \mathrm{m\,s^{-2}}$. The difference between the means is statistically significant with t-test P value = 2.01% as seen in Table 1.

The correlation between projected speed and width of CMEs for solar cycle rise phase and decay phase events is shown in Figure 8. It is difficult to find the true width of halo and partial halo CMEs. Therefore, we have excluded the full halo CMEs and wider CMEs with width ≥200° from the present analysis. There exists correlation for solar cycle decay phase events with a correlation coefficient of $R = 0.79$ and solar cycle decay phase events with $R = 0.60$ with the removal of outliers (the outliers are the extreme/error in the data points which affects the correlation, so they are removed from the analysis) shown by red diamond in Figure 8, and these results are in agreement with the earlier reports [12, 18, 19].

The correlation between speed and acceleration of CMEs for solar cycle rise and decay phase events is shown in Figure 9. Out of 229 events, 171 events are having 4 data points and are considered for the analysis. Out of 171 events, 105 events (61%) are decelerating and the remaining 66 events (39%) accelerate in the LASCO FOV. In 127 solar cycle rise phase events, only 99 events are found to have 4 data points, 59% (58/99) are decelerating events, and the remaining 41% (41/99) are accelerating events, while, in 102 solar cycle decay phase events, only 72 events are found to have 4 data points, 65% (47/72) are decelerating events, and the remaining 35% (25/72) are accelerating events. The faster events decelerate more with negative correlation between speed and acceleration of CMEs. There exists a good correlation for solar cycle decay phase events with a correlation coefficient of $R = -0.80$, while a weak correlation with $R = -0.57$ exists for solar cycle rise phase events with the removal of outliers (the outliers are the extreme/error in the data points which affects the correlation, so they are removed from the analysis) shown by red diamond in Figure 9. The acceleration decreases with increase in speed of CMEs. The faster CMEs decelerate more than slower CMEs, because of the drag force faced by the CMEs in the interplanetary medium [12, 18, 19, 21, 31]. Earlier reports [12, 21] suggest that the CMEs faster than $900\ \mathrm{km\,s^{-1}}$ have tendency to decelerate in the LASCO FOV.

The correlation between CME speed and DH-type II end frequency for solar cycle rise and decay phase events is shown in Figure 10. The correlations between ending frequency and CME speed are very weak for both data sets ($R = -0.43$ and $R = -0.36$ for solar cycle rise and decay phase events, resp.).

4. Conclusions

The properties of flares and CMEs associated with DH-type II bursts during the period 1997–2008 in the 23rd solar cycle are investigated. The collected events are divided into two groups as (i) solar cycle rise phase events and (ii) solar cycle decay phase events. The main results of our study are summarized as follows.

(i) There is no notable/significant difference in the properties of type II bursts and flares like start frequency,

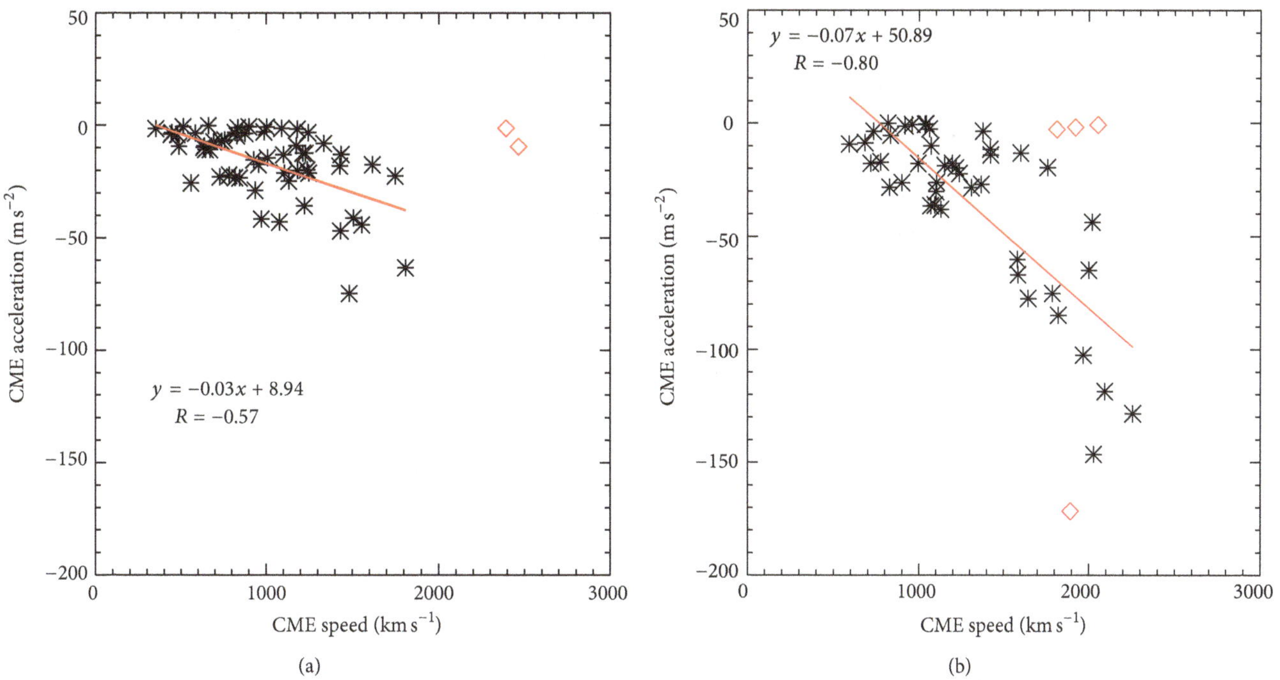

FIGURE 9: Correlation between speed and acceleration of CMEs for (a) rise phase events and (b) decay phase events (red diamonds are outlier points).

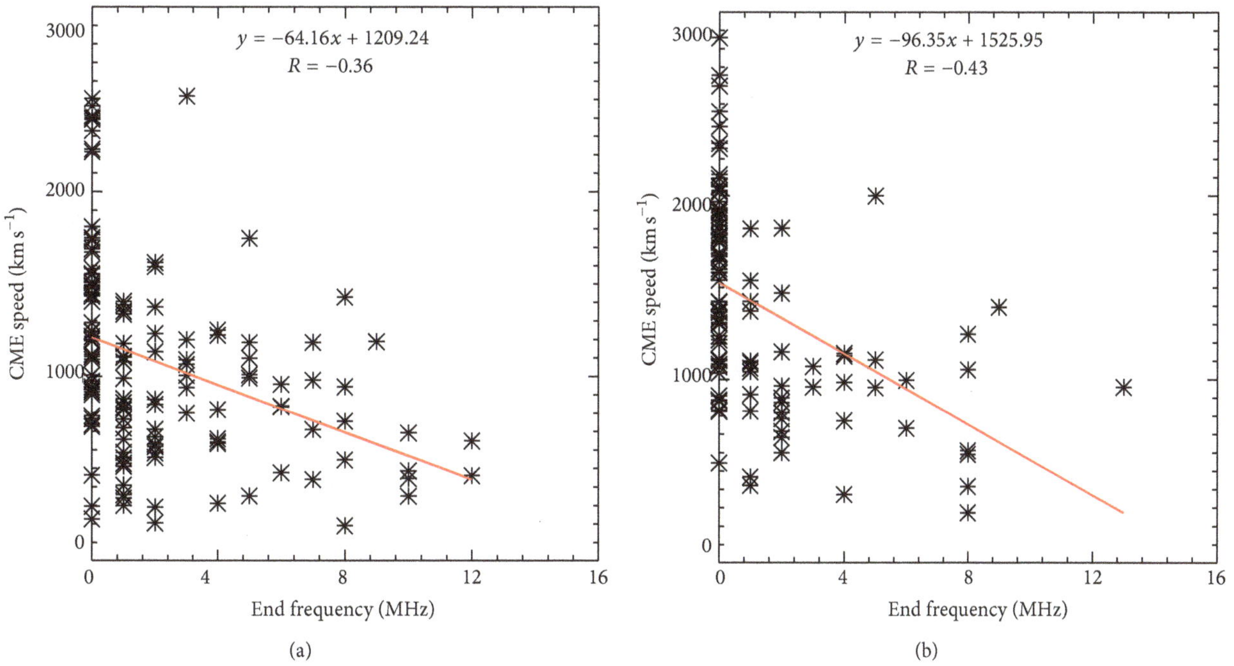

FIGURE 10: Correlation between end frequency and speed of CMEs for (a) rise phase events and (b) decay phase events.

end frequency, flare duration, rise time, and decay time in solar cycle rise phase events compared to solar cycle decay phase events, which indicates that these parameters of type II bursts do not depend on the phase of solar cycle.

(ii) The correlation between mean start frequency of type II bursts and drift rate is better for solar cycle decay phase events (R = 0.65) compared to solar cycle rise phase events (R = 0.50).

(iii) The mean CME speed of solar cycle decay phase events (1373 km s^{-1}) is found to be slightly higher than the mean CME speed of solar cycle rise phase events (1058 km s^{-1}).

(iv) Only a small difference was found between the mean widths of solar cycle decay phase events (276°) compared to solar cycle rise phase events (260°).

(v) The mean acceleration in solar cycle decay phase events ($-15.18\ \mathrm{m\,s^{-2}}$) is found to be higher than solar cycle rise phase events ($-1.32\ \mathrm{m\,s^{-2}}$).

(vi) The shock speed of the DH-type II bursts in solar cycle decay phase events ($1260\ \mathrm{km\,s^{-1}}$) is slightly larger than in solar cycle rise phase events ($1135\ \mathrm{km\,s^{-1}}$).

(vii) The correlation between end frequency of the type II bursts and CME speed is very weak for both groups ($R = 0.43$ and $R = 0.36$ for solar cycle rise phase and decay phase events, resp.).

(viii) There exists good correlation in solar cycle decay phase events between (i) CME speed and width ($R = 0.79$) and (ii) CME speed and acceleration ($R = -0.80$), while the correlations are weak and nonexisting for solar cycle rise phase events with $R = 0.60$ and $R = -0.57$.

About 229 DH-type II bursts associated with flares and CMEs are statistically studied in the 23rd solar cycle during the period 1997–2008. The number of events is larger for solar cycle rise phase (127/232) as a maximum number of events are evolved during solar maximum and lower for solar cycle decay phase (102/229) as expected in a solar cycle. There is no notable/significant difference in the properties of type II bursts and flares like start frequency, end frequency, flare duration, rise time, and decay time in solar cycle rise phase events compared to solar cycle decay phase events. But the properties of CMEs (like CME speed and CME acceleration and correlation between (i) CME speed and width and (ii) CME speed and acceleration) associated with DH-type II bursts are better for solar cycle decay phase events compared to solar cycle rise phase events. Therefore, this clearly shows that more energetic CMEs are also produced during the solar cycle decay phase events and this may produce intense/severe geomagnetic storms at earth. Only the number of events follows the solar cycle, but more energetic CMEs may be produced at all time throughout the solar cycle and may produce stronger geomagnetic storms. The DH-type II bursts are driven by energetic CMEs [12]. Therefore the parameters of DH-type II bursts do not depend on the time of appearance (phase) in the solar cycle.

Conflict of Interests

The authors declare that there is no conflict of interests regarding the publication of this paper.

Acknowledgments

The authors thank the referee for his fruitful comments for the improvement of their paper. The authors greatly acknowledge the various online data centers of NOAA and NASA. Also, they thank the Wind/WAVES team for providing the type II catalog. The CME catalog used is provided by the Center for Solar Physics and Space Weather, the Catholic University of America, in cooperation with Naval Research Laboratory and NASA.

References

[1] M. J. Reiner, "High-frequency Type II radio emissions associated with shocks driven by coronal mass ejections," *Journal of Geophysical Research A: Space Physics*, vol. 104, no. 8, pp. 16979–16991, 1999.

[2] J. P. Wild and L. L. McCready, "Observations of the spectrum of high intensity solar radiations at metrewavelengths," *Australian Journal of Scientific Research A*, vol. 3, no. 4, pp. 541–557, 1950.

[3] R. Payne-Scott, D. E. Yabsley, and J. G. Bolton, "Relative times of arrival of bursts of solar noise on different radio frequencies," *Nature*, vol. 160, no. 4060, pp. 256–257, 1947.

[4] H. H. Malitson, J. Fainberg, and R. G. Stone, "Observation of a Type II solar radio burst to 37 Rsun," *Astrophysical Journal*, vol. 14, pp. 111–114, 1973.

[5] A. Boischot, A. C. Riddle, J. B. Pearce, and J. W. Warwick, "Shock waves and Type II radiobursts in the interplanetary medium," *Solar Physics*, vol. 65, no. 2, pp. 397–404, 1980.

[6] H. V. Cane, N. R. Sheeley, and R. A. Howard, "Energetic interplanetary shocks, radio emission, and coronal mass ejections," *Journal of Geophysical Research*, vol. 92, no. 9, pp. 9869–9874, 1987.

[7] D. Lengyel-Frey and R. G. Stone, "Characteristics of interplanetary Type II radio emission and the relationship to shock and plasma properties," *Journal of Geophysical Research*, vol. 94, no. 1, pp. 159–167, 1989.

[8] J.-L. Bougeret, M. L. Kaiser, P. J. Kellogg et al., "WAVES: the radio and plasma wave investigation on the wind spacecraft," *Space Science Reviews*, vol. 71, no. 1–4, pp. 231–263, 1995.

[9] M. L. Kaiser, M. J. Reiner, N. Gopalswamy et al., "Type II radio emissions in the frequency range from 1- MHz associated with the April 7, 1997 solar event," *Geophysical Research Letters*, vol. 25, no. 14, pp. 2501–2504, 1998.

[10] M. J. Reiner, M. L. Kaiser, J. Fainberg, J.-L. Bougeret, and R. G. Stone, "On the origin of radio emissions associated with the January 6–11, 1997, CME," *Geophysical Research Letters*, vol. 25, no. 14, pp. 2493–2496, 1998.

[11] V. Vasanth, S. Umapathy, B. Vršnak, and M. Anna Lakshmi, "Characteristics of Type-II radio bursts associated with flares and CMEs," *Solar Physics*, vol. 273, no. 1, pp. 143–162, 2011.

[12] N. Gopalswamy, S. Yashiro, M. L. Kaiser, R. A. Howard, and J.-L. Bougeret, "Characteristics of coronal mass ejections associated with long-wavelength Type II radio bursts," *Journal of Geophysical Research A: Space Physics*, vol. 106, no. 12, pp. 29219–29229, 2001.

[13] B. Vršnak and E. W. Cliver, "Origin of coronal shock waves: iIlnvited review," *Solar Physics*, vol. 253, no. 1-2, pp. 215–235, 2008.

[14] J. Magdalenić, B. Vršnak, S. Pohjolainen, M. Temmer, H. Aurass, and N. J. Lehtinen, "A flare-generated shock during a coronal mass ejection on 24 December 1996," *Solar Physics*, vol. 253, no. 1-2, pp. 305–317, 2008.

[15] A. Nindos, C. E. Alissandrakis, A. Hillaris, and P. Preka-Papadema, "On the relationship of shock waves to flares and coronal mass ejections," *Astronomy and Astrophysics*, vol. 531, article A31, 12 pages, 2011.

[16] A. Lara, N. Gopalswamy, S. Nunes, G. Munoz, and S. Yashiro, "A statistical study of CMEs associated with metric Type II bursts," *Geophysical Research Letters*, vol. 30, no. 12, pp. SEP 4-1–SEP 4-4, 2003.

[17] N. Gopalswamy, A. Lara, R. P. Lepping, M. L. Kaiser, D. Berdichevsky, and O. C. St. Cyr, "Interplanetary acceleration of coronal mass ejections," *Geophysical Research Letters*, vol. 27, no. 2, pp. 145–148, 2000.

[18] V. Vasanth and S. Umapathy, "A statistical study on CMEs associated with DH-Type-II radio bursts based on their source location (limb and disk events)," *Solar Physics*, vol. 282, no. 1, pp. 239–247, 2013.

[19] V. Vasanth and S. Umapathy, "A statistical study on DH CMEs and its geo-effectiveness," *ISRN Astronomy and Astrophysics*, vol. 2013, Article ID 129426, 13 pages, 2013.

[20] N. Gopalswamy, S. Yashiro, G. Michalek et al., "The SOHO/LASCO CME catalog," *Earth, Moon and Planets*, vol. 104, no. 1–4, pp. 295–313, 2009.

[21] S. Yashiro, N. Gopalswamy, G. Michalek et al., "A catalog of white light coronal mass ejections observed by the SOHO spacecraft," *Journal of Geophysical Research A: Space Physics*, vol. 109, no. 7, article 105, 2004.

[22] G. E. Brueckner, J.-P. Delaboudiniere, R. A. Howard et al., "Geomagnetic storms caused by coronal mass ejections (CMEs): March 1996 through June 1997," *Geophysical Research Letters*, vol. 25, no. 15, pp. L3019–L3022, 1998.

[23] N. Gopalswamy, E. Aguilar-Rodriguez, S. Yashiro, S. Nunes, M. L. Kaiser, and R. A. Howard, "Type II radio bursts and energetic solar eruptions," *Journal of Geophysical Research A: Space Physics*, vol. 110, no. 12, article S07, 2005.

[24] N. Gopalswamy, "Coronal mass ejections and Type II bursts," in *Solar Eruptions and Energetic Particles*, N. Gopalswamy, R. Mewaldt, and J. Torsti, Eds., vol. 165 of *Geophysical Monograph Series*, p. 207, 2006.

[25] Y. Leblanc, G. A. Dulk, and J.-L. Bougeret, "Tracing the electron density from the corona to 1 AU," *Solar Physics*, vol. 183, no. 1, pp. 165–180, 1998.

[26] N. Gopalswamy, W. T. Thompson, J. M. Davila et al., "Relation between Type II bursts and CMEs inferred from STEREO observations," *Solar Physics*, vol. 259, no. 1-2, pp. 227–254, 2009.

[27] B. Vršnak, J. Magdalenić, and H. Aurass, "Comparative analysis of Type II bursts and of thermal and non-thermal flare signatures," *Solar Physics*, vol. 202, no. 2, pp. 319–335, 2001.

[28] B. Vršnak, "Processes and mechanisms governing the initiation and propagation of CMEs," *Annales Geophysicae*, vol. 26, no. 10, pp. 3089–3101, 2008.

[29] J. T. Gosling, E. Hildner, R. M. MacQueen, R. H. Munro, A. I. Poland, and C. L. Ross, "The speeds of coronal mass ejection events," *Solar Physics*, vol. 48, no. 2, pp. 389–397, 1976.

[30] Y.-J. Moon, G. S. Choe, H. Wang et al., "A statistical study of two classes of coronal mass ejections," *Astrophysical Journal Letters*, vol. 581, no. 1, pp. 694–702, 2002.

[31] B. Vršnak, "Deceleration of coronal mass ejections," *Solar Physics*, vol. 202, no. 1, pp. 173–189, 2001.

2

The Total Solar Irradiance, UV Emission and Magnetic Flux during the Last Solar Cycle Minimum

E. E. Benevolenskaya[1,2] and I. G. Kostuchenko[3]

[1] *Pulkovo Astronomical Observatory, Pulkovskoe sh. 65, Saint Petersburg 196140, Russia*
[2] *Saint Petersburg State University, Saint Petersburg 198504, Russia*
[3] *Karpov Institute of Physical Chemistry, Ul. Vorontsovo Pole 10, Moscow 105064, Russia*

Correspondence should be addressed to E. E. Benevolenskaya; benevolenskayae@mail.ru

Academic Editors: A. Caliandro, A. Cellino, A. Meli, J. F. Valdés-Galicia, and S. Wedemeyer-Bohm

We have analyzed the total solar irradiance (TSI) and the spectral solar irradiance as ultraviolet emission (UV) in the wavelength range 115–180 nm, observed with the instruments TIM and SOLSTICE within the framework of SORCE (the solar radiation and climate experiment) during the long solar minimum between the 23rd and 24th cycles. The wavelet analysis reveals an increase in the magnetic flux in the latitudinal zone of the sunspot activity, accompanied with an increase in the TSI and UV on the surface rotation timescales of solar activity complexes. In-phase coherent structures between the midlatitude magnetic flux and TSI/UV appear when the long-lived complexes of the solar activity are present. These complexes, which are related to long-lived sources of magnetic fields under the photosphere, are maintained by magnetic fluxes reappearing in the same longitudinal regions. During the deep solar minimum (the period of the absence of sunspots), a coherent structure has been found, in which the phase between the integrated midlatitude magnetic flux is ahead of the total solar irradiance on the timescales of the surface rotation.

1. Introduction

The minimum of the solar activity separating the activity cycles 23 and 24 is often called "an unusual minimum." For comparison, during the previous minimum, the lowest annual sunspot number was 8.6 in 1996. During the latest minimum, the annual sunspot number reached lower values: 7.5, 2.5, and 3.1 in 2007, 2008, and 2009, respectively. A typical minimum lasts for about 486 spotless days (http://spaceweather.com/), but since 2004, as many as 821 spotless days were observed. Woods [1] studied the irradiance during the latest solar cycle minimum and compared it with the previous minimum in 1996. He found that the solar irradiance was lower during the latest minimum. Thus, the total solar irradiance (TSI) was smaller in 2008 than that in 1996 by about 200 ppm. The irradiance measured with the SOHO Solar EUV Monitor (SEM) at 26 to 34 nm was by about 15% lower in 2008 than that in 1996. This EUV decrease could be explained by the abundance of low-latitude coronal holes during the latest cycle minimum.

It is known that variations of the sunspot activity during a solar cycle occur because of the generation of the magnetic flux in the convective zone. The magnetic flux is produced by dynamo processes and moves to the solar surface on account of magnetic buoyancy. The emerged magnetic flux impacts the solar luminosity, causing its cyclic behavior. However, the relationship between the magnetic flux and solar irradiance is rather complicated. Various magnetic features such as sunspots, flares, network elements, faculae, and prominences contribute to TSI in different ways. The major contribution to TSI is made by the photosphere. On the other hand, lines formed in the chromosphere and in the corona also impact the total solar irradiance. The development of the sunspot activity is accompanied with total irradiance reduction. This happens only because sunspots on the solar disk in the visible wavelength are dark. And the sunspots, in turn, are dark due to the suppression of the convection in the presence of strong magnetic fields. It is the so-called TSI blocking effect, which makes it difficult to determine the relationship between the magnetic flux and the total solar irradiance.

Fortunately, bright plages or faculae surrounding sunspots contribute to the total solar irradiance, which varies along with the solar cycle.

In a certain way, the brightness of magnetic elements is a function of wavelength [2]. If sunspots display negative fluctuations in the visible wavelengths, they are typically bright in the UV and EUV. It is a consequence of the fact that the complex of solar activity affects all layers of the solar atmosphere. Above the complex of solar activity, there exists a loop structure with a closed configuration of the magnetic field strength, filled by hot plasma emitting in UV, EUV, and X-ray wavelengths.

The irradiance associated with the network elements displays even more complicated behavior. According to the HINODE observations [3], weak magnetic fields enhance the brightness of the quiet sun. However, inner network fields, which are traced by weak horizontal fields, do not indicate the brightness effect. We can suggest, therefore, that the impact of magnetic fields on brightness in quiet regions is mostly due to strong, vertical, small-scale (subpixel) fields.

In this paper, we study the impact of the magnetic flux of the sunspot activity on the solar irradiance, with the purpose to understand the role of the long-lived complexes of the solar activity. In this aspect, the long and deep solar minimum gives a unique opportunity to analyze in detail variations of the total solar irradiance (TSI) and the spectral solar irradiance as a function of the magnetic flux and sunspot area. We can estimate the contribution of each complex of the solar activity to the TSI and UV, separately. We can also compare these results with those for the time when the sunspots do not contribute to the TSI at all.

In addition, the distribution of the sources of the TSI and UV is particularly important for the development of empirical and semiempirical irradiance models [4], for prediction of the solar luminosity.

2. Data Analysis

Here, we use the data for TSI and UV (115–180 nm) obtained by the SORCE instruments (http://lasp.colorado.edu/sorce/). The sunspot areas are taken from the webpage (http://solarscience.msfc.nasa.gov/greenwch.shtml). The magnetic data are represented by synoptic maps of the Wilcox Solar Observatory (WSO, http://wso.stanford.edu/).

The solar radiation and climate experiment (SORCE) is a NASA-sponsored satellite mission that is providing state-of-the-art measurements of the incoming X-ray, ultraviolet, visible, near-infrared, and total solar radiation. The SORCE spacecraft was launched on 25 January, 2003. SORCE carries four instruments, including the spectral irradiance monitor (SIM), solar stellar irradiance comparison experiment (SOLSTICE), total irradiance monitor (TIM), and the XUV photometer system (XPS). In our study, we used the daily data of TIM and SOLSTICE from 10 March, 2007, to 23 January, 2010. TIM measures the total solar irradiance (TSI) with an estimated absolute accuracy of 350 ppm (0.035%). Relative variations in the solar irradiance were measured with the accuracy less than 0.001%/yr [5]. The SOLSTICE measurements provide coverage from 115 nm to 320 nm with the spectral resolution of 1 nm, the absolute accuracy better than 5%, and the relative accuracy of 0.5% per year [6].

The Wilcox Solar observatory at Stanford provides us with large-scale low-resolution synoptic maps of the line-of-sight component of the magnetic field strength $B_{||}$ (measured in microTesla, μT). The resolution of these maps is 5° of Carrington longitude (73 points from 0° to 360°). Each synoptic map consists of 30 data points in equal steps along the latitude sine from −0.97 to 0.97. Using these magnetic data, we calculated the magnetic flux as the absolute values of $B_{||}$. The magnetic flux (F_{mag}) was integrated over a box with a width of ±90° in the longitude and with a height of ±40° in the latitude. F_{mag} was calculated as a function of Julian date, moving the box along the Carrington longitude. This procedure makes it possible to take into account the evolution of the magnetic flux (F_{mag}) from east to west on the solar limb relatively to the center of the solar disk.

We analyzed the variable fluxes of the total solar irradiance (TSI), ultraviolet emission (UV) (115–180 nm), integrated daily sunspot area, and integrated midlatitudinal magnetic flux (F_{mag}) during the time 10 March 2007 to 23 January 2010. Figure 1 presents all the studied variables as a function of time. Here, time is in Julian dates, in years, and in days since the first day of the time series. Figure 1(a) (upper plot) depicts daily values of the TSI. Figure 1(a) (bottom plot) displays variations of the ultraviolet emission integrated over the wavelength range 115–180 nm. The integrated midlatitude magnetic flux, F_{mag}, is represented in Figure 1(b) (upper plot). Bottom plot (Figure 1(b)) shows the daily integrated sunspot area in thousandths of the solar hemisphere. The daily integrated sunspot area is estimated by summing the area of all sunspots per day.

A simple comparison of the variables TSI and F_{mag} indicates their consistency, except for the times of strong TSI negative variations, which occurred due to the sunspot blocking effect. The increase in the daily integrated sunspot area corresponds to that in the integrated midlatitude magnetic flux. There is a strong correlation between the F_{mag} and the intensity of the UV. As we can see in Figure 1, during the deep minimum, the integrated midlatitude magnetic flux corresponds to low luminosity in the total and UV irradiance.

3. A Nonaxisymmetric Pattern of the Solar Irradiance

We analyzed the TSI and magnetic data ($B_{||}$) in the form of Carrington maps using the method described in [7]. First, the time series (TSI) is interpolated with the reference to the beginning (to_i) and the end (tn_i) of each Carrington rotation "i" with the step (tn_i − to_i)/27. The synodic period of the Carrington rotation is 27.2753 days. Thus, the first day of the interpolated time series corresponds to 360° of Carrington longitude for Carrington rotation CR2055, and the last day

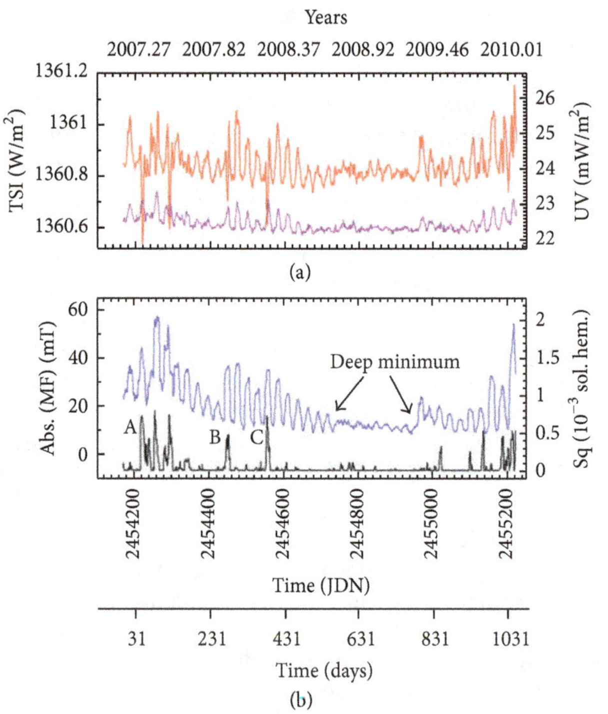

FIGURE 1: The daily values of the total irradiance (TSI) during the time 10 March 2007–23 January 2010 ((a), upper red line) in W/m^2; the variations of the UV in the range of 115–180 nm ((a), bottom purple line) in mW/m^2; the WSO absolute values of the magnetic field strength of the line-of-sight component (magnetic flux, F_{mag}) integrated over the latitude from −40° to 40° and from −90° to 90° over the longitude ((b), upper blue line) in milliTesla (mT); the daily integrated sunspot area is represented as thousandths of the solar hemisphere ((b), bottom black lines). The beginning and the end of the deep minimum are marked by arrows.

of this time series corresponds to 0° of Carrington longitude CR2091. TSI is represented in the form of 2D matrix, one dimension of which is the Carrington longitude with the step 360°/27 and the second is the time measured in the numbers of Carrington rotations (CR2055 to CR2091). This approach makes it possible to study features with the lifetime longer than one Carrington period. Figure 2 presents the 2D distribution of TSI for CR2055–CR2091 (b). The stacked WSO synoptic maps of the line-of-sight component of the magnetic field strength during the studied time are shown in Figure 2(a). This figure reveals the nonuniform longitudinal distribution of the magnetic flux and TSI related to the occurrence and development of solar activity complexes. The complexes of the solar activity are marked by "A," "B," and "C." The solar complexes "B" and "C" are related to the strong fluctuations of the sunspot area within the defined longitudinal zone about 180°–270°.

The solar complex "A" corresponds to the multiple strong variations of the sunspot area (Figure 1(b), bottom plot), which occur during several Carrington rotations and are spread along the longitude. But only one of these fluctuations, the one related to the complexes of solar activity, exists inside the aforementioned longitudinal zone.

It is clearly seen that, during CR2055–2076 (March 2007–November 2008), a strong magnetic activity exists in the longitudinal zone 200°–300° and rotates with the velocity rate slightly exceeding that of the Carrington rotation. This magnetic structure, combined from "A," "B," and "C" activity complexes, contributes to the TSI due to the surrounding bright faculae. The longitudinal zones in which a strong solar activity occurs during several Carrington rotations are called active longitudinal zones. They are related to a source of the magnetic field under the photosphere.

It is interesting that a long-lived longitudinal activity occurred at the ascending phase of the cycle 24 during the Carrington rotation CR2107–CR2115 at the longitude 300°–360° and latitude 13°–22° [8]. This complex produced a strong X-flare (X6.9) in September, 2011. It was the first long-lived complex in the cycle 24. At the beginning of the solar cycle 23, a similar longitudinal pattern existed from June, 1996, to June, 1998 [7]. During this time, the TSI distribution displayed an increase at longitudes 200°–300°. The EIT/SOHO data for the extreme UV irradiance indicated the increase in the coronal temperature associated with the activity complex at the beginning of the cycle 23. Therefore, the heating of the solar corona is closely related to long-lived complexes of the solar activity and to a corresponding source of the magnetic field under the photosphere.

The complexes of the solar activity "A," "B," and "C" are seen during CR2055–CR2063 (31 March, 2007–1 December, 2007), CR2064–CR2067 (1 December, 2007–20 March, 2008), and CR2068–CR2072 (20 March, 2008–3 August, 2008), respectively (Figure 2). They are maintained by the magnetic flux emerging from a subsurface source. The complex "A" consists of five fluctuations of the daily summarized area (Figure 1(b), bottom plot), but only three of them are strong. They exist mainly due to long-lived sunspots. Moreover, the blocking effect is related only to the 1st and the 3rd strong fluctuations of the sunspot area. What sunspots are responsible for these negative TSI fluctuations?

During the time of evolution of the complex "A," five long-lived sunspots were observed. The first long-lived large sunspot in the region 0953 NOAA appeared on 26 April, 2007 (S14° E73°, CL308°), and lasted for 14 days. Here, CL indicates the Carrington longitude. This region reached its maximum area of about 520 millionths of the solar hemisphere on 29 April, 2007 (according to daily sunspot summaries from the NOAA Space Environment Center). The second sunspot 0956 NOAA began to be observed on 15 May, 2007 (N02° E61°, CL070°), and disappeared on 24 May, 2007. On 18 May, 2007, its maximum area was about 300 millionths of the solar hemisphere. The third sunspot 0960 NOAA appeared on 2 June, 2007 (S06°, CL181°), and went behind the solar limb on June, 14. Its area reached about 540 millionths of the solar hemisphere on 4 June, 2007. The fourth region 0961 NOAA appeared on 26 June, 2007 (S09° E77°, CL219°), and moved behind the solar limb on 8 July, 2007. On 1 July, 2007, the maximum area was about 210 millionths of the solar hemisphere. The last, fifth region 0963 NOAA appeared on 8 July, 2007 (S09° E83°, CL054°), and moved behind the solar limb on 20 July, 2007 (its maximum area was about 530 millionths of the solar hemisphere on 11 July, 2007). The three

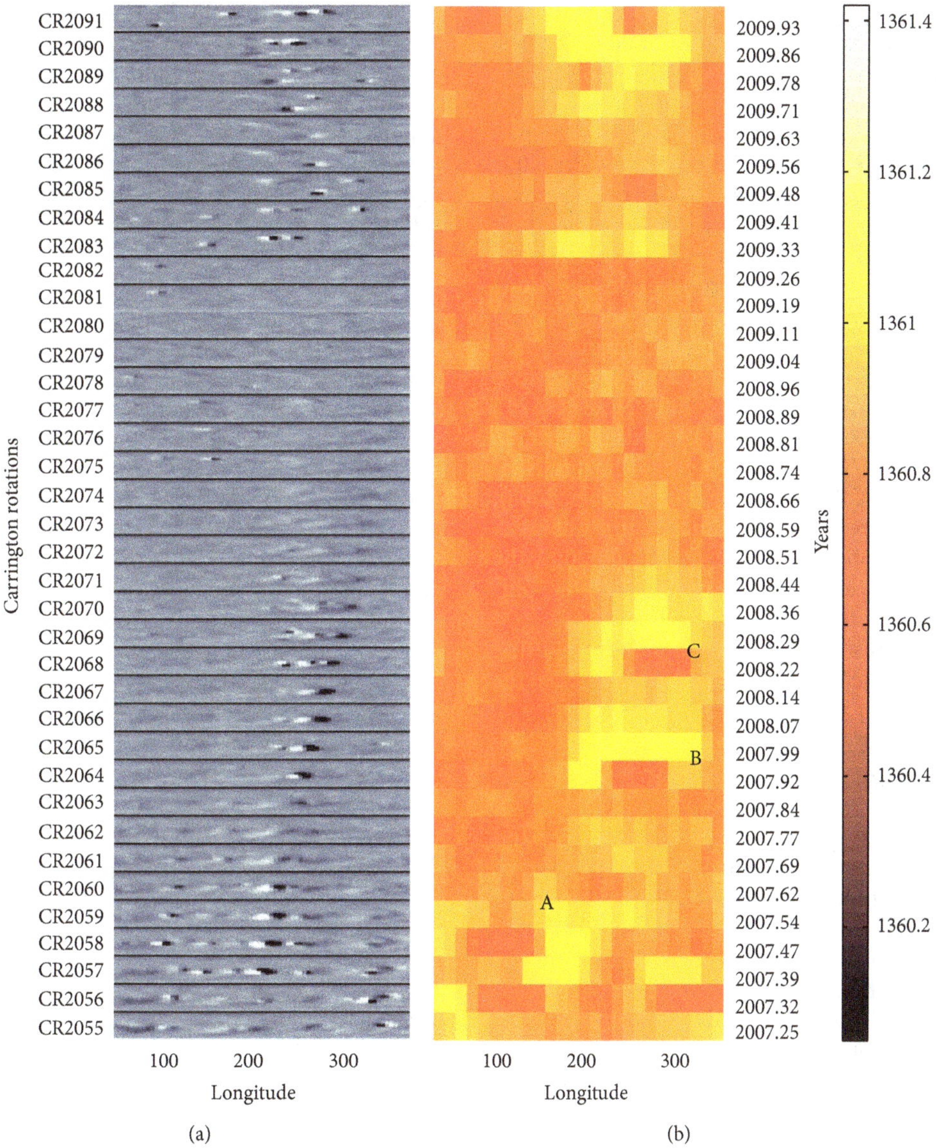

FIGURE 2: (a) Stacked WSO synoptic maps, 10 March, 2007, to 23 January, 2010, in gray scale from −250 to 250 microTesla; (b) TSI as a function of the Carrington number and the Carrington longitude. Complexes of solar activity are marked with A, B, and C. The time scale on the right indicates the beginnings of Carrington rotations. The color bar shows the TSI intensity in W/m^2.

strongest fluctuations of the sunspot area that belong to the complex "A" (Figure 1(b), bottom plot) correspond to the long-lived sunspots 0953 NOAA, 0960 NOAA, and 0963 NOAA. However, the strong blocking effect of the total solar irradiance occurs only due to the regions 0953 NOAA and 0963 NOAA. Why the long-lived sunspot 0960 NOAA with the area of about 540 millionths of the solar hemisphere does not block the total solar irradiance is still unclear.

The activity of B and C complexes is of special importance, as they belong to the overlapping cycles 23 and 24. According to the Hale's law, the polarity of the preceding and following parts in bipolar complexes of the solar activity alters from one solar cycle to the next. However, within the time interval of the so-called "cycles' overlap," the activity complexes of both polarities coexist.

During 2008-2009, the "old" magnetic flux (which belonged to the cycle 23) was concentrated in longitudinal zones, and the largest part of the "new" flux (which belonged to the cycle 24) with reversed magnetic polarity emerged in the same zones, within the longitude interval 180°–270° [9]. The complex marked by "B" appeared in the region 0978 NOAA and lasted from 7 to 19 December, 2007. The sunspot 0981 NOAA (N30°, CL246°) emerged in the same longitudinal zone as the sunspot 0980 NOAA with the "old" polarity (S06°, CL239°). The complex marked by "C" began since 28 March, 2008, when the cycle 23 returned and three big sunspots (0987 NOAA, 0988 NOAA, and 0989 NOAA) appeared. Their magnetic flux displayed the polarity of the "old" solar cycle 23. After that, the sun was practically blank, without sunspots (see Figure 2(a)).

Generally, at the beginning of a "new" cycle, sunspots with the "new" polarity appear at latitudes 25° to 35°, while those with the "old" polarity, located close to the equator, disappear. However, as we mentioned, there is a time of the cycles' overlap, when sunspots with both "old" and "new" polarities coexist. On 5 October, 2008, the sunspot 1003 NOAA of the new cycle appeared in the southern hemisphere at 23° of latitude and at 222° of Carrington longitude. It disappeared rapidly, and during the following several days we observed plages instead of the sunspot in the same location. On 11 October, 2008, a small sunspot (1004 NOAA) emerged at the latitude S08° and Carrington longitude 188° and another (1005 NOAA) at a higher latitude (N26°, CL116°). During the following day, two plages, 1003 NOAA (S23°, CL222°) and 1004 NOAA (S08°, CL188°), coexisted with the northern sunspot 1005 NOAA with the coordinates N26° and CL116°. In CR2075-CR2076 (September-October, 2008), "new" magnetic flux spread over longitude, but it was weak and short lived. Therefore, during the years 2008-2009, the sunspot activity was low, and magnetic regions with the "old" and "new" polarity coexisted.

To study the relationship between the nonuniform distribution of the solar magnetic flux, TSI, and UV, we applied the wavelet analysis.

4. Wavelet Analysis of TSI, UV, and F_{mag}

We used MATLAB package to perform cross-wavelet and wavelet coherence analysis developed by Grinsted et al. [10]. This software includes a code originally written by C. Torrence and G. Compo (available at http://paos.colorado.edu/research/wavelets/) and by E. Breitenberger from the University of Alaska, which was adapted from the SSA-MTM Toolkit freeware (http://www.atmos.ucla.edu/tcd/ssa/).

Wavelet analysis solves problems by decomposing a time series into time and frequency spaces simultaneously. We obtain the information on both the amplitude of any "periodic" signal within the series and the variation of this amplitude with the time. The wavelet transformation is generally used to analyze time series that contain nonstationary power at many different frequencies. Grinsted and his colleagues [10] recommend using the Morlet wavelet with $\omega_0 = 6$, since it provides a good balance between time and frequency localization. In this case, the Morlet wavelet scale is almost equal to Fourier period. According to the description presented in [10], Morlet wavelet which consists of a plane wave modulated by a Gaussian is defined as

$$\psi_0(\eta) = \pi^{-1/4} \exp^{i\omega_0 \eta} \exp^{-\eta^2/2}, \quad (1)$$

where ω_0 is the dimensionless frequency, s is the scale, and $\eta = st$. The continuous wavelet transformation of a discrete sequence X_n is defined as the convolution of X_n with a scaled and normalized wavelet:

$$W_n^X(s) = \sqrt{\frac{\delta t}{s}} \sum_{n'=0}^{N-1} x_{n'} \psi^* \left[\frac{(n' - n)\,\delta t}{s} \right], \quad (2)$$

where "$*$" indicates the complex conjugation. Finally, the wavelet power spectrum is defined as $|W_n(s)|^2$.

Figure 3 presents the Morlet continuous wavelet power spectrum for three sets of normalized time series (magnetic flux, ultraviolet emission, and total solar irradiance). Here, we indicate the cone of influence to display the statistical significance of the wavelet power (marked by solid lines in Figure 3). The continuous wavelet transformation displays edge artifacts, because the wavelet is not completely localized in time. It is therefore useful to introduce a cone of influence (COI), in which the edge effects cannot be ignored. Here, we take COI as the area in which the wavelet power caused by a discontinuity at the edge drops with the factor of e^2 [10].

The Morlet continuous wavelet transformation reveals common features with some differences in three time series. In the case of F_{mag} (Figure 3(a)), significant peaks occur on the timescale of the solar activity rotation during the descending phase of the solar cycle 23 and at the beginning of the solar cycle 24. The wavelet power shows separation between the complexes "A" and "B&C" within the same time intervals (Figure 3(b)). The separation increases for the wavelet power of TSI. Significant peaks for "A" are expanded to smaller time intervals of about 10 days. The TSI and UV series also display high values of the power spectrum in the band with the maximum at about 120–130 days during 200–500 days. This time interval coincides with the lifetime of the solar activity complexes "B" and "C."

5. Cross-Wavelet Transformation and Coherence Structure of TSI, UV, and F_{mag}

If a cross-wavelet power transformation reveals areas with high common power spectrum in the time-frequency space, then the wavelet coherence transformation finds local phase-locked behavior in this space. The cross-wavelet transformation (XWT) of two series X and Y can be defined as $W_n^{XY}(s) = W_n^X(s) W_n^{Y*}(s)$, where "$*$" denotes complex conjugation. The wavelet transformations of the series X and Y are $W_n^X(s)$ and $W_n^Y(s)$, where s is the scale, so that $\eta = s \cdot t$; η is the dimensionless time. The cross-wavelet spectrum is complex and can be defined as the cross-wavelet power spectrum $|W_n^{XY}(s)|$. Another useful parameter derived from the wavelet analysis is the wavelet transformation coherence (WTC) defined as the square of the cross-spectrum (XWT) normalized to the individual power spectrum. Phase coherence is defined as $\tan^{-1}[\mathrm{Im}[|W_n^{XY}(s)|]/\mathrm{Re}[|W_n^{XY}(s)|]]$ [10].

Figure 4(a) presents the cross-wavelet power spectrum of the TSI and F_{mag}. This figure indicates strong correlation between the irradiance and the magnetic flux during the descending phase of the cycle 23 and during the overlap of cycles 23 an 24 (up to 500th day of the time series) with the maximum at periods close to that of the solar activity surface rotation. The relative phase relationship of the two data series is shown with arrows. Arrows pointing to the right show the in-phase behavior of both selected data sets in the time-frequency space; those pointing to the left indicate the antiphase behavior of the series. Both time series are in-phase during the evolution of the complexes of solar activity "B" and "C" (arrows pointing to the right). But there is no in-phase

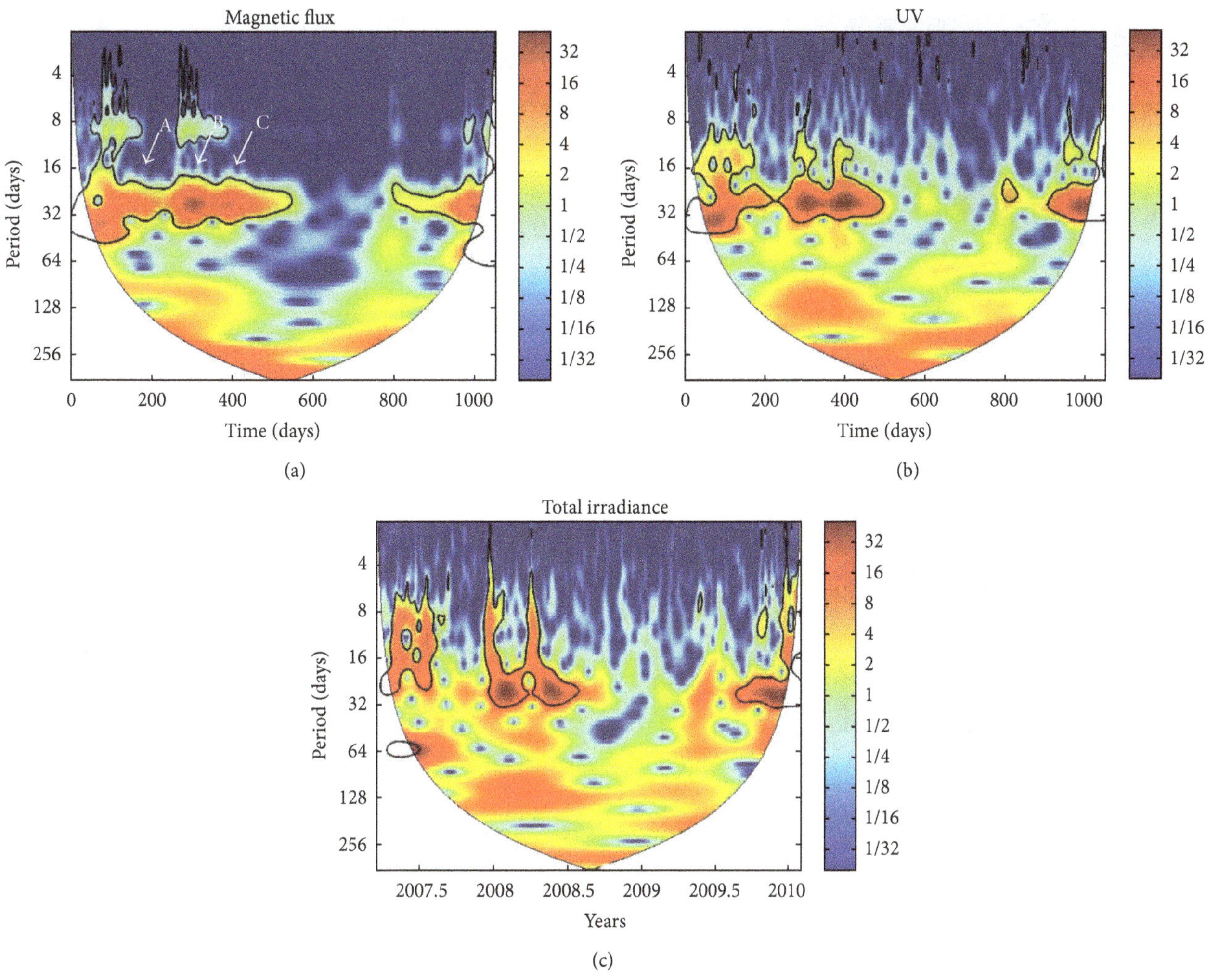

FIGURE 3: Morlet continuous power spectrum of the magnetic flux (a), ultraviolet emission (b), and total irradiance (c). The thick black contour designates the 5% significance level against the red noise. The cone of influence (COI), where the edge effects might distort the picture, is not shown by color. Complexes of solar activity contributing to TSI and UV are marked by A, B, and C in (a).

behavior of the TSI and F_{mag} in the presence of the complex activity marked by "A." Note that this complex consists of several activity complexes separated in the Carrington longitude. Then, during the deep minimum (500–800 days of the time series), the cross-wavelet power spectrum of the TSI and F_{mag} does not reveal strong common areas in the time-frequency space (Figure 4(a)). Further, significant peaks appear again due to the development of the solar cycle 24.

On the periods of about 120 days, cross-wavelet transformation of the TSI and F_{mag} does not show in-phase behavior of the power spectrum in the time-frequency space. This may be related to the fact that strong magnetic flux coincides with dark sunspots. An increase in the magnetic flux of the sunspots results in the reduction of the TSI.

The cross-wavelet transformation power spectrum for UV and F_{mag} is slightly different (Figure 4(c)). The cross-wavelet of the UV and magnetic flux indicates the in-phase behavior for all three complexes of solar activity on the rotational timescale (the synodic period is about 27-28 days). Moreover, we observe the in-phase behavior of the cross-wavelet transformation power spectrum between the UV and F_{mag} on timescales of about 100–140 days.

The wavelet transformation coherence (WTC) makes it possible to find structures in the time-frequency space, which display locally phase-locked behavior. Such a coherence structure indicates that variations of both time series are in the same phase in the local time-frequency space.

In Figures 4(b) and 4(d), the arrows pointing to the right indicate an in-phase behavior of TSI and UV, respectively. Arrows pointing to the left indicate the antiphase behavior; those pointing downward show that the first series of data is by 90° second; those pointing upward indicate that the second data series is by 90° ahead of the first.

The wavelet transformation coherence of the TSI and F_{mag} and that of the UV and F_{mag} display a detailed picture of coherent structures in time-frequency space. There is a significant difference between the TSI, F_{mag} series, on the one hand, and the UV, F_{mag} series, on the other hand, in

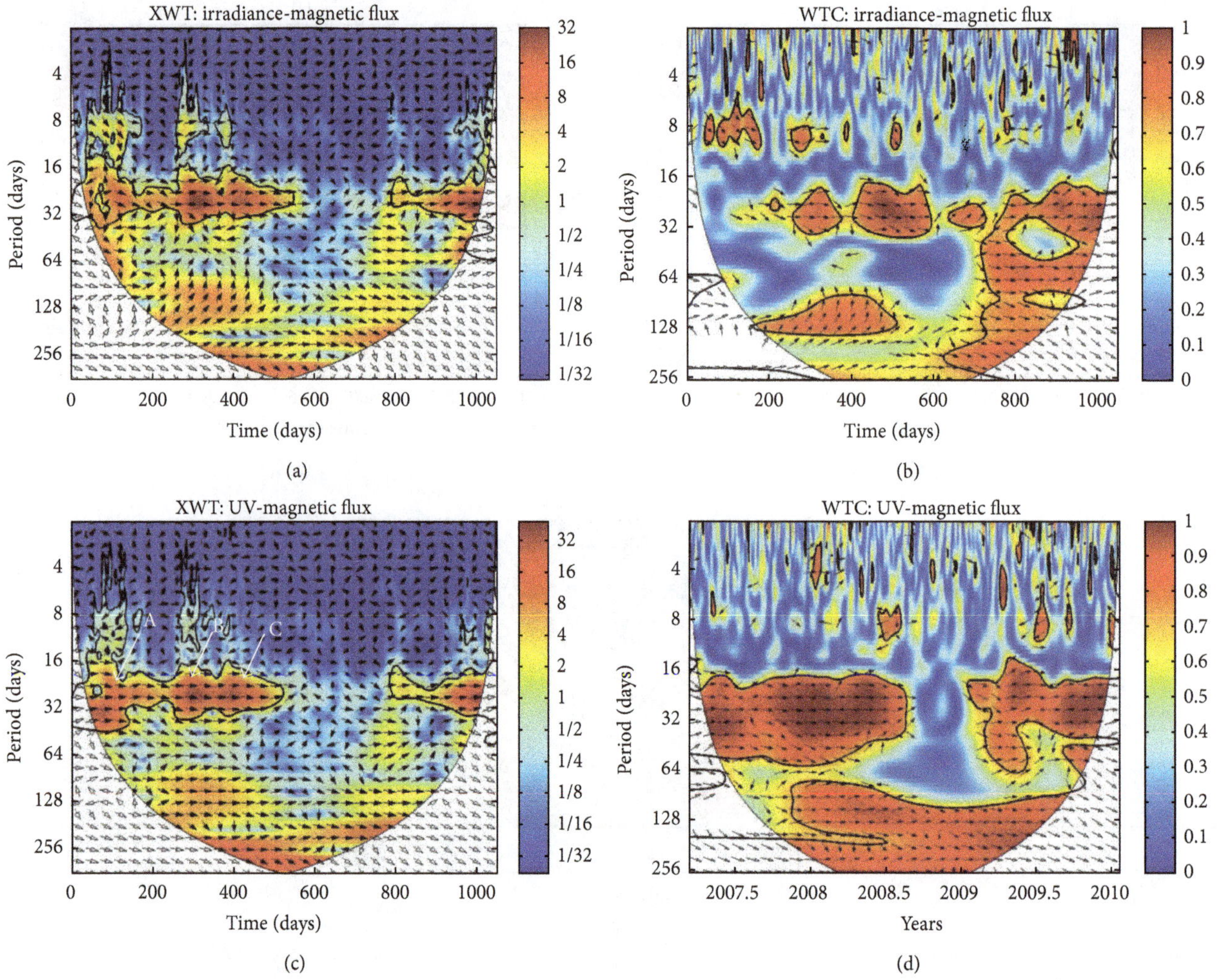

FIGURE 4: (a) Cross-wavelet transformation of the irradiance and the magnetic flux; (b) coherence of the irradiance and the magnetic flux; (c) cross-wavelet transformation of the ultraviolet and the magnetic flux; (d) coherence of the ultraviolet and the magnetic flux. Arrows indicate the phase relationship between the two data series in time-frequency space: (1) arrows pointing to the right show the in-phase behavior; (2) arrows pointing to the left indicate the antiphase behavior; (3) arrows pointing downward show that the first data series is by 90° ahead of the second series; (4) arrows pointing upward indicate that the second data series is by 90° ahead of the first series. Complexes of solar activity contributing to the TSI and UV are marked by A, B, and C in (c).

the coherence areas on periods of the surface rotation. The UV and magnetic flux coherence structures associated with the magnetic activity (Figure 4(d)) are in-phase (arrows point to the right). The stronger magnetic flux causes stronger UV emission. Also, we observe the extended coherence structure on time intervals from about 20 to 50 days during the descending phase of the cycle 23. The coherence structure related to the beginning of the cycle 24 appears slightly earlier then the development of sunspot activity begins (see Figures 4(d) and 1(b)).

The coherence structure of the TSI and the F_{mag} on periods of the surface activity rotation is decomposed into several structures during the time of overlapping of cycles 23 and 24 (Figure 4(b)) and during the solar cycle minimum. Thus, complex "A" does not create a coherent structure, but "B" and "C" display two coherent regions on periods of the surface activity rotation. The separation is related to the sunspot blocking effect. The second coherent region in frequency-time space shifts to the region of the deep minimum. The third coherent structure during the deep minimum shows that the phase of the F_{mag} is ahead of the total solar irradiance. This effect requires explanation. It may be a result of the influence of the polar irradiance (due to bright polar faculae), which is in the antiphase with the magnetic flux at midlatitude during the solar minimum. In the case of the UV, the relationship between the polar irradiance and the midlatitude magnetic flux is in-phase due to the coronal holes visible in X-ray, UV, and EUV emissions. On the periods of about 120 days, we see the coherent structure, in which the phase of the F_{mag} is ahead of the TSI phase.

Therefore, long-lived complexes of the solar activity are maintained by the magnetic flux, which appears in the same longitudinal zones. These emerged fluxes create the in-phase coherent structure of the magnetic flux and the total solar irradiance due to bright faculae. The complex "A," consisted of regions, which spread over the Carrington longitude, does not produce a strong coherent structure on the surface rotational timescales. However, the complexes "B" and "C" display strong coherence structures on surface rotational timescales.

6. Discussion

The total solar irradiance (TSI) describes the total radiant energy, in the form of electromagnetic radiation emitted by the sun at all wavelengths, that falls for each second on 1 square meter outside the Earth's atmosphere, a value proportional to the "solar constant" introduced earlier in the last century. TSI is a measure of the solar energy flux. The radiative flux decreases when dark sunspots are present on the disk and due to bright faculae or plages [11].

Fligge and Solanki [12], using a reconstruction of the total solar irradiance since 1700, found that long-term TSI variations exist due to the evolution of the solar network or other processes. Therefore, connections between the solar magnetic activity and total irradiance are based on the following factors: irradiance reductions that result from the passage of dark sunspots across the solar disk; irradiance increases due to facular brightening; and long-term irradiance variations related to changes in the solar network. Moreover, the long-term and solar cycle variations are obliged to the existence of long-lived complexes of solar activity with extended bright plages or faculae. There is a good consistency between the TSI, UV, and the irradiance reconstructed from Ca K faculae area [13]. Usually, a large and long-lived complex of the solar activity is surrounded by a large area of faculae. Moreover, sometimes a part of an emerging bipolar magnetic flux is represented by bright faculae rather than by dark sunspots in continuum. Bright and expended plages can explain the brightening in 2D maps of TSI (Figure 2(b)).

Long-lived longitudinal nonuniformity in the solar activity (active longitudes) has been known for a long time, but generally, this phenomenon is related to active phases of solar cycles, when there are a number of (several) sunspot complexes and solar magnetic field has a complicated structure. Here, we observed a pronounced and stable longitudinal nonuniformity for the relatively weak solar magnetic flux during the long solar minimum and two cycles overlapping.

7. Conclusions

A pronounced longitudinal nonuniform emerging of faculae and sunspots and, as a result, a nonaxisymmetrical distribution of the total and spectral solar irradiance were observed during the long minimum between cycles 23 and 24.

During the descending phase of the solar cycle 23, sunspots reappeared in the same plages, in a limited longitudinal zone, during 22 Carrington rotations. The Carrington longitude of these spots and faculae consecutively shifted from 180° to 270°, demonstrating the rotation rate slightly greater than the Carrington's. After the long minimum, sunspots of the cycle 24 tended to appear in the same longitudinal zone, while the spots of the "old" cycle still exist.

The revealed tendencies make it possible to suppose that a long-lived local source of the emerging magnetic flux exists under solar photosphere. Our results also suggest that conditions for such a source to appear are stable relative to the process of cycle changing. According to the solar dynamo theory, the solar magnetic field during the minimum is expected to be a pure dipole. However, the existence of such a local long-lived subsurface source of magnetic field apparently indicates generation of a nonaxisymmetrical component of the solar magnetic field due to dynamo process.

Our comparative and cross-wavelet analyses show that the increase in the magnetic flux in the latitudinal zone of the sunspot activity is accompanied by the increase in the TSI and UV on the timescale of the rotation of the solar activity complexes. The coherent structures between the midlatitude magnetic flux and TSI/UV occurred in the case of the existence of the long-lived complexes of the solar activity. Therefore, the nonuniform longitudinal distribution of the long-lived solar magnetic activity affects the solar irradiance. Indeed, the found coherent structures are associated with the development of such complexes of the solar activity. Moreover, coherent structures between the TSI and the magnetic flux confirm the idea about the interrelation of the activity processes going on the sun in total.

A similar longitudinal pattern was observed at the beginning of the solar cycle 23 from June 1996 to June 1998 [7]. During this time, the longitudinal distribution of the total solar irradiance, EUV irradiance from the transition region and corona, and solar magnetic flux integrated over solar disk increase in the same longitudinal zone (200°–300°). We concluded that precisely the long-living complexes of sunspot activity associated with this longitudinal zone make a significant contribution to the variation of the total solar irradiance and to the heating of the solar corona.

Therefore, the revealed nonaxisymmetrical character of the solar magnetic field turned out to be quite stable during the solar minimum, and the solar irradiance is closely related to the nature of the solar magnetic field.

Acknowledgments

The authors thank Dr. Grinsted, Professor Moore, Dr. Jevrejeva, Dr. Torrence, Dr. Combo, and Dr. Breitenberger for the development of the Wavelet MATLAB package. Also, they are thankful to the science team of SORCE and WSO for making their data available for free download. Part of this work is supported by the Program 22 of the Russian Academy of Science.

References

[1] T. N. Woods, "Irradiance variations during this solar cycle minimum," in *SOHO-23: Understanding a Peculiar Solar Minimum*, S. Cranmer, T. Hoeksema, and J. Kohl, Eds., vol. 428 of *ASP Conference Series*, p. 63, 2010.

[2] M. Haberreiter, Y. Yamaguchi, and T. Yokoyama, "Influence of the Schwabe/Hale solar cycles on climate change during the

Maunder Minimum," in *Solar and Stellar Variability: Impact on Earth and Planets Proceedings IAU Symposium No. 264*, Kosovichev, A. H. Andrei, and J.-P. Rozelot, Eds., p. 231, 2009.

[3] R. S. Schnerr and H. C. Spruit, "The total solar irradiance and small scale magnetic fields," in *Solar Polarization Workshop 6*, K. Berdyugina, H. Keil, L. Rimmele, and Trujillo-Bueno, Eds., vol. 437 of *ASP Conference Series*, p. 167, 2011.

[4] S. K. Solanki and Y. C. Unruh, "Solar irradiance variability," *Astronomische Nachrichten*, vol. 334, no. 1-2, pp. 145–150, 2012.

[5] G. Kopp, J. L. Lean, and A. New, "Lower value of total solar irradiance: evidence and climate significance," *Geophysical Research Letters*, vol. 38, no. 1, Article ID L01706, 2011.

[6] R. C. Bless, A. D. Code, and E. T. Fairchild, "Ultraviolet photometry from the orbiting astronomical observatory. XXI. Absolute energy distribution of stars in the ultraviolet," *The Astrophysical Journal*, vol. 203, pp. 410–416, 1976.

[7] E. E. Benevolenskaya, "Non-axisymmetrical distributions of solar magnetic activity and irradiance," *Advances in Space Research*, vol. 29, no. 12, pp. 1941–1946, 2002.

[8] D. Batmunh, A. A. Golovko, V. D. Trifonov, and S. A. Yazev, "Observations of the solar activity complex which produced strong solar flares in August-September 2011 made in Irkutst and Ulan Bator," in *Proceedings of the 3rd Selected Astronomical Problems of National Russian Astronomical Conference*, S. A. Yazev, Ed., vol. 75, Irkutsk, Russia, November 2011.

[9] E. E. Benevolenskaya and D. Y. Ponyavin, "Synoptic magnetic field in cycle 23 and in the beginning of the cycle 24 ," *Advances in Space Research*, vol. 50, no. 6, pp. 656–661, 2012.

[10] A. Grinsted, J. C. Moore, and S. Jevrejeva, "Application of the cross wavelet transform and wavelet coherence to geophysical times series," *Nonlinear Processes in Geophysics*, vol. 11, no. 5-6, pp. 561–566, 2004.

[11] J. Lean and C. Frohlich, "Solar total irradiance variations," in *Synoptic Solar Physics*, K. Blasubramaniam, J. Harvey, and D. Rabin, Eds., vol. 140 of *ASP Conference Series*, pp. 281–292, 1998.

[12] M. Fligge and S. K. Solanki, "The solar spectral irradiance since 1700," *Geophysical Research Letters*, vol. 27, no. 14, pp. 2157–2160, 2000.

[13] P. Foukal, "A new look at solar irradiance variation," *Solar Physics*, vol. 279, no. 2, pp. 365–381, 2012.

3

Dark Energy from the Gas of Wormholes

A. A. Kirillov and E. P. Savelova

Dubna International University of Nature, Society and Man, Universitetskaya Street 19, Dubna 141980, Russia

Correspondence should be addressed to A. A. Kirillov; ka98@mail.ru

Academic Editors: C. Ghag, M. Jamil, and S. Profumo

We assume the space-time foam picture in which the vacuum is filled with a gas of virtual wormholes. It is shown that virtual wormholes form a finite (of the Planckian order) value of the energy density of zero-point fluctuations. However such a huge value is compensated by the contribution of virtual wormholes to the mean curvature and the observed value of the cosmological constant is close to zero. A nonvanishing value appears due to the polarization of vacuum in external classical fields. In the early Universe some virtual wormholes may form actual ones. We show that in the case of actual wormholes vacuum polarization effects are negligible while their contribution to the mean curvature is apt to form the observed dark energy phenomenon. Using the contribution of wormholes to dark matter and dark energy we find estimates for characteristic parameters of the gas of wormholes.

1. Introduction

As is well known modern astrophysics (and, even more generally, theoretical physics) faces two key problems. Those are the nature of dark matter and dark energy. Recall that more than 90% of matter of the Universe has a nonbaryonic dark (to say, mysterious) form, while lab experiments still show no evidence for the existence of such matter. Both dark components are intrinsically incorporated in the most successful ΛCDM (Lambda cold dark matter) model which reproduces correctly properties of the Universe at very large scales (e.g., see [1] and references therein). We point out that ΛCDM predicts also the presence of cusps ($\rho_{\rm DM} \sim 1/r$) in centers of galaxies [2] and a too large number of galaxy satellites. Therefore other models are proposed, for example, like axions [3], which may avoid these. To be successful such models should involve a periodic self-interaction and therefore require a fine tuning, while in general the presence of standard nonbaryonic particles cannot solve the problem of cusps. Indeed, if we admit the existence of a self-interaction in the dark matter component, or some coupling to baryonic matter (which should be sufficiently strong to remove cusps), then we completely change properties of the dark matter component at the moment of recombination and destroy all successful predictions at very large scales. Recall that both warm and self-interacting dark matter candidates are rejected by the observing $\Delta T/T$ spectrum [1]. In other words, the two key observational phenomena (cores of dark matter in centers of galaxies [4–6] and $\Delta T/T$ spectrum) give a very narrow gap for dark matter particles which seems to require attracting some exotic objects in addition to standard nonbaryonic particles.

As it was demonstrated recently [7] the problem of cusps can be cured, if some part of nonbaryon particles is replaced by wormholes. Wormholes represent extremely heavy (in comparison to particles) objects which at very large scales behave exactly like nonbaryon cold particles, while at smaller scales (in galaxies) they strongly interact with baryons and form the observed [4–6] cored ($\rho_{\rm DM} \sim$ const) distribution. We note that stable wormholes violate necessarily the averaged null energy conditions which gives the basic argument against the existence of such objects. Without exotic matter stable wormholes may however exist in modified theories, for example, see [8] and references therein. In the case when the energy conditions hold a wormhole collapses into a couple of conjugated (of equal masses) blackholes which almost impossible to distinguish from standard primordial blackholes. However, the topological nontriviality of such objects retains and gravitational effects of a gas of wormholes considered in [9] and some results of [7] still remain valid which means that nontraversable wormholes can be used to smooth cusps in centers of galaxies. Thus, it worth expecting that wormholes may play an important role in the explanation of the dark matter phenomenon.

Saving the dark matter component ΛCDM requires the presence (~70%) of dark energy (of the cosmological constant). Moreover, there is evidence for the start of an acceleration phase in the evolution of the Universe [10–14]. In the present paper we use virtual wormholes to estimate the contribution of zero-point fluctuations in the value of the cosmological constant. The idea to relate virtual wormholes (or baby universes) and the cosmological constant is not new, in somewhat different context it was used by Coleman in [15] and developed in [16]. Our basic aim is to demonstrate that virtual wormholes form a finite value of the energy density of zero-point fluctuations.

It is necessary to point here out to the principle difference between actual and virtual wormholes. The principle difference is that a virtual wormhole exists only for a very small period of time and at very small scales and does not necessarily obey to the Einstein equations. It represents tunnelling event and therefore, the averaged null energy condition (ANEC) cannot forbid the origin of such an object. For the future we also note that a set of virtual wormholes may work as an actual wormhole opening thus the way for an artificial construction of wormhole-type objects in lab experiments.

In the present paper we describe a virtual wormhole as follows. From the very beginning we use the Euclidean approach (e.g., see [17] and the standard textbooks [18]). Then the simplest virtual wormhole is described by the metric ($\alpha = 1, 2, 3, 4$)

$$ds^2 = h^2(r)\,\delta_{\alpha\beta}dx^\alpha dx^\beta, \tag{1}$$

where

$$h(r) = 1 + \theta(a-r)\left(\frac{a^2}{r^2} - 1\right) \tag{2}$$

and $\theta(x)$ is the step function. Such a wormhole has vanishing throat length, while the step function at the junction may cause a problem in Einstein's equation or when a topological Euler term is involved. A more careful analysis needs to consider distributional curvature and so forth, see [19]. To avoid these difficulties we may consider from the very beginning a wormhole of a finite throat length $\sim 1/\beta$ where the step function is replaced with a smooth function (e.g., $\theta(x,\beta) = (\exp(\beta x)+1)^{-1}$). Then where it is necessary one may consider the limit $\beta \to \infty$ only in final expressions. This insures that the Bianchi identity holds and that the above metric remains inside of the domain of usual gravity. (Unexpectedly our approach (the use of step function) seems to irritate an essential part of physicists working with wormholes (e.g., PRD reviewers, moreover there is a claim that we study some other exotic objects which are not wormholes [20]). Therefore, it is necessary to clarify our position here. We are quite aware that at present state the physics of wormholes is certainly on the most speculative side. Therefore, we do not see the difference between specific forms of a smooth metric that one may use. For example, the simplest choice $h = (1 + a^2/r^2)$ gives the well-known metric which in 3-dimensions is called as the Bronnikov-Ellis metric, or the metric $h = (1 + (b/r) + (a^2/r^2))$ which includes an additional parameter (an arbitrary length of the handle) which is not called by any name but is not less trivial. In the case of actual wormholes the exact and correct form of metric may be established only upon understanding the nature and properties of exotic matter which may support such a metric as a stable solution to the Einstein equations. For this time has not come yet. In the case of virtual wormholes even this is not important; for in the complete theory one has to sum over all possible metrics which formally may be included in φ in (5). Our basic aim is to present sufficiently clear and simple model (let it be far from realistic) which retains basic qualitative features related to a nontrivial topology.)

In the region $r > a$, $h = 1$ and the metric (2) is flat, while the region $r < a$, with the obvious transformation $y^\alpha = (a^2/r^2)x^\alpha$, is also flat for $y > a$. Therefore, the regions $r > a$ and $r < a$ represent two Euclidean spaces glued at the surface of a sphere S^3 with the centre at the origin $r = 0$ and radius $r - a$. Such a space can be described with the ordinary double-valued flat metric in the region $r_\pm > a$ by

$$ds^2 = \delta_{\alpha\beta}dx_\pm^\alpha dx_\pm^\beta, \tag{3}$$

where the coordinates $x_\pm^\alpha$ describe two different sheets of space. Now, identifying the inner and outer regions of the sphere S^3 allows the construction of a wormhole which connects regions in the same space (instead of two independent spaces). This is achieved by gluing the two spaces in (3) by motions of the Euclidean space (the Poincare motions). If $R_\pm$ is the position of the sphere in coordinates $x_\pm^\mu$, then the gluing is the rule

$$x_+^\mu = R_+^\mu + \Lambda_\nu^\mu\left(x_-^\nu - R_-^\nu\right), \tag{4}$$

where $\Lambda_\nu^\mu \in O(4)$, which represents the composition of a translation and a rotation of the Euclidean space (Lorentz transformation). In terms of common coordinates such a wormhole represents the standard flat space in which the two spheres $S_\pm^3$ (with centers at positions $R_\pm$) are glued by the rule (4). We point out that the physical region is the outer region of the two spheres. Thus, in general, the wormhole is described by a set of parameters: the throat radius a, positions of throats $R_\pm$, and rotation matrix $\Lambda_\nu^\mu \in O(4)$.

In the present paper we assume the space-time foam picture in which the vacuum is filled with a gas of virtual wormholes. We show that virtual wormholes form a finite (of the Planckian order) value of the energy density of zero-point fluctuations. However such a huge value is compensated by the contribution of virtual wormholes to the mean curvature and the observed value of the cosmological constant should be close to zero.

To achieve our aim we, in Section 2, present the construction of the generating functional in quantum field theory. The main idea is that the partition function includes the sum over field configurations and the sum over topologies.

Where the sum over topologies is the sum over virtual wormholes described above. Such an approach gives a rather good leading approximation for calculation of the partition function and corresponds to the standard methods (e.g., Ritza method, etc.). In Section 3 we investigate properties of the two-point Green function. We show that the presence of the gas of virtual wormholes can be described by the topological bias exactly as it happens in the presence of actual wormholes [7, 9]. For limiting topologies when the density of virtual wormholes becomes infinite the Green function shows a good ultraviolet behavior which means that there exists a class of such systems when quantum field theories are free of divergencies. We demonstrate how the sum over topologies defines the mean value for the bias which takes the sense of a cutoff function in the space of modes. In Section 4 we explicitly demonstrate that for a particular set of virtual wormholes the bias defines not more than the projection operator on the subspace of functions obeying to the proper boundary conditions at wormhole throats. The projective nature of the bias means that wormholes merely cut some portion of degrees of freedom (modes). Phenomenologically it means that wormholes can be described by the presence of ghost fields which compensate the extra (cut by wormholes) modes. In Section 5 we show how the cutoff expresses via some dynamic parameters of wormholes. The exact definition of such parameters we leave it for the future investigation. In Section 6 we consider the origin of the cosmological constant. We demonstrate that the cosmological constant is determined by the contribution of the energy density of zero-point fluctuations and by the contribution of virtual wormholes to the mean curvature. We estimate contribution of virtual wormholes to the mean curvature and show also that wormholes lead to a finite (of the planckian order) value of $\langle T_{\mu\nu}\rangle$ which requires considering the contribution from the smaller and smaller wormholes with divergent density $n \to \infty$. We also present arguments of why in the absence of external classical fields the total value of the cosmological constant is exactly zero, while it acquires a nonvanishing value due to vacuum polarization effects (i.e., due to an additional distribution of virtual wormholes) in external fields. We also speculate the possibility of the formation of actual wormholes and in Section 7 we estimate their contribution to the dark energy. Finally in Section 8 we repeat basic results an discuss some perspectives.

2. Generating Function

The basic aim of this section is to construct the generating functional which can be used to get all possible correlation functions. Consider the partition function which includes the sum over topologies and the sum over field configurations

$$Z_{\text{total}} = \sum_{\tau}\sum_{\varphi} e^{-S}. \tag{5}$$

For the sake of simplicity we use from the very beginning the Euclidean approach. The action has the form

$$S = -\frac{1}{2}\left(\varphi\widehat{A}\varphi\right) + \left(J\varphi\right) \tag{6}$$

and we use the notions $(J\varphi) = \int J(x)\varphi(x)d^4x$. If we fix the topology of space by placing a set of wormholes with parameters ξ_i, then the sum over field configurations φ gives the well-known result

$$Z^*(J) = Z_0\left(\widehat{A}\right)e^{-(1/2)(J\widehat{A}^{-1}J)}, \tag{7}$$

where $Z_0(\widehat{A}) = \int [D\varphi]e^{(1/2)(\varphi\widehat{A}\varphi)}$ is the standard expression and $\widehat{A}^{-1} = A^{-1}(\xi)$ is the Green function for a fixed topology, that is, for a fixed set of wormholes $\xi_1, \ldots, \xi_N$.

Consider now the sum over topologies τ. To this end we restrict the sum over the number of wormholes and integrals over parameters of wormholes:

$$\sum_{\tau} \longrightarrow \sum_{N}\int\prod_{i=1}^{N} d\xi_i = \int [DF], \tag{8}$$

where

$$F(\xi, N) = \frac{1}{N}\sum_{i=1}^{N}\delta(\xi - \xi_i) \tag{9}$$

and NF is the density of wormholes in the configuration space ξ. We also point out that in general the integration over parameters is not free (e.g., it obeys the obvious restriction $|\vec{R}_i^+ - \vec{R}_i^-| \geq 2a_i$). This defines the generating function as

$$Z_{\text{total}}(J) = \int [DF]\, Z_0\left(\widehat{A}\right)e^{-(1/2)(J\widehat{A}^{-1}J)}. \tag{10}$$

The sum over topologies assumes an additional averaging out for all mean values with the measure $d\mu_N = \rho(\xi, N)d^N\xi$, where

$$\rho(\xi, N) = \frac{Z_0\left(\widehat{A}(\xi, N)\right)}{Z_{\text{total}}(0)}, \tag{11}$$

which obey the obvious normalization condition $\sum_N \int d\mu_N = \sum_N \rho_N = 1$. The averaging out over topologies assumes the two stages. First we fix the total number of wormholes N and average over the parameters of wormholes ξ (i.e., over parameters of a static gas of wormholes in R^4). Then we sum over the number of wormholes N (the so-called big canonical ensemble).

The basic difficulty of the standard field theory is that the perturbation scheme based on (7) leads to divergent expressions. This remains true for every particular topology of space (i.e., for any particular finite set of wormholes), since there always exists a scale below which the space looks like the ordinary Euclidean space. What we expect is that the sum over all possible topologies will remove such a difficulty.

And indeed, the above measure (11) has the structure

$$Z_0\left(\widehat{A}(\xi, N)\right) = \exp\left(-\int \Lambda(\xi, N)\, d^4x\right), \tag{12}$$

where $\Lambda(\xi, N)$ is the cosmological constant related to the energy density of zero-point fluctuations calculated for a particular distribution of wormholes. (We recall that the total cosmological constant should include also the contribution from the mean curvature (55).) Any finite distribution of wormholes leads to the divergent expression $\Lambda(\xi, N) \rightarrow \infty$ and is suppressed (i.e., $\rho(\xi, N) \rightarrow 0$). However, the sum over all possible topologies assumes also the limiting topologies $n \rightarrow \infty$, where $n = N/V$ is the density of wormholes. In this limit wormhole throats degenerate into points and the minimal scale below which the space looks like the Euclidean space is merely absent. We point out that from the rigorous mathematical standpoint such limiting topologies cannot be described in terms of smooth manifolds, since they are not locally Euclidean and does not possess a finite set of maps. In mathematics similar objects are well known, for example, fractal sets. However, if a fractal set is obtained by cutting (by means of a specific rule or iterations) portions of space, our limiting topologies are obtained by gluing (identifying) some portions (or in the limit couples of points) of the Euclidean space. The basic feature of such topologies is that QFT becomes finite on such a set. Indeed, as we shall see a particular infinite distribution of wormholes can always be chosen in such a way that the energy of zero-point fluctuations becomes a finite $0 \leq \Lambda_{\infty}(\xi) < \infty$ (e.g., see the next section or the second term in (67)). In the sum over topologies only such limiting topologies do survive (i.e., $\rho_{\infty}(\xi) \neq 0$).

3. The Two-Point Green Function

From (7) we see that the very basic role in QFT plays the two point Green function. Such a Green function can be found from the equation

$$\widehat{A}G(x, x') = -\delta(x - x') \tag{13}$$

with proper boundary conditions at wormholes, which gives $G = A^{-1}$. Now let us introduce the bias function $N(x, x')$ as

$$G(x, y) = \int G_0(x, x') N(x', y) dx', \tag{14}$$

where $G_0(x, x')$ is the ballistic (or the standard Euclidean Green function) and the bias can be presented as

$$N(x, x') = \delta(x - x') + \sum_i b_i \delta(x - x_i), \tag{15}$$

where b_i are fictitious sources at positions x_i which should be added to obey the proper boundary conditions. We point out that the bias can be explicitly expressed via parameters of wormholes; that is, $N(x, x') = N(x, x', \xi_1, \ldots \xi_N)$. For the sake of illustration we consider first a particular example.

3.1. The Bias for a Particular Distribution of Wormholes (Rarefied Gas Approximation). Consider now the bias for a particular set of wormholes. For the sake of simplicity we consider the case when $m = 0$. The Green function obeys the Laplace equation

$$-\Delta G(x, x') = \delta(x - x') \tag{16}$$

with proper boundary conditions at throats (we require G and $\partial G/\partial n$ to be continual at throats). The Green function for the Euclidean space is merely $G_0(x, x') = 1/4\pi^2(x - x')^2$ (and $G_0(k) = 1/k^2$ for the Fourier transform). In the presence of a single wormhole which connects two Euclidean spaces this equation admits the exact solution. For outer region of the throat S^3 the source $\delta(x - x')$ generates a set of multipoles placed in the center of sphere which gives the corrections to the Green function G_0 in the form (we suppose the center of the sphere at the origin)

$$\delta G = -\frac{1}{4\pi^2 x^2} \sum_{n=1}^{\infty} \frac{1}{n+1} \left(\frac{a}{x'}\right)^{2n} \left(\frac{x'}{x}\right)^{n-1} Q_n, \tag{17}$$

where $Q_n = (4\pi^2/2n) \sum_{l=0}^{n-1} \sum_{m=-l}^{l} Q^{*\prime}_{nlm} Q_{nlm}$ and $Q_{nlm}(\Omega)$ are four-dimensional spherical harmonics, for example, see [21]. In the present section we shall consider a dilute gas approximation and, therefore, it is sufficient to retain the lowest (monopole) term only. A single wormhole which connects two regions in the same space is a couple of conjugated spheres $S^3_{\pm}$ of the radius a with a distance $\vec{X} = \vec{R}_+ - \vec{R}_-$ between centers of spheres. So the parameters of the wormhole are (The additional parameter (rotation matrix Λ) is important only for multipoles of higher orders.) $\xi = (a, R_+, R_-)$. The interior of the spheres is removed and surfaces are glued together. Then the proper boundary conditions (the actual topology) can be accounted for by adding the bias of the source

$$\delta(x - x') \longrightarrow \delta(x - x') + b(x, x'). \tag{18}$$

In the approximation $a/X \ll 1$ (e.g., see for details [9]) the bias for a single wormhole takes the form

$$b_1(x, x', \xi) = \frac{a^2}{2} \left(\frac{1}{(R_- - x')^2} - \frac{1}{(R_+ - x')^2} \right) \times \left[\delta(\vec{x} - \vec{R}_+) - \delta(\vec{x} - \vec{R}_-)\right]. \tag{19}$$

This form for the bias is convenient when constructing the true Green function and considering the long-wave limit; however it is not acceptable in considering the short-wave behavior and vacuum polarization effects. Indeed, the positions of additional sources are in the physically nonadmissible region of space (the interior of spheres $S^3_{\pm}$). To account for the finite value of the throat size we should replace in (19) the point-like source with the surface density (induced on the throat); that is,

$$\delta(\vec{x} - \vec{R}_{\pm}) \longrightarrow \frac{1}{2\pi^2 a^3} \delta(|\vec{x} - \vec{R}_{\pm}| - a). \tag{20}$$

Such a replacement does not change the value of the true Green function; however, now all extra sources are in the physically admissible region of space.

In the rarefied gas approximation the total bias is additive; that is,

$$b_{\text{total}}(x, x') = \sum b_1(x, x', \xi_i) = N \int b_1(x, x', \xi) F(\xi) d\xi, \tag{21}$$

where NF is given by (9). For a homogeneous and isotropic distribution $F(\xi) = F(a, X)$, and then for the bias we find

$$b_{\text{total}}\left(x - x'\right) = \int \frac{1}{2\pi^2 a}\left(\frac{1}{R_-^2} - \frac{1}{R_+^2}\right)\delta\left(\left|\vec{x} - \vec{x}' - \vec{R}_+\right| - a\right) NF(\xi)\, d\xi. \quad (22)$$

Consider the Fourier transform $F(a,X) = \int F(a,k)e^{-ikX} \cdot (d^4k/(2\pi)^4)$ and using the integral $1/x^2 = \int (4\pi^2/k^2)e^{-ikx} \cdot (d^4k/(2\pi)^4)$ we find for $b(k) = \int b(x)e^{ikx}d^4x$ the expression

$$b_{\text{total}}(k) = N\int a^2 \frac{4\pi^2}{k^2}\left(F(a,k) - F(a,0)\right)\frac{J_1(ka)}{ka/2}da. \quad (23)$$

(1) Example of a Finite Density of Wormholes. Consider now a particular (of a finite density) distribution of wormholes $F(a, X)$, for example,

$$NF(a,X) = \frac{n}{2\pi^2 r_0^3}\delta(a - a_0)\,\delta(X - r_0), \quad (24)$$

where $n = N/V$ is the density of wormholes. In the case $N = 1$ this function corresponds to a single wormhole with the throat size a_0 and the distance between throats $r_0 = |R_+ - R_-|$. We recall that the action (6) remains invariant under translations and rotations which straightforwardly leads to the above function. Then $NF(a,k) = \int NF(a,X)e^{ikx}d^4x$ reduces to $NF(a,k) = n(J_1(kr_0)/(kr_0/2))\delta(a - a_0)$. Thus from (23) we find

$$b(k) = -na^2\frac{4\pi^2}{k^2}\left(1 - \frac{J_1(kr_0)}{kr_0/2}\right)\frac{J_1(ka_0)}{ka_0/2}. \quad (25)$$

And for the true Green function we get

$$G_{\text{true}} = G_0(k)\,N(k) = G_0(k)\,(1 + b(k)). \quad (26)$$

In the short-wave limit (ka, $kr_0 \gg 1$) $b(k) \to 0$ and therefore $N(k) \to 1$. This means that at very small scales the space filled with a finite density of wormholes looks like the ordinary Euclidean space. In the long-wave limit $k \to 0$ we get $J_1(kr_0)/(kr_0/2) \approx 1 - (1/2)\,(kr_0/2)^2 + \cdots$ which gives $b(k) \approx -\pi^2 na^2 r_0^2/2$, while in a more general case we find $b(k) \approx -\int(\pi^2/2)a^2r_0^2 n(a,r_0)da\,dr_0$, where $n(a, r_0)$ is the density of wormholes with a particular values of a and r_0, and for the bias function (15) we get

$$N(k) \longrightarrow 1 - \frac{\pi^2}{2}\int a^4 n(a,r_0)\frac{r_0^2}{a^2}da\,dr_0 \le 1. \quad (27)$$

In other words, in the long-wave limit (ka, $kr_0 \ll 1$) the presence of a particular set of virtual wormholes diminishes merely the value of the charge values.

(2) Limiting Topologies or Infinite Densities of Wormholes. Consider now the limiting distribution when the density of wormholes $n \to \infty$. Since every throat cuts the finite portion of the volume $(\pi^2/2)a^4$, this case requires considering the limit $a \to 0$. We assume that in this limit $a^2NF(a,X) \to \delta(a)\nu(X)$, where $\nu(X)$ is a finite specific distribution. Then (23) reduces to $b_{\text{total}}(k) = (4\pi^2/k^2)\,(\widetilde{\nu}(k) - \widetilde{\nu}(0))$ where $\widetilde{\nu}(k) = \int \nu(X)e^{ikX}d^4X$ and the bias (15), (18) $N(k)$ becomes

$$N(k) = 1 - \frac{4\pi^2}{k^2}\left(\widetilde{\nu}(0) - \widetilde{\nu}(k)\right). \quad (28)$$

It is important that this limit still agrees with the rarefied gas approximation, for the basic gas parameter (i.e., the portion of volume cut by wormholes) tends to $\xi = \int(\pi^2/2)a^4F(a,X)d^4X da \to 0$. The above expression is obtained in the linear approximation only. Taking into account next orders (e.g., see [22] where the case of a dense gas is also considered) we find $N(k) = 1 + b_{\text{total}}(k) + b_{\text{total}}^2(k) + \cdots$ and the true Green function in a gas of wormholes becomes

$$G_{\text{true}} = G_0(k)\,N(k) = \frac{1}{k^2 + 4\pi^2\left(\widetilde{\nu}(0) - \widetilde{\nu}(k)\right)}. \quad (29)$$

In the long-wave limit $k \to 0$ the function $\widetilde{\nu}(k)$ can be expanded as $\widetilde{\nu}(k) \approx \widetilde{\nu}(0) + (1/2)\widetilde{\nu}''(0)k^2$ which also defines a renormalization of charge values $N(k) \to 1/(1 - 2\pi^2\widetilde{\nu}''(0))$ which coincides with (27). However in this limiting case we have some freedom in the choice of $\widetilde{\nu}(k)$, which we can use to assign $N(k)$ an arbitrary function of k. In other words we may get here a class of limiting topologies where Green functions $G_{\text{true}} = G_0(k)N(k)$ have a good ultraviolet behavior and quantum field theories in such spaces turn out to be finite. In particular, this will certainly be the case when

$$G_{\text{true}}\left(x - x' = 0\right) = \int \frac{1}{k^2 + 4\pi^2\left(\widetilde{\nu}(0) - \widetilde{\nu}(k)\right)}\frac{d^4k}{(2\pi)^4} < \infty. \quad (30)$$

This possibility however requires the further and more deep investigation.

3.2. Green Function, General Consideration. The action (6) remains invariant under translations $\vec{x}' = \vec{x} + \vec{c}$ with an arbitrary $\vec{c}$ which means that the measure (11) does not actually depend on the position of the center of mass of the gas of wormholes and, therefore, we may restrict ourself with homogeneous distributions $F(\xi)$ of wormholes in space only. Indeed, we may define $d^N\xi = d^N\xi' d^4c$, while the integration over d^4c gives the volume of R^4; that is, $\int d^4c = L^4 = V$ which disappears from (11) due to the denominator. (Technically, we may first restrict a portion of R^4 in (6) to a finite volume V and then in final expressions consider the limit $V \to \infty$ (which represents the standard tool in thermodynamics and QFT).) In what follows we shall omit the prime from ξ'.

Let us consider the Fourier representation $N(x, x', \xi) \to N(k, k', \xi)$ which in the case of a homogeneous distribution of wormholes gives $N(k,k') = N(k,\xi)\,\delta(k - k')$; then we find

$$G(k) = G_0(k)\,N(k,\xi) \quad (31)$$

and the Green function can be taken as

$$G = \frac{N(k,\xi)}{k^2 + m^2}. \quad (32)$$

Then for the total partition function we find

$$Z_{\text{total}}(J) = \int [DN(k)]\, e^{-I(N)} e^{-(1/2)(L^4/(2\pi)^4)\int((N(k)/(k^2+m^2))|J_k|^2)dk}, \quad (33)$$

where $[DN] = \prod_k dN_k$ and $\sigma(N)$ comes from the integration measure (i.e., from the Jacobian of transformation from $F(\xi)$ to $N(k)$)

$$e^{-I(N)} = \int [DF]\, Z_0(N(k,\xi))\, \delta(N(k) - N(k,\xi)). \quad (34)$$

We point out that $I(N)$ can be changed by means of adding to the action (6) of an arbitrary "nondynamical" constant term which depends only on topology $S \to S + \Delta S(N(k))$ (e.g., a topological Euler term). The multiplier $Z_0(N)$ defines the simplest measure for topologies. Now by means of using the expression (32) and (33) we find the two-point Green function in the form

$$G(k) = \frac{\overline{N}(k)}{k^2 + m^2}, \quad (35)$$

where $\overline{N}(k)$ is the cutoff function (the mean bias) which is given by (We recall that in this integral contribute only limiting topologies in which density of wormholes diverges $(n \to \infty)$)

$$\overline{N}(k) = \frac{1}{Z_{\text{total}}(0)} \int [DN]\, e^{-I(N)} N(k). \quad (36)$$

At the present stage we still cannot evaluate the exact form for the cutoff function $\overline{N}(k)$ in virtue of the ambiguity of $\Delta S(N(k))$ pointed out. Such a term may include two parts. First part $\Delta_1 S$ describes the proper dynamics of wormholes and should be considered separately. Indeed, in general wormholes are dynamical self-gravitating objects which require considering the gravitational contribution to the action. Some part of such a contribution (mean curvature induced by wormholes) is discussed in Section 6. However, since a wormhole represents an extended nonlocal object, it possesses a rather complex dynamics and this problem requires the further investigation. The second part $\Delta_2 S$ may describe "external conditions" (e.g., an external classical field in (33)) for the mean topology. Actually the last term can be used to prescribe an arbitrary particular value for the cutoff function $\overline{N}(k) = f(k)$. Indeed, the "external conditions" can be accounted for by adding the term $\Delta_2 S = (\lambda, N) = \int \lambda(k) N(k) d^4k$, where $\lambda(k)$ plays the role of a specific chemical potential which implicitly depends on $f(k)$ through the equation

$$f(k) = \frac{1}{Z_{\text{total}}(\lambda, 0)} \int [DN]\, e^{(\lambda,N) - I(N)} N(k). \quad (37)$$

From (33) we see that the role of such a chemical potential may play the external current $\lambda(k) = -(1/2)(L^4/(2\pi)^4) \cdot G_0(k)|J_k^{\text{ext}}|^2$ or equivalently an external classical field $\varphi^{\text{ext}} = G_0(k)J_k^{\text{ext}}$. In quantum field theory such a term leads merely to a renormalization of the cosmological constant. By other words the mean topology (i.e., the cutoff function or mean distribution of wormholes) is driven by the cosmological constant Λ and vice versa.

4. Topological Bias as a Projection Operator

By the construction the topological bias $N(x, x')$ plays the role of a projection operator onto the space of functions (a subspace of functions on R^4) which obey the proper boundary conditions at throats of wormholes. This means that for any particular topology (for a set of wormholes) there exists the basis $\{f_i(x)\}$ in which it takes the diagonal form $N(x, x') = \sum N_i f_i(x) f_i^*(x')$ with eigenvalues $N_i = 0, 1$ (since $N_i^2 = N_i$). In this section we illustrate this simple fact (which is probably not obvious for readers) by the explicit construction of the reference system for a single wormhole when physical functions become (due to the boundary conditions) periodic functions of one of coordinates.

Indeed, consider a single wormhole with parameters ξ (i.e., $\xi = (a, R^+, R^-)$, where a is the throat radius and $R^\pm$ are positions of throats in space. (In general, there exists an additional parameter Λ_β^α which defines a rotation of one of throats before gluing. However, it does not change the subsequent construction. There always exists a diffeomorphic map of coordinates $x' = h(x)$ which sets such a matrix to unity).) Consider now a particular solution ϕ_0 to the equation $\Delta\phi_0 = 0$ (harmonic function) for R^4 in the presence of the wormhole, which corresponds to the situation when throats possess a unit charge/mass but those have the opposite signs. Now define the family of lines of force $x(s, x_0)$ which obey the equation $dx/ds = -\nabla\phi_0(x)$ with initial conditions $x(0) = x_0$. Physically, such lines correspond to lines of force for a two charged particles in positions $R^\pm$ with charges ± 1. We note that all points which lay on the trajectory $x(s, x_0)$ may be taken as initial conditions and they define the same line of force with the obvious redefinition $s \to s - s_0$. By other words we may take as a new coordinates the parameter s and portion of the coordinates orthogonal to the family of lines $x_0^\perp$. Coordinates $x_0^\perp$ can be taken as laying in the hyperplane R^3 which is orthogonal to the vector $\vec{d} = \vec{R}^- - \vec{R}^+$ and goes through the point $\vec{X}_0 = (\vec{R}^- + \vec{R}^+)/2$. (Instead of the construction used here one may use also another method. Indeed, consider two point charges, and then the function $\phi_0 = 1/(x - x_+)^2 - 1/(x - x_-)^2$ can be taken as a new coordinate. Wormhole appears when we identify (glue) surfaces $\phi_0 = \pm\omega$. We point out that such surfaces are not spheres, though they reduce to spheres in the limit $|x_+ - x_-| \to \infty$ or $\omega \to \infty$.)

Let $s^\pm(x_0^\perp)$ be the values of the parameter s at which the line intersects the throats $R^\pm$. Then instead of s we may consider a new parameter θ as $s(\theta) = s^- + (s^+ - s^-)\theta/2\pi$, so that when $\theta = 0, 2\pi$ the parameter s takes the values

$s = s^-$, s^+ respectively. The gluing procedure at throats means merely that we identify points at $\theta = 0$ and $\theta = 2\pi$ and all physical functions in the space R^4 with a single wormhole ξ become periodic functions of θ. Thus, the coordinate transformation $x = x(\theta, x_0^\perp)$ gives the map of the above space onto the cylinder with a specific metric $dl^2 = (d\vec{x}(\theta, x_0^\perp))^2 = g_{\alpha\beta}dy^\alpha dy^\beta$ (where $y = (\theta, x_0^\perp)$) whose components are also periodic in terms of θ. Now we can continue the coordinates to the whole space R^4 (to construct a cover of the fundamental region $\theta \in [0, 2\pi]$) simply admitting all values $-\infty < \theta < +\infty$ this, however, requires to introduce the bias

$$\frac{1}{\sqrt{g}}\delta(\theta - \theta') \longrightarrow N(\theta - \theta') = \sum_{n=-\infty}^{+\infty} \frac{1}{\sqrt{g}}\delta(\theta - \theta' + 2\pi n), \tag{38}$$

since every point and every source in the fundamental region acquires a countable set of images in the nonphysical region (inside of wormhole throats). Considering now the Fourier transforms for θ we find

$$N(k, k') = \sum_{n=-\infty}^{+\infty} \delta(k - n)\,\delta(k - k'). \tag{39}$$

We point out that the above bias gives the unit operator in the space of periodic functions of θ. From the standpoint of all possible functions on R^4 it represents the projection operator $\widehat{N}^2 = \widehat{N}(\xi)$ (taking an arbitrary function f we find that upon the projection $f_N = \widehat{N}f$ f_N becomes a periodic function of θ; that is, only periodic functions survive).

The above construction can be easily generalized to the presence of a set of wormholes. In the approximation of a dilute gas of wormholes we may neglect the influence of wormholes on each other (at least there always exists a sufficiently smooth map which transforms the family of lines of force for "independent" wormholes onto the actual lines). Then the total bias (projection) may be considered as the product

$$N_{\text{total}}(x, x') = \int \left(\prod_i \sqrt{g_i} d^4 y_i\right) N(\xi_1, x, y_1) \times N(\xi_2, y_1, y_2) \cdots N(\xi_N, y_{N-1}, x'), \tag{40}$$

where $N(\xi_i, x, x')$ is the bias for a single wormhole with parameters ξ_i. Every such a particular bias $N(\xi_i, x, x')$ realizes projection on a subspace of functions which are periodic with respect to a particular coordinate $\theta_i(x)$, while the total bias gives the projection onto the intersection of such particular subspaces (functions which are periodic with respect to every parameter θ_i).

5. Cutoff

The projective nature of the bias operator $N(x, x')$ allows us to express the cutoff function $\overline{N}(k)$ via dynamic parameters of wormholes. Indeed, consider a box L^4 in R^4 and periodic boundary conditions which gives $k = 2\pi n/L$ (in final expressions we consider the limit $L \to \infty$, which gives $\sum_k \to (L^4/(2\pi)^4)\int d^4k$). And let us consider the decomposition for the integration measure in (33) as

$$I = I_0 + \sum \lambda_1(k)\,N(k) + \frac{1}{2}\sum \lambda_2(k, k')\,N(k)\,N(k') + \cdots, \tag{41}$$

where $\lambda_1(k)$ includes also the contribution from $Z_0(k)$. We point out that this measure plays the role of the action for the bias $N(k)$. Indeed, the variation of the above expression gives the equation of motions for the bias in the form

$$\sum_{k'} \lambda_2(k, k')\,N(k') = -\lambda_1(k), \tag{42}$$

which can be found by considering the proper dynamics of wormholes. We however do not consider the problem of the dynamic description of wormholes here and leave this for the future research. Moreover, we may expect that in the first approximation one may retain the linear term only. Indeed, this takes place when $N(k)$ is not a dynamic variable, or if we take into account that $N(k)$ is a collective variable. Then the projective nature of the bias $N(k) = 0, 1$ means that it can be phenomenologically expressed via some Fermionic ghost field $\Psi(k)$ (e.g., $N(k, k') = \Psi(k)\,\Psi^+(k')$) where the negative and positive frequency parts of the operator $\Psi(k)$ obey the anticommutation relations $\Psi^+(k)\,\Psi(k') + \Psi(k')\,\Psi^+(k) = \delta(k - k')$. In the absence of ghost particles $\Psi(k)|0\rangle = 0$ we get $N(k, k') = \delta(k - k')$; that is, $N(k) = 1$ and wormholes are absent. In the term of the ghost field the action becomes $I(N) = I_0 + (\Psi, \widehat{\lambda}_1\Psi) + \cdots$. Therefore in the leading approximation equations of motion take the linear form $\widehat{\lambda}_1\Psi = 0$.

Thus taken into account that $N(k) = 0, 1$ ($N^2 = N$) we find

$$\overline{N}(k) = \frac{1}{Z_{\text{total}}(k)} \sum_{N=0,1} e^{-\lambda_1(k)N(k)} N(k) = \frac{e^{-\lambda_1(k)}}{1 + e^{-\lambda_1(k)}}. \tag{43}$$

The simplest choice gives merely $\lambda_1(k) = -\sum \ln Z_0(k)$, where the sum is taken over the number of fields and $Z_0(k)$ is given by $Z_0(k) = \sqrt{\pi/(k^2 + m^2)}$. In the case of a set of massless fields we find $\overline{N}(k) = Z(k)/(1 + Z(k))$ where $Z(k) = (\sqrt{\pi}/k)^\alpha$ and α is the effective number of degrees of freedom (the number of boson minus fermion fields). To ensure the absence of divergencies one has to consider the number of fields $\alpha > 4$ [23]. However, such a choice gives the simplest estimate which, in general, cannot be correct. Indeed, while its behavior at very small scales (i.e., when exceeding the Planckian scales $Z(k) \lesssim 1$ and $\overline{N}(k) = Z(k)$) may be physically accepted, since it produces some kind of a cutoff, on the mass-shell $k^2 + m^2 \to 0$ it gives the behavior $\overline{N}(k) \to 1$ which is merely incorrect (e.g., from (27) we see that the true behavior should be $\overline{N}(k) \to \text{const} < 1$).

One may expect that the true cutoff function has a much more complex behavior. Indeed, some theoretical models in particle physics (e.g., string theory) have the property to be lower-dimensional at very small scales. The mean cutoff

$\overline{N}(k)$ gives the natural tool to describe a scale-dependant dimensional reduction [24, 25]. In fact, this function defines the spectral number of modes in the interval between k and $k + dk$ as

$$\int \overline{N}(k) \frac{d^4k}{(2\pi)^4} = \int \frac{\overline{N}(k) k^4}{(2\pi)^2} \frac{dk}{k}. \quad (44)$$

Hence we can define the effective spectral dimension D of space as follows:

$$k^4 \overline{N}(k) \sim k^D. \quad (45)$$

From the empirical standpoint the dimension $D = 4$ is verified at laboratory scales only, while the rigorous tool to define the spectral density of states (or the mean cutoff) can give the lattice quantum gravity, for example, see [26, 27] and references therein. And indeed, the spectral dimension for nonperturbative quantum gravity defined via Euclidean dynamical triangulations was calculated recently in [28]. It turns out that it runs from a value of $D = 3/2$ at short distance to $D = 4$ at large distance scales. We also point out that all observed dark matter phenomena can be explained by the fractal dimension $D \approx 2$ starting from scales $L \gtrsim (1 \div 5)$ Kpc, for example, [29–33].

6. Cosmological Constant

Let us consider the total Euclidean action [17]

$$I_E = -\frac{1}{16\pi G} \int (R - 2\Lambda_0) \sqrt{g} d^4x - \int L_m \sqrt{g} d^4x. \quad (46)$$

The variation of the above action leads to the Einstein equations

$$R_{ab} - \frac{1}{2} g_{ab} R + g_{ab} \Lambda_0 = 8\pi G T_{ab}, \quad (47)$$

where $T^{ab} = (1/2)(g)^{-1/2}(\delta L_m / \delta g_{ab})$ is the stress energy tensor and Λ_0 is a naked cosmological constant. In cosmology such equations are considered from the classical standpoint, which means that they involve characteristic scales $\ell \gg \ell_{\rm pl}$. However, the presence of virtual wormholes at Planckian scales defines some additional contribution in both parts of these equations which can be adsorbed into the cosmological constant. Therefore the total cosmological constant can be defined as

$$\Lambda_{\rm tot} = \Lambda_0 + \Lambda_m + \Lambda_R = \Lambda_0 + 2\pi G \langle T \rangle + \frac{1}{4} \langle R \rangle_w, \quad (48)$$

where $\langle T \rangle$ is the energy of zero-point fluctuations. (It includes also the contribution of zero-point fluctuations of gravitons.) That is, the mean vacuum value (we recall that in the standard QFT Λ_m is infinite, while wormholes form a finite value) and $\langle R \rangle_w = \Lambda_R$ is a contribution of wormholes into the mean curvature due to gluing (1).

6.1. Contribution of Virtual Wormholes into Mean Curvature. Consider a single wormhole whose metric is given by (1) $ds^2 = h^2(r)\delta_{\alpha\beta} dx^\alpha dx^\beta$; then the components of the Ricci tensor are (We are much obliged to the Referee who pointed out to the subtleties when working with a step function in the metric.)

$$R_{\alpha\beta} = 7 - \left(\frac{h'}{h}\right)' \left(\delta_{\alpha\beta} + (N-2) n_\alpha n_\beta\right) - \frac{1}{r}\frac{h'}{h} \left((N-2) \Delta_{\alpha\beta} + \delta_{\alpha\beta}(N-1)\right) - \left(\frac{h'}{h}\right)^2 (N-2) \Delta_{\alpha\beta}, \quad (49)$$

where $n^\nu = x^\nu / a$ is the unite normal vector to the throat surface, $\Delta_{\alpha\beta} = \delta_{\alpha\beta} - n_\alpha n_\beta$, $h' = \partial h / \partial r$, and N is the number of dimensions. The curvature scalar is

$$R = -\frac{(N-1)}{h^2} \left[2\frac{h''}{h} + 2\frac{1}{r}\frac{h'}{h}(N-1) + \left(\frac{h'}{h}\right)^2 (N-4) \right]. \quad (50)$$

Then substituting $h = 1 + (a^2/r^2 - 1)\theta$ in the above equation we find

$$-\frac{h^4}{N-1} R = 4h\frac{a^2}{r^3}\delta + \frac{2h}{r^{N-1}}\left(r^{N-1}\lambda\right)' + (N-4)\left(\lambda^2 - 4\theta\frac{a^2}{r^3}\lambda\right) + 4(N-4)\frac{a^2}{r^4}\theta(\theta - 1), \quad (51)$$

where θ is a smooth function which only in a limit becomes a step function, $\lambda = (1 - a^2/r^2)\delta$, and $\delta = -\theta'$. In the limit $\theta \to \theta(a - r)$ we find $\delta \to \delta(r - a)$ and the last two terms in (51) are negligible as compared to the first two terms, while in four dimensions the last two terms vanish. Then the curvature is concentrated on the throat of the wormhole where θ differs from the step function. In the limit of a vanishing throat size we have $h \to 1$ at the throat and we get

$$-R = \frac{4(N-1)}{a}\delta(r-a) + \frac{2(N-1)}{r^{N-1}}\left(r^{N-3}\left(r^2 - a^2\right)\delta\right)'. \quad (52)$$

In general all the terms together in (51) and analogous terms in the Ricci tensor provide that the Bianchi identity holds and the energy is conserved [19]. Therefore none of them can be dropped out. However it is easy to see the leading contribution to the integral over space comes from the first term only and in the limit of a vanishing throat size we find for a single wormhole

$$\frac{1}{4}\int R\sqrt{g} d^4x = -6\pi^2 a^2. \quad (53)$$

In the case of a set of wormholes (24) we find

$$\frac{1}{4}\int R\sqrt{g}d^4x = -6\pi^2\sum_j a_j^2$$
$$= -12\pi^2\int n(a)\,a^2 da\,d^4x = \int \Lambda_R d^4x, \quad (54)$$

where $2n(a) = \int n(a,r_0)dr_0$ is the density of wormhole throats with a fixed value of the throat size a. This defines the contribution to the cosmological constant from the mean curvature as

$$\Lambda_R = -12\pi^2\int n(a)\,a^2 da < 0. \quad (55)$$

We see that this quantity is always negative.

6.2. Stress Energy Tensor. In this section we consider the contribution from matter fields. In the case of a scalar field the stress energy tensor has the form

$$-T_{\alpha\beta}(x) = \partial_\alpha\varphi\partial_\beta\varphi - \frac{1}{2}g_{\alpha\beta}\left(\partial^\mu\varphi\partial_\mu\varphi + m^2\varphi^2\right). \quad (56)$$

Then the mean vacuum value of the stress energy tensor can be obtained directly from the two-point green function (32), (35) as

$$-\left\langle T_{\alpha\beta}(x)\right\rangle$$
$$= \lim_{x'\to x}\left(\partial_\alpha\partial'_\beta - \frac{1}{2}g_{\alpha\beta}\left(\partial^\mu\partial'_\mu + m^2\right)\right)\left\langle G\left(x,x',\xi\right)\right\rangle. \quad (57)$$

By means of using the Fourier transform $G(x,x',\xi) = \int e^{-ik(x-x')}G(k,\xi)\,(d^4k/(2\pi)^4)$ and the expressions (32), (35) we arrive at

$$\left\langle T_{\alpha\beta}(x)\right\rangle = \frac{1}{4}g_{\alpha\beta}\int\left(1+\frac{m^2}{k^2+m^2}\right)\overline{N}(k)\frac{d^4k}{(2\pi)^4}, \quad (58)$$

where the property $\int k_\alpha k_\beta f(k^2)d^4k = (1/4)g_{\alpha\beta}\int k^2 f(k^2)d^4k$ has been used and $\overline{N}(k) = \langle N(k,\xi)\rangle$ is the cutoff function (36).

For the sake of simplicity we consider the massless case. Then by the use of the cutoff $\overline{N}(k) = \pi^{\alpha/2}/(\pi^{\alpha/2}+k^\alpha)$ from the previous section we get the finite estimate ($\alpha > 4$ is the effective number of the field helicity states)

$$\Lambda_m = 2\pi G\sum\int\frac{\pi^{\alpha/2}}{\pi^{\alpha/2}+k^\alpha}\frac{d^4k}{(2\pi)^4} = \frac{\pi G}{4}\Gamma\left(\frac{\alpha-4}{\alpha}\right)\Gamma\left(\frac{4}{\alpha}\right) \sim 1, \quad (59)$$

where the sum is taken over the number of fields. Since the leading contribution comes here from very small scales, we may hope that this value will not essentially change if the true cutoff function changes the behavior on the mass-shell as $k \to 0$ (e.g., if we take $\lambda_1(k) = -\sum\ln Z_0(k) + \delta\lambda(k)$ with $\delta\lambda(k) \ll \ln Z_0(k)$ as $k \gg 1$).

To understand how wormholes remove divergencies, it will be convenient to split the bias function into two parts $N(k,\xi) = 1 + b(k,\xi)$, where 1 corresponds to the standard Euclidean contribution, while $b(k,\xi)$ is the contribution of wormholes. The first part gives the well-known divergent contribution of vacuum field fluctuations $8\pi G\langle T^0_{\alpha\beta}\rangle = \Lambda_* g_{\alpha\beta}$ with $\Lambda_* \to +\infty$, while the second part remains finite for any finite number of wormholes and, due to the projective nature of the bias described in the previous section, it partially compensates (reduces) the value of the cosmological constant; that is, $8\pi G\langle\Delta T_{\alpha\beta}\rangle = \delta\Lambda g_{\alpha\beta}$, where $\delta\Lambda = \sum_N \rho_N\delta\Lambda(N)$ and $\delta\Lambda(N)$ is a negative finite contribution of a finite set of wormholes.

Consider now the particular distribution of virtual wormholes (24) and evaluate their contribution to the cosmological constant which is given by $\delta\Lambda(N) = 2\pi G\int b(k)\,(d^4k/(2\pi)^4) = 2\pi G b_{\text{total}}(0)$. Then from the expressions (22) and (24) we get

$$b_{\text{total}}(0) = -\frac{n}{4\pi^4 a^3 r_0^3}\int\left(1-\frac{a^2}{R_-^2}\right)\delta\left(R_+ - a\right)$$
$$\times\,\delta\left(\left|R_+ - R_-\right| - r_0\right)d^4R_- d^4R_+, \quad (60)$$

which gives

$$b_{\text{total}}(0) = -n\left(1 - f\left(\frac{a}{r_0}\right)\right), \quad (61)$$

where

$$f\left(\frac{a}{r_0}\right) = \frac{2}{\pi}\int_0^\pi\frac{a^2\sin^2\theta d\theta}{a^2+2ar_0\cos\theta+r_0^2}. \quad (62)$$

For $a/r_0 \ll 1$ (we recall that by the construction $a/r_0 \le 1/2$) this function has the value $f(a/r_0) \approx a^2/r_0^2$. Thus, for the contribution of wormholes we find

$$\delta\Lambda_m = -2\pi G\int n(a,r_0)\left(1 - f\left(\frac{a}{r_0}\right)\right)da\,dr_0$$
$$= -2\pi G n\left(1 - \langle f\rangle\right). \quad (63)$$

6.3. Vacuum Value of the Cosmological Constant. From the above expression we see that to get the finite value of the cosmological constant $\Lambda_m = \Lambda_* + \delta\Lambda_m < \infty$ one should consider the limit $n \to \infty$ (infinite density of virtual wormholes) which requires considering the smaller and smaller wormholes. From the other hand we have the obvious restriction $\int 2n(a,r_0)(\pi^2/2)a^4 da\,dr_0 < 1$, where $(\pi^2/2)a^4$ is the volume of one throat (wormholes cannot cut more than the total volume of space). (We also point out that in removing divergencies the leading role plays the zero-point energy. Indeed $\delta\Lambda_m \sim -2\pi G n$, while the mean curvature has the order $\Lambda_R \sim -a^2 n$ and for $a \ll \ell_{\text{pl}}$ we have $\delta\Lambda_m \gg \Lambda_R$. Moreover in the limit $n \to \infty$, one gets $a \to 0$ and therefore $\Lambda_R/\delta\Lambda_m \to 0$.) Therefore, in the leading order it seems to be sufficient to retain point-like wormholes only (i.e., consider the limit $a \to 0$). Then instead of (24) we may assume the vacuum distribution of virtual wormholes in the form

$$NF(a, X) = \frac{1}{a^2}\delta(a)\,\nu(X), \tag{64}$$

where $\nu(X) = \int a^2 NF(a,X)da$ and $\int (1/a^2)\nu(X)d^4X = n \to \infty$ has the meaning of the infinite density of point-like wormholes, while $\nu \sim a^2 n$ remains a finite. In this case the volume cut by wormholes vanishes $\int 2n(\pi^2/2)a^4 da\, dr_0 = a^2 \int \nu(X)d^4X \to 0$ and the rarefied gas approximation works well. This defines the bias and the mean cutoff (here we define the Fourier transform $\widetilde{\nu}(k) = \int \nu(X)e^{ikX}d^4X$) as

$$\overline{N}(k) = 1 - \frac{4\pi^2}{k^2}\left(\widetilde{\nu}(0) - \widetilde{\nu}(k)\right). \tag{65}$$

The contribution to the mean curvature (55) can be expressed via the same function $\nu(k)$ as

$$\begin{aligned}\Lambda_R &= -12\pi^2 \int n(a, r_0)\, a^2 da\, dr_0 \\ &= -12\pi^2 \int \nu(X)\, d^4X = -12\pi^2 \widetilde{\nu}(0).\end{aligned} \tag{66}$$

Thus, for the total cosmological constant we get the expression

$$\begin{aligned}\Lambda_{\text{tot}} &= \Lambda_0 \\ &+ 2\pi G \int \left(1 - \frac{4\pi^2}{k^2}\left(\widetilde{\nu}(0) - \widetilde{\nu}(k)\right)\right)\frac{d^4k}{(2\pi)^4} - 12\pi^2\widetilde{\nu}(0).\end{aligned} \tag{67}$$

We stress that all these terms should be finite. Indeed all distributions of virtual wormholes $\widetilde{\nu}(k)$ which lead to an infinite value of Λ_{tot} are suppressed in (5) by the factor $\sim e^{-\int \Lambda_{\text{tot}} d^4x}$, while the minimal value is reached when wormholes cut all of the volume of space and the action is merely $S = 0$. (Frankly speaking this statement is not rigorous. At first look the two last terms in (67) are independent and one may try to take $\widetilde{\nu}(0)$ an arbitrary big. If this were the case then the action would not possess the minimum at all. However $\widetilde{\nu}(0)$ cannot be arbitrary big, since it will violate the rarefied gas approximation and the linear expression (67) brakes down. Moreover, fermions give here a contribution of the opposite sign. The rigorous investigation of this problem requires the further studying and we present it elsewhere.) We also point out that here we considered the real scalar field as the matter source, while in the general case the stress energy tensor should include all existing Bose and Fermi fields (Fermi fields give a negative contribution to Λ_m).

The value of Λ_0 looks like a free parameter, which in quantum gravity runs with scales [27]. However at large scales its asymptotic value may be uniquely fixed by the simple arguments as follows. Indeed in quantum field theory properties of the ground state (vacuum) change when we imply an external classical fields. The same is true for the distribution of virtual wormholes $\widetilde{\nu}(k, J)$, for example, see (33) and (37) and, therefore, $\Lambda_{\text{tot}} = \Lambda_{\text{tot}}(J)$ which we describe in the next subsection. We recall that in gravity the role of the external current plays the stress energy tensor of matter fields $J = T_{ab}$. However one believes that in the absence of all classical fields the vacuum state should represent the most symmetric (Lorentz invariant) state which in our case corresponds to the Euclidean space. In order to be consistent with the Einstein equations this requires $\Lambda_{\text{tot}}(J = 0) = 0$ which uniquely fixes the value of Λ_0 in (67). We point out that from somewhat different considerations such a choice was advocated earlier in [15, 34, 35].

6.4. Vacuum Polarization in an External Field. Consider now topology fluctuations in the presence of an external current. In the presence of an external current J^{ext} the distribution of virtual wormholes changes $\widetilde{\nu}(k, J) = \widetilde{\nu}(k) + \delta\widetilde{\nu}(k, J)$. Indeed in (33) for the case of a weak external field the contribution of the external current into the action can be expanded as $\exp(-V(J)) \simeq 1 - V$, where

$$\begin{aligned}V &= -\frac{1}{2}\int J(x)\,G(x,y)\,J(y)\,d^4x\,d^4y \\ &= -\frac{1}{2}\frac{L^4}{(2\pi)^4}\int G_0(k)\,N(k)\left|J_k\right|^2 d^4k.\end{aligned} \tag{68}$$

Then using (36) we find $\overline{N}(k, J) = \overline{N}(k, 0) + \delta N(k, J)$, where

$$\delta N(k, J) = \delta b(J) \simeq -\frac{1}{2}\frac{L^4}{(2\pi)^4}\int \sigma^2(k,p)\,G_0(p)\left|J_p\right|^2 d^4p \tag{69}$$

is the bias related to an additional distribution of virtual wormholes and $\sigma^2(k, p) = \overline{\Delta N^*(k)\Delta N(p)}$

$$\sigma^2(k,p) = \frac{1}{Z_{\text{total}}(0)}\int [DN]\, e^{-I(N)}\Delta N^*(k)\,\Delta N(p) \tag{70}$$

defines the dispersion of vacuum topology fluctuations (here $\Delta N = N - \overline{N}$). The exact definition of $\sigma^2(k, p)$ requires the further development of a fundamental theory. In particular, it can be numerically calculated in lattice quantum gravity [28]. However it can be shown that at scales $k, p \gg k_{\text{pl}}$ it reduces to $\sigma^2(k, p) \to \sigma_k^2\delta(k - p)$ and therefore

$$\delta b(J) = -\sigma_k^2 \frac{4\pi^2}{2k^2}\left|J_k^{\text{ext}}\right|^2. \tag{71}$$

Now comparing this function with (23) we relate the additional distribution of virtual wormholes and the external classical field as

$$\frac{4\pi^2}{k^2}\left(\delta\widetilde{\nu}(0, J) - \delta\widetilde{\nu}(k, J)\right) = \frac{1}{2}\sigma_k^2\frac{4\pi^2}{k^2}\left|J_k^{\text{ext}}\right|^2, \tag{72}$$

where $\delta\widetilde{\nu}(k, J) = \int a^2 \delta NF(a, k)da$. We point out that the above expression does not define the value $\delta\widetilde{\nu}(0, J)$ which requires an additional consideration. Moreover, in general the external field J does not possess a symmetry and therefore the correction $\langle \delta T_{\alpha\beta}(x)\rangle$ does not reduce to a single cosmological constant. However, such corrections always violate the averaged null energy condition [36, 37] and may be considered as some kind of dark energy or, by other words, it represents an exotic matter. Some portion of dark energy still has the form of the cosmological constant which defines a nonvanishing present day value (we recall that fermions give a contribution of the opposite sign)

$$\delta\Lambda_{\text{tot}} = -2\pi G \int \frac{4\pi^2}{k^2} \langle \delta\widetilde{\nu}(0,J) - \delta\widetilde{\nu}(k,J)\rangle \frac{d^4k}{(2\pi)^4} - 12\pi^2\delta\widetilde{\nu}(0,J), \tag{73}$$

where $\langle \delta\widetilde{\nu}(k,J)\rangle$ denotes an averaging over rotations.

The only unknown parameter in (72) is the dispersion σ_k^2 which defines the intensity of topology fluctuations in the vacuum. It has also the sense of the efficiency coefficient which defines the portion of the energy of the external field spent on the formation of additional wormholes. Though the evaluation of σ_k^2 requires the further development of a fundamental theory, one may expect that $\sigma_k^2 = \overline{N}(k)(1 - \overline{N}(k))$, where $\overline{N}(k)$ is the mean cutoff. It is expected that $\overline{N}(k) \to 0$ as $k \gg k_{\text{pl}}$ and $\overline{N}(k) \to \overline{N}_0 \leq 1$. This means that $\sigma \to 0$ as $k \gg k_{\text{pl}}$ and $\sigma \to \sigma_0 \ll 1$ as $k \ll k_{\text{pl}}$, while it takes the maximum value $\sigma_{\max} \sim 1$ at Planckian scales $k \sim k_{\text{pl}}$. By other words, the most efficient transmission of the energy into wormholes takes place for wormholes of the Planckian size. In the case when external classical fields have characteristic scales $\lambda = 2\pi/k \gg \ell_{\text{pl}}$ in (72) the efficiency coefficient σ_k^2 and the cutoff $\overline{N}(k)$ become constant $\sigma \simeq \sigma_0$, $\overline{N}(k) \simeq \overline{N}_0$, while their ratio may be estimated as $\alpha\sigma_0^2/\overline{N}_0 = \Omega_{\text{DE}}/\Omega_b$, where α is the effective number of fundamental fields which contribute to $\delta\Lambda_{\text{tot}}$ and Ω_{DE}, Ω_b are dark energy and baryon energy densities, respectively. According to the modern picture this ratio gives $\Omega_{\text{DE}}/\Omega_b \approx 0.75/0.05 = 15$, while $\sigma_0^2/\overline{N}_0 \sim 1$ (as $\overline{N}_0 \ll 1$) and therefore this defines the estimate for the effective number of fundamental fields (helicity states) as $\alpha \sim 15$.

6.5. Speculations on the Formation of Actual Wormholes. As we already pointed out the additional distribution of virtual wormholes (72) reflects the symmetry of external classical fields and therefore it forms a homogeneous and isotropic background and perturbations. We recall that virtual wormholes represent an exotic form of matter. In the early Universe such perturbations start to develop and may form actual wormholes. The rigorous description of such a process represents an extremely complex and interesting problem which requires the further study. Some aspects of the behavior of the exotic density perturbations were considered in [7], while the simplest example of the formation of a wormhole-type object was discussed recently by us in [38]. Therefore we may expect that some portion of such a form of dark energy is reserved now in actual wormholes which we consider in the next section.

7. Dark Energy from Actual Wormholes

Consider now the contribution to the dark energy from the gas of actual wormholes. Unlike the virtual wormholes, actual wormholes do exist at all times and, therefore, a single wormhole can be viewed as a couple of conjugated cylinders $T_{\pm}^3 = S_{\pm}^2 \times R^1$. So that the number of parameters of an actual wormhole is less $\eta = (a, r_+, r_-)$, where a is the radius of $S_{\pm}^2$ and $r_{\pm} \in R^3$ is a spatial part of $R_{\pm}$.

Actual wormholes also produce two kinds of contribution to the dark energy. One comes from their contribution to the mean curvature which corresponds to an exotic stress energy momentum tensor. Such a stress energy momentum tensor reflects the dark energy reserved by additional virtual wormholes discussed in the previous section. Such energy is necessary to support actual wormholes as a solution to the Einstein equations. The second part comes from vacuum polarization effects by actual wormholes. The consideration in the previous section shows that for macroscopic wormholes the second part has the order $\langle \Delta T_{\alpha\beta}\rangle \sim 8\pi Gn$ and is negligible as compared to the curvature $R \sim a^2 n$ (since macroscopic wormholes have throats $a \gg \ell_{\text{pl}}$). However, for the sake of completeness and for methodological aims we describe it as well.

For rigorous evaluation of dark energy of the second type we, first, have to find the bias $b_1(x, x', \eta)$ analogous to (19) for the topology $R^4/(T_+^3 \cup T_-^3)$. There are many papers treating different wormholes in this respect (e.g., see [36, 37] and references therein). However, in the present paper for an estimation we shall use a more simple trick.

7.1. Beads of Virtual Wormholes (Quantum Wormhole). Indeed, instead of the cylinders $T_{\pm}^3$ we consider a couple of chains (beads of virtual wormholes $T_{\pm}^3 \to \cup_n S_{\pm,n}^3$). Such an idea was first suggested in [39] and one may call such an object as quantum wormhole. Then the bias can be written straightforwardly

$$b_1(x, x', \eta) = \sum_{n=-\infty}^{+\infty} \frac{1}{4\pi^2 a} \left(\frac{1}{(R_{-,n} - x')^2} - \frac{1}{(R_{+,n} - x')^2} \right) \times \left[\delta\left(\left|\vec{x} - \vec{R}_{+,n}\right| - a\right) - \delta\left(\left|\vec{x} - \vec{R}_{-,n}\right| - a\right)\right], \tag{74}$$

where $R_{\pm,n} = (t_n, r_{\pm})$ with $t_n = t_0 + 2\ell n$ and $\ell \geq a$ is the step. We may expect that upon averaging over the position $t_0 \in [-\ell, \ell]$ the bias for the beads will reproduce the bias for cylinders $T_{\pm}^3$ (at least it looks like a very good approximation). We point out that the averaging out $(1/2\ell)\int_{-\ell}^{\ell} dt_0$ and the sum $\sum_{n=-\infty}^{+\infty}$ reduces to a single integral $(1/2\ell)\int_{-\infty}^{\infty} dt$ of the zero term in (74). And moreover, the resulting total bias corresponds merely to a specific choice of the distribution function $F(\xi)$ in (21). Namely, we may take

$$NF(\xi) = \frac{1}{2\ell}\delta(t_+ - t_-) f(|r_+ - r_-|, a), \tag{75}$$

where $R_{\pm} = (t_{\pm}, r_{\pm})$ and $f(s,a)$ is the distribution of cylinders, which can be taken as ($\tilde{n}$ is 3-dimensional density)

$$f(\eta) = \frac{\tilde{n}(a)}{4\pi r_0^2}\delta(s - r_0). \tag{76}$$

Using the normalization condition $\int NF(\xi)d\xi = N$ we find the relation $N = (1/2\ell)\tilde{n}V = nV$, where n is a 4-dimensional density of wormholes and $1/(2\ell)$ is the effective number of wormholes on the unit length of the cylinder (i.e., the frequency with which the virtual wormhole appears at the positions $r_{\pm}$). This frequency is uniquely fixed by the requirement that the volume which cuts the bead is equal to that which cuts the cylinder $(4/3)\pi a^3 = (\pi^2/2)a^4(1/2\ell)$ (i.e., $2\ell = (3\pi/8)a$ and $n = (8/3\pi a)\tilde{n}$). Thus, we can use directly expression (23) and find (compare to (25))

$$b(k) = -\int n(a)\,a^2\frac{4\pi^2}{k^2}\left(1 - \frac{\sin|\mathbf{k}|\,r_0}{|\mathbf{k}|\,r_0}\right)\frac{J_1(ka)}{ka/2}da, \tag{77}$$

where $k = (k_0, \mathbf{k})$. Here the first term merely coincides with that in (25) and, therefore, it gives the contribution to the cosmological constant $\delta\Lambda/(8\pi G) = -n/4 = -2\tilde{n}/(3\pi a)$, while the second term describes a correction which does not reduce to the cosmological constant and requires a separate consideration.

7.2. Stress Energy Tensor. From (57) we find that the stress energy tensor

$$-\left\langle \Delta T_{\alpha\beta}(x)\right\rangle = \int \frac{k_\beta k_\alpha - (1/2)\,g_{\alpha\beta}k^2}{k^2}b(k,\xi)\frac{d^4k}{(2\pi)^4} \tag{78}$$

reduces to the two functions

$$T_{00} = \varepsilon = \lambda_1 - \frac{1}{2}\mu,$$
$$T_{ij} = p\delta_{ij}, \qquad p = \frac{1}{3}\lambda_2 - \frac{1}{2}\mu, \tag{79}$$

where $\varepsilon + 3p = -\mu$ and $\lambda_1 + \lambda_2 = \mu$ and these functions are

$$\lambda_1 = -\int \frac{k_0^2}{k^2}b\frac{d^4k}{(2\pi)^4}, \qquad \lambda_2 = -\int \frac{|\mathbf{k}|^2}{k^2}b\frac{d^4k}{(2\pi)^4}. \tag{80}$$

By means of the use of the spherical coordinates $k_0^2/k^2 = \cos^2\theta$, $|\mathbf{k}|^2/k^2 = \sin^2\theta$, and $d^4k = 4\pi\sin^2\theta k^3 dk d\theta$ we get

$$\lambda_i = \frac{n(a,r_0)}{4\beta_i}\left(1 - 2\beta_i\left(\frac{a}{r_0}\right)^2 f_i\left(\frac{a}{r_0}\right)\right), \tag{81}$$

where $\beta_1 = 1$, $\beta_2 = 1/3$, and f_i is given by

$$f_{\binom{1}{2}}(y) = \frac{2}{\pi}\int_{-1}^{1}\int_0^{\infty}\sin(x\sin\theta)\frac{J_1(yx)}{yx/2}\begin{pmatrix}\cos^2\theta\\ \sin^2\theta\end{pmatrix}dx\,d\cos\theta. \tag{82}$$

For $a/r_0 \ll 1$ we find

$$f_{1,2}\left(\frac{a}{r_0}\right) \approx \left(1 + o_{1,2}\left(\frac{a}{r_0}\right)\right). \tag{83}$$

Thus, finally we find

$$\varepsilon \simeq -\frac{n}{4} = -\frac{2\tilde{n}}{3\pi a}, \qquad p \simeq \varepsilon\left(1 - \frac{4}{3}\left(\frac{a}{r_0}\right)^2\right), \tag{84}$$

which upon the continuation to the Minkowsky space gives the equation of state in the form (An arbitrary gas of wormholes splits in fractions with a fixed a and r_0.)

$$p = -\left(1 - \frac{4}{3}\left(\frac{a}{r_0}\right)^2\right)\varepsilon, \tag{85}$$

which in the case when $a/r_0 \ll 1$ behaves like a cosmological constant. However when $a \gg \ell_{pl}$ such a constant is extremely small and can be neglected, while the leading contribution comes from the mean curvature.

7.3. Mean Curvature. In this subsection we consider the Minkowsky space. Then the simplest actual wormhole can be described by the metric analogous to (1), for example, see [7]

$$ds^2 = c^2dt^2 - h^2(r)\,\delta_{\alpha\beta}dx^\alpha dx^\beta, \tag{86}$$

where $h(r) = 1 + \theta(a - r)(a^2/r^2 - 1)$. To avoid problems with the Bianchi identity and the conservation of energy the step function should be also smoothed as in (1). The stress energy tensor which produces such a wormhole can be found from the Einstein equation $8\pi GT_\alpha^\beta = R_\alpha^\beta - (1/2)\delta_\alpha^\beta R$. Both regions $r > a$ and $r < a$ represent portions of the ordinary flat Minkowsky space and therefore the curvature is $R_i^k \equiv 0$. However on the boundary $r = a$ it has the singularity. Since the metric (86) does not depend on time we find

$$R_0^0 = R_\alpha^0 = 0, \qquad R_\alpha^\beta = \frac{2}{a}\delta(a - r)\left\{n_\alpha n^\beta + \delta_\alpha^\beta\right\} + \lambda_\alpha^\beta, \tag{87}$$

where $n^\alpha = n_\alpha = x^\alpha/r$ is the outer normal to the throat S^2, and λ_α^β are additional terms (e.g., see (49) and (51)) which in the leading order are negligible upon averaging over some portion of space $\Delta V \gtrsim a^3$. In the case of a set of wormholes this gives in the leading order

$$R_0^0 = R_\alpha^0 = 0, \qquad R_\alpha^\beta = \sum \frac{2}{a_i}\delta(a_i - |r - r_i|)\left\{n_{i\alpha}n_i^\beta + \delta_\alpha^\beta\right\}, \tag{88}$$

where a_i is the radius of a throat and r_i is the position of the center of the throat in space and $n_i^\alpha = (x^\alpha - r_i^\alpha)/|r - r_i|$. In the case of a homogeneous and isotropic distribution of such throats we find $R_\alpha^\beta = (1/3)R\delta_\alpha^\beta$ (averaging over spatial directions gives $\langle n_\alpha n^\beta\rangle = (1/3)\delta_\alpha^\beta$) where

$$R = -8\pi GT = \sum \frac{8}{a_i}\delta(a_i - |r - R_i|) = 32\pi\int a\tilde{n}(a)\,da, \tag{89}$$

where T stands for the trace of the stress energy momentum tensor which one has to add to the Einstein equations to support such a wormhole. It is clear that such a source violates the weak energy condition and, therefore, it reproduces the form of dark energy (i.e., $T = \varepsilon + 3p < 0$). If the density of such sources (and, resp., the density of wormholes) is sufficiently high, then this results in the observed [10–14] acceleration of the scale factor for the Friedmann space as $\sim t^{\alpha}$ with $\alpha = 2\varepsilon/3(\varepsilon + p) = 2\varepsilon/(2\varepsilon + (\varepsilon + 3p)) > 1$, for example, see also [40–42]. In terms of the 4-dimensional density of wormholes $n = (8/3\pi a)\widetilde{n}$ we get $R \sim a^2 n \gg 8\pi G n$ as $a \gg \ell_{\text{pl}}$ and, therefore, the leading contribution indeed comes from the mean curvature.

8. Estimates and Concluding Remarks

Now consider the simplest estimates. Actual wormholes seem to be responsible for the dark matter [7, 9]. Therefore, to get the estimate to the number density of wormholes is rather straightforward. First wormholes appear at scales when dark matter effects start to display themselves; that is, at scales of the order $L \sim (1 \div 5)$ Kpc, which gives in that range the number density

$$\widetilde{n} \sim \frac{1}{L^3} \sim (3 \div 0.024) \times 10^{-65} \text{cm}^{-3}. \tag{90}$$

The characteristic size of throats can be estimated from (89) $\varepsilon_{\text{DE}} \sim (G)^{-1}\widetilde{n}\overline{a}$. Since the density of dark energy is $\varepsilon_{\text{DE}}/\varepsilon_0 = \Omega_{\text{DE}} \sim 0.75$, where ε_0 is the critical density, then we find the estimate

$$\overline{a} \sim \frac{2}{3}(1 \div 125) \times 10^{-3} R_{\odot}\Omega_{\text{DE}}h_{75}^2, \tag{91}$$

where $R_{\odot}$ is the Solar radius, $h_{75} = H/(75\,\text{km}/(\text{sec Mpc}))$ and H is the Hubble constant. We also recall that the background density of baryons ε_b generates a nonvanishing wormhole rest mass $M_w = (4/3)\pi\overline{a}^3 R^3 \varepsilon_b$ (where $R(t)$ is the scale factor of the Universe and therefore M_w remains constant), for example, see [7]. It produces the dark matter density related to the wormholes as $\varepsilon_{\text{DM}} \simeq M_w \widetilde{n}$. The typical mass of a wormhole M_w is estimated as

$$M_w \sim 1{,}7 \times (1 \div 125) \times 10^2 M_{\odot}\Omega_{\text{DM}}h_{75}^2, \tag{92}$$

where $M_{\odot}$ is the Solar mass. We point out that this mass has not the direct relation to the parameters of the gas of wormholes. However it defines the moment when wormhole throats separated from the cosmological expansion. The above estimate shows that if wormholes form due to the development of perturbations in the exotic matter, then this process should start much earlier than the formation of galaxies.

Thus, we see that virtual wormholes should indeed lead to the regularization of all divergencies in QFT which agrees with recent results [28]. Therefore, they form the local finite value of the cosmological constant. In the absence of external classical fields such a value should be exactly zero at macroscopic scales. A some nonvanishing value for the cosmological constant appears as the result of vacuum polarization effects in external fields. Indeed, external fields form an additional distribution of virtual wormholes which possess an exotic stress energy tensor (some kind of dark energy). Only some part of it forms the cosmological constant, while the rest reflects the symmetry of external fields and possesses inhomogeneities. We assume that during the evolution of our Universe inhomogeneities in the exotic matter develop and may form actual wormholes. Although this problem requires the further and more deep investigation we refer to [38] where the formation of a simplest wormhole-like object has been considered. In other words, such polarization energy is reserved now in a gas of actual wormholes. We estimated parameters of such a gas and believe that such a gas may indeed be responsible for both, dark matter and dark energy phenomena.

References

[1] J. R. Primack, "The nature ofdark matter," http://arxiv.org/abs/astro-ph/0112255.

[2] J. Diemand, M. Zemp, B. Moore, J. Stadel, and M. Carollo, "Cusps in cold dark matter haloes," *Monthly Notices of the Royal Astronomical Society*, vol. 364, no. 2, pp. 665–673, 2005.

[3] E. W. Mielke and J. A. V. Vélez Pérez, "Axion condensate as a model for dark matter halos," *Physics Letters B*, vol. 671, no. 1, pp. 174–178, 2009.

[4] G. Gentile, P. Salucci, U. Klein, D. Vergani, and P. Kalberla, "The cored distribution of dark matter in spiral galaxies," *Monthly Notices of the Royal Astronomical Society*, vol. 351, no. 3, pp. 903–922, 2004.

[5] D. T. F. Weldrake, W. J. G. de Blok, and F. Walter, "A high-resolution rotation curve of NGC 6822: a test-case for cold dark matter," *Monthly Notices of the Royal Astronomical Society*, vol. 340, no. 1, pp. 12–28, 2003.

[6] W. J. G. de Blok and A. Bosma, "High-resolution rotation curves of low surface brightness galaxies," *Astronomy and Astrophysics*, vol. 385, pp. 816–846, 2002.

[7] A. Kirillov and E. P. Savelova, "Density perturbations in a gas of wormholes," *Monthly Notices of the Royal Astronomical Society*, vol. 412, no. 3, pp. 1710–1720, 2011.

[8] P. Kanti, B. Kleihaus, and J. Kunz, "Stable Lorentzian wormholes in dilatonic Einstein-Gauss-Bonnet theory," *Physical Review D*, vol. 85, no. 4, Article ID 044007, 2012.

[9] A. A. Kirillov and E. P. Savelova, "Dark matter from a gas of wormholes," *Physics Letters B*, vol. 660, no. 3, pp. 93–99, 2008.

[10] A. G. Riess, A. V. Filippenko, P. Challis et al., "Observational evidence from supernovae for an accelerating universe and a cosmological constant," *The Astronomical Journal*, vol. 116, no. 3, p. 1009, 1998.

[11] S. Perlmutter, G. Aldering, G. Goldhaber et al., "Measurements of Ω and Λ from 42 high-redshift supernovae," *The Astrophysical Journal*, vol. 517, no. 2, p. 565, 1999.

[12] J. L. Tonry, B. P. Schmidt, B. Barris et al., "Cosmological results from high-z supernovae," *The Astrophysical Journal*, vol. 594, no. 1, p. 594, 2003.

[13] N. W. Halverson, E. M. Leitch, C. Pryke et al., "Degree angular scale interferometer first results: a measurement of the cosmic microwave background angular power spectrum," *The Astrophysical Journal*, vol. 568, no. 1, p. 38, 2002.

[14] C. B. Netterfield, P. A. R. Ade, J. J. Bock et al., "A measurement by BOOMERANG of multiple peaks in the angular power spectrum of the cosmic microwave background," *The Astrophysical Journal*, vol. 571, no. 2, p. 604, 2002.

[15] S. Coleman, "Why there is nothing rather than something: a theory of the cosmological constant," *Nuclear Physics B*, vol. 310, no. 3-4, pp. 643–668, 1988.

[16] I. Klebanov, L. Susskind, and T. Banks, "Wormholes and the cosmological constant," *Nuclear Physics B*, vol. 317, no. 3, pp. 665–692, 1989.

[17] S. W. Hawking, "Spacetime foam," *Nuclear Physics B*, vol. 114, p. 349, 1978.

[18] T. P. Cheng and L. F. Li, *Gauge Theory of Elementary Particles*, Clarendon Press, Oxford, UK, 1984.

[19] A. H. Taub, "Space-times with distribution valued curvature tensors," *Journal of Mathematical Physics*, vol. 21, no. 6, pp. 1423–1431, 1980.

[20] A. Shatsky, privite communications.

[21] V. A. Fock, "Zur Theorie des Wasserstoffatoms," *Zeitschrift für Physik*, vol. 98, no. 3-4, pp. 145–154, 1935.

[22] E. P. Savelova, "Gas of wormholes in Euclidean quantum field theory," http://arxiv.org/abs/1211.5106.

[23] A. A. Kirillov and E. P. Savelova, "On the behavior of bosonic systems in the presence of topology fluctuations," http://arxiv.org/abs/0808.2478.

[24] A. A. Kirillov, "Violation of the Pauli principle and dimension of the Universe at very large distances," *Physics Letters B*, vol. 555, no. 1-2, pp. 13–21, 2003.

[25] A. A. Kirillov, "Dark matter, dark charge, and the fractal structure of the Universe," *Physics Letters B*, vol. 535, no. 1–4, pp. 22–24, 2002.

[26] J. Ambjorn, J. Jurkiewicz, and R. Loll, "The spectral dimension of the universe is scale dependent," *Physical Review Letters*, vol. 95, no. 17, Article ID 171301, 4 pages, 2005.

[27] E. Manrique, S. Rechenberger, and F. Saueressig, "Asymptotically safe Lorentzian gravity," *Physical Review Letters*, vol. 106, no. 25, Article ID 251302, 4 pages, 2011.

[28] J. Laiho and D. Coumbe, "Evidence for asymptotic safety from lattice quantum gravity," *Physical Review Letters*, vol. 107, no. 16, Article ID 161301, 2011.

[29] A. A. Kirillov and D. Turaev, "On modification of the Newton's law of gravity at very large distances," *Physics Letters B*, vol. 532, no. 3-4, pp. 185–192, 2002.

[30] A. A. Kirillov and D. Turaev, "The universal rotation curve of spiral galaxies," *Monthly Notices of the Royal Astronomical Society*, vol. 371, no. 1, pp. L31–L35, 2006.

[31] A. A. Kirillov and D. Turaev, "Foam-like structure of the universe," *Physics Letters B*, vol. 656, pp. 1–8, 2007.

[32] A. A. Kirillov, "The nature of dark matter," *Physics Letters B*, vol. 632, no. 4, pp. 453–462, 2006.

[33] A. A. Kirillov and E. P. Savelova, "Astrophysical effects of spacetime foam," *Gravitation and Cosmology*, vol. 14, no. 3, pp. 256–261, 2008.

[34] S. W. Hawking and N. Turok, "Open inflation without false vacua," *Physics Letters B*, vol. 425, no. 1-2, pp. 25–32, 1998.

[35] Z. C. Wu, "The cosmological constant is probably zero, and a proof is possibly right," *Physics Letters B*, vol. 659, no. 5, pp. 891–893, 2008.

[36] R. Garattini, "Entropy and the cosmological constant: a spacetime-foam approach," *Nuclear Physics B*, vol. 88, no. 1–3, Article ID 991003, pp. 297–300, 2000.

[37] M. Visser, *Lorentzian Wormholes*, Springer, New York, NY, USA, 1996.

[38] A. A. Kirillov and E. P. Savelova, "Artificial wormhole," http://arxiv.org/abs/1204.0351.

[39] E. W. Mielke, "Knot wormholes in geometrodynamics?" *General Relativity and Gravitation*, vol. 8, no. 3, pp. 175–196, 1977.

[40] E. B. Gliner, "Algebraic properties of the energy-momentum tensor and vacuum-like states of matter," *Journal of Experimental and Theoretical Physics*, vol. 22, p. 378, 1966.

[41] A. H. Guth, "Inflationary universe: a possible solution to the horizon and flatness problems," *Physical Review D*, vol. 23, no. 2, pp. 347–356, 1981.

[42] A. A. Linde, "A new inflationary universe scenario: a possible solution of the horizon, flatness, homogeneity, isotropy and primordial monopole problems," *Physics Letters B*, vol. 108, no. 6, pp. 389–393, 1982.

4

Growth of Accreting Supermassive Black Hole Seeds and Neutrino Radiation

Gagik Ter-Kazarian

Division of Theoretical Astrophysics, Ambartsumian Byurakan Astrophysical Observatory, Byurakan, 378433 Aragatsotn, Armenia

Correspondence should be addressed to Gagik Ter-Kazarian; gago_50@yahoo.com

Academic Editor: Gary Wegner

In the framework of *microscopic theory of black hole* (MTBH), which explores the most important processes of rearrangement of vacuum state and spontaneous breaking of gravitation gauge symmetry at huge energies, we have undertaken a large series of numerical simulations with the goal to trace an evolution of the mass assembly history of 377 plausible accreting supermassive black hole seeds in active galactic nuclei (AGNs) to the present time and examine the observable signatures today. Given the redshifts, masses, and luminosities of these black holes at present time collected from the literature, we compute the initial redshifts and masses of the corresponding seed black holes. For the present masses $M_{\mathrm{BH}}/M_{\odot} \simeq 1.1 \times 10^{6}$ to 1.3×10^{10} of 377 black holes, the computed intermediate seed masses are ranging from $M_{\mathrm{BH}}^{\mathrm{Seed}}/M_{\odot} \simeq 26.4$ to 2.9×10^{5}. We also compute the fluxes of ultrahigh energy (UHE) neutrinos produced via simple or modified URCA processes in superdense protomatter nuclei. The AGNs are favored as promising pure UHE neutrino sources, because the computed neutrino fluxes are highly beamed along the plane of accretion disk, peaked at high energies, and collimated in smaller opening angle ($\theta \ll 1$).

1. Introduction

With typical bolometric luminosities $\sim 10^{45-48}\ \mathrm{erg\,s^{-1}}$, the AGNs are amongst the most luminous emitters in the universe, particularly at high energies (gamma-rays) and radio wavelengths. From its historical development, up to current interests, the efforts in the AGN physics have evoked the study of a major unsolved problem of how efficiently such huge energies observed can be generated. This energy scale severely challenges conventional source models. The huge energy release from compact regions of AGN requires extremely high efficiency (typically ≥10 per cent) of conversion of rest mass to other forms of energy. This serves as the main argument in favour of supermassive black holes, with masses of millions to billions of times the mass of the Sun, as central engines of massive AGNs. The astrophysical black holes come in a wide range of masses, from $\geq 3M_{\odot}$ for stellar mass black holes [1] to $\sim 10^{10}M_{\odot}$ for supermassive black holes [2, 3]. Demography of local galaxies suggests that most galaxies harbour quiescent supermassive black holes in their nuclei at the present time and that the mass of the hosted black hole is correlated with properties of the host bulge. The visible universe should therefore contain at least 100 billion supermassive black holes. A complex study of evolution of AGNs requires an answer to the key questions such as how did the first black holes form, how did massive black holes get to the galaxy centers, and how did they grow in accreting mass, namely, an understanding of the important phenomenon of mass assembly history of accreting supermassive black hole seeds. The observations support the idea that black holes grow in tandem with their hosts throughout cosmic history, starting from the earliest times. While the exact mechanism for the formation of the first black holes is not currently known, there are several prevailing theories [4]. However, each proposal towards formation and growth of initial seed black holes has its own advantage and limitations in proving the whole view of the issue. In this report we review the mass assembly history of 377 plausible accreting supermassive black hole seeds in AGNs and their neutrino radiation in the framework of gravitation theory, which explores the most important processes of rearrangement of vacuum state and a spontaneous breaking of gravitation gauge symmetry at huge energies. We will proceed according to the following structure. Most observational, theoretical, and computational aspects of the growth of black hole seeds are summarized in Section 2. The other important phenomenon of ultrahigh energy cosmic rays, in relevance to AGNs, is discussed in Section 3. The objectives of suggested approach are outlined in Section 4. In Section 5 we review the spherical accretion on superdense protomatter nuclei, in use. In Section 6 we discuss the growth of the seed black hole at accretion and derive its intermediate mass, initial redshift, and neutrino *preradiation time* (PRT). Section 7 is devoted to the neutrino radiation produced in superdense protomatter nuclei. The simulation results of the seed black hole intermediate masses, PRTs, seed redshifts, and neutrino fluxes for 377 AGN black holes are brought in Section 8. The concluding remarks are given in Section 9. We will refrain from providing lengthy details of the proposed gravitation theory at huge energies and neutrino flux computations. For these the reader is invited to visit the original papers and appendices of the present paper. In the latter we also complete the spacetime deformation theory, in the model context of gravitation, by new investigation of building up the complex of *distortion* (DC) of spacetime continuum and showing how it restores the *world-deformation tensor*, which still has been put in by hand. Finally, note that we regard the considered black holes only as the potential neutrino sources. The obtained results, however, may suffer if not all live black holes at present reside in final stage of their growth driven by the formation of protomatter disk at accretion and they radiate neutrino. We often suppress the indices without notice. Unless otherwise stated, we take geometrized units throughout this paper.

2. A Breakthrough in Observational and Computational Aspects on Growth of Black Hole Seeds

Significant progress has been made in the last few years in understanding how supermassive black holes form and grow. Given the current masses of $10^{6-9}M_{\odot}$, most black hole growth happens in the AGN phase. A significant fraction of the total black hole growth, 60% [6], happens in the most luminous AGN, quasars. In an AGN phase, which lasts ~10^8 years, the central supermassive black hole can gain up to ~$10^{7-8}M_{\odot}$, so even the most massive galaxies will have only a few of these events over their lifetime. Aforesaid gathers support especially from a breakthrough made in recent observational, theoretical, and computational efforts in understanding of evolution of black holes and their host galaxies, particularly through self-regulated growth and feedback from accretion-powered outflows; see, for example, [4, 7–18]. Whereas the multiwavelength methods are used to trace the growth of seed BHs, the prospects for future observations are reviewed. The observations provide strong support for the existence of a correlation between supermassive black holes and their hosts out to the highest redshifts. The observations of the quasar luminosity function show that the most supermassive black holes get most of their mass at high redshift, while at low redshift only low mass black holes are still growing [19]. This is observed in both optical [20] and hard X-ray luminosity functions [19, 21], which indicates that this result is independent of obscuration. Natarajan [13] has reported that the initial black hole seeds form at extremely high redshifts from the direct collapse of pregalactic gas discs. Populating dark matter halos with seeds formed in this fashion and using a Monte-Carlo merger tree approach, he has predicted the black hole mass function at high redshifts and at the present time. The most aspects of the models that describe the growth and accretion history of supermassive black holes and evolution of this scenario have been presented in detail by [9, 10]. In these models, at early times the properties of the assembling black hole seeds are more tightly coupled to properties of the dark matter halo as their growth is driven by the merger history of halos. While a clear picture of the history of black hole growth is emerging, significant uncertainties still remain [14], and in spite of recent advances [6, 13], the origin of the seed black holes remains an unsolved problem at present. The NuSTAR deep high-energy observations will enable obtaining a nearly complete AGN survey, including heavily obscured Compton-thick sources, up to $z \sim 1.5$ [22]. A similar mission, ASTRO-H [23], will be launched by Japan in 2014. These observations in combination with observations at longer wavelengths will allow for the detection and identification of most growing supermassive black holes at $z \sim 1$. The ultradeep X-ray and near-infrared surveys covering at least ~1 $\deg^2$ are required to constrain the formation of the first black hole seeds. This will likely require the use of the next generation of space-based observatories such as the James Webb Space Telescope and the International X-ray Observatory. The superb spatial resolution and sensitivity of the Atacama Large Millimeter Array (ALMA) [24] will revolutionize our understanding of galaxy evolution. Combining these new data with existing multiwavelength information will finally allow astrophysicists to pave the way for later efforts by pioneering some of the census of supermassive black hole growth, in use today.

3. UHE Cosmic-Ray Particles

The galactic sources like supernova remnants (SNRs) or microquasars are thought to accelerate particles at least up to energies of 3×10^{15} eV. The ultrahigh energy cosmic-ray (UHECR) particles with even higher energies have since been detected (comprehensive reviews can be found in [25–29]). The accelerated protons or heavier nuclei up to energies exceeding 10^{20} eV are firstly observed by [30]. The cosmic-ray events with the highest energies so far detected have energies of 2×10^{11} GeV [31] and 3×10^{11} GeV [32]. These energies are 10^7 times higher than the most energetic man-made accelerator, the LHC at CERN. These highest energies are believed to be reached in extragalactic sources like AGNs or gamma-ray bursts (GRBs). During propagation of such energetic particles through the universe, the threshold for pion photoproduction on the microwave background is ~2×10^{10} GeV, and at ~3×10^{11} GeV the energy-loss distance is about 20 Mpc. Propagation of cosmic rays over substantially larger distances gives rise to a cutoff in the spectrum at ~ 10^{11} GeV as was first shown by [33, 34], the GZK cutoff. The recent confirmation [35, 36] of GZK suppression in the cosmic-ray energy spectrum indicates that the cosmic rays with energies above the GZK cutoff, $E_{GZK} \sim 40$ EeV, mostly come from relatively close (within the GZK radius, $r_{GZK} \sim 100$ Mpc) extragalactic sources. However, despite the detailed measurements of the cosmic-ray spectrum, the identification of the sources of the cosmic-ray particles is still an open question as they are deflected in the galactic and extragalactic magnetic fields and hence have lost all information about their origin when reaching Earth. Only at the highest energies beyond $\sim 10^{19.6}$ GeV cosmic-ray particles may retain enough directional information to locate their sources. The latter must be powerful enough to sustain the energy density in extragalactic cosmic rays of about 3×10^{-19} erg cm^{-3} which is equivalent to $\sim 8 \times 10^{44}$ erg Mpc^{-3} yr^{-1}. Though it has not been possible up to now to identify the sources of galactic or extragalactic cosmic rays, general considerations allow limiting potential source classes. For example, the existing data on the cosmic-ray spectrum and on the isotropic 100 MeV gamma-ray background limit significantly the parameter space in which topological defects can generate the flux of the highest energy cosmic rays and rule out models with the standard X-particle mass of 10^{16} GeV and higher [37]. Eventually, the neutrinos will serve as unique astronomical messengers, and they will significantly enhance and extend our knowledge on galactic and extragalactic sources of the UHE universe. Indeed, except for oscillations induced by transit in a vacuum Higgs field, neutrinos can penetrate cosmological distances and their trajectories are not deflected by magnetic fields as they are neutral, providing powerful probes of high energy astrophysics in ways which no other particle can. Moreover, the flavor composition of neutrinos originating at astrophysical sources can serve as a probe of new physics in the electroweak sector. Therefore, an appealing possibility among the various hypotheses of the origin of UHECR is so-called Z-burst scenario [38–51]. This suggests that if ZeV astrophysical neutrino beam is sufficiently strong, it can produce a large fraction of observed UHECR particles within 100 Mpc by hitting local light relic neutrinos clustered in dark halos and form UHECR through the hadronic Z (s-channel production) and W-bosons (t-channel production) decays by weak interactions. The discovery of UHE neutrino sources would also clarify the production mechanism of the GeV-TeV gamma rays observed on Earth [43, 52, 53] as TeV photons are also produced in the up-scattering of photons in reactions to accelerated electrons (inverse-Compton scattering). The direct link between TeV gamma-ray photons and neutrinos through the charged and neutral pion production, which is well known from particle physics, allows for a quite robust prediction of the expected neutrino fluxes provided that the sources are transparent and the observed gamma rays originate from pion decay. The weakest link in the Z-burst hypothesis is probably both unknown boosting mechanism of the primary neutrinos up to huge energies of hundreds ZeV and their large flux required at the resonant energy $E_\nu \simeq M_Z^2/(2m_\nu) \simeq 4.2 \times 10^{21}$ eV (eV$/m_\nu$) well above the GZK cutoff. Such a flux severely challenges conventional source models. Any concomitant photon flux should not violate existing upper limits [37, 48, 49, 54]. The obvious question is then raised: where in the Cosmos are these neutrinos coming from? It turns out that currently, at energies in excess of 10^{19} eV, there are only two good candidate source classes for UHE neutrinos: AGNs and GRBs. The AGNs as significant point sources of neutrinos were analyzed in [50, 55, 56]. While hard to detect, neutrinos have the advantage of representing aforesaid unique fingerprints of hadron interactions and, therefore, of the sources of cosmic rays. Two basic event topologies can be distinguished: track-like patterns of detected Cherenkov light (hits) which originate from muons produced in charged-current interactions of muon neutrinos (muon channel); spherical hit patterns which originate from the hadronic cascade at the vertex of neutrino interactions or the electromagnetic cascade of electrons from charged-current interactions of electron neutrinos (cascade channel). If the charged-current interaction happens inside the detector or in case of charged-current tau-neutrino interactions, these two topologies overlap which complicates the reconstruction. At the relevant energies, the neutrino is approximately collinear with the muon and, hence, the muon channel is the prime channel for the search for point-like sources of cosmic neutrinos. On the other hand, cascades deposit all of their energy inside the detector and therefore allow for a much better energy reconstruction with a resolution of a few 10%. Finally, numerous reports are available at present in literature on expected discovery potential and sensitivity of experiments to neutrino point-like sources. Currently operating high energy neutrino telescopes attempt to detect UHE neutrinos, such as ANTARES [57, 58] which is the most sensitive neutrino telescope in the Northern Hemisphere, IceCube [35, 59–64] which is worldwide largest and hence most sensitive neutrino telescope in the Southern Hemisphere, BAIKAL [65], as well as the CR extended experiments of The Telescope Array [66], Pierre Auger Observatory [67, 68], and JEM-EUSO mission [69]. The JEM-EUSO mission, which is planned to be launched by a H2B rocket around 2015-2016, is designed to explore the extremes in the universe and fundamental physics through the detection of the extreme energy ($E > 10^{20}$ eV) cosmic rays. The possible origins of the soon-to-be famous 28 IceCube neutrino-PeV events [59–61] are the first hint for astrophysical neutrino signal. Aartsen et al. have published an observation of two ~1 PeV neutrinos, with a P value 2.8σ beyond the hypothesis that these events were atmospherically generated [59]. The analysis revealed an additional 26 neutrino candidates depositing "electromagnetic equivalent energies" ranging from about 30 TeV up to 250 TeV [61]. New results were presented at the IceCube Particle Astrophysics Symposium (IPA 2013) [62–64]. If cosmic neutrinos are primarily of extragalactic origin, then the 100 GeV gamma ray flux observed by Fermi-LAT constrains the normalization at PeV energies at injection, which in turn demands a neutrino spectral index $\Gamma < 2.1$ [70].

4. MTBH, Revisited: Preliminaries

For the benefit of the reader, a brief outline of the key ideas behind the *microscopic theory of black hole*, as a guiding principle, is given in this section to make the rest of the paper understandable. There is a general belief reinforced by statements in textbooks that, according to general relativity (GR), a long-standing standard phenomenological black hole model (PBHM)—namely, the most general Kerr-Newman black hole model, with parameters of mass (M), angular momentum (J), and charge (Q), *still has to be put in by hand*—can describe the growth of accreting supermassive black hole seed. However, such beliefs are suspect and should be critically reexamined. The PBHM cannot be currently accepted as convincing model for addressing the aforementioned problems, because in this framework the very source of gravitational field of the black hole is a kind of curvature singularity at the center of the stationary black hole. A meaningless central singularity develops which is hidden behind the event horizon. The theory breaks down inside the event horizon which is causally disconnected from the exterior world. Either the Kruskal continuation of the Schwarzschild ($J = 0, Q = 0$) metric, or the Kerr ($Q = 0$) metric, or the Reissner-Nordstrom ($J = 0$) metric, shows that the static observers fail to exist inside the horizon. Any object that collapses to form a black hole will go on to collapse to a singularity inside the black hole. Thereby any timelike worldline must strike the central singularity which wholly absorbs the infalling matter. Therefore, the ultimate fate of collapsing matter once it has crossed the black hole surface is unknown. This, in turn, disables any accumulation of matter in the central part and, thus, neither the growth of black holes nor the increase of their mass-energy density could occur at accretion of outside matter, or by means of merger processes. As a consequence, the mass and angular momentum of black holes will not change over the lifetime of the universe. But how can one be sure that some hitherto unknown source of pressure does not become important at huge energies and halt the collapse? To fill the void which the standard PBHM presents, one plausible idea to innovate the solution to alluded key problems would appear to be the framework of *microscopic theory of black hole*. This theory has been originally proposed by [71] and references therein and thoroughly discussed in [72–75]. Here we recount some of the highlights of the MTBH, which is the extension of PBHM and rather completes it by exploring the most important processes of rearrangement of vacuum state and a spontaneous breaking of gravitation gauge symmetry at huge energies [71, 74, 76]. We will not be concerned with the actual details of this framework but only use it as a backdrop to validate the theory with some observational tests. For details, the interested reader is invited to consult the original papers. Discussed gravitational theory is consistent with GR up to the limit of neutron stars. But this theory manifests its virtues applied to the physics of internal structure of galactic nuclei. In the latter a significant change of properties of spacetime continuum, so-called *inner distortion* (ID), arises simultaneously with the strong gravity at huge energies (see Appendix A). Consequently the matter undergoes phase transition of second kind, which supplies a powerful pathway to form a stable superdense protomatter core (SPC) inside the event horizon. Due to this, the stable equilibrium holds in outward layers too and, thus, an accumulation of matter is allowed now around the SPC. The black hole models presented in phenomenological and microscopic frameworks have been schematically plotted in Figure 1, to guide the eye. A crucial point of the MTBH is that a central singularity cannot occur, which is now replaced by finite though unbelievably extreme conditions held in the SPC, where the static observers existed. The SPC surrounded by the accretion disk presents the microscopic model of AGN. The SPC accommodates the highest energy scale up to hundreds of ZeV in central protomatter core which accounts for the spectral distribution of the resulting radiation of galactic nuclei. External physics of accretion onto the black hole in earlier part of its lifetime is identical to the processes in Schwarzschild's model. However, a strong difference in the model context between the phenomenological black hole and the SPC is arising in the second part of its lifetime (see Section 6). The seed black hole might grow up driven by the accretion of outside matter when it was getting most of its mass. An infalling matter with time forms the protomatter disk around the protomatter core tapering off faster at reaching out the thin edge of the event horizon. At this, metric singularity inevitably disappears (see appendices) and the neutrinos may escape through vista to outside world, even after the neutrino trapping. We study the growth of protomatter disk and derive the intermediate mass and initial redshift of seed black hole and examine luminosities, neutrino surfaces for the disk. In this framework, we have computed the fluxes of UHE neutrinos [75], produced in the medium of the SPC via simple (quark and pionic reactions) or modified URCA processes, even after the neutrino trapping *(G. Gamow was inspired to name the process URCA after the name of a casino in Rio de Janeiro, when M. Schenberg remarked to him that "the energy disappears in the nucleus of the supernova as quickly as the money disappeared at that roulette table")*. The "trapping" is due to the fact that as the neutrinos are formed in protomatter core at superhigh densities they experience greater difficulty escaping from the protomatter core before being dragged along with the matter; namely, the neutrinos are "trapped" comove with matter. The part of neutrinos annihilates to produce, further, the secondary particles of expected ultrahigh energies. In this model, of course, a key open question is to enlighten the mechanisms that trigger the activity, and how a large amount of matter can be steadily funneled to the central regions to fuel this activity. In high luminosity AGNs the large-scale internal gravitational instabilities drive gas towards the nucleus which trigger big starbursts, and the coeval compact cluster just formed. It seemed they have some connection to the nuclear fueling through mass loss of young stars as well as their tidal disruption and supernovae. Note that we regard the UHECR particles as a signature of existence of *superdence protomatter* sources in the universe. Since neutrino events are expected to be of sufficient intensity, our estimates can be used to guide investigations of neutrino detectors for the distant future.

5. Spherical Accretion onto SPC

As alluded to above, the MTBH framework supports the idea of accreting supermassive black holes which link to AGNs. In order to compute the mass accretion rate $\dot{M}$, in use, it is

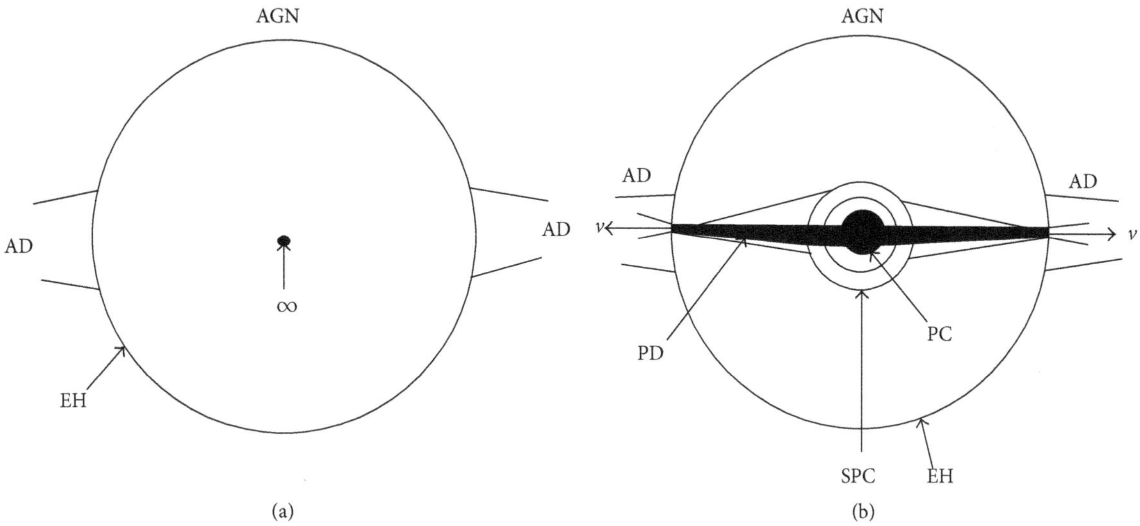

FIGURE 1: (a) The phenomenological model of AGN with the central stationary black hole. The meaningless singularity occurs at the center inside the black hole. (b) The microscopic model of AGN with the central stable SPC. In due course, the neutrinos of huge energies may escape through the vista to outside world. Accepted notations: EH = event horizon, AD = accretion disk, SPC = superdense protomatter core, PC = protomatter core.

necessary to study the accretion onto central supermassive SPC. The main features of spherical accretion can be briefly summed up in the following three idealized models that illustrate some of the associated physics [72].

5.1. Free Radial Infall. We examine the motion of freely moving test particle by exploring the external geometry of the SPC, with the line element (A.7), at $x = 0$. Let us denote the 4-vector of velocity of test particle $v^\mu = d\tilde{x}^\mu/d\tilde{s}$, $\tilde{x}^\mu = (\tilde{t}, \tilde{r}, \tilde{\theta}, \tilde{\varphi})$, and consider it initially for simplest radial infall $v^2 = v^3 = 0$. We determine the value of local velocity $v^{\tilde{r}} < 0$ of the particle for the moment of crossing the EH sphere, as well as at reaching the surface of central stable SPC. The equation of geodesics is derived from the variational principle $\delta \int dS = 0$, which is the extremum of the distance along the wordline for the Lagrangian at hand

$$2L = (1 - x_0)^2 \dot{\tilde{t}}^2 - (1 + x_0)^2 \dot{\tilde{r}}^2 - \tilde{r}^2 \sin^2\tilde{\theta}\dot{\tilde{\varphi}}^2 - \tilde{r}^2\dot{\tilde{\theta}}^2, \qquad (1)$$

where $\dot{\tilde{t}} \equiv d\tilde{t}/d\lambda$ is the t-component of 4-momentum and λ is the affine parameter along the worldline. We are using an affine parametrization (by a rescaling $\lambda \rightarrow \lambda(\lambda'))$ such that $L = \text{const}$ is constant along the curve. A static observer makes measurements with local orthonormal tetrad:

$$\vec{e}_{\tilde{t}} = \left|1 - x_0\right|^{-1} \vec{e}_t, \qquad \vec{e}_{\tilde{r}} = \left(1 + x_0\right)^{-1} \vec{e}_r,$$
$$\vec{e}_{\tilde{\theta}} = \tilde{r}^{-1} \vec{e}_\theta, \qquad \vec{e}_{\tilde{\varphi}} = \left(\tilde{r} \sin\tilde{\theta}\right)^{-1} \vec{e}_\theta. \qquad (2)$$

The Euler-Lagrange equations for $\tilde{\theta}$, $\tilde{\phi}$, and $\tilde{t}$ can be derived from the variational principle. A local measurement of the particle's energy made by a static observer in the equatorial plane gives the time component of the 4-momentum as measured in the observer's local orthonormal frame. This is the projection of the 4-momentum along the time basis vector. The Euler-Lagrange equations show that if we orient the coordinate system as initially the particle is moving in the equatorial plane (i.e., $\tilde{\theta} = \pi/2, \dot{\tilde{\theta}} = 0$), then the particle always remains in this plane. There are two constants of the motion corresponding to the ignorable coordinates $\tilde{t}$ and $\tilde{\phi}$, namely, the E-"energy-at-infinity" and the l-angular momentum. We conclude that the free radial infall of a particle from the infinity up to the moment of crossing the EH sphere, as well as at reaching the surface of central body, is absolutely the same as in the Schwarzschild geometry of black hole (Figure 2(a)). We clear up a general picture of orbits just outside the event horizon by considering the Euler-Lagrange equation for radial component with "effective potential." The circular orbits are stable if V is concave up, namely, at $\tilde{r} > 4\widetilde{M}$, where $\widetilde{M}$ is the mass of SPC. The binding energy per unit mass of a particle in the last stable circular orbit at $\tilde{r} = 4\widetilde{M}$ is $\widetilde{E}_{\text{bind}} = (m - E)/\widetilde{M} \simeq 1 - (27/32)^{1/2}$. Namely, this is the fraction of rest-mass energy released when test particle originally at rest at infinity spirals slowly toward the SPC to the innermost stable circular orbit and then plunges into it. Thereby one of the important parameters is the capture cross section for particles falling in from infinity: $\sigma_{\text{capt}} = \pi b_{\max}^2$, where $b_{\max}$ is the maximum impact parameter of a particle that is captured.

5.2. Collisionless Accretion. The distribution function for a collisionless gas is determined by the collisionless Boltzmann equation or Vlasov equation. For the stationary and spherical flow we obtain then

$$\dot{M}(E > 0) = 16\pi \left(G\widetilde{M}\right)^2 \rho_\infty v_\infty^{-1} c^{-2}, \qquad (3)$$

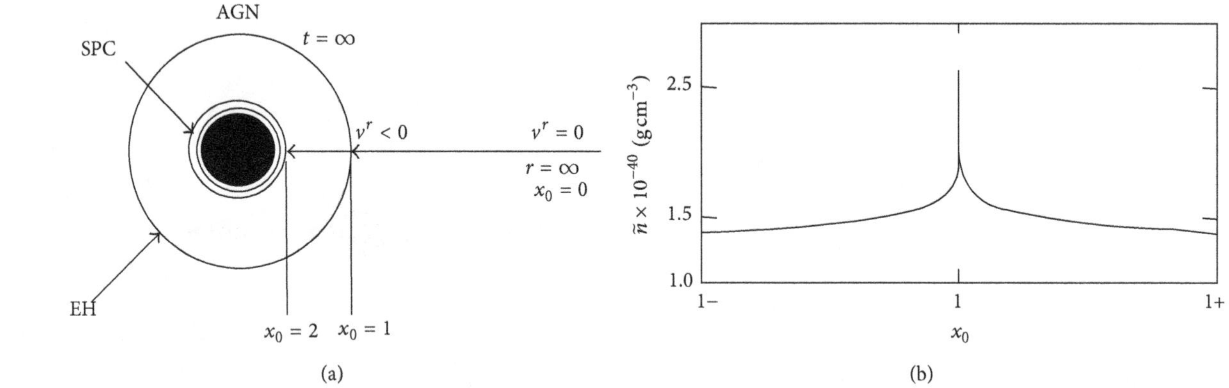

FIGURE 2: (a) The free radial infall of a particle from the infinity to EH sphere ($x_0 = 1$), which is similar to the Schwarzschild geometry of BH. Crossing the EH sphere, a particle continues infall reaching finally the surface ($x_0 = 2$) of the stable SPC. (b) Approaching the EH sphere ($x_0 = 1$), the particle concentration increases asymptotically until the threshold value of protomatter. Then, due to the action of cutoff effect, the metric singularity vanishes and the particles well pass EH sphere.

where the particle density ρ_∞ is assumed to be uniform at far from the SPC and the particle speed is $v_\infty \ll 1$. During the accretion process the particles approaching the EH become relativistic. Approaching event horizon, the particle concentration increases asymptotically as $(\tilde{n}(\tilde{r})/n_\infty)_{x_0 \to 1} \approx -(\ln \tilde{g}_{00})/2\, v_\infty$, up to the ID threshold value $\tilde{n}_d(\tilde{r})^{-1/3} = 0.4\,\text{fm}$ (Figure 2(b)). Due to the action of cutoff effect, the metric singularity then vanishes and the particles well pass EH sphere ($x_0 = 1$) and in the sequel form the protomatter disk around the protomatter core.

5.3. Hydrodynamic Accretion. For real dynamical conditions found in considered superdense medium, it is expected that the mean free path for collisions will be much shorter than the characteristic length scale; that is, the accretion of ambient gas onto a stationary, nonrotating compact SPC will be hydrodynamical in nature. For any equation of state obeying the causality constraint the sound speed implies $a^2 < 1$ and the flow must pass through a critical sonic point r_s outside the event horizon. The locally measured particle velocity reads $v^{\hat{r}} = (1 - \tilde{g}_{00}/E^2)$, where $E = E_\infty/m = (\tilde{g}_{00}/(1-u^2))^{1/2}$ and E_∞ is the energy at infinity of individual particle of the mass m. Thus, the proper flow velocity $v^{\hat{r}} = u \to 0$ and is subsonic. At $\tilde{r} = R_g/2$, the proper velocity equals the speed of light $|v^{\hat{r}}| = u = 1 > a$ and the flow is supersonic. This condition is independent of the magnitude of u and is not sufficient by itself to guarantee that the flow passes through a critical point outside EH. For large $\tilde{r} \geq r_s$, it is expected that the particles be nonrelativistic with $a \leq a_s \ll 1$ (i.e., $T \ll mc^2/K = 10^{13}\,K$), as they were nonrelativistic at infinity ($a_\infty \ll 1$). Considering the equation of accretion onto superdense protomatter core, which is an analogue of Bondi equations for spherical, steady-state adiabatic accretion onto the SPC, we determine a mass accretion rate

$$\dot{M} = 2\pi m n_s r_s^{5/2} \left(\ln \tilde{g}_{00}\right)'_s, \tag{4}$$

where prime $(')_s$ denotes differentiation with respect to $\tilde{r}$ at the point r_s. The gas compression can be estimated as

$$\frac{\tilde{n}}{n_\infty} \approx \frac{r_s^{5/2}}{2r^2}\left[\frac{(\ln \tilde{g}_{00})'_s}{1 + \tilde{g}^{rr}(\tilde{r})}\right]^{1/2}. \tag{5}$$

The approximate equality between the sound speed and the mean particle speed implies that the hydrodynamic accretion rate is larger than the collisionless accretion rate by the large factor $\approx 10^9$.

6. The Intermediate Mass, PRT, and Initial Redshift of Seed Black Hole

The key objectives of the MTBH framework are then an increase of the mass, $M_{\text{BH}}^{\text{Seed}}$, gravitational radius, R_g^{Seed}, and of the seed black hole, BH^{Seed}, at accretion of outside matter. Thereby an infalling matter forms protomatter disk around protomatter core tapering off faster at reaching the thin edge of event horizon. So, a practical measure of growth $\text{BH}^{\text{Seed}} \to$ BH may most usefully be the increase of gravitational radius or mass of black hole:

$$\begin{aligned}
\Delta R_g &= R_g^{\text{BH}} - R_g^{\text{Seed}} = \frac{2G}{c^2} M_d = \frac{2G}{c^2}\rho_d V_d, \\
\Delta M_{\text{BH}} &= M_{\text{BH}} - M_{\text{BH}}^{\text{Seed}} = M_{\text{BH}}^{\text{Seed}} \frac{\Delta R_g}{R_g^{\text{Seed}}},
\end{aligned} \tag{6}$$

where M_d, ρ_d, and V_d, respectively, are the total mass, density, and volume of protomatter disk. At the value $\widehat{R}_g^{\text{BH}}$ of gravitational radius, when protomatter disk has finally reached the event horizon of grown-up supermassive black

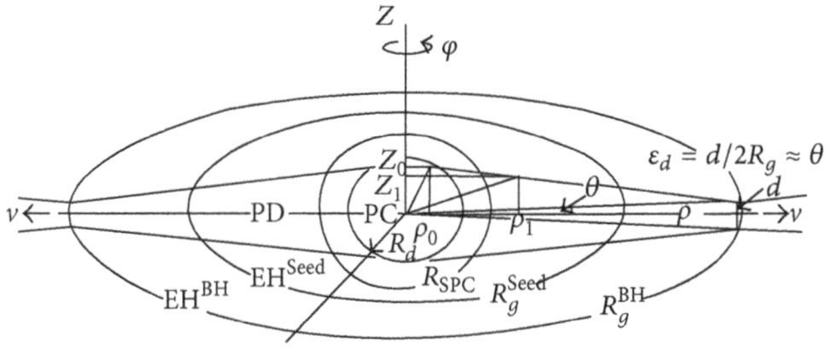

FIGURE 3: A schematic cross section of the growth of supermassive black hole driven by the formation of protomatter disk at accretion, when protomatter disk has finally reached the event horizon of grown-up supermassive black hole.

hole, the volume $\widehat{V}_d$ can be calculated in polar coordinates (ρ, z, φ) from Figure 3:

$$\begin{aligned}\widehat{V}_d &= \int_{\rho_0}^{\widehat{R}_g^{BH}} d\rho \int_0^{2\pi} \rho d\phi \int_{-z_1(\rho)}^{z_1(\rho)} dz \\ &\quad - \int_{\rho_0}^{R_d} d\rho \int_0^{2\pi} \rho d\phi \int_{-z_0(\rho)}^{z_0(\rho)} dz \\ &\overset{(R_d \ll \widehat{R}_g^{BH})}{\simeq} \frac{\sqrt{2}\pi}{3} R_d \left(\widehat{R}_g^{BH}\right)^2,\end{aligned} \tag{7}$$

where $z_1(\rho) \simeq z_0 - z_0(\rho - \rho_0)/(\widehat{R}_g^{BH} - \rho_0)$, $z_0(\rho) = \sqrt{R_d^2 - \rho^2}$, and in approximation $R_d \ll \widehat{R}_g^{BH}$ we set $z_0(\rho_0) \simeq \rho_0 \simeq R_d/\sqrt{2}$.

6.1. The Intermediate Mass of Seed Black Hole. From the first line of (6), by virtue of (7), we obtain

$$\widehat{R}_g^{BH} = k\left(1 \pm \sqrt{1 - \frac{2}{k} R_g^{Seed}}\right), \tag{8}$$

where $2/k = 8.73\,[\text{km}] R_d \rho_d / M_\odot$. The (8) is valid at $(2/k) R_g^{Seed} \le 1$; namely,

$$\frac{R_\odot}{R_d} \ge 2.09 \frac{[\text{km}]}{R_\odot} \frac{\rho_d}{\overline{\rho}_\odot} \frac{R_g^{Seed}}{R_\odot}. \tag{9}$$

For the values $\rho_d = 2.6 \times 10^{16}\,[\text{g cm}]^{-3}$ (see below) and $R_g^{Seed} \simeq 2.95\,[\text{km}](10^3 \text{ to } 10^6)$, inequality (9) is reduced to $R_\odot/R_d \ge 2.34 \times 10^8 (1 \text{ to } 10^3)$ or $[\text{cm}]/R_d \ge 0.34(10^{-2} \text{ to } 10)$. This condition is always satisfied, because for considered 377 black holes, with the masses $M_{BH}/M_\odot \simeq 1.1 \times 10^6$ to 1.3×10^{10}, we approximately have $R_d/r_{OV} \simeq 10^{-10}$ to 10^{-7} [71]. Note that Woo and Urry [5] collect and compare all the AGN/BH mass and luminosity estimates from the literature. According to (6), the intermediate mass of seed black hole reads

$$\frac{M_{BH}^{Seed}}{M_\odot} \simeq \frac{M_{BH}}{M_\odot}\left(1 - 1.68 \times 10^{-6} \frac{R_d}{[\text{cm}]} \frac{M_{BH}}{M_\odot}\right). \tag{10}$$

6.2. PRT. The PRT is referred to as a lapse of time T_{BH} from the birth of black hole till neutrino radiation, the earlier part of the lifetime. That is, $T_{BH} = M_d/\dot{M}$, where $\dot{M}$ is the accretion rate. In approximation at hand $R_d \ll R_g$, the PRT reads

$$T_{BH} = \rho_d \frac{V_d}{\dot{M}} \simeq 9.33 \cdot 10^{15}\,\left[\text{g cm}^{-3}\right] \frac{R_d R_g^2}{\dot{M}}. \tag{11}$$

In case of collisionless accretion, (3) and (11) give

$$T_{BH} \simeq 2.6 \cdot 10^{16} \frac{R_d}{\text{cm}} \frac{10^{-24}\,\text{g cm}^{-3}}{\rho_\infty} \frac{v_\infty}{10\,\text{km}\,s^{-1}}\,\text{yr}. \tag{12}$$

In case of hydrodynamic accretion, (4) and (11) yield

$$T_{BH} \simeq 8.8 \cdot 10^{38} \frac{R_d R_g^2\,\text{cm}^{-3}}{n_s r_s^{5/2} \left(\ln g_{00}\right)_s'}. \tag{13}$$

Note that the spherical accretion onto black hole, in general, is not necessarily an efficient mechanism for converting rest-mass energy into radiation. Accretion onto black hole may be far from spherical accretion, because the accreted gas possesses angular momentum. In this case, the gas will be thrown into circular orbits about the black hole when centrifugal forces will become significant before the gas plunges through the event horizon. Assuming a typical mass-energy conversion efficiency of about $\epsilon \sim 10\%$, in approximation $R_d \ll R_g$, according to (12) and (13), the resulting relationship of typical PRT versus bolometric luminosity becomes

$$T_{BH} \simeq 0.32 \frac{R_d}{r_{OV}} \left(\frac{M_{BH}}{M_\odot}\right)^2 \frac{10^{39}\,W}{L_{bol}}\,[\text{yr}]. \tag{14}$$

We supplement this by computing neutrino fluxes in the next section.

6.3. Redshift of Seed Black Hole. Interpreting the redshift as a cosmological Doppler effect and that the Hubble law could most easily be understood in terms of expansion of the universe, we are interested in the purely academic question of principle to ask what could be the initial redshift, z^{Seed}, of seed black hole if the mass, the luminosity, and the redshift, z, of black hole at present time are known. To follow the history of seed black hole to the present time, let us place ourselves at the origin of coordinates $r = 0$ (according to the Cosmological Principle, this is mere convention) and consider a light traveling to us along the $-r$ direction, with angular variables fixed. If the light has left a seed black hole, located at r_s, θ_s, and φ_s, at time t_s, and it has to reach us at a time t_0, then a power series for the redshift as a function of the time of flight is $z^{Seed} = H_0(t_0 - t_s) + \cdots$, where t_0 is the present moment and H_0 is Hubble's constant. Similar expression, $z = H_0(t_0 - t_1) + \cdots$, can be written for the current black hole, located at r_1, θ_1, and φ_1, at time t_1, where $t_1 = t_s + T_{BH}$, as seed black hole is an object at early times. Hence, in the first-order approximation by Hubble's constant, we may obtain the following relation between the redshifts of seed and present black holes: $z^{\text{Seed}} \simeq z + H_0 T_{\text{BH}}$. This relation is in agreement with the scenario of a general recession of distant galaxies away from us in all directions, the furthest naturally being those moving the fastest. This relation, incorporating with (14), for the value $H_0 = 70$ [km]/[s Mpc] favored today yields

$$z^{\text{Seed}} \simeq z + 2.292 \times 10^{28} \frac{R_d}{r_{\text{OV}}} \left(\frac{M_{\text{BH}}}{M_\odot}\right)^2 \frac{W}{L_{\text{bol}}}. \quad (15)$$

7. UHE Neutrino Fluxes

The flux can be written in terms of luminosity as $J_{\nu\varepsilon} = \widetilde{L}_{\nu\varepsilon}/4\pi D_L^2(z)(1+z)$, where z is the redshift and $D_L(z)$ is the luminosity distance depending on the cosmological model. The $(1+z)^{-1}$ is due to the fact that each neutrino with energy $\widetilde{E}_\nu'$ if observed near the place and time of emission t' will be red-shifted to energy $\widetilde{E}_\nu = \widetilde{E}_\nu' R(t_1)/R(t_0) = \widetilde{E}_\nu'(1+z)^{-1}$ of the neutrino observed at time t after its long journey to us, where $R(t)$ is the cosmic scale factor. Computing the UHE neutrino fluxes in the framework of MTBH, we choose the cosmological model favored today, with a flat universe, filled with matter $\Omega_M = \rho_M/\rho_c$ and vacuum energy densities $\Omega_V = \rho_V/\rho_c$, thereby $\Omega_V + \Omega_M = 1$, where the critical energy density $\rho_c = 3H_0^2/(8\pi G_N)$ is defined through the Hubble parameter H_0 [77]:

$$D_L(z) = \frac{(1+z)c}{H_0\sqrt{\Omega_M}} \int_1^{1+z} \frac{dx}{\sqrt{\Omega_V/\Omega_M + x^3}} = 2.4 \times 10^{28} I(z) \text{ cm}. \quad (16)$$

Here $I(z) = (1+z)\int_1^{1+z} dx/\sqrt{2.3 + x^3}$, we set the values $H_0 = 70$ km/s Mpc, $\Omega_V = 0.7$ and $\Omega_M = 0.3$.

7.1. URCA Reactions. The neutrino luminosity of SPC of given mass, $\widetilde{M}$, by modified URCA reactions with no muons reads [75]

$$\widetilde{L}_{\nu\varepsilon}^{\text{URCA}} = 3.8 \times 10^{50} \varepsilon_d \left(\frac{M_\odot}{\widetilde{M}}\right)^{1.75} \left[\text{erg s}^{-1}\right], \quad (17)$$

where $\varepsilon_d = d/2R_g$ and d is the thickness of the protomatter disk at the edge of even horizon. The resulting total UHE neutrino flux of cooling of the SPC can be obtained as

$$J_{\nu\varepsilon}^{\text{URCA}} \simeq 5.22 \times 10^{-8} \times \frac{\varepsilon_d}{I^2(z)(1+z)} \left(\frac{M_\odot}{\widetilde{M}}\right)^{1.75} \left[\text{erg cm}^{-2}\,\text{s}^{-1}\,\text{sr}^{-1}\right], \quad (18)$$

where the neutrino is radiated in a cone with the beaming angle $\theta \sim \varepsilon_d \ll 1$, $I(z) = (1+z)\int_1^{1+z} dx/\sqrt{2.3+x^3}$. As it is seen, the nucleon modified URCA reactions can contribute efficiently only to extragalactic objects with enough small redshift $z \ll 1$.

7.2. Pionic Reactions. The pionic reactions, occurring in the superdense protomatter medium of SPC, allow both the distorted energy and momentum to be conserved. This is the analogue of the simple URCA processes:

$$\pi^- + n \longrightarrow n + e^- + \overline{\nu}_e, \qquad \pi^- + n \longrightarrow n + \mu^- + \overline{\nu}_\mu \quad (19)$$

and the two inverse processes. As in the modified URCA reactions, the total rate for all four processes is essentially four times the rate of each reaction alone. The muons are already present when pions appear. The neutrino luminosity of the SPC of given mass, $\widetilde{M}$, by pionic reactions reads [75]

$$\widetilde{L}_{\nu\varepsilon}^{\pi} = 5.78 \times 10^{58} \varepsilon_d \left(\frac{M_\odot}{\widetilde{M}}\right)^{1.75} \left[\text{erg s}^{-1}\right]. \quad (20)$$

Then, the UHE neutrino total flux is

$$J_{\nu\varepsilon}^{\pi} \simeq 7.91 \frac{\varepsilon_d}{I^2(z)(1+z)} \left(\frac{M_\odot}{\widetilde{M}}\right)^{1.75} \left[\text{erg cm}^{-2}\,\text{s}^{-1}\,\text{sr}^{-1}\right]. \quad (21)$$

The resulting total energy-loss rate will then be dramatically larger due to the pionic reactions (19) rather than the modified URCA processes.

7.3. Quark Reactions. In the superdense protomatter medium the distorted quark Fermi energies are far below the charmed c-, t-, and b-quark production thresholds. Therefore, only up-, down-, and strange quarks are present. The β equilibrium is maintained by reactions like

$$d \longrightarrow u + e^- + \overline{\nu}_e, \qquad u + e^- \longrightarrow d + \nu_e, \quad (22)$$

$$s \longrightarrow u + e^- + \overline{\nu}_e, \qquad u + e^- \longrightarrow s + \nu_e, \quad (23)$$

which are β decay and its inverse. These reactions constitute simple URCA processes, in which there is a net loss of a $\nu_l\overline{\nu}_l$ pair at nonzero temperatures. In this application a sufficient accuracy is obtained by assuming β-equilibrium and that the neutrinos are not retained in the medium of Λ-like protomatter. The quark reactions (22) and (23) proceed at equal rates in β equilibrium, where the participating quarks must reside close to their Fermi surface. Hence, the total energy of flux due to simple URCA processes is rather twice than that of (22) or (23) alone. For example, the spectral fluxes of the UHE antineutrinos and neutrinos for different redshifts from quark reactions are plotted, respectively, in Figures 4 and 5 [75]. The total flux of UHE neutrino can be written as

$$J_{\nu\varepsilon}^{q} \simeq 70.68 \frac{\varepsilon_d}{I^2(z)(1+z)} \left(\frac{M_\odot}{\widetilde{M}}\right)^{1.75} \left[\text{erg cm}^{-2}\,\text{s}^{-1}\,\text{sr}^{-1}\right]. \quad (24)$$

8. Simulation

For simulation we use the data of AGN/BH mass and luminosity estimates for 377 black holes presented by [5]. These masses are mostly based on the virial assumption for the broad emission lines, with the broad-line region size

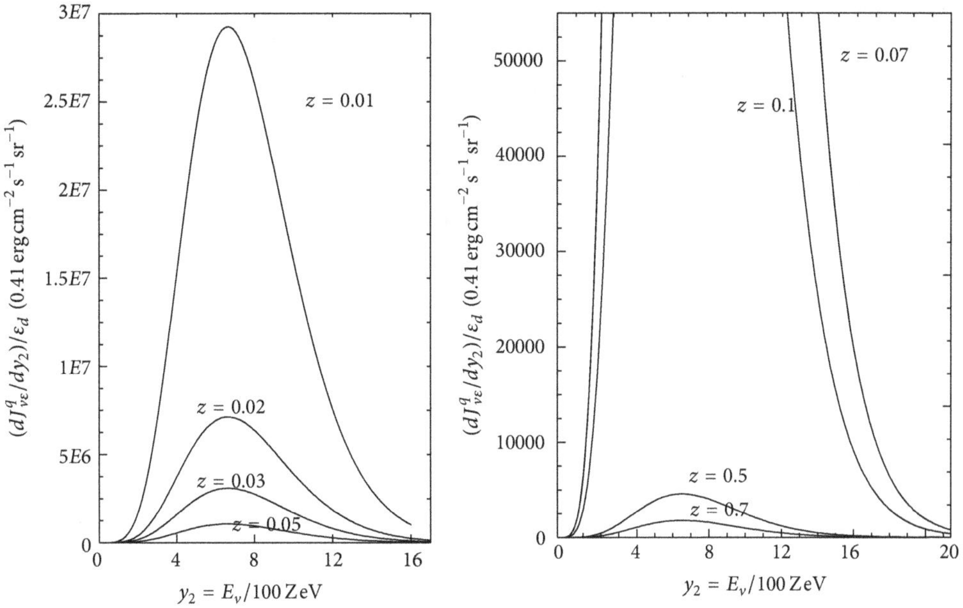

FIGURE 4: The spectral fluxes of UHE antineutrinos for different redshifts from quark reactions.

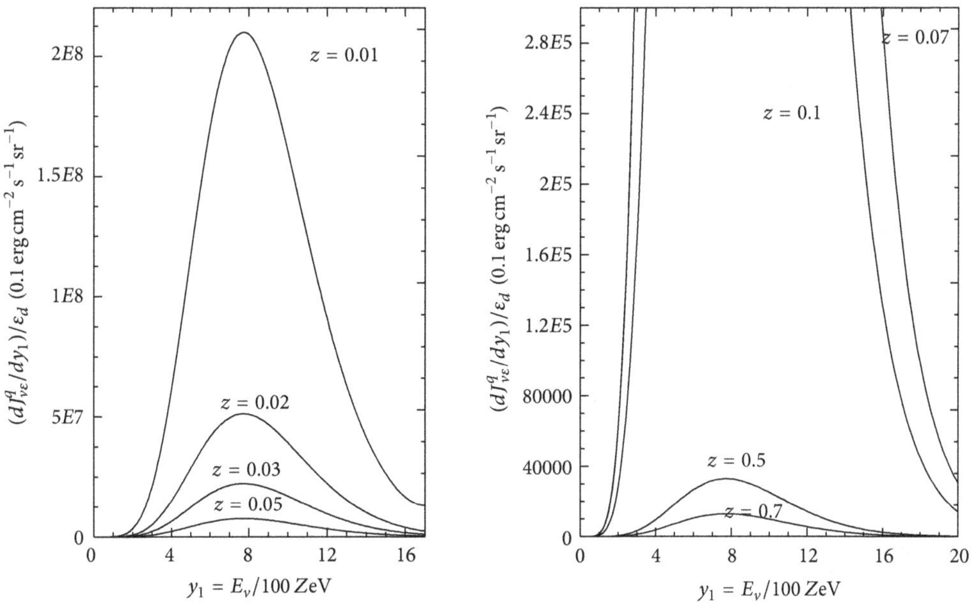

FIGURE 5: The spectral fluxes of UHE neutrinos for different redshifts from quark reactions.

determined from either reverberation mapping or optical luminosity. Additional black hole mass estimates based on properties of the host galaxy bulges, either using the observed stellar velocity dispersion or using the fundamental plane relation. Since the aim is to have more than a thousand of realizations, each individual run is simplified, with a use of previous algorithm of the SPC-configurations [71] as a working model, given in Appendix G. Computing the corresponding PRTs, seed black hole intermediate masses, and total neutrino fluxes, a main idea comes to solving an inverse problem. Namely, by the numerous reiterating integrations of the state equations of SPC-configurations we determine those required central values of particle concentration $\tilde{n}(0)$ and ID-field $x(0)$, for which the integrated total mass of configuration has to be equal to the black hole mass M_{BH} given from observations. Along with all integral characteristics, the radius R_d is also computed, which is further used in (10), (14), (15), (18), (21), and (24) for calculating $M^{\mathrm{Seed}}_{\mathrm{BH}}$, T_{BH}, z^{Seed}, and $J^i_{\nu\varepsilon}$, respectively. The results are summed up in Tables 1, 2, 3, 4 and 5. Figure 6 gives the intermediate seed masses $M^{\mathrm{Seed}}_{\mathrm{BH}}/M_\odot$ versus the present masses $M_{\mathrm{BH}}/M_\odot$ of 337 black holes, on logarithmic scales. For the present masses $M_{\mathrm{BH}}/M_\odot \simeq 1.1 \times 10^6$ to 1.3×10^{10}, the computed intermediate seed masses

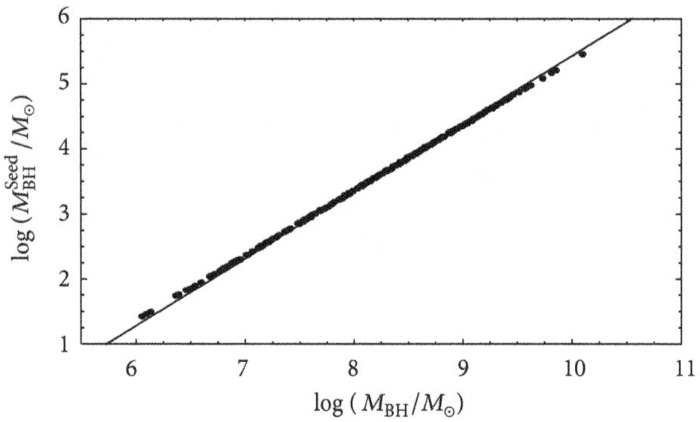

FIGURE 6: The $M_{BH}^{Seed}/M_{\odot}$-$M_{BH}/M_{\odot}$ relation on logarithmic scales of 337 black holes from [5]. The solid line is the best fit to data of samples.

are ranging from $M_{BH}^{Seed}/M_{\odot} \simeq 26.4$ to 2.9×10^5. The computed neutrino fluxes are ranging from (1) (quark reactions)—$J_{\nu\varepsilon}^{q}/\varepsilon_d$ [erg cm^{-2} s^{-1} sr^{-1}] $\simeq 8.29\times10^{-16}$ to 3.18×10^{-4}, with the average $\overline{J}_{\nu\varepsilon}^{q} \simeq 5.53\times10^{-10}\,\varepsilon_d$ [erg cm^{-2} s^{-1} sr^{-1}]; (2) (pionic reactions)—$J_{\nu\varepsilon}^{\pi} \simeq 0.112 J_{\nu\varepsilon}^{q}$, with the average $\overline{J}_{\nu\varepsilon}^{\pi} \simeq 3.66 \times 10^{-11}\,\varepsilon_d$ [erg cm^{-2} s^{-1} sr^{-1}]; and (3) (modified URCA processes)—$J_{\nu\varepsilon}^{URCA} \simeq 7.39\times10^{-11} J_{\nu\varepsilon}^{q}$, with the average $\overline{J}_{\nu\varepsilon}^{URCA} \simeq 2.41 \times 10^{-20}\,\varepsilon_d$ [erg cm^{-2} s^{-1} sr^{-1}]. In accordance, the AGNs are favored as promising pure neutrino sources because the computed neutrino fluxes are highly beamed along the plane of accretion disk and peaked at high energies and collimated in smaller opening angle $\theta \sim \varepsilon_d = d/2r_g \ll 1$. To render our discussion here a bit more transparent and to obtain some feeling for the parameter ε_d we may estimate $\varepsilon_d \simeq 1.69\times10^{-10}$, just for example, only, for the suppermassive black hole of typical mass $\sim 10^9 M_{\odot}(2R_g = 5.9\times10^{14}$ cm), and so $d \sim 1$ km. But the key problem of fixing the parameter ε_d more accurately from experiment would be an important topic for another investigation elsewhere.

9. Conclusions

The growth of accreting supermassive black hole seeds and their neutrino radiation are found to be common phenomena in the AGNs. In this report, we further expose the assertions made in the framework of microscopic theory of black hole via reviewing the mass assembly history of 377 plausible accreting supermassive black hole seeds. After the numerous reiterating integrations of the state equations of SPC-configurations, we compute their intermediate seed masses, M_{BH}^{Seed}, PRTs, initial redshifts, z^{Seed}, and neutrino fluxes. All the results are presented in Tables 1–5. Figure 6 gives the intermediate seed masses $M_{BH}^{Seed}/M_{\odot}$ versus the present masses $M_{BH}/M_{\odot}$ of 337 black holes, on logarithmic scales. In accordance, the AGNs are favored as promising pure UHE neutrino sources. Such neutrinos may reveal clues on the puzzle of origin of UHE cosmic rays. We regard the considered black holes only as the potential neutrino sources. The obtained results, however, may suffer and that would be underestimated if not all 377 live black holes in the $M_{BH}/M_{\odot} \simeq 1.1 \times 10^6$ to 1.3×10^{10} mass range at present reside in final stage of their growth, when the protomatter disk driven by accretion has reached the event horizon.

Appendices

A. Outline of the Key Points of Proposed Gravitation Theory at Huge Energies

The proposed gravitation theory explores the most important processes of rearrangement of vacuum state and a spontaneous breaking of gravitation gauge symmetry at huge energies. From its historical development, the efforts in gauge treatment of gravity mainly focus on the quantum gravity and microphysics, with the recent interest, for example, in the theory of the quantum superstring or, in the very early universe, in the inflationary model. The papers on the gauge treatment of gravity provide a unified picture of gravity modified models based on several Lie groups. However, currently no single theory has been uniquely accepted as the convincing gauge theory of gravitation which could lead to a consistent quantum theory of gravity. They have evoked the possibility that the treatment of spacetime might involve non-Riemannian features on the scale of the Planck length. This necessitates the study of dynamical theories involving post-Riemannian geometries. It is well known that the notions of space and connections should be separated; see, for example, [78–81]. The curvature and torsion are in fact properties of a connection, and many different connections are allowed to exist in the same spacetime. Therefore, when considering several connections with different curvature and torsion, one takes spacetime simply as a manifold and connections as additional structures. From this view point in a recent paper [82] we tackle the problem of spacetime deformation. This theory generalizes and, in particular cases, fully recovers the results of the conventional theory. Conceptually and techniquewise this theory is versatile and powerful and manifests its practical and technical virtue in the fact that through a nontrivial choice of explicit form of the *world-deformation tensor*, which we have at our disposal, in general, we have a way to deform the spacetime which displayed different connections, which may reveal different post-Riemannian spacetime structures as corollary. All the fundamental gravitational structures in fact—the metric as much as the coframes and connections—acquire a spacetime deformation induced theoretical interpretation. There is another line of reasoning which supports the side of this method. We address the theory of teleparallel gravity and construct a consistent Einstein-Cartan (EC) theory with the *dynamical torsion*. We show that the equations of the standard EC theory, in which the equation defining torsion is the algebraic type and, in fact, no propagation of torsion is allowed, can be equivalently replaced by the set of modified Einstein-Cartan equations in which the torsion, in general, is *dynamical*. Moreover, the special physical constraint imposed upon the spacetime deformations yields the short-range propagating spin-spin interaction. For the self-contained arguments in Appendix A.1 and Appendices B and C we complete the

TABLE 1: Seed black hole intermediate masses, preradiation times, redshifts, and neutrino fluxes from spatially resolved kinematics. Columns: (1) name, (2) redshift, (3) AGN type: SY2: Seyfert 2, (4) log of the bolometric luminosity (ergs^{-1}), (5) log of the radius of protomatter core in special unit r_{OV} = 13.68 km, (6) log of the black hole mass in solar masses, (7) log of the seed black hole intermediate mass in solar masses, (8) log of the neutrino preradiation time (yrs), (9) redshift of seed black hole, (10) $J^{i=q}$, (11) $J^{i=\mathrm{URCA}}$, and (12) $J^{i=\pi}$, where $J^i \equiv \log(J^i_{\nu\varepsilon}/\varepsilon_d \,\mathrm{erg\,cm^{-2}\,s^{-1}\,sr^{-1}})$.

Name	z	Type	$\log L_{\mathrm{bol}}$	$\log\left(\frac{R_d}{r_{\mathrm{ov}}}\right)$	$\log\left(\frac{M_{\mathrm{BH}}}{M_\odot}\right)$	$\log\left(\frac{M^{\mathrm{Seed}}_{\mathrm{BH}}}{M_\odot}\right)$	$\log T_{\mathrm{BH}}$	z^{Seed}	J^q	J^{URCA}	J^π
NGC 1068	0.004	SY2	44.98	−7.59201	7.23	2.5922	8.02	0.006	−5.49	−15.62	−6.44
NGC 4258	0.001	SY2	43.45	−7.98201	7.62	2.9822	10.33	0.148	−4.97	−15.10	−5.92

TABLE 2: Seed black hole intermediate masses, preradiation times, redshifts, and neutrino fluxes from reverberation mapping. Columns: (1) name, (2) redshift, (3) AGN type: SY2: Seyfert 2, (4) log of the bolometric luminosity (ergs^{-1}), (5) log of the radius of protomatter core in special unit r_{OV} = 13.68 km, (6) log of the black hole mass in solar masses, (7) log of the seed black hole intermediate mass in solar masses, (8) log of the neutrino preradiation time (yrs), (9) redshift of seed black hole, (10) $J^{i=q}$, (11) $J^{i=\mathrm{URCA}}$, and (12) $J^{i=\pi}$, where $J^i \equiv \log(J^i_{\nu\varepsilon}/\varepsilon_d \,\mathrm{erg\,cm^{-2}\,s^{-1}\,sr^{-1}})$.

Name	z	Type	$\log L_{\mathrm{bol}}$	$\log\left(\frac{R_d}{r_{\mathrm{ov}}}\right)$	$\log\left(\frac{M_{\mathrm{BH}}}{M_\odot}\right)$	$\log\left(\frac{M^{\mathrm{Seed}}_{\mathrm{BH}}}{M_\odot}\right)$	$\log T_{\mathrm{BH}}$	z^{Seed}	J^q	J^{URCA}	J^π
3C 120	0.033	SY1	45.34	−7.78201	7.42	2.7822	8.04	0.034	−7.69	−17.82	−8.64
3C 390.3	0.056	SY1	44.88	−8.91201	8.55	3.9122	10.76	0.103	−10.15	−20.28	−11.10
Akn 120	0.032	SY1	44.91	−8.63201	8.27	3.6322	10.17	0.055	−9.15	−19.28	−10.10
F9	0.047	SY1	45.23	−8.27201	7.91	3.2722	9.13	0.052	−8.87	−19.00	−9.82
IC 4329A	0.016	SY1	44.78	−7.13201	6.77	2.1322	7.30	0.017	−5.91	−16.04	−6.86
Mrk 79	0.022	SY1	44.57	−8.22201	7.86	3.2222	9.69	0.041	−8.10	−18.23	−9.05
Mrk 110	0.035	SY1	44.71	−7.18201	6.82	2.1822	7.47	0.036	−6.69	−16.82	−7.64
Mrk 335	0.026	SY1	44.69	−7.05201	6.69	2.0522	7.23	0.027	−6.20	−16.33	−7.15
Mrk 509	0.034	SY1	45.03	−8.22201	7.86	3.2222	9.23	0.041	−8.49	−18.62	−9.44
Mrk 590	0.026	SY1	44.63	−7.56201	7.20	2.5622	8.31	0.030	−7.09	−17.22	−8.04
Mrk 817	0.032	SY1	44.99	−7.96201	7.60	2.9622	8.75	0.036	−7.98	−18.11	−8.93
NGC 3227	0.004	SY1	43.86	−8.00201	7.64	3.0022	9.96	0.064	−6.21	−16.34	−7.16
NGC 3516	0.009	SY1	44.29	−7.72201	7.36	2.7222	8.97	0.021	−6.43	−16.56	−7.38
NGC 3783	0.010	SY1	44.41	−7.30201	6.94	2.3022	8.01	0.013	−5.79	−15.92	−6.74
NGC 4051	0.002	SY1	43.56	−6.49201	6.13	1.4922	7.24	0.006	−2.96	−13.10	−3.91
NGC 4151	0.003	SY1	43.73	−7.49201	7.13	2.4922	9.07	0.028	−5.07	−15.20	−6.02
NGC 4593	0.009	SY1	44.09	−7.27201	6.91	2.2722	8.27	0.016	−5.64	−15.77	−6.59
NGC 5548	0.017	SY1	44.83	−8.39201	8.03	3.3922	9.77	0.033	−8.16	−18.30	−9.11
NGC 7469	0.016	SY1	45.28	−7.20201	6.84	2.2022	6.94	0.016	−6.03	−16.16	−6.98
PG 0026 + 129	0.142	RQQ	45.39	−7.94201	7.58	2.9422	8.31	0.144	−9.35	−19.48	−10.30
PG 0052 + 251	0.155	RQQ	45.93	−8.77201	8.41	3.7722	9.43	0.158	−10.89	−21.02	−11.84
PG 0804 + 761	0.100	RQQ	45.93	−8.60201	8.24	3.6022	9.09	0.102	−10.16	−20.29	−11.11
PG 0844 + 349	0.064	RQQ	45.36	−7.74201	7.38	2.7422	7.94	0.065	−8.23	−18.36	−9.18
PG 0953 + 414	0.239	RQQ	46.16	−8.60201	8.24	3.6022	8.86	0.240	−11.04	−21.17	−11.99
PG 1211 + 143	0.085	RQQ	45.81	−7.85201	7.49	2.8522	7.71	0.085	−8.69	−18.82	−9.64
PG 1229 + 204	0.064	RQQ	45.01	−8.92201	8.56	3.9222	10.65	0.099	−10.29	−20.42	−11.24
PG 1307 + 085	0.155	RQQ	45.83	−8.26201	7.90	3.2622	8.51	0.156	−9.99	−20.12	−10.94
PG 1351 + 640	0.087	RQQ	45.50	−8.84201	8.48	3.8422	10.00	0.097	−10.44	−20.57	−11.39
PG 1411 + 442	0.089	RQQ	45.58	−7.93201	7.57	2.9322	8.10	0.090	−8.87	−19.00	−9.82
PG 1426 + 015	0.086	RQQ	45.19	−8.28201	7.92	3.2822	9.19	0.091	−9.45	−19.58	−10.40
PG 1613 + 658	0.129	RQQ	45.66	−8.98201	8.62	3.9822	10.12	0.138	−11.07	−21.20	−12.02
PG 1617 + 175	0.114	RQQ	45.52	−8.24201	7.88	3.2422	8.78	0.116	−9.65	−19.78	−10.60
PG 1700 + 518	0.292	RQQ	46.56	−8.67201	8.31	3.6722	8.60	0.293	−11.38	−21.51	−12.33
PG 2130 + 099	0.061	RQQ	45.47	−8.10201	7.74	3.1022	8.55	0.063	−8.81	−18.94	−9.76
PG 1226 + 023	0.158	RLQ	47.35	−7.58201	7.22	2.5822	5.63	0.158	−8.82	−18.95	−9.77
PG 1704 + 608	0.371	RLQ	46.33	−8.59201	8.23	3.5922	8.67	0.372	−11.50	−21.64	−12.45

TABLE 3: Seed black hole intermediate masses, preradiation times, redshifts, and neutrino fluxes from optimal luminosity. Columns: (1) name, (2) redshift, (3) AGN type: SY2: Seyfert 2, (4) log of the bolometric luminosity (ergs^{-1}), (5) log of the radius of protomatter core in special unit r_{OV} = 13.68 km, (6) log of the black hole mass in solar masses, (7) log of the seed black hole intermediate mass in solar masses, (8) log of the neutrino preradiation time (yrs), (9) redshift of seed black hole, (10) $J^{i=q}$, (11) $J^{i=\mathrm{URCA}}$, and (12) $J^{i=\pi}$, where $J^i \equiv \log(J^i_{\nu\varepsilon}/\varepsilon_d \text{ erg cm}^{-2}\text{ s}^{-1}\text{ sr}^{-1})$.

Name	z	Type	$\log L_{\mathrm{bol}}$	$\log\left(\frac{R_d}{r_{\mathrm{ov}}}\right)$	$\log\left(\frac{M_{\mathrm{BH}}}{M_\odot}\right)$	$\log\left(\frac{M^{\mathrm{Seed}}_{\mathrm{BH}}}{M_\odot}\right)$	$\log T_{\mathrm{BH}}$	z^{Seed}	J^q	J^{URCA}	J^π
Mrk 841	0.036	SY1	45.84	−8.46201	8.10	3.4622	8.90	0.038	−8.96	−19.09	−9.91
NGC 4253	0.013	SY1	44.40	−6.90201	6.54	1.9022	7.22	0.014	−5.32	−15.45	−6.27
NGC 6814	0.005	SY1	43.92	−7.64201	7.28	2.6422	9.18	0.028	−5.78	−15.91	−6.73
0054 + 144	0.171	RQQ	45.47	−9.26201	8.90	4.2622	10.87	0.198	−11.84	−21.97	−12.79
0157 + 001	0.164	RQQ	45.62	−8.06201	7.70	3.0622	8.32	0.165	−9.70	−19.83	−10.65
0204 + 292	0.109	RQQ	45.05	−7.03201	6.67	2.0322	6.83	0.109	−7.49	−17.62	−8.44
0205 + 024	0.155	RQQ	45.45	−8.22201	7.86	3.2222	8.81	0.158	−9.92	−20.05	−10.87
0244 + 194	0.176	RQQ	45.51	−8.39201	8.03	3.3922	9.09	0.179	−10.35	−20.48	−11.30
0923 + 201	0.190	RQQ	46.22	−9.30201	8.94	4.3022	10.20	0.195	−12.02	−22.15	−12.97
1012 + 008	0.185	RQQ	45.51	−8.15201	7.79	3.1522	8.61	0.187	−9.98	−20.11	−10.93
1029 − 140	0.086	RQQ	46.03	−9.44201	9.08	4.4422	10.67	0.097	−11.48	−21.61	−12.43
1116 + 215	0.177	RQQ	46.02	−8.57201	8.21	3.5722	8.94	0.179	−10.67	−20.80	−11.62
1202 + 281	0.165	RQQ	45.39	−8.65201	8.29	3.6522	9.73	0.173	−10.74	−20.87	−11.69
1309 + 355	0.184	RQQ	45.39	−8.36201	8.00	3.3622	8.91	0.186	−10.34	−20.47	−11.29
1402 + 261	0.164	RQQ	45.13	−7.65201	7.29	2.6522	7.99	0.165	−8.98	−19.11	−9.93
1444 + 407	0.267	RQQ	45.93	−8.42201	8.06	3.4222	8.73	0.268	−10.84	−20.97	−11.79
1635 + 119	0.146	RQQ	45.13	−8.46201	8.10	3.4622	9.61	0.155	−10.28	−20.41	−11.23
0022 − 297	0.406	RLQ	44.98	−8.27201	7.91	3.2722	9.38	0.414	−11.05	−21.18	−12.00
0024 + 348	0.333	RLQ	45.31	−6.73201	6.37	1.7322	5.97	0.333	−8.13	−18.26	−9.08
0056 − 001	0.717	RLQ	46.54	−9.07201	8.71	4.0722	9.42	0.718	−13.13	−23.26	−14.08
0110 + 495	0.395	RLQ	45.78	−8.70201	8.34	3.7022	9.44	0.399	−11.77	−21.90	−12.72
0114 + 074	0.343	RLQ	44.02	−7.16201	6.80	2.1622	8.12	0.349	−8.91	−19.04	−9.86
0119 + 041	0.637	RLQ	45.57	−8.74201	8.38	3.7422	9.73	0.643	−12.41	−22.54	−13.36
0133 + 207	0.425	RLQ	45.83	−9.88201	9.52	4.8822	11.75	0.474	−13.92	−24.05	−14.87
0133 + 476	0.859	RLQ	46.69	−9.09201	8.73	4.0922	9.31	0.860	−13.40	−23.53	−14.35
0134 + 329	0.367	RLQ	46.44	−9.10201	8.74	4.1022	9.58	0.369	−12.38	−22.52	−13.33
0135 − 247	0.831	RLQ	46.64	−9.49201	9.13	4.4922	10.16	0.834	−14.06	−24.19	−15.01
0137 + 012	0.258	RLQ	45.22	−8.93201	8.57	3.9322	10.46	0.280	−11.70	−21.83	−12.65
0153 − 410	0.226	RLQ	44.74	−7.92201	7.56	2.9222	8.92	0.233	−9.79	−19.92	−10.74
0159 − 117	0.669	RLQ	46.84	−9.63201	9.27	4.6322	10.24	0.672	−14.03	−24.16	−14.98
0210 + 860	0.186	RLQ	44.92	−6.90201	6.54	1.9022	6.70	0.186	−7.80	−17.93	−8.75
0221 + 067	0.510	RLQ	44.94	−7.65201	7.29	2.6522	8.18	0.512	−10.23	−20.36	−11.18
0237 − 233	2.224	RLQ	47.72	−8.88201	8.52	3.8822	7.86	2.224	−14.39	−24.52	−15.34
0327 − 241	0.888	RLQ	46.01	−8.96201	8.60	3.9622	9.73	0.892	−13.22	−23.35	−14.17
0336 − 019	0.852	RLQ	46.32	−9.34201	8.98	4.3422	10.18	0.857	−13.83	−23.96	−14.78
0403 − 132	0.571	RLQ	46.47	−9.43201	9.07	4.4322	10.21	0.575	−13.48	−23.61	−14.43
0405 − 123	0.574	RLQ	47.40	−9.83201	9.47	4.8322	10.08	0.575	−14.19	−24.32	−15.14
0420 − 014	0.915	RLQ	47.00	−9.39201	9.03	4.3922	9.60	0.916	−14.01	−24.14	−14.96
0437 + 785	0.454	RLQ	46.15	−9.15201	8.79	4.1522	9.97	0.458	−12.72	−22.85	−13.67
0444 + 634	0.781	RLQ	46.12	−8.89201	8.53	3.8922	9.48	0.784	−12.93	−23.06	−13.88
0454 − 810	0.444	RLQ	45.32	−8.49201	8.13	3.4922	9.48	0.450	−11.54	−21.67	−12.49
0454 + 066	0.405	RLQ	45.12	−7.78201	7.42	2.7822	8.26	0.407	−10.19	−20.32	−11.14
0502 + 049	0.954	RLQ	46.36	−9.24201	8.88	4.2422	9.94	0.957	−13.80	−23.93	−14.75
0514 − 459	0.194	RLQ	45.36	−7.91201	7.55	2.9122	8.28	0.196	−9.61	−19.74	−10.56
0518 + 165	0.759	RLQ	46.34	−8.89201	8.53	3.8922	9.26	0.761	−12.89	−23.02	−13.84
0538 + 498	0.545	RLQ	46.43	−9.94201	9.58	4.9422	11.27	0.559	−14.32	−24.45	−15.27
0602 − 319	0.452	RLQ	45.69	−9.38201	9.02	4.3822	10.89	0.473	−13.11	−23.25	−14.07

TABLE 3: Continued.

Name	z	Type	$\log L_{\text{bol}}$	$\log\left(\frac{R_d}{r_{\text{ov}}}\right)$	$\log\left(\frac{M_{\text{BH}}}{M_\odot}\right)$	$\log\left(\frac{M_{\text{BH}}^{\text{Seed}}}{M_\odot}\right)$	$\log T_{\text{BH}}$	z^{Seed}	J^q	J^{URCA}	J^π
0607 − 157	0.324	RLQ	46.30	−9.04201	8.68	4.0422	9.60	0.326	−12.14	−22.27	−13.09
0637 − 752	0.654	RLQ	47.16	−9.77201	9.41	4.7722	10.20	0.656	−14.24	−24.37	−15.19
0646 + 600	0.455	RLQ	45.58	−9.10201	8.74	4.1022	10.44	0.469	−12.63	−22.76	−13.58
0723 + 679	0.846	RLQ	46.41	−9.03201	8.67	4.0322	9.47	0.848	−13.27	−23.41	−14.23
0736 + 017	0.191	RLQ	46.41	−8.36201	8.00	3.3622	8.57	0.192	−10.38	−20.51	−11.33
0738 + 313	0.631	RLQ	46.94	−9.76201	9.40	4.7622	10.40	0.634	−14.18	−24.31	−15.13
0809 + 483	0.871	RLQ	46.54	−8.32201	7.96	3.3222	7.92	0.871	−12.07	−22.20	−13.02
0838 + 133	0.684	RLQ	46.23	−8.88201	8.52	3.8822	9.35	0.686	−12.74	−22.87	−13.69
0906 + 430	0.668	RLQ	45.99	−8.26201	7.90	3.2622	8.35	0.669	−11.63	−21.76	−12.58
0912 + 029	0.427	RLQ	45.26	−8.08201	7.72	3.0822	8.72	0.430	−10.77	−20.90	−11.72
0921 − 213	0.052	RLQ	44.63	−8.50201	8.14	3.5022	10.19	0.084	−9.36	−19.50	−10.32
0923 + 392	0.698	RLQ	46.26	−9.64201	9.28	4.6422	10.84	0.708	−14.10	−24.23	−15.05
0925 − 203	0.348	RLQ	46.35	−8.82201	8.46	3.8222	9.11	0.349	−11.83	−21.97	−12.78
0953 + 254	0.712	RLQ	46.59	−9.36201	9.00	4.3622	9.95	0.715	−13.63	−23.76	−14.58
0954 + 556	0.901	RLQ	46.54	−8.43201	8.07	3.4322	8.14	0.901	−12.31	−22.44	−13.26
1004 + 130	0.240	RLQ	46.21	−9.46201	9.10	4.4622	10.53	0.248	−12.55	−22.68	−13.50
1007 + 417	0.612	RLQ	46.71	−9.15201	8.79	4.1522	9.41	0.613	−13.08	−23.21	−14.03
1016 − 311	0.794	RLQ	46.63	−9.25201	8.89	4.2522	9.69	0.796	−13.58	−23.71	−14.53
1020 − 103	0.197	RLQ	44.87	−8.72201	8.36	3.7222	10.39	0.228	−11.04	−21.18	−11.99
1034 − 293	0.312	RLQ	46.20	−9.11201	8.75	4.1122	9.84	0.316	−12.22	−22.35	−13.17
1036 − 154	0.525	RLQ	44.55	−8.16201	7.80	3.1622	9.59	0.543	−11.16	−21.29	−12.11
1045 − 188	0.595	RLQ	45.80	−7.19201	6.83	2.1922	6.40	0.595	−9.61	−19.74	−10.56
1100 + 772	0.311	RLQ	46.49	−9.67201	9.31	4.6722	10.67	0.318	−13.20	−23.33	−14.15
1101 − 325	0.355	RLQ	46.33	−8.97201	8.61	3.9722	9.43	0.357	−12.12	−22.25	−13.07
1106 + 023	0.157	RLQ	44.97	−7.86201	7.50	2.8622	8.57	0.160	−9.31	−19.44	−10.26
1107 − 187	0.497	RLQ	44.25	−7.26201	6.90	2.2622	8.09	0.501	−9.52	−19.65	−10.47
1111 + 408	0.734	RLQ	46.26	−10.18201	9.82	5.1822	11.92	0.770	−15.11	−25.24	−16.06
1128 − 047	0.266	RLQ	44.08	−7.08201	6.72	2.0822	7.90	0.270	−8.49	−18.62	−9.44
1136 − 135	0.554	RLQ	46.78	−9.14201	8.78	4.1422	9.32	0.555	−12.94	−23.07	−13.89
1137 + 660	0.656	RLQ	46.85	−9.72201	9.36	4.7222	10.41	0.659	−14.16	−24.29	−15.11
1150 + 497	0.334	RLQ	45.98	−9.09201	8.73	4.0922	10.02	0.340	−12.26	−22.39	−13.21
1151 − 348	0.258	RLQ	45.56	−9.38201	9.02	4.3822	11.02	0.287	−12.26	−22.39	−13.21
1200 − 051	0.381	RLQ	46.41	−8.77201	8.41	3.7722	8.95	0.382	−12.26	−22.39	−13.21
1202 − 262	0.789	RLQ	45.81	−9.36201	9.00	4.3622	10.73	0.804	−13.76	−23.89	−14.71
1217 + 023	0.240	RLQ	45.83	−8.77201	8.41	3.7722	9.53	0.244	−11.34	−21.47	−12.29
1237 − 101	0.751	RLQ	46.63	−9.64201	9.28	4.6422	10.47	0.755	−14.19	−24.32	−15.14
1244 − 255	0.633	RLQ	46.48	−9.40201	9.04	4.4022	10.14	0.637	−13.55	−23.69	−14.51
1250 + 568	0.321	RLQ	45.61	−8.78201	8.42	3.7822	9.77	0.327	−11.67	−21.81	−12.62
1253 − 055	0.536	RLQ	46.10	−8.79201	8.43	3.7922	9.30	0.538	−12.28	−22.42	−13.24
1254 − 333	0.190	RLQ	45.52	−9.19201	8.83	4.1922	10.68	0.210	−11.83	−21.96	−12.78
1302 − 102	0.286	RLQ	45.86	−8.66201	8.30	3.6622	9.28	0.289	−11.34	−21.47	−12.29
1352 − 104	0.332	RLQ	45.81	−8.51201	8.15	3.5122	9.03	0.334	−11.24	−21.37	−12.19
1354 + 195	0.720	RLQ	47.11	−9.80201	9.44	4.8022	10.31	0.722	−14.42	−24.55	−15.37
1355 − 416	0.313	RLQ	46.48	−10.09201	9.73	5.0922	11.52	0.331	−13.94	−24.07	−14.89
1359 − 281	0.803	RLQ	46.19	−8.43201	8.07	3.4322	8.49	0.804	−12.16	−22.29	−13.11
1450 − 338	0.368	RLQ	43.94	−6.82201	6.46	1.8222	7.52	0.371	−8.40	−18.53	−9.35
1451 − 375	0.314	RLQ	46.16	−9.18201	8.82	4.1822	10.02	0.319	−12.35	−22.48	−13.30
1458 + 718	0.905	RLQ	46.93	−9.34201	8.98	4.3422	9.57	0.906	−13.91	−24.04	−14.86
1509 + 022	0.219	RLQ	44.54	−8.35201	7.99	3.3522	9.98	0.247	−10.51	−20.64	−11.46
1510 − 089	0.361	RLQ	46.38	−9.01201	8.65	4.0122	9.46	0.363	−12.21	−22.34	−13.16

Table 3: Continued.

Name	z	Type	$\log L_{\rm bol}$	$\log\left(\frac{R_d}{r_{\rm ov}}\right)$	$\log\left(\frac{M_{\rm BH}}{M_\odot}\right)$	$\log\left(\frac{M_{\rm BH}^{\rm Seed}}{M_\odot}\right)$	$\log T_{\rm BH}$	$z^{\rm Seed}$	J^q	$J^{\rm URCA}$	J^π
1545 + 210	0.266	RLQ	45.86	−9.29201	8.93	4.2922	10.54	0.278	−12.36	−22.49	−13.31
1546 + 027	0.412	RLQ	46.00	−9.08201	8.72	4.0822	9.98	0.417	−12.48	−22.61	−13.43
1555 − 140	0.097	RLQ	44.94	−7.61201	7.25	2.6122	8.10	0.099	−8.39	−18.53	−9.34
1611 + 343	1.401	RLQ	46.99	−9.93201	9.57	4.9322	10.69	1.405	−15.54	−25.67	−16.49
1634 + 628	0.988	RLQ	45.47	−7.64201	7.28	2.6422	7.63	0.989	−11.05	−21.18	−12.00
1637 + 574	0.750	RLQ	46.68	−9.54201	9.18	4.5422	10.22	0.753	−14.01	−24.14	−14.96
1641 + 399	0.594	RLQ	46.89	−9.78201	9.42	4.7822	10.49	0.597	−14.14	−24.27	−15.09
1642 + 690	0.751	RLQ	45.78	−8.12201	7.76	3.1222	8.28	0.752	−11.53	−21.66	−12.48
1656 + 053	0.879	RLQ	47.21	−9.98201	9.62	4.9822	10.57	0.882	−14.99	−25.12	−15.94
1706 + 006	0.449	RLQ	44.01	−6.99210	6.63	2.9922	7.79	0.453	−8.92	−19.06	−9.87
1721 + 343	0.206	RLQ	45.63	−8.40201	8.04	3.4022	8.99	0.209	−10.53	−20.66	−11.48
1725 + 044	0.293	RLQ	46.07	−8.43201	8.07	3.4322	8.61	0.294	−10.96	−21.09	−11.91

spacetime deformation theory [82] by new investigation of building up the *distortion-complex* of spacetime continuum and showing how it restores the *world-deformation tensor*, which still has been put in by hand. We extend necessary geometrical ideas of spacetime deformation in concise form, without going into the subtleties, as applied to the gravitation theory which underlies the MTBH framework. I have attempted to maintain a balance between being overly detailed and overly schematic. Therefore the text in the appendices should resemble a "hybrid" of a new investigation and some issues of proposed gravitation theory.

A.1. A First Glance at Spacetime Deformation. Consider a smooth deformation map $\Omega : M_4 \to \widetilde{\mathcal{M}}_4$, written in terms of the *world-deformation* tensor (Ω), the general ($\widetilde{\mathcal{M}}_4$), and flat ($M_4$) smooth differential 4D-manifolds. The following notational conventions will be used throughout the appendices. All magnitudes related to the space, $\widetilde{\mathcal{M}}_4$, will be denoted by an over "~". We use the Greek alphabet ($\mu, \nu, \rho, \ldots = 0, 1, 2, 3$) to denote the holonomic world indices related to $\widetilde{\mathcal{M}}_4$ and the second half of Latin alphabet ($l, m, k, \ldots = 0, 1, 2, 3$) to denote the world indices related to M_4. The tensor, Ω, can be written in the form $\Omega = \widetilde{D}\widetilde{\psi}$ ($\Omega^m{}_l = \widetilde{D}^m_\mu \widetilde{\psi}^\mu_l$), where the DC-members are the invertible distortion matrix $\widetilde{D}(\widetilde{D}^m_\mu)$ and the tensor $\widetilde{\psi}(\widetilde{\psi}^\mu_l \equiv \partial_l \widetilde{x}^\mu$ and $\partial_l = \partial/\partial x^l)$. The principle foundation of the *world-deformation* tensor (Ω) comprises the following two steps: (1) the basis vectors e_m at given point ($p \in M_4$) undergo the *distortion* transformations by means of $\widetilde{D}$; and (2) the diffeomorphism $\widetilde{x}^\mu(x) : M_4 \to \widetilde{M}_4$ is constructed by seeking new holonomic coordinates $\widetilde{x}^\mu(x)$ as the solutions of the first-order partial differential equations. Namely,

$$\widetilde{e}_\mu = \widetilde{D}^l_\mu e_l, \qquad \widetilde{e}_\mu \widetilde{\psi}^\mu_l = \Omega^m{}_l e_m, \tag{A.1}$$

where the conditions of integrability, $\partial_k \widetilde{\psi}^\mu_l = \partial_l \widetilde{\psi}^\mu_k$, and nondegeneracy, $\|\psi\| \neq 0$, necessarily hold [83, 84]. For reasons that will become clear in the sequel, next we write the norm $\widetilde{ds} \equiv \widetilde{id}$ (see Appendix B) of the infinitesimal displacement $d\widetilde{x}^\mu$ on the $\widetilde{\mathcal{M}}_4$ in terms of the spacetime structures of M_4

$$\widetilde{id} = \widetilde{e}\widetilde{\vartheta} = \widetilde{e}_\mu \otimes \widetilde{\vartheta}^\mu = \Omega^m{}_l e_m \otimes \vartheta^l \in \widetilde{\mathcal{M}}_4. \tag{A.2}$$

A deformation $\Omega : M_4 \to \widetilde{\mathcal{M}}_4$ comprises the following two 4D deformations $\overset{\circ}{\Omega} : M_4 \to V_4$ and $\breve{\Omega} : V_4 \to \widetilde{M}_4$, where V_4 is the semi-Riemannian space and $\overset{\circ}{\Omega}$ and $\breve{\Omega}$ are the corresponding *world deformation* tensors. The key points of the theory of spacetime deformation are outlined further in Appendix B. Finally, to complete this theory we need to determine $\widetilde{D}$ and $\widetilde{\psi}$, figured in (A.1). In the standard theory of gravitation they can be determined from the standard field equations by means of the general linear frames (C.10). However, it should be emphasized that the standard Riemannian space interacting quantum field theory cannot be a satisfactory ground for addressing the most important processes of rearrangement of vacuum state and gauge symmetry breaking in gravity at huge energies. The difficulties arise there because Riemannian geometry, in general, does not admit a group of isometries, and it is impossible to define energy-momentum as Noether local currents related to exact symmetries. This, in turn, posed severe problem of nonuniqueness of the physical vacuum and the associated Fock space. A definition of positive frequency modes cannot, in general, be unambiguously fixed in the past and future, which leads to $|\text{in}\rangle \neq |\text{out}\rangle$, because the state $|\text{in}\rangle$ is unstable against decay into many particle $|\text{out}\rangle$ states due to interaction processes allowed by lack of Poincaré invariance. A nontrivial Bogolubov transformation between past and future positive frequency modes implies that particles are created from the vacuum and this is one of the reasons for $|\text{in}\rangle \neq |\text{out}\rangle$.

A.2. General Gauge Principle. Keeping in mind the aforesaid, we develop the alternative framework of the *general gauge principle* (GGP), which is the *distortion gauge induced* fiber-bundle formulation of gravitation. As this principle was in use as a guide in constructing our theory, we briefly discuss its general implications in Appendix D. The interested reader

TABLE 4: Seed black hole intermediate masses, preradiation times, redshifts, and neutrino fluxes from observed stellar velocity dispersions. Columns: (1) name, (2) redshift, (3) AGN type: SY2: Seyfert 2, (4) log of the bolometric luminosity (ergs^{-1}), (5) log of the radius of protomatter core in special unit r_{OV} = 13.68 km, (6) log of the black hole mass in solar masses, (7) log of the seed black hole intermediate mass in solar masses, (8) log of the neutrino preradiation time (yrs), (9) redshift of seed black hole, (10) $J^{i=q}$, (11) $J^{i=\mathrm{URCA}}$, and (12) $J^{i=\pi}$, where $J^i \equiv \log(J^i_{\nu\varepsilon}/\varepsilon_d \,\mathrm{erg\,cm^{-2}\,s^{-1}\,sr^{-1}})$.

Name	z	Type	$\log L_{\mathrm{bol}}$	$\log\left(\frac{R_d}{r_{\mathrm{ov}}}\right)$	$\log\left(\frac{M_{\mathrm{BH}}}{M_\odot}\right)$	$\log\left(\frac{M^{\mathrm{Seed}}_{\mathrm{BH}}}{M_\odot}\right)$	$\log T_{\mathrm{BH}}$	z^{Seed}	J^q	J^{URCA}	J^π
NGC 1566	0.005	SY1	44.45	−7.28201	6.92	2.2822	7.93	0.008	−5.15	−15.28	−6.10
NGC 2841	0.002	SY1	43.67	−8.57201	8.21	3.5722	11.29	0.347	−6.60	−16.74	−7.55
NGC 3982	0.004	SY1	43.54	−6.45201	6.09	1.45220	7.18	0.008	−3.50	−13.63	−4.45
NGC 3998	0.003	SY1	43.54	−9.31201	8.95	4.3122	12.90	2.561	−8.25	−18.38	−9.20
Mrk 10	0.029	SY1	44.61	−7.83291	7.47	4.7908	8.87	0.036	−7.66	−17.79	−8.61
UGC 3223	0.016	SY1	44.27	−7.38201	7.02	2.3822	8.31	0.022	−6.34	−16.47	−7.29
NGC 513	0.002	SY2	42.52	−8.01201	7.65	3.0122	11.32	1.345	−5.62	−15.76	−6.57
NGC 788	0.014	SY2	44.33	−7.87201	7.51	2.8722	9.23	0.029	−7.08	−17.21	−8.03
NGC 1052	0.005	SY2	43.84	−8.55201	8.19	3.5522	11.08	0.228	−7.37	−17.50	−8.32
NGC 1275	0.018	SY2	45.04	−8.87201	8.51	3.8722	10.52	0.047	−9.05	−19.19	−10.01
NGC 1320	0.009	SY2	44.02	−7.54201	7.18	2.5422	8.88	0.023	−6.12	−16.25	−7.07
NGC 1358	0.013	SY2	44.37	−8.24201	7.88	3.2422	9.93	0.045	−7.66	−17.80	−8.62
NGC 1386	0.003	SY2	43.38	−7.60201	7.24	2.6022	9.64	0.075	−0.020	−15.39	−6.21
NGC 1667	0.015	SY2	44.69	−8.24201	7.88	3.2422	9.61	0.030	−7.79	−17.92	−8.74
NGC 2110	0.008	SY2	44.10	−8.66201	8.30	3.6622	11.04	0.166	−7.97	−18.10	−8.92
NGC 2273	0.006	SY2	44.05	−7.66201	7.30	2.6622	9.09	0.024	−5.97	−16.10	−6.92
NGC 2992	0.008	SY2	43.92	−8.08201	7.72	3.0822	10.06	0.071	−6.96	−17.09	−7.91
NGC 3185	0.004	SY2	43.08	−6.42201	6.06	1.4222	7.58	0.014	−3.45	−13.58	−4.40
NGC 3362	0.028	SY2	44.27	−7.13201	6.77	2.1322	7.81	0.031	−6.40	−16.54	−7.36
NGC 3786	0.009	SY2	43.47	−7.89201	7.53	2.8922	10.13	0.123	−6.73	−16.86	−7.68
NGC 4117	0.003	SY2	43.64	−7.19201	6.83	2.1922	8.56	0.018	−4.54	−14.67	−5.49
NGC 4339	0.004	SY2	43.38	−7.76201	7.40	2.7622	9.96	0.108	−5.79	−15.92	−6.74
NGC 5194	0.002	SY2	43.79	−7.31201	6.95	2.3122	8.65	0.016	−4.40	−14.53	−5.35
NGC 5252	0.023	SY2	45.39	−8.40201	8.04	3.4022	9.23	0.027	−8.45	−18.58	−9.40
NGC 5273	0.004	SY2	43.03	−6.87201	6.51	1.8722	8.53	0.034	−4.23	−14.36	−5.18
NGC 5347	0.008	SY2	43.81	−7.15201	6.79	2.1522	8.31	0.018	−5.33	−15.46	−6.28
NGC 5427	0.009	SY2	44.12	−6.75201	6.39	1.7522	7.20	0.011	−4.73	−14.86	−5.68
NGC 5929	0.008	SY2	43.04	−7.61201	7.25	2.6122	10.00	0.169	−6.13	−16.27	−7.09
NGC 5953	0.007	SY2	44.05	−7.30201	6.94	2.3022	8.37	0.015	−5.48	−15.61	−6.43
NGC 6104	0.028	SY2	43.60	−7.96201	7.60	2.9622	10.14	0.128	−7.86	−17.99	−8.81
NGC 7213	0.006	SY2	44.30	−8.35201	7.99	3.3522	10.22	0.055	−7.18	−17.31	−8.13
NGC 7319	0.023	SY2	44.19	−7.74201	7.38	2.7422	9.11	0.038	−7.30	−17.43	−8.25
NGC 7603	0.030	SY2	44.66	−8.44201	8.08	3.4422	10.04	0.056	−8.76	−18.89	−9.71
NGC 7672	0.013	SY2	43.86	−7.24201	6.88	2.2422	8.44	0.023	−5.91	−16.05	−6.87
NGC 7682	0.017	SY2	43.93	−7.64201	7.28	2.6422	9.17	0.039	−6.85	−16.98	−7.80
NGC 7743	0.006	SY2	43.60	−6.95201	6.59	1.9522	8.12	0.016	−4.73	−14.86	−5.68
Mrk 1	0.016	SY2	44.20	−7.52201	7.16	2.5222	8.66	0.025	−6.59	−16.72	−7.54
Mrk 3	0.014	SY2	44.54	−9.01201	8.65	4.0122	11.30	0.142	−9.08	−19.21	−10.03
Mrk 78	0.037	SY2	44.59	−8.23201	7.87	3.2322	9.69	0.056	−8.58	−18.71	−9.53
Mrk 270	0.010	SY2	43.37	−7.96201	7.60	2.9622	10.37	0.179	−6.94	−17.07	−7.89
Mrk 348	0.015	SY2	44.27	−7.57201	7.21	2.5722	8.69	0.024	−6.62	−16.75	−7.57
Mrk 533	0.029	SY2	45.15	−7.92201	7.56	2.9222	8.51	0.032	−7.82	−17.95	−8.77
Mrk 573	0.017	SY2	44.44	−7.64201	7.28	2.6422	8.66	0.024	−6.85	−16.98	−7.80
Mrk 622	0.023	SY2	44.52	−7.28201	6.92	2.2822	7.86	0.026	−6.49	−16.62	−7.44
Mrk 686	0.014	SY2	44.11	−7.92201	7.56	2.9222	9.55	0.042	−7.17	−17.30	−8.12
Mrk 917	0.024	SY2	44.75	−7.98201	7.62	2.9822	9.03	0.031	−7.75	−17.89	−8.70
Mrk 1018	0.042	SY2	44.39	−8.45201	8.09	3.4522	10.33	0.092	−9.08	−19.21	−10.03

TABLE 4: Continued.

Name	z	Type	$\log L_{\text{bol}}$	$\log\left(\frac{R_d}{r_{\text{ov}}}\right)$	$\log\left(\frac{M_{\text{BH}}}{M_\odot}\right)$	$\log\left(\frac{M_{\text{BH}}^{\text{Seed}}}{M_\odot}\right)$	$\log T_{\text{BH}}$	z^{Seed}	J^q	J^{URCA}	J^π
Mrk 1040	0.017	SY2	44.53	−8.00201	7.64	3.0022	9.29	0.030	−7.48	−17.61	−8.43
Mrk 1066	0.012	SY2	44.55	−7.37201	7.01	2.3722	8.01	0.015	−6.07	−16.20	−7.02
Mrk 1157	0.015	SY2	44.27	−7.19201	6.83	2.1922	7.93	0.019	−5.95	−16.08	−6.90
Akn 79	0.018	SY2	45.24	−7.90201	7.54	2.9022	8.38	0.020	−7.36	−17.49	−8.31
Akn 347	0.023	SY2	44.84	−8.36201	8.00	3.3622	9.70	0.037	−8.38	−18.51	−9.33
IC 5063	0.011	SY2	44.53	−8.10201	7.74	3.1022	9.49	0.027	−7.27	−17.40	−8.22
II ZW55	0.025	SY2	44.54	−8.59201	8.23	3.5922	10.46	0.074	−8.86	−18.99	−9.81
F 341	0.016	SY2	44.13	−7.51201	7.15	2.5122	8.71	0.026	−6.57	−16.70	−7.52
UGC 3995	0.016	SY2	44.39	−8.05201	7.69	3.0522	9.53	0.036	−7.52	−17.65	−8.47
UGC 6100	0.029	SY2	44.48	−8.06201	7.70	3.0622	9.46	0.046	−8.06	−18.20	−9.01
1ES 1959 + 65	0.048	BLL	—	−10.39501	8.09	3.4522	7.79	0.052	−9.20	−19.34	−10.15
Mrk 180	0.045	BLL	—	−10.51501	8.21	3.5722	7.91	0.051	−9.35	−19.49	−10.31
Mrk 421	0.031	BLL	—	−10.59501	8.29	3.6522	7.99	0.038	−9.16	−19.29	−10.11
Mrk 501	0.034	BLL	—	−11.51501	9.21	4.5722	8.91	0.092	−10.85	−20.98	−11.80
I Zw 187	0.055	BLL	—	−10.16501	7.86	3.2222	7.56	0.058	−8.93	−19.06	−9.88
3C 371	0.051	BLL	—	−10.81501	8.51	3.8722	8.21	0.063	−9.99	−20.13	−10.95
1514 − 241	0.049	BLL	—	−10.40501	8.10	3.4622	7.80	0.054	−9.24	−19.37	−10.19
0521 − 365	0.055	BLL	—	−10.95501	8.65	4.0122	8.35	0.071	−10.31	−20.44	−11.26
0548 − 322	0.069	BLL	—	−10.45501	8.15	3.5122	7.85	0.074	−9.65	−19.78	−10.60
0706 + 591	0.125	BLL	—	−10.56501	8.26	3.6222	7.96	0.132	−10.41	−20.54	−11.36
2201 + 044	0.027	BLL	—	−10.40501	8.10	3.4622	7.80	0.032	−8.70	−18.83	−9.65
2344 + 514	0.044	BLL	—	−11.10501	8.80	4.1622	8.50	0.067	−10.37	−20.50	−11.32
3C 29	0.045	RG	—	−10.50501	8.20	3.5622	7.90	0.051	−9.34	−19.47	−10.29
3C 31	0.017	RG	—	−10.80501	8.50	3.8622	8.20	0.028	−8.99	−19.12	−9.94
3C 33	0.059	RG	—	−10.68501	8.38	3.7422	8.08	0.068	−9.90	−20.03	−10.85
3C 40	0.018	RG	—	−10.16501	7.86	3.2222	7.56	0.021	−7.92	−18.05	−8.87
3C 62	0.148	RG	—	−10.97501	8.67	4.0322	8.37	0.165	−11.29	−21.43	−12.25
3C 76.1	0.032	RG	—	−10.43501	8.13	3.4922	7.83	0.037	−8.90	−19.04	−9.86
3C 78	0.029	RG	—	−10.90501	8.60	3.9622	8.30	0.043	−9.64	−19.77	−10.59
3C 84	0.017	RG	—	−10.79501	8.49	3.8522	8.19	0.028	−8.97	−19.10	−9.92
3C 88	0.030	RG	—	−10.33501	8.03	3.3922	7.73	0.034	−8.67	−18.80	−9.62
3C 89	0.139	RG	—	−10.82501	8.52	3.8822	8.22	0.151	−10.97	−21.10	−11.92
3C 98	0.031	RG	—	−10.18501	7.88	3.2422	7.58	0.034	−8.44	−18.57	−9.39
3C 120	0.033	RG	—	−10.43501	8.13	3.4922	7.83	0.038	−8.93	−19.06	−9.88
3C 192	0.060	RG	—	−10.36501	8.06	3.4222	7.76	0.064	−9.36	−19.49	−10.31
3C 196.1	0.198	RG	—	−10.51501	8.21	3.5722	7.91	0.204	−10.79	−20.92	−11.74
3C 223	0.137	RG	—	−10.45501	8.15	3.5122	7.85	0.142	−10.31	−20.44	−11.26
3C 293	0.045	RG	—	−10.29501	7.99	3.3522	7.69	0.048	−8.97	−19.10	−9.92
3C 305	0.041	RG	—	−10.22501	7.92	3.2822	7.62	0.044	−8.76	−18.89	−9.71
3C 338	0.030	RG	—	−11.08501	8.78	4.1422	8.48	0.052	−9.98	−20.12	−10.93
3C 388	0.091	RG	—	−11.48501	9.18	4.5422	8.88	0.145	−11.71	−21.84	−12.66
3C 444	0.153	RG	—	−9.98501	7.68	3.0422	7.38	0.155	−9.60	−19.73	−10.55
3C 449	0.017	RG	—	−10.63501	8.33	3.6922	8.03	0.025	−8.69	−18.82	−9.64
gin 116	0.033	RG	—	−11.05501	8.75	4.1122	8.45	0.053	−10.02	−20.15	−10.97
NGC 315	0.017	RG	—	−11.20501	8.90	4.2622	8.60	0.045	−9.69	−19.82	−10.64
NGC 507	0.017	RG	—	−11.30501	9.00	4.3622	8.70	0.053	−9.86	−19.99	−10.81
NGC 708	0.016	RG	—	−10.76501	8.46	3.8222	8.16	0.026	−8.86	−18.99	−9.81
NGC 741	0.018	RG	—	−11.02501	8.72	4.0822	8.42	0.037	−9.42	−19.55	−10.37
NGC 4839	0.023	RG	—	−10.78501	8.48	3.8422	8.18	0.034	−9.22	−19.35	−10.17
NGC 4869	0.023	RG	—	−10.42501	8.12	3.4822	7.82	0.028	−8.59	−18.72	−9.54

TABLE 4: Continued.

Name	z	Type	$\log L_{\rm bol}$	$\log\left(\frac{R_d}{r_{\rm ov}}\right)$	$\log\left(\frac{M_{\rm BH}}{M_\odot}\right)$	$\log\left(\frac{M^{\rm Seed}_{\rm BH}}{M_\odot}\right)$	$\log T_{\rm BH}$	$z^{\rm Seed}$	J^q	$J^{\rm URCA}$	J^π
NGC 4874	0.024	RG	—	−10.93501	8.63	3.9922	8.33	0.039	−9.52	−19.65	−10.47
NGC 6086	0.032	RG	—	−11.26501	8.96	4.3222	8.66	0.065	−10.36	−20.49	−11.31
NGC 6137	0.031	RG	—	−11.11501	8.81	4.1722	8.51	0.054	−10.07	−20.20	−11.02
NGC 7626	0.025	RG	—	−11.27501	8.97	4.3322	8.67	0.058	−10.15	−20.28	−11.10
0039 − 095	0.000	RG	—	−11.02501	8.72	4.0822	8.42	0.019	−2.89	−13.02	−3.84
0053 − 015	0.038	RG	—	−11.12501	8.82	4.1822	8.52	0.062	−10.27	−20.40	−11.22
0053 − 016	0.043	RG	—	−10.81501	8.51	3.8722	8.21	0.055	−9.84	−19.97	−10.79
0055 − 016	0.045	RG	—	−11.15501	8.85	4.2122	8.55	0.070	−10.47	−20.61	−11.43
0110 + 152	0.044	RG	—	−10.39501	8.09	3.4522	7.79	0.048	−9.12	−19.26	−10.07
0112 − 000	0.045	RG	—	−10.83501	8.53	3.8922	8.23	0.057	−9.91	−20.05	−10.87
0112 + 084	0.000	RG	—	−11.48501	9.18	4.5422	8.88	0.054	−3.70	−13.83	−4.65
0147 + 360	0.018	RG	—	−10.76501	8.46	3.8222	8.16	0.028	−3.70	−13.83	−4.65
0131 − 360	0.030	RG	—	−10.83501	8.53	3.8922	8.23	0.042	−9.55	−19.68	−10.50
0257 − 398	0.066	RG	—	−10.59501	8.29	3.6522	7.99	0.073	−9.85	−19.98	−10.80
0306 + 237	0.000	RG	—	−10.81501	8.51	3.8722	8.21	0.012	−2.52	−12.66	−3.48
0312 − 343	0.067	RG	—	−10.87501	8.57	3.9322	8.27	0.080	−10.35	−20.48	−11.30
0325 + 024	0.030	RG	—	−10.59501	8.29	3.6522	7.99	0.037	−9.13	−19.26	−10.08
0431 − 133	0.033	RG	—	−10.95501	8.65	4.0122	8.35	0.049	−9.84	−19.97	−10.79
0431 − 134	0.035	RG	—	−10.61501	8.31	3.6722	8.01	0.042	−9.30	−19.43	−10.25
0449 − 175	0.031	RG	—	−10.02501	7.72	3.0822	7.42	0.033	−8.16	−18.29	−9.11
0546 − 329	0.037	RG	—	−11.59501	9.29	4.6522	8.99	0.107	−11.07	−21.20	−12.02
0548 − 317	0.034	RG	—	−9.58501	7.28	2.6422	6.98	0.035	−7.47	−17.60	−8.42
0634 − 206	0.056	RG	—	−10.39501	8.09	3.4522	7.79	0.060	−9.35	−19.48	−10.30
0718 − 340	0.029	RG	—	−11.31501	9.01	4.3722	8.71	0.066	−10.36	−20.49	−11.31
0915 − 118	0.054	RG	—	−10.99501	8.69	4.0522	8.39	0.072	−10.36	−20.49	−11.31
0940 − 304	0.038	RG	—	−11.59501	9.29	4.6522	8.99	0.108	−11.09	−21.22	−12.04
1043 − 290	0.060	RG	—	−10.67501	8.37	3.7322	8.07	0.068	−9.90	−20.03	−10.85
1107 − 372	0.010	RG	—	−11.11501	8.81	4.1722	8.51	0.033	−9.06	−19.19	−10.01
1123 − 351	0.032	RG	—	−11.83501	9.53	4.8922	9.23	0.153	−11.35	−21.49	−12.31
1258 − 321	0.015	RG	—	−10.91501	8.61	3.9722	8.31	0.030	−9.07	−19.20	−10.02
1333 − 337	0.013	RG	—	−11.07501	8.77	4.1322	8.47	0.034	−9.22	−19.35	−10.17
1400 − 337	0.014	RG	—	−11.19501	8.89	4.2522	8.59	0.042	−9.50	−19.63	−10.45
1404 − 267	0.022	RG	—	−11.11505	8.81	4.8798	8.51	0.045	−9.76	−19.89	−10.71
1510 + 076	0.053	RG	—	−11.33501	9.03	4.3922	8.73	0.091	−10.94	−21.07	−11.89
1514 + 072	0.035	RG	—	−10.95501	8.65	4.0122	8.35	0.051	−9.90	−20.03	−10.85
1520 + 087	0.034	RG	—	−10.59501	8.29	3.6522	7.99	0.041	−9.24	−19.37	−10.19
1521 − 300	0.020	RG	—	−10.10501	7.80	3.1622	7.50	0.022	−7.91	−18.04	−8.86
1602 + 178	0.041	RG	—	−10.54501	8.24	3.6022	7.94	0.047	−9.32	−19.45	−10.27
1610 + 296	0.032	RG	—	−11.26501	8.96	4.3222	8.66	0.065	−10.36	−20.49	−11.31
2236 − 176	0.070	RG	—	−10.79501	8.49	3.8522	8.19	0.081	−10.25	−20.39	−11.20
2322 + 143	0.045	RG	—	−10.47501	8.17	3.5322	7.87	0.050	−9.28	−19.42	−10.24
2322 − 122	0.082	RG	—	−10.63501	8.33	3.6922	8.03	0.090	−10.12	−20.25	−11.07
2333 − 327	0.052	RG	—	−10.95501	8.65	4.0122	8.35	0.068	−10.26	−20.39	−11.21
2335 + 267	0.030	RG	—	−11.38501	9.08	4.4422	8.78	0.073	−10.51	−20.64	−11.46

is invited to consult the original paper [74] for details. In this, we restrict ourselves to consider only the simplest spacetime deformation map, $\widetilde{\Omega} : M_4 \longrightarrow V_4$ ($\breve{\Omega}^{\mu}{}_{\nu} \equiv \delta^{\mu}_{\nu}$). This theory accounts for the *gravitation gauge group* G_V generated by the hidden local internal symmetry $U^{\rm loc}$. We assume that a *distortion* massless gauge field $a(x)$ ($\equiv a_n(x)$) has to act on the external spacetime groups. This field takes values in the Lie algebra of the abelian group $U^{\rm loc}$. We pursue a principle goal of building up the *world-deformation* tensor, $\widetilde{\Omega}(F) = \widetilde{D}(a)\widetilde{\psi}(a)$, where F is the

TABLE 5: Seed black hole intermediate masses, preradiation times, redshifts, and neutrino fluxes from fundamental plane-derived velocity dispersions. Columns: (1) name, (2) redshift, (3) AGN type: SY2: Seyfert 2, (4) log of the bolometric luminosity (ergs^{-1}), (5) log of the radius of protomatter core in special unit $r_{OV} = 13.68$ km, (6) log of the black hole mass in solar masses, (7) log of the seed black hole intermediate mass in solar masses, (8) log of the neutrino preradiation time (yrs), (9) redshift of seed black hole, (10) $J^{i=q}$, (11) $J^{i=\mathrm{URCA}}$, and (12) $J^{i=\pi}$, where $J^i \equiv \log(J^i_{\nu\varepsilon}/\varepsilon_d\ \mathrm{erg\,cm^{-2}\,s^{-1}\,sr^{-1}})$.

Name	z	Type	$\log\left(\frac{R_d}{r_{OV}}\right)$	$\log\left(\frac{M_{BH}}{M_\odot}\right)$	$\log\left(\frac{M_{BH}^{Seed}}{M_\odot}\right)$	$\log T_{BH}$	z^{Seed}	J^q	J^{URCA}	J^π
0122 + 090	0.339	BLL	−11.12501	8.82	4.1822	8.52	0.363	−12.43	−22.57	−13.39
0145 + 138	0.124	BLL	−10.72501	8.42	3.7822	8.12	0.133	−10.68	−20.81	−11.63
0158 + 001	0.229	BLL	−10.38501	8.08	3.4422	7.78	0.233	−10.71	−20.84	−11.66
0229 + 200	0.139	BLL	−11.54501	9.24	4.6022	8.94	0.201	−12.23	−22.36	−13.18
0257 + 342	0.247	BLL	−10.96501	8.66	4.0222	8.36	0.263	−11.81	−21.94	−12.76
0317 + 183	0.190	BLL	−10.25501	7.95	3.3122	7.65	0.193	−10.29	−20.42	−11.24
0331 − 362	0.308	BLL	−11.05501	8.75	4.1122	8.45	0.328	−12.21	−22.34	−13.16
0347 − 121	0.188	BLL	−10.95501	8.65	4.0122	8.35	0.204	−11.50	−21.63	−12.45
0350 − 371	0.165	BLL	−11.12501	8.82	4.1822	8.52	0.189	−11.67	−21.80	−12.62
0414 + 009	0.287	BLL	−10.86501	8.56	3.9222	8.26	0.300	−11.80	−21.93	−12.75
0419 + 194	0.512	BLL	−10.91501	8.61	3.9722	8.31	0.527	−12.54	−22.68	−13.50
0506 − 039	0.304	BLL	−11.05501	8.75	4.1122	8.45	0.324	−12.19	−22.32	−13.14
0525 + 713	0.249	BLL	−11.33501	9.03	4.3922	8.73	0.287	−12.46	−22.60	−13.41
0607 + 710	0.267	BLL	−10.95501	8.65	4.0122	8.35	0.283	−12.46	−22.60	−13.41
0737 + 744	0.315	BLL	−11.24501	8.94	4.3022	8.64	0.346	−12.56	−22.69	−13.51
0922 + 749	0.638	BLL	−11.91501	9.61	4.9722	9.31	0.784	−14.56	−24.69	−15.51
0927 + 500	0.188	BLL	−10.64501	8.34	3.7022	8.04	0.196	−10.96	−21.09	−11.91
0958 + 210	0.344	BLL	−11.33501	9.03	4.3922	8.73	0.382	−12.82	−22.95	−13.77
1104 + 384	0.031	BLL	−11.69501	9.39	4.7522	9.09	0.119	−11.08	−21.21	−12.03
1133 + 161	0.460	BLL	−10.62501	8.32	3.6822	8.02	0.467	−11.91	−22.04	−12.86
1136 + 704	0.045	BLL	−11.25501	8.95	4.3122	8.65	0.077	−10.65	−20.78	−11.60
1207 + 394	0.615	BLL	−11.40501	9.10	4.4622	8.80	0.660	−13.62	−23.76	−14.58
1212 + 078	0.136	BLL	−11.29501	8.99	4.3522	8.69	0.171	−11.77	−21.90	−12.72
1215 + 303	0.130	BLL	−10.42501	8.12	3.4822	7.82	0.135	−10.20	−20.33	−11.15
1218 + 304	0.182	BLL	−10.88501	8.58	3.9422	8.28	0.196	−11.35	−21.48	−12.30
1221 + 245	0.218	BLL	−10.27501	7.97	3.3322	7.67	0.221	−10.47	−20.60	−11.42
1229 + 643	0.164	BLL	−11.71501	9.41	4.7722	9.11	0.256	−12.69	−22.82	−13.64
1248 − 296	0.370	BLL	−11.31501	9.01	4.3722	8.71	0.407	−12.87	−23.00	−13.82
1255 + 244	0.141	BLL	−10.88501	8.58	3.9422	8.28	0.155	−11.09	−21.22	−12.04
1407 + 595	0.495	BLL	−11.60501	9.30	4.6622	9.00	0.566	−13.71	−23.84	−14.66
1418 + 546	0.152	BLL	−11.33501	9.03	4.3922	8.73	0.190	−11.95	−22.08	−12.90
1426 + 428	0.129	BLL	−11.43501	9.13	4.4922	8.83	0.177	−11.96	−22.09	−12.91
1440 + 122	0.162	BLL	−10.74501	8.44	3.8022	8.14	0.172	−10.98	−21.11	−11.93
1534 + 014	0.312	BLL	−11.10501	8.80	4.1622	8.50	0.335	−12.31	−22.44	−13.26
1704 + 604	0.280	BLL	−11.07501	8.77	4.1322	8.47	0.301	−12.14	−22.27	−13.09
1728 + 502	0.055	BLL	−10.43501	8.13	3.4922	7.83	0.060	−9.40	−19.53	−10.35
1757 + 703	0.407	BLL	−11.05501	8.75	4.1122	8.45	0.427	−12.52	−22.65	−13.47
1807 + 698	0.051	BLL	−12.40501	10.10	5.4622	9.80	0.502	−12.78	−22.91	−13.73
1853 + 671	0.212	BLL	−10.53501	8.23	3.5922	7.93	0.218	−10.89	−21.02	−11.84
2005 − 489	0.071	BLL	−11.33501	9.03	4.3922	8.73	0.109	−11.21	−21.34	−12.16
2143 + 070	0.237	BLL	−10.76501	8.46	3.8222	8.16	0.247	−11.41	−21.54	−12.36
2200 + 420	0.069	BLL	−10.53501	8.23	3.5922	7.93	0.075	−9.79	−19.92	−10.74
2254 + 074	0.190	BLL	−10.92501	8.62	3.9822	8.32	0.205	−11.46	−21.59	−12.41
2326 + 174	0.213	BLL	−11.04501	8.74	4.1022	8.44	0.233	−11.79	−21.92	−12.74
2356 − 309	0.165	BLL	−10.90501	8.60	3.9622	8.30	0.179	−11.28	−21.41	−12.23
0230 − 027	0.239	RG	−10.27501	7.97	3.3322	7.67	0.242	−10.56	−20.70	−11.52
0307 + 169	0.256	RG	−10.96501	8.66	4.0222	8.36	0.272	−11.85	−21.98	−12.80

TABLE 5: Continued.

Name	z	Type	$\log\left(\frac{R_d}{r_{ov}}\right)$	$\log\left(\frac{M_{BH}}{M_\odot}\right)$	$\log\left(\frac{M_{BH}^{Seed}}{M_\odot}\right)$	$\log T_{BH}$	z^{Seed}	J^q	J^{URCA}	J^π
0345 + 337	0.244	RG	−9.42501	7.12	2.4822	6.82	0.244	−9.10	−19.23	−10.05
0917 + 459	0.174	RG	−10.51501	8.21	3.5722	7.91	0.180	−10.65	−20.78	−11.60
0958 + 291	0.185	RG	−10.23501	7.93	3.2922	7.63	0.188	−10.23	−20.36	−11.18
1215 − 033	0.184	RG	−10.23501	7.93	3.2922	7.63	0.187	−10.22	−20.35	−11.17
1215 + 013	0.118	RG	−10.50501	8.20	3.5622	7.90	0.124	−10.25	−20.38	−11.20
1330 + 022	0.215	RG	−10.12501	7.82	3.1822	7.52	0.217	−10.19	−20.32	−11.14
1342 − 016	0.167	RG	−10.71501	8.41	3.7722	8.11	0.176	−10.96	−21.09	−11.91
2141 + 279	0.215	RG	−10.12501	7.82	3.1822	7.52	0.217	−10.19	−20.32	−11.14
0257 + 024	0.115	RQQ	−11.05501	8.75	4.1122	8.45	0.135	−11.18	−21.32	−12.13
1549 + 203	0.250	RQQ	−9.22501	6.92	2.2822	6.62	0.250	−8.78	−18.91	−9.73
2215 − 037	0.241	RQQ	−10.50501	8.20	3.5622	7.90	0.247	−10.98	−21.11	−11.93
2344 + 184	0.138	RQQ	−9.37501	7.07	2.4322	6.77	0.138	−8.42	−18.56	−9.38
0958 + 291	0.185	RG	−10.23501	7.93	3.2922	7.63	0.188	−10.23	−20.36	−11.18
1215 − 033	0.184	RG	−10.23501	7.93	3.2922	7.63	0.187	−10.22	−20.35	−11.17
1215 + 013	0.118	RG	−10.50501	8.20	3.5622	7.90	0.124	−10.25	−20.38	−11.20
1330 + 022	0.215	RG	−10.12501	7.82	3.1822	7.52	0.217	−10.19	−20.32	−11.14
1342 − 016	0.167	RG	−10.71501	8.41	3.7722	8.11	0.176	−10.96	−21.09	−11.91
2141 + 279	0.215	RG	−10.12501	7.82	3.1822	7.52	0.217	−10.19	−20.32	−11.14
0257 + 024	0.115	RQQ	−11.05501	8.75	4.1122	8.45	0.135	−11.18	−21.32	−12.13
1549 + 203	0.250	RQQ	−9.22501	6.92	2.2822	6.62	0.250	−8.78	−18.91	−9.73
2215 − 037	0.241	RQQ	−10.50501	8.20	3.5622	7.90	0.247	−10.98	−21.11	−11.93
2344 + 184	0.138	RQQ	−9.37501	7.07	2.4322	6.77	0.138	−8.42	−18.56	−9.38

differential form of gauge field $F = (1/2)F_{nm}\vartheta^n \wedge \vartheta^m$. We connect the structure group G_V, further, to the nonlinear realization of the Lie group G_D of *distortion* of extended space $M_6(\to \overset{\frown}{M}_6)$ (E.1), underlying the M_4. This extension appears to be indispensable for such a realization. In using the 6D language, we will be able to make a necessary reduction to the conventional 4D space. The laws guiding this redaction are given in Appendix E. The nonlinear realization technique or the method of phenomenological Lagrangians [85–91] provides a way to determine the transformation properties of fields defined on the quotient space. In accordance, we treat the distortion group G_D and its stationary subgroup $H = SO(3)$, respectively, as the dynamical group and its algebraic subgroup. The fundamental field is distortion gauge field (a) and, thus, all the fundamental gravitational structures in fact—the metric as much as the coframes and connections—acquire a *distortion-gauge induced* theoretical interpretation. We study the geometrical structure of the space of parameters in terms of Cartan's calculus of exterior forms and derive the Maurer-Cartan structure equations, where the distortion fields (a) are treated as the Goldstone fields.

A.3. A Rearrangement of Vacuum State. Addressing the rearrangement of vacuum state, in realization of the group G_V we implement the abelian local group [74]

$$U^{loc} = U(1)_Y \times \overline{U}(1) \equiv U(1)_Y \times \text{diag}\,[SU(2)], \quad \text{(A.3)}$$

on the space M_6 (spanned by the coordinates η), with the group elements of $\exp[i(Y/2)\theta_Y(\eta)]$ of $U(1)_Y$ and $\exp[iT^3\theta_3(\eta)]$ of $\overline{U}(1)$. This group leads to the renormalizable theory, because gauge invariance gives a conservation of charge, and it also ensures the cancelation of quantum corrections that would otherwise result in infinitely large amplitudes. This has two generators, the third component T^3 of isospin $\vec{T}$ related to the Pauli spin matrix $\vec{\tau}/2$, and hypercharge Y implying $Q^d = T^3 + Y/2$, where Q^d is the *distortion charge* operator assigning the number −1 to particles, but +1 to antiparticles. The group (A.3) entails two neutral gauge bosons of $\overline{U}(1)$, or that coupled to T^3, and of $U(1)_Y$, or that coupled to the hypercharge Y. Spontaneous symmetry breaking can be achieved by introducing the neutral complex scalar Higgs field. Minimization of the vacuum energy fixes the nonvanishing vacuum expectation value (VEV), which spontaneously breaks the theory, leaving the $U(1)_d$ subgroup intact, that is, leaving one Goldstone boson. Consequently, the left Goldstone boson is gauged away from the scalar sector, but it essentially reappears in the gauge sector providing the longitudinally polarized spin state of one of gauge bosons which acquires mass through its coupling to Higgs scalar. Thus, the two neutral gauge bosons were mixed to form two physical orthogonal states of the massless component of *distortion* field, (a) $(M_a = 0)$, which is responsible for gravitational interactions, and its massive component, $(\overline{a})$ $(M_{\overline{a}} \neq 0)$, which is responsible for the ID-regime. Hence, a substantial change of the properties of the spacetime continuum besides the curvature may arise at huge energies. This theory is renormalizable, because gauge invariance gives conservation of charge and also ensures the cancelation of quantum corrections that would otherwise result in infinitely large amplitudes. Without careful thought we expect that in this framework the renormalizability of the theory will not be spoiled in curved space-time too, because the infinities arise from ultraviolet properties of Feynman integrals in momentum space which, in coordinate space, are short distance properties, and locally (over short distances) all the curved spaces look like *maximally symmetric* (flat) space.

A.4. Model Building: Field Equations. The field equations follow at once from the total gauge invariant Lagrangian in terms of Euler-Lagrange variations, respectively, on both curved and flat spaces. The Lagrangian of distortion gauge field (a) defined on the flat space is undegenerated Killing form on the Lie algebra of the group U^{loc} in adjoint representation, which yields the equation of distortion field (F.1). We are interested in the case of a spherical-symmetric gravitational field $a_0(r)$ in presence of one-dimensional space-like ID-field $\bar{a}$ (F.6). In the case at hand, one has the group of motions $SO(3)$ with 2D space-like orbits S^2 where the standard coordinates are $\tilde{\theta}$ and $\tilde{\varphi}$. The stationary subgroup of $SO(3)$ acts isotropically upon the tangent space at the point of sphere S^2 of radius $\tilde{r}$. So, the bundle $p : V_4 \to \tilde{R}^2$ has the fiber $S^2 = p^{-1}(\tilde{x})$, $\tilde{x} \in V_4$, with a trivial connection on it, where $\tilde{R}^2$ is the quotient-space $V_4/SO(3)$. Considering the equilibrium configurations of degenerate barionic matter, we assume an absence of transversal stresses and the transference of masses in V_4

$$T_1^1 = T_2^2 = T_3^3 = -\tilde{P}(\tilde{r}), \qquad T_0^0 = -\tilde{\rho}(\tilde{r}), \tag{A.4}$$

where $\tilde{P}(\tilde{r})$ and $\tilde{\rho}(\tilde{r})$ ($\tilde{r} \in \tilde{R}^3$) are taken to denote the internal pressure and macroscopic density of energy defined in proper frame of reference that is being used. The equations of gravitation (a_0) and ID ($\bar{a}$) fields can be given in Feynman gauge [71] as

$$\begin{aligned}\Delta a_0 &= \frac{1}{2}\left\{\tilde{g}_{00}\frac{\partial\tilde{g}^{00}}{\partial a_0}\tilde{\rho}(\tilde{r}) - \left[\tilde{g}_{33}\frac{\partial\tilde{g}^{33}}{\partial a_0} + \tilde{g}_{11}\frac{\partial\tilde{g}^{11}}{\partial a_0} + \tilde{g}_{22}\frac{\partial\tilde{g}^{22}}{\partial a_0}\right]\tilde{P}(\tilde{r})\right\}, \\ \left(\Delta - \lambda_a^{-2}\right)\bar{a} &= \frac{1}{2}\left\{\tilde{g}_{00}\frac{\partial\tilde{g}^{00}}{\partial\bar{a}}\tilde{\rho}(\tilde{r}) - \left[\tilde{g}_{33}\frac{\partial\tilde{g}^{33}}{\partial\bar{a}} + \tilde{g}_{11}\frac{\partial\tilde{g}^{11}}{\partial\bar{a}} + \tilde{g}_{22}\frac{\partial\tilde{g}^{22}}{\partial\bar{a}}\right]\tilde{P}(\tilde{r})\right\} \\ &\quad \times\theta\left(\lambda_a - \tilde{n}^{-1/3}\right),\end{aligned} \tag{A.5}$$

where $\tilde{n}$ is the concentration of particles and $\lambda_a = \hbar/m_a c \simeq 0.4$ fm is the Compton lenghth of the ID-field (but substantial ID-effects occur far below it), and a diffeomorphism $\tilde{r}(r) : M_4 \to V_4$ is given as $r = \tilde{r} - R_g/4$. A distortion of the basis $\tilde{e}$ in the ID regime, in turn, yields the transformations of Poincaré generators of translations. Given an explicit form of distorted basis vectors (F.7), it is straightforward to derive the laws of phase transition for individual particle found in the ID-region ($x_0 = 0$, $x \neq 0$) of the space-time continuum $\tan\tilde{\theta}_3 = -x$, $\tilde{\theta}_1 = \tilde{\theta}_1 = 0$. The Poincaré generators P_μ of translations are transformed as follows [71]:

$$\begin{aligned}\tilde{E} &= E, \qquad \tilde{P}_{1,2} = P_{1,2}\cos\tilde{\theta}_3, \\ \tilde{P}_3 &= P_3 - \tan\tilde{\theta}_3 mc, \\ \tilde{m} &= \left|\left(m - \tan\tilde{\theta}_3\frac{P_3}{c}\right)^2 + \sin^2\tilde{\theta}_3\frac{P_1^2 + P_2^2}{c^2} - \tan^2\tilde{\theta}_3\frac{E^2}{c^4}\right|^{1/2},\end{aligned} \tag{A.6}$$

where E, $\vec{P}$, and m and $\tilde{E}$, $\tilde{\vec{P}}$, and $\tilde{m}$ are ordinary and distorted energy, momentum, and mass at rest. Hence the matter found in the ID-region ($\bar{a} \neq 0$) of space-time continuum has undergone phase transition of II-kind; that is, each particle goes off from the mass shell—a shift of mass and energy-momentum spectra occurs upwards along the energy scale. The matter in this state is called *protomatter* with the thermodynamics differing strongly from the thermodynamics of ordinary compressed matter. The resulting deformed metric on V_4 in holonomic coordinate basis takes the form

$$\begin{aligned}\tilde{g}_{00} &= (1 - x_0)^2 + x^2, \qquad \tilde{g}_{\mu\nu} = 0 \quad (\mu \neq \nu), \\ \tilde{g}_{33} &= -\left[(1 + x_0)^2 + x^2\right], \qquad \tilde{g}_{11} = -\tilde{r}^2, \\ \tilde{g}_{22} &= -\tilde{r}^2\sin^2\tilde{\theta}.\end{aligned} \tag{A.7}$$

As a working model we assume the SPC-configurations given in Appendix G, which are composed of spherical-symmetric distribution of matter in many-phase stratified states. This is quick to estimate the main characteristics of the equilibrium degenerate barionic configurations and will guide us toward first look at some of the associated physics. The simulations confirm in brief the following scenario [71]: the energy density and internal pressure have sharply increased in protomatter core of SPC-configuration (with respect to corresponding central values of neutron star) proportional to gravitational forces of compression. This counteracts the collapse and equilibrium holds even for the masses $\sim 10^9 M_\odot$. This feature can be seen, for example, from Figure 7 where the state equation of the II-class $\mathrm{SPC_{II}}$ configuration, with the quark protomatter core, is plotted.

B. A Hard Look at Spacetime Deformation

The holonomic metric on $\mathcal{M}_4$ can be recast in the form $\tilde{g} = \tilde{g}_{\mu\nu}\tilde{\vartheta}^\mu \otimes \tilde{\vartheta}^\nu = \tilde{g}(\tilde{e}_\mu, \tilde{e}_\nu)\tilde{\vartheta}^\mu \otimes \tilde{\vartheta}^\nu$, with components $\tilde{g}_{\mu\nu} = \tilde{g}(\tilde{e}_\mu, \tilde{e}_\nu)$ in dual holonomic base $\{\tilde{\vartheta}^\mu \equiv d\tilde{x}^\mu\}$. In order to relate local Lorentz symmetry to more general deformed

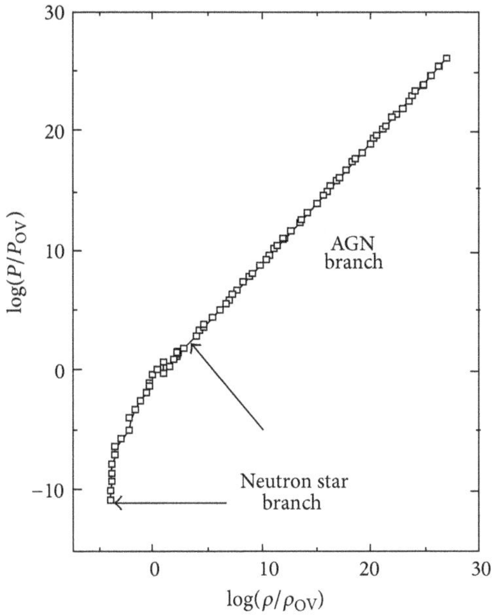

FIGURE 7: The state equation of SPC_{II} on logarithmic scales, where P and ρ are the internal pressure and density, given in special units $P_{OV} = 6.469 \times 10^{36}$ [erg cm^{-3}] and $\rho_{OV} = 7.194 \times 10^{15}$ [g cm^{-3}], respectively.

spacetime, there is, however, a need to introduce the soldering tools, which are the linear frames and forms in tangent fiber-bundles to the external smooth differential manifold, whose components are so-called tetrad (vierbein) fields. The $\widetilde{\mathcal{M}}_4$ has at each point a tangent space, $\widetilde{T}_{\widetilde{x}}\widetilde{M}_4$, spanned by the anholonomic orthonormal frame field, $\widetilde{e}$, as a shorthand for the collection of the 4-tuplet $(\widetilde{e}_0, \ldots, \widetilde{e}_3)$, where $\widetilde{e}_a = \widetilde{e}_a{}^{\mu}\widetilde{\partial}_{\mu}$. We use the first half of Latin alphabet $(a, b, c, \ldots = 0, 1, 2, 3)$ to denote the anholonomic indices related to the tangent space. The frame field, $\widetilde{e}$, then defines a dual vector, $\widetilde{\vartheta}$, of differential forms, $\widetilde{\vartheta} = \begin{pmatrix} \widetilde{\vartheta}^0 \\ \vdots \\ \widetilde{\vartheta}^3 \end{pmatrix}$, as a shorthand for the collection of the $\widetilde{\vartheta}^b = \widetilde{e}^b{}_{\mu} d\widetilde{x}^{\mu}$, whose values at every point form the dual basis, such that $\widetilde{e}_a \rfloor \widetilde{\vartheta}^b = \delta^b_a$, where $\rfloor$ denotes the interior product; namely, this is a C^{∞}-bilinear map $\rfloor : \Omega^1 \to \Omega^0$ with Ω^p denoting the C^{∞}-modulo of differential p-forms on $\widetilde{\mathcal{M}}_4$. In components, we have $\widetilde{e}_a{}^{\mu}\widetilde{e}^b{}_{\mu} = \delta^b_a$. On the manifold, $\widetilde{\mathcal{M}}_4$, the tautological tensor field, $\widetilde{id}$, of type $(1,1)$ can be defined which assigns to each tangent space the identity linear transformation. Thus, for any point $\widetilde{x} \in \widetilde{\mathcal{M}}_4$ and any vector $\widetilde{\xi} \in \widetilde{T}_{\widetilde{x}}\widetilde{\mathcal{M}}_4$, one has $\widetilde{id}(\widetilde{\xi}) = \widetilde{\xi}$. In terms of the frame field, the $\widetilde{\vartheta}^a$ give the expression for $\widetilde{id}$ as $\widetilde{id} = \widetilde{e}\widetilde{\vartheta} = \widetilde{e}_0 \otimes \widetilde{\vartheta}^0 + \cdots \widetilde{e}_3 \otimes \widetilde{\vartheta}^3$, in the sense that both sides yield $\widetilde{\xi}$ when applied to any tangent vector $\widetilde{\xi}$ in the domain of definition of the frame field. One can also consider general transformations of the linear group, $GL(4, R)$, taking any base into any other set of four linearly independent fields. The notation, $\{\widetilde{e}_a, \widetilde{\vartheta}^b\}$, will be used below for general linear frames. Let us introduce so-called *first deformation matrices*, $(\pi(x)^m{}_k$ and $\widetilde{\pi}^a{}_l(\widetilde{x})) \in GL(4, \widetilde{M})$ for all $\widetilde{x}$, as follows:

$$\widetilde{D}^m_{\mu} = \widetilde{e}_{\mu}{}^k \pi^m{}_k, \qquad \widetilde{\psi}^{\mu}_l = \widetilde{e}^{\mu}{}_k \pi^k{}_l,$$
$$\widetilde{e}_{\mu}{}^k \widetilde{e}^{\mu}{}_m = \delta^k_m, \qquad \widetilde{\pi}_a{}^m = \widetilde{e}_a{}^{\mu}\widetilde{D}^m_{\mu}, \tag{B.1}$$
$$\widetilde{\pi}^a{}_l = \widetilde{e}^a{}_{\mu}\widetilde{\psi}^{\mu}_l,$$

where $\widetilde{g}_{\mu\nu}\widetilde{e}_k{}^{\mu}\widetilde{e}_s{}^{\nu} = \eta_{ks}$; η_{ks} is the metric on M_4. A deformation tensor, $\Omega^m{}_l = \pi^m{}_k \pi^k{}_l$, yields local tetrad deformations

$$\widetilde{e}_a = \widetilde{\pi}_a{}^m e_m, \qquad \widetilde{\vartheta}^a = \widetilde{\pi}^a{}_l \vartheta^l,$$
$$\widetilde{e}_k = \pi^m{}_k e_m, \qquad \widetilde{\vartheta}^k = \pi^k{}_l \vartheta^l, \tag{B.2}$$

and $\widetilde{id} = \widetilde{e}_a \otimes \widetilde{\vartheta}^a = \widetilde{e}_k \otimes \widetilde{\vartheta}^k \in \widetilde{\mathcal{M}}_4$. The *first deformation matrices* π and $\widetilde{\pi}$, in general, give rise to the right cosets of the Lorentz group; that is, they are the elements of the quotient group $GL(4, \widetilde{M})/SO(3,1)$. If we deform the cotetrad according to (B.2), we have two choices to recast metric as follows: either writing the deformation of the metric in the space of tetrads or deforming the tetrad field:

$$\widetilde{g} = o_{ab}\widetilde{\vartheta}^a \otimes \widetilde{\vartheta}^b = o_{ab}\widetilde{\pi}^a{}_l \widetilde{\pi}^b{}_m \vartheta^l \otimes \vartheta^m$$
$$= \gamma_{lm}\vartheta^l \otimes \vartheta^m, \tag{B.3}$$

where the *second deformation matrix*, γ_{lm}, reads $\gamma_{lm} = o_{ab}\widetilde{\pi}^a{}_l \widetilde{\pi}^b{}_m$. The deformed metric splits as

$$\widetilde{g}_{\mu\nu} = \Upsilon^2 \eta_{\mu\nu} + \widetilde{\gamma}_{\mu\nu}, \tag{B.4}$$

provided that $\Upsilon = \widetilde{\pi}^a{}_a = \pi^k{}_k$ and

$$\widetilde{\gamma}_{\mu\nu} = \left(\gamma_{al} - \Upsilon^2 o_{al}\right)\widetilde{e}^a{}_{\mu}\widetilde{e}^l{}_{\nu}$$
$$= \left(\gamma_{ks} - \Upsilon^2 \eta_{ks}\right)\widetilde{e}^k{}_{\mu}\widetilde{e}^s{}_{\nu}. \tag{B.5}$$

The anholonomic orthonormal frame field, $\widetilde{e}$, relates $\widetilde{g}$ to the tangent space metric, $o_{ab} = \mathrm{diag}(+ - - -)$, by $o_{ab} = \widetilde{g}(\widetilde{e}_a, \widetilde{e}_b) = \widetilde{g}_{\mu\nu}\widetilde{e}_a{}^{\mu}\widetilde{e}_b{}^{\nu}$, which has the converse $\widetilde{g}_{\mu\nu} = o_{ab}\widetilde{e}^a{}_{\mu}\widetilde{e}^b{}_{\nu}$ because $\widetilde{e}_a{}^{\mu}\widetilde{e}^a{}_{\nu} = \delta^{\mu}_{\nu}$. With this provision, we build up a *world-deformation* tensor Ω yielding a deformation of the flat space M_4. The γ_{lm} can be decomposed in terms of symmetric $\widetilde{\pi}_{(al)}$ and antisymmetric $\widetilde{\pi}_{[al]}$ parts of the matrix $\widetilde{\pi}_{al} = o_{ac}\widetilde{\pi}^c{}_l$ (or, resp., in terms of $\pi_{(kl)}$ and $\pi_{[kl]}$, where $\pi_{kl} = \eta_{ks}\pi^s{}_l$) as

$$\gamma_{al} = \widetilde{\Upsilon}^2 o_{al} + 2\widetilde{\Upsilon}\widetilde{\Theta}_{al} + o_{cd}\widetilde{\Theta}^c{}_a \widetilde{\Theta}^d{}_l$$
$$+ o_{cd}\left(\widetilde{\Theta}^c{}_a \widetilde{\varphi}^d{}_l + \widetilde{\varphi}^c{}_a \widetilde{\Theta}^d{}_l\right) + o_{cd}\widetilde{\varphi}^c{}_a \widetilde{\varphi}^d{}_l, \tag{B.6}$$

where

$$\widetilde{\pi}_{al} = \widetilde{\Upsilon} o_{al} + \widetilde{\Theta}_{al} + \widetilde{\varphi}_{al}, \tag{B.7}$$

$\widetilde{\Upsilon} = \widetilde{\pi}^a{}_a$, $\widetilde{\Theta}_{al}$ is the traceless symmetric part, and $\widetilde{\varphi}_{al}$ is the skew symmetric part of the first deformation matrix.

The anholonomy objects defined on the tangent space, $\widetilde{T}_{\widetilde{x}}\widetilde{M}_4$, read

$$\widetilde{C}^a := d\widetilde{\vartheta}^a = \frac{1}{2}\widetilde{C}^a{}_{bc}\widetilde{\vartheta}^b \wedge \widetilde{\vartheta}^c, \quad \text{(B.8)}$$

where the anholonomy coefficients, $\widetilde{C}^a{}_{bc}$, which represent the curls of the base members, are

$$\begin{aligned}\widetilde{C}^c{}_{ab} &= -\widetilde{\vartheta}^c\left([\widetilde{e}_a, \widetilde{e}_b]\right) = \widetilde{e}_a{}^{\mu}\widetilde{e}_b{}^{\nu}\left(\widetilde{\partial}_{\mu}\widetilde{e}^c{}_{\nu} - \widetilde{\partial}_{\nu}\widetilde{e}^c{}_{\mu}\right)\\ &= -\widetilde{e}^c{}_{\mu}\left[\widetilde{e}_a\left(\widetilde{e}_b{}^{\mu}\right) - \widetilde{e}_b\left(\widetilde{e}_a{}^{\mu}\right)\right] \quad \text{(B.9)}\\ &= 2\pi^c{}_l\widetilde{e}_m{}^{\mu}\left(\pi^{-1m}{}_{[a}\widetilde{\partial}_{\mu}\pi^{-1l}{}_{b]}\right).\end{aligned}$$

In particular case of constant metric in the tetradic space, the deformed connection can be written as

$$\widetilde{\Gamma}^a{}_{bc} = \frac{1}{2}\left(\widetilde{C}^a{}_{bc} - o^{aa'}o_{bb'}\widetilde{C}^{b'}{}_{a'c} - o^{aa'}o_{cc'}\widetilde{C}^{c'}{}_{a'b}\right). \quad \text{(B.10)}$$

All magnitudes related to the V_4 will be denoted by an over "$\circ$". According to (A.1), we have $\mathring{\Omega}_l^m = \mathring{D}^m_{\mu}\mathring{\psi}^{\mu}_l$ and $\breve{\Omega}^{\mu}{}_{\nu} = \breve{D}^{\mu}_{\rho}\breve{\psi}^{\rho}_{\nu}$, provided

$$\begin{aligned}\mathring{e}_{\mu} &= \mathring{D}^l_{\mu}e_l, &\quad \mathring{e}_{\mu}\mathring{\psi}^{\mu}_l &= \mathring{\Omega}^m_l e_m,\\ \widetilde{e}_{\rho} &= \breve{D}^{\mu}_{\rho}\mathring{e}_{\mu}, &\quad \widetilde{e}_{\rho}\breve{\psi}^{\rho}_{\nu} &= \breve{\Omega}^{\mu}{}_{\nu}\mathring{e}_{\mu}.\end{aligned} \quad \text{(B.11)}$$

In analogy with (B.1), the following relations hold:

$$\begin{aligned}\mathring{D}^m_{\mu} &= \mathring{e}_{\mu}{}^k\mathring{\pi}^m{}_k, &\quad \mathring{\psi}^{\mu}_l &= \mathring{e}^{\mu}{}_k\mathring{\pi}^k{}_l,\\ \mathring{e}_{\mu}{}^k\mathring{e}^{\mu}{}_m &= \delta^k_m, &\quad \mathring{\pi}_a{}^m &= \mathring{e}_a{}^{\mu}\mathring{D}^m_{\mu},\end{aligned} \quad \text{(B.12)}$$
$$\mathring{\pi}^a{}_l = \mathring{e}^a{}_{\mu}\mathring{\psi}^{\mu}_l,$$

where $\mathring{\Omega}^m_l = \mathring{\pi}^m_{\rho}\mathring{\pi}^{\rho}{}_l$ and $\breve{\Omega}^{\mu}{}_{\nu} = \breve{\pi}^{\mu}{}_{\rho}\breve{\pi}^{\rho}{}_{\nu}$. We also have $\mathring{g}_{\mu\nu}\mathring{e}_k{}^{\mu}\mathring{e}_s{}^{\nu} = \eta_{ks}$ and

$$\begin{aligned}\breve{D}^{\mu}_{\rho} &= \breve{e}_{\nu}{}^{\mu}\breve{\pi}^{\nu}{}_{\rho}, &\quad \breve{\psi}^{\rho}_{\nu} &= \breve{e}^{\rho}{}_{\mu}\breve{\pi}^{\mu}{}_{\nu},\\ \breve{e}_{\nu}{}^{\mu}\breve{e}^{\nu}{}_{\rho} &= \delta^{\mu}_{\rho}, &\quad \breve{\pi}_a{}^{\mu} &= \breve{e}_a{}^{\rho}\breve{D}^{\mu}_{\rho},\end{aligned} \quad \text{(B.13)}$$
$$\breve{\pi}^a{}_{\nu} = \breve{e}^a{}_{\rho}\breve{\psi}^{\rho}_{\nu}.$$

The norm $d\mathring{s} \equiv \mathring{id}$ of the displacement $d\mathring{x}^{\mu}$ on V_4 can be written in terms of the spacetime structures of M_4 as

$$\mathring{id} = \mathring{e}\mathring{\vartheta} = \mathring{\Omega}^m_l e_m \otimes \vartheta^l \in V_4. \quad \text{(B.14)}$$

The holonomic metric can be recast in the form

$$\mathring{g} = \mathring{g}_{\mu\nu}\mathring{\vartheta}^{\mu} \otimes \mathring{\vartheta}^{\nu} = \mathring{g}\left(\mathring{e}_{\mu}, \mathring{e}_{\nu}\right)\mathring{\vartheta}^{\mu} \otimes \mathring{\vartheta}^{\nu}. \quad \text{(B.15)}$$

The anholonomy objects defined on the tangent space, $\mathring{T}_{\mathring{x}}V_4$, read

$$\mathring{C}^a := d\mathring{\vartheta}^a = \frac{1}{2}\mathring{C}^a{}_{bc}\mathring{\vartheta}^b \wedge \mathring{\vartheta}^c, \quad \text{(B.16)}$$

where the anholonomy coefficients, $\mathring{C}^a{}_{bc}$, which represent the curls of the base members, are

$$\begin{aligned}\mathring{C}^c{}_{bc} &= -\mathring{\vartheta}^c\left([\mathring{e}_a, \mathring{e}_b]\right)\\ &= \mathring{e}_a{}^{\mu}\mathring{e}_b{}^{\nu}\left(\mathring{\partial}_{\mu}\mathring{e}^c{}_{\nu} - \mathring{\partial}_{\nu}\mathring{e}^c{}_{\mu}\right) \quad \text{(B.17)}\\ &= -\mathring{e}^c{}_{\mu}\left[\mathring{e}_a\left(\mathring{e}_b{}^{\mu}\right) - \mathring{e}_b\left(\mathring{e}_a{}^{\mu}\right)\right].\end{aligned}$$

The (anholonomic) Levi-Civita (or Christoffel) connection can be written as

$$\mathring{\Gamma}_{ab} := \mathring{e}_{[a}\rfloor d\mathring{\vartheta}_{b]} - \frac{1}{2}\left(\mathring{e}_a\rfloor\mathring{e}_b\rfloor d\mathring{\vartheta}_c\right) \wedge \mathring{\vartheta}^c, \quad \text{(B.18)}$$

where $\mathring{\vartheta}_c$ is understood as the down indexed 1-form $\mathring{\vartheta}_c = o_{cb}\mathring{\vartheta}^b$. The norm $i\widetilde{d}$ (A.2) can then be written in terms of the spacetime structures of V_4 and M_4 as

$$\begin{aligned}i\widetilde{d} &= \widetilde{e}\widetilde{\vartheta} = \widetilde{e}_{\rho} \otimes \widetilde{\vartheta}^{\rho} = \widetilde{e}_a \otimes \widetilde{\vartheta}^a = \breve{\Omega}^{\mu}{}_{\nu}\mathring{e}_{\mu} \otimes \mathring{\vartheta}^{\nu}\\ &= \breve{\Omega}^a{}_b\breve{e}_a\breve{\vartheta}^b = \Omega^m{}_l e_m \otimes \vartheta^l \in \widetilde{\mathscr{M}}_4,\end{aligned} \quad \text{(B.19)}$$

provided

$$\begin{aligned}&\breve{\Omega}^a{}_b = \breve{\pi}^a{}_c\breve{\pi}^c{}_b = \breve{\Omega}^{\mu}{}_{\nu}\mathring{e}^a{}_{\mu}\mathring{e}_b{}^{\nu}, \qquad \widetilde{e}_{\rho} = \breve{\pi}^{\nu}{}_{\rho}\mathring{e}_{\nu},\\ &\widetilde{\vartheta}^{\rho} = \breve{\pi}_{\mu}{}^{\rho}\mathring{\vartheta}^{\mu}, \qquad \widetilde{e}_c = \breve{\pi}_c{}^a\mathring{e}_a, \qquad \widetilde{\vartheta}^c = \breve{\pi}^c{}_b\mathring{\vartheta}^b.\end{aligned} \quad \text{(B.20)}$$

Under a local tetrad deformation (B.20), a general spin connection transforms according to

$$\widetilde{\omega}^a{}_{b\mu} = \breve{\pi}_c{}^a\mathring{\omega}^c{}_{d\mu}\breve{\pi}^d{}_b + \breve{\pi}_c{}^a\widetilde{\partial}_{\mu}\breve{\pi}^c{}_b = \pi_l{}^a\widetilde{\partial}_{\mu}\pi^l{}_b. \quad \text{(B.21)}$$

We have then two choices to recast metric as follows:

$$\begin{aligned}\widetilde{g} &= o_{ab}\widetilde{\vartheta}^a \otimes \widetilde{\vartheta}^b = o_{ab}\breve{\pi}^a{}_c\breve{\pi}^b{}_d\mathring{\vartheta}^c \otimes \mathring{\vartheta}^d\\ &= \breve{\gamma}_{cd}\mathring{\vartheta}^c \otimes \mathring{\vartheta}^d.\end{aligned} \quad \text{(B.22)}$$

In the first case, the contribution of the Christoffel symbols constructed by the metric $\breve{\gamma}_{ab} = o_{cd}\breve{\pi}^c{}_a\breve{\pi}^d{}_b$ reads

$$\begin{aligned}\widetilde{\Gamma}^a{}_{bc} &= \frac{1}{2}\left(\mathring{C}^a{}_{bc} - \breve{\gamma}^{aa'}\breve{\gamma}_{bb'}\mathring{C}^{b'}{}_{a'c} - \breve{\gamma}^{aa'}\breve{\gamma}_{cc'}\mathring{C}^{c'}{}_{a'b}\right)\\ &\quad + \frac{1}{2}\breve{\gamma}^{aa'}\left(\mathring{e}_c\rfloor d\breve{\gamma}_{ba'} - \mathring{e}_b\rfloor d\breve{\gamma}_{ca'} - \mathring{e}_{a'}\rfloor d\breve{\gamma}_{bc}\right).\end{aligned} \quad \text{(B.23)}$$

As before, the second deformation matrix, $\breve{\gamma}_{ab}$, can be decomposed in terms of symmetric, $\breve{\pi}_{(ab)}$, and antisymmetric, $\breve{\pi}_{[ab]}$, parts of the matrix $\breve{\pi}_{ab} = o_{ac}\breve{\pi}^c{}_b$. So,

$$\breve{\pi}_{ab} = \breve{\Upsilon}o_{ab} + \breve{\Theta}_{ab} + \breve{\varphi}_{ab}, \quad \text{(B.24)}$$

where $\breve{\Upsilon} = \breve{\pi}^a{}_a$, $\breve{\Theta}_{ab}$ is the traceless symmetric part, and $\breve{\varphi}_{ab}$ is the skew symmetric part of the first deformation matrix. In analogy with (B.4), the deformed metric can then be split as

$$\widetilde{g}_{\mu\nu}\left(\breve{\pi}\right) = \breve{\Upsilon}^2\left(\breve{\pi}\right)\mathring{g}_{\mu\nu} + \breve{\gamma}_{\mu\nu}\left(\breve{\pi}\right), \quad \text{(B.25)}$$

where

$$\breve{\gamma}_{\mu\nu}(\breve{\pi}) = \left[\breve{\gamma}_{ab} - \breve{\Upsilon}^2 o_{ab}\right] \overset{\circ}{e}^a{}_\mu \overset{\circ}{e}^b{}_\nu. \quad \text{(B.26)}$$

The inverse deformed metric reads

$$\widetilde{g}^{\mu\nu}(\breve{\pi}) = o^{cd} \breve{\pi}^{-1a}{}_c \breve{\pi}^{-1b}{}_d \overset{\circ}{e}_a{}^\mu \overset{\circ}{e}_b{}^\nu, \quad \text{(B.27)}$$

where $\breve{\pi}^{-1a}{}_c \breve{\pi}^c{}_b = \breve{\pi}^c{}_b \breve{\pi}^{-1a}{}_c = \delta^a_b$. The (anholonomic) Levi-Civita (or Christoffel) connection is

$$\widetilde{\Gamma}_{ab} := \widetilde{e}_{[a} \rfloor d\widetilde{\vartheta}_{b]} - \frac{1}{2}\left(\widetilde{e}_a \rfloor \widetilde{e}_b \rfloor d\widetilde{\vartheta}_c\right) \wedge \widetilde{\vartheta}^c, \quad \text{(B.28)}$$

where $\widetilde{\vartheta}_c$ is understood as the down indexed 1-form $\widetilde{\vartheta}_c = o_{cb}\widetilde{\vartheta}^b$. Hence, the usual Levi-Civita connection is related to the original connection by the relation

$$\widetilde{\Gamma}^\mu{}_{\rho\sigma} = \breve{\Gamma}^\mu{}_{\rho\sigma} + \breve{\Pi}^\mu{}_{\rho\sigma}, \quad \text{(B.29)}$$

provided

$$\begin{aligned}\Pi^\mu{}_{\rho\sigma} &= 2\widetilde{g}^{\mu\nu}\breve{g}_{\nu(\rho}\breve{\nabla}_{\sigma)}\breve{\Upsilon} - \breve{g}_{\rho\sigma}g^{\mu\nu}\breve{\nabla}_\nu\breve{\Upsilon} \\ &+ \frac{1}{2}\widetilde{g}^{\mu\nu}\left(\breve{\nabla}_\rho\breve{\gamma}_{\nu\sigma} + \breve{\nabla}_\sigma\breve{\gamma}_{\rho\nu} - \breve{\nabla}_\nu\breve{\gamma}_{\rho\sigma}\right),\end{aligned} \quad \text{(B.30)}$$

where $\breve{\nabla}$ is the covariant derivative. The contravariant deformed metric, $\widetilde{g}^{\nu\rho}$, is defined as the inverse of $\widetilde{g}_{\mu\nu}$, such that $\widetilde{g}_{\mu\nu}\widetilde{g}^{\nu\rho} = \delta^\rho_\mu$. Hence, the connection deformation $\Pi^\mu{}_{\rho\sigma}$ acts like a force that deviates the test particles from the geodesic motion in the space, V_4. A metric-affine space $(\widetilde{M}_4, \widetilde{g}, \widetilde{\Gamma})$ is defined to have a metric and a linear connection that need not be dependent on each other. In general, the lifting of the constraints of metric-compatibility and symmetry yields the new geometrical property of the spacetime, which are the *nonmetricity* 1-form $\widetilde{N}_{ab}$ and the affine *torsion* 2-form $\widetilde{T}^a$ representing a translational misfit (for a comprehensive discussion see [92–95]). These, together with the *curvature* 2-form $\widetilde{R}_a{}^b$, symbolically can be presented as [96, 97]

$$\left(\widetilde{N}_{ab}, \widetilde{T}^a, \widetilde{R}_a{}^b\right) \sim \widetilde{\mathscr{D}}\left(\widetilde{g}_{ab}, \widetilde{\vartheta}^a, \widetilde{\Gamma}_a{}^b\right), \quad \text{(B.31)}$$

where $\widetilde{\mathscr{D}}$ is the *covariant exterior derivative*. If the nonmetricity tensor $\widetilde{N}_{\lambda\mu\nu} = -\widetilde{\mathscr{D}}_\lambda\widetilde{g}_{\mu\nu} \equiv -\widetilde{g}_{\mu\nu;\lambda}$ does not vanish, the general formula for the affine connection written in the spacetime components is

$$\widetilde{\Gamma}^\rho{}_{\mu\nu} = \overset{\circ}{\Gamma}{}^\rho{}_{\mu\nu} + \widetilde{K}^\rho{}_{\mu\nu} - \widetilde{N}^\rho{}_{\mu\nu} + \frac{1}{2}\widetilde{N}^\rho{}_{(\mu\nu)}, \quad \text{(B.32)}$$

where $\overset{\circ}{\Gamma}{}^\rho{}_{\mu\nu}$ is the Riemann part and $\widetilde{K}^\rho{}_{\mu\nu} := 2\widetilde{Q}_{(\mu\nu)}{}^\rho + \widetilde{Q}^\rho{}_{\mu\nu}$ is the non-Riemann part, the affine *contortion tensor*. The torsion, $\widetilde{Q}^\rho{}_{\mu\nu} = (1/2)\widetilde{T}^\rho{}_{\mu\nu} = \widetilde{\Gamma}^\rho{}_{[\mu\nu]}$ given with respect to a holonomic frame, $d\widetilde{\vartheta}^\rho = 0$, is the third-rank tensor, antisymmetric in the first two indices, with 24 independent components. In a presence of curvature and torsion, the coupling prescription of a general field carrying an arbitrary representation of the Lorentz group will be

$$\widetilde{\partial}_\mu \longrightarrow \widetilde{\mathscr{D}}_\mu = \widetilde{\partial}_\mu - \frac{i}{2}\left(\widetilde{\omega}^{ab}{}_\mu - \widetilde{K}^{ab}{}_\mu\right) J_{ab}, \quad \text{(B.33)}$$

with J_{ab} denoting the corresponding Lorentz generator. The Riemann-Cartan manifold, U_4, is a particular case of the general metric-affine manifold $\widetilde{\mathscr{M}}_4$, restricted by the metricity condition $\widetilde{N}_{\lambda\mu\nu} = 0$, when a nonsymmetric linear connection is said to be metric compatible. The Lorentz and diffeomorphism invariant scalar curvature, $\widetilde{R}$, becomes either a function of $\widetilde{e}^a{}_\mu$ only, or $\widetilde{g}_{\mu\nu}$:

$$\begin{aligned}\widetilde{R}(\widetilde{\omega}) &\equiv \widetilde{e}_a{}^\mu \widetilde{e}_b{}^\nu \widetilde{R}_{\mu\nu}{}^{ab}(\widetilde{\omega}) = \widetilde{R}\left(\widetilde{g}, \widetilde{\Gamma}\right) \\ &\equiv \widetilde{g}^{\rho\nu}\widetilde{R}^\mu{}_{\rho\mu\nu}\left(\widetilde{\Gamma}\right).\end{aligned} \quad \text{(B.34)}$$

C. Determination of $\widetilde{D}$ and $\widetilde{\psi}$ in Standard Theory of Gravitation

Let $\widetilde{\omega}^{ab} = \widetilde{\omega}^{ab}_\mu \wedge d\widetilde{x}^\mu$ be the 1-forms of corresponding connections assuming values in the Lorentz Lie algebra. The action for gravitational field can be written in the form

$$\widetilde{S}_g = \overset{\circ}{S} + \widetilde{S}_Q, \quad \text{(C.1)}$$

where the integral

$$\begin{aligned}\overset{\circ}{S} &= -\frac{1}{4æ}\int \star\overset{\circ}{R} = -\frac{1}{4æ}\int \star\overset{\circ}{R}_{cd} \wedge \widetilde{\vartheta}^c \wedge \widetilde{\vartheta}^d \\ &= -\frac{1}{2æ}\int \overset{\circ}{R}\sqrt{-\widetilde{g}}d\Omega\end{aligned} \quad \text{(C.2)}$$

is the usual Einstein action, with the coupling constant relating to the Newton gravitational constant $æ = 8\pi G_N/c^4$, S_Q is the phenomenological action of the spin-torsion interaction, and $\star$ denotes the Hodge dual. This is a C^∞-linear map $\star : \Omega^p \to \Omega^{n-p}$, which acts on the wedge product monomials of the basis 1-forms as $\star(\widetilde{\vartheta}^{a_1\cdots a_p}) = \varepsilon^{a_1\cdots a_n}\widetilde{e}_{a_{p+1}\cdots a_n}$. Here we used the abbreviated notations for the wedge product monomials, $\widetilde{\vartheta}^{a_1\cdots a_p} = \widetilde{\vartheta}^{a_1} \wedge \widetilde{\vartheta}^{a_2} \wedge \cdots \wedge \widetilde{\vartheta}^{a_p}$, defined on the U_4 space, the $\widetilde{e}_{a_i}$ $(i = p+1, \ldots, n)$ are understood as the down indexed 1-forms $\widetilde{e}_{a_i} = o_{a_i b}\widetilde{\vartheta}^b$, and $\varepsilon^{a_1\cdots a_n}$ is the total antisymmetric pseudotensor. The variation of the connection 1-form $\widetilde{\omega}^{ab}$ yields

$$\delta\widetilde{S}_Q = \frac{1}{æ}\int \star\widetilde{\mathscr{T}}_{ab} \wedge \delta\widetilde{\omega}^{ab}, \quad \text{(C.3)}$$

where

$$\begin{aligned}\star\widetilde{\mathscr{T}}_{ab} &:= \frac{1}{2}\star\left(\widetilde{Q}_a \wedge \widetilde{e}_b\right) = \widetilde{Q}^c \wedge \widetilde{\vartheta}^d \varepsilon_{cdab} \\ &= \frac{1}{2}\widetilde{Q}^c{}_{\mu\nu} \wedge \widetilde{e}^d{}_\alpha \varepsilon_{abcd}\widetilde{\vartheta}^{\mu\nu\alpha},\end{aligned} \quad \text{(C.4)}$$

and also

$$\widetilde{Q}^a = \widetilde{D}\widetilde{\vartheta}^a = d\widetilde{\vartheta}^a + \widetilde{\omega}^a{}_b \wedge \widetilde{\vartheta}^b. \quad (C.5)$$

The variation of the action describing the macroscopic matter sources $\widetilde{S}_m$ with respect to the coframe ϑ^a and connection 1-form $\widetilde{\omega}^{ab}$ reads

$$\begin{aligned} \delta\widetilde{S}_m &= \int \delta\widetilde{L}_m \\ &= \int \left(- \star \widetilde{\theta}_a \wedge \delta\widetilde{\vartheta}^a + \frac{1}{2} \star \widetilde{\Sigma}_{ab} \wedge \delta\widetilde{\omega}^{ab} \right), \end{aligned} \quad (C.6)$$

where $\star\widetilde{\theta}_a$ is the dual 3-form relating to the canonical energy-momentum tensor, $\widetilde{\theta}^{\mu}_{a}$, by

$$\star\widetilde{\theta}_a = \frac{1}{3!}\widetilde{\theta}^{\mu}{}_a \varepsilon_{\mu\nu\alpha\beta}\widetilde{\vartheta}^{\nu\alpha\beta} \quad (C.7)$$

and $\star\widetilde{\Sigma}_{ab} = - \star \widetilde{\Sigma}_{ba}$ is the dual 3-form corresponding to the canonical spin tensor, which is identical with the dynamical spin tensor $\widetilde{S}_{abc}$; namely,

$$\star\widetilde{\Sigma}_{ab} = \widetilde{S}^{\mu}{}_{ab}\varepsilon_{\mu\nu\alpha\beta}\widetilde{\vartheta}^{\nu\alpha\beta}. \quad (C.8)$$

The variation of the total action, $\widetilde{S} = \widetilde{S}_g + \widetilde{S}_m$, with respect to the $\widetilde{e}_a$, $\widetilde{\omega}^{ab}$ and $\widetilde{\Phi}$ gives the following field equations:

$$\begin{aligned} &(1)\ \frac{1}{2}\mathring{R}_{cd} \wedge \widetilde{\vartheta}^c = æ\widetilde{\theta}_d = 0, \\ &(2)\ \star \widetilde{\mathcal{T}}_{ab} = -\frac{1}{2}æ \star \widetilde{\Sigma}_{ab}, \\ &(3)\ \frac{\delta\widetilde{L}_m}{\delta\widetilde{\Phi}} = 0, \qquad \frac{\delta\widetilde{L}_m}{\delta\overline{\widetilde{\Phi}}} = 0. \end{aligned} \quad (C.9)$$

In the sequel, the DC-members $\widetilde{D}$ and $\widetilde{\psi}$ can readily be determined as follows:

$$\widetilde{D}^l_a = \eta^{lm}\left\langle \widetilde{e}_a, e_m \right\rangle, \qquad \widetilde{\psi}^a_l = \eta_{lm}\widetilde{\vartheta}^a\left(\vartheta^{-1}\right)^m. \quad (C.10)$$

D. The GGP in More Detail

Note that an invariance of the Lagrangian $L_{\widetilde{\Phi}}$ under the infinite-parameter group of general covariance (A.5) in V_4 implies an invariance of $L_{\widetilde{\Phi}}$ under the G_V group and vice versa if and only if the generalized local gauge transformations of the fields $\widetilde{\Phi}(\widetilde{x})$ and their covariant derivative $\widetilde{\nabla}_\mu\widetilde{\Phi}(\widetilde{x})$ are introduced by finite local $U_V \in G_V$ gauge transformations:

$$\begin{aligned} \widetilde{\Phi}'(\widetilde{x}) &= U_V(\widetilde{x})\widetilde{\Phi}(\widetilde{x}), \\ \left[\widetilde{\gamma}^\mu(\widetilde{x})\widetilde{\nabla}_\mu\widetilde{\Phi}(\widetilde{x})\right]' &= U_V(\widetilde{x})\left[\widetilde{\gamma}^\mu(\widetilde{x})\widetilde{\nabla}_\mu\widetilde{\Phi}(\widetilde{x})\right]. \end{aligned} \quad (D.1)$$

Here $\widetilde{\nabla}_\mu$ denotes the covariant derivative agreeing with the metric, $\widetilde{g}^{\mu\nu} = (1/2)(\widetilde{\gamma}^\mu\widetilde{\gamma}^\nu + \widetilde{\gamma}^\nu\widetilde{\gamma}^\mu)$: $\widetilde{\nabla}_\mu = \widetilde{\partial}_\mu + \widetilde{\Gamma}_\mu$, where $\widetilde{\Gamma}_\mu(\widetilde{x}) = (1/2)J^{ab}\widetilde{e}_a{}^\nu(\widetilde{x})\widetilde{\partial}_\mu\widetilde{e}_{b\nu}(\widetilde{x})$ is the connection and J_{ab} are the generators of Lorentz group Λ. The tetrad components $\widetilde{e}_a{}^\mu(\widetilde{x})$ associate with the chosen representation $D(\Lambda)$ by which the $\widetilde{\Phi}(\widetilde{x})$ is transformed as $[D(\Lambda)]^{l'\cdots k'}_{l\cdots k}\widetilde{\Phi}(\widetilde{x})$, where $D(\Lambda) = I + (1/2)\widetilde{\omega}^{ab}J_{ab}$, $\widetilde{\omega}_{ab} = -\widetilde{\omega}_{ba}$ are the parameters of the Lorentz group. One has, for example, to set $\widetilde{\gamma}^\mu(\widetilde{x}) \rightarrow \widetilde{e}^\mu(\widetilde{x})$ for the fields of spin ($j = 0, 1$); for vector field $[J_{ab}]^l_k = \delta^l_a\eta_{bk} - \delta^l_b\eta_{ak}$; but $\widetilde{\gamma}^\mu(\widetilde{x}) = \widetilde{e}_a{}^\mu(\widetilde{x})\gamma^a$ and $J_{ab} = -(1/4)[\gamma_a, \gamma_b]$ for the spinor field ($j = 1/2$), where γ^a are the Dirac matrices.

Given the principal fiber bundle $\widetilde{P}(V_4, G_V; \widetilde{s})$ with the structure group G_V, the local coordinates $\widetilde{p} \in \widetilde{P}$ are $\widetilde{p} = (\widetilde{x}, U_V)$, where $\widetilde{x} \in V_4$ and $U_V \in G_V$, the total bundle space $\widetilde{P}$ is a smooth manifold, and the surjection $\widetilde{s}$ is a smooth map $\widetilde{s} : \widetilde{P} \rightarrow V_4$. A set of open coverings $\{\widetilde{\mathcal{U}}_i\}$ of V_4 with $\widetilde{x} \in \{\widetilde{\mathcal{U}}_i\} \subset V_4$ satisfy $\bigcup_\alpha \widetilde{\mathcal{U}}_\alpha = V_4$. The collection of matter fields of arbitrary spins $\widetilde{\Phi}(\widetilde{x})$ take values in standard fiber over $\widetilde{x} : \widetilde{s}^{-1}(\widetilde{\mathcal{U}}_i) = \widetilde{\mathcal{U}}_i \times \widetilde{F}_{\widetilde{x}}$. The fibration is given as $\bigcup_{\widetilde{x}}\widetilde{s}^{-1}(\widetilde{x}) = \widetilde{P}$. The local gauge will be the diffeomorphism map $\widetilde{\chi}_i$: $\widetilde{\mathcal{U}}_i \times_{V_4} G_V \rightarrow \widetilde{s}^{-1}(\widetilde{\mathcal{U}}_i) \in \widetilde{P}$, since $\widetilde{\chi}_i^{-1}$ maps $\widetilde{s}^{-1}(\widetilde{\mathcal{U}}_i)$ onto the direct (Cartesian) product $\widetilde{\mathcal{U}}_i \times_{V_4} G_V$. Here $\times_{V_4}$ represents the fiber product of elements defined over space V_4 such that $\widetilde{s}(\widetilde{\chi}_i(\widetilde{x}, U_V)) = \widetilde{x}$ and $\widetilde{\chi}_i(\widetilde{x}, U_V) = \widetilde{\chi}_i(\widetilde{x}, (id)_{G_V})U_V = \widetilde{\chi}_i(\widetilde{x})U_V$ for all $\widetilde{x} \in \{\widetilde{\mathcal{U}}_i\}$, where $(id)_{G_V}$ is the identity element of the group G_V. The fiber $\widetilde{s}^{-1}$ at $\widetilde{x} \in V_4$ is diffeomorphic to $\widetilde{F}$, where $\widetilde{F}$ is the fiber space, such that $\widetilde{s}^{-1}(\widetilde{x}) \equiv \widetilde{F}_{\widetilde{x}} \approx \widetilde{F}$. The action of the structure group G_V on $\widetilde{P}$ defines an isomorphism of the Lie algebra $\widetilde{\mathfrak{g}}$ of G_V onto the Lie algebra of vertical vector fields on $\widetilde{P}$ tangent to the fiber at each $\widetilde{p} \in \widetilde{P}$ called fundamental. To involve a drastic revision of the role of gauge fields in the physical concept of the spacetime deformation, we generalize the standard gauge scheme by exploring a new special type of *distortion* gauge field, (a), which is assumed to act on the external spacetime groups. Then, we also consider the principle fiber bundle, $P(M_4, U^{\text{loc}}; s)$, with the base space M_4, the structure group U^{loc}, and the surjection s. The matter fields $\Phi(x)$ take values in the standard fiber which is the Hilbert vector space where a linear representation $U(x)$ of group U^{loc} is given. This space can be regarded as the Lie algebra of the group U^{loc} upon which the Lie algebra acts according to the law of the adjoint representation: $a \leftrightarrow \text{ad}\, a\Phi \rightarrow [a, \Phi]$.

The GGP accounts for the *gravitation gauge group* G_V generated by the hidden local internal symmetry U^{loc}. *The physical system of the fields* $\widetilde{\Phi}(\widetilde{x})$ *defined on* V_4 *must be invariant under the finite local gauge transformations* U_V (D.1) *of the Lie group of gravitation* G_V (see Scheme 1), where $R_\psi(a)$ is the matrix of unitary map:

$$\begin{aligned} R_\psi(a) &: \Phi \longrightarrow \widetilde{\Phi}, \\ S(a)R_\psi(a) &: \left(\gamma^k D_k\Phi\right) \longrightarrow \left(\widetilde{\gamma}^\nu(\widetilde{x})\nabla_\nu\widetilde{\Phi}\right). \end{aligned} \quad (D.2)$$

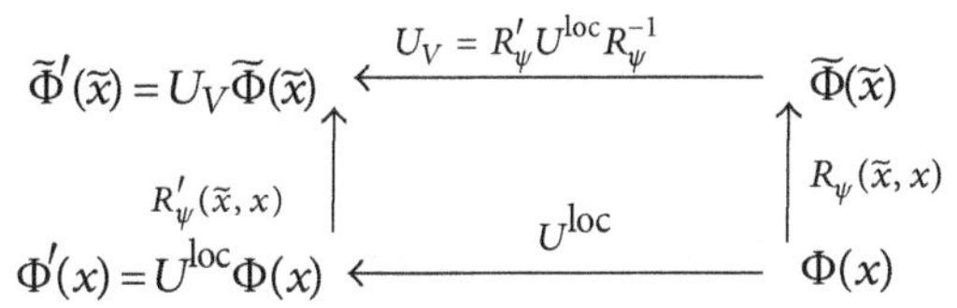

Scheme 1: The GGP.

Here $S(F)$ is the gauge invariant scalar function $S(F) \equiv (1/4)\widetilde{\pi}^{-1}(F) = (1/4)\widetilde{\psi}^l_\mu \widetilde{D}^\mu_l$, $D_k = \partial_k - i æ, a_k$. In an illustration of the point at issue, the (D.2) explicitly may read

$$\begin{aligned} \widetilde{\Phi}^{\mu\cdots\delta}(\widetilde{x}) &= \widetilde{\psi}^\mu_l \cdots \widetilde{\psi}^\delta_m R(a)\, \Phi^{l\cdots m}(x) \\ &\equiv \left(R_\psi\right)^{\mu\cdots\delta}_{l\cdots m} \Phi^{l\cdots m}(x), \end{aligned} \tag{D.3}$$

and also

$$\begin{aligned} &\widetilde{g}^\nu(\widetilde{x}) \nabla_\nu \widetilde{\Phi}^{\mu\cdots\delta}(\widetilde{x}) \\ &\quad = S(F)\, \widetilde{\psi}^\mu_l \cdots \widetilde{\psi}^\delta_m R(a)\, \gamma^k D_k \Phi^{l\cdots m}(x). \end{aligned} \tag{D.4}$$

In case of zero curvature, one has $\psi^\mu_l = D^\mu_l = e^\mu{}_l = (\partial x^\mu / \partial X^l)$, $\|D\| \neq 0$, where X^l are the inertial coordinates. In this, the conventional gauge theory given on the M_4 is restored in both curvilinear and inertial coordinates. Although the distortion gauge field (a_A) is a vector field, only the gravitational attraction is presented in the proposed theory of gravitation.

E. A Lie Group of Distortion

The extended space M_6 reads 0

$$\begin{aligned} M_6 &= R^3_+ \oplus R^3_- = R^3 \oplus T^3, \\ \operatorname{sgn}\left(R^3\right) &= (+++), \qquad \operatorname{sgn}\left(T^3\right) = (---). \end{aligned} \tag{E.1}$$

The $e_{(\lambda\alpha)} = O_\lambda \times \sigma_\alpha$ $(\lambda = \pm, \alpha = 1, 2, 3)$ are linearly independent unit basis vectors at the point (p) of interest of the given three-dimensional space R^3_λ. The unit vectors O_λ and σ_α imply

$$\langle O_\lambda, O_\tau \rangle = {}^*\delta_{\lambda\tau}, \qquad \langle \sigma_\alpha, \sigma_\beta \rangle = \delta_{\alpha\beta}, \tag{E.2}$$

where $\delta_{\alpha\beta}$ is the Kronecker symbol and ${}^*\delta_{\lambda\tau} = 1 - \delta_{\lambda\tau}$. Three spatial $e_\alpha = \xi \times \sigma_\alpha$ and three temporal $e_{0\alpha} = \xi_0 \times \sigma_\alpha$ components are the basis vectors, respectively, in spaces R^3 and T^3, where $O_\pm = (1/\sqrt{2})(\xi_0 \pm \xi)$, $\xi_0^2 = -\xi^2 = 1$, $\langle \xi_0, \xi \rangle = 0$. The 3D space $R^3_\pm$ is spanned by the coordinates $\eta_{(\pm\alpha)}$. In using this language it is important to consider a reduction to the space M_4 which can be achieved in the following way.

(1) In case of free flat space M_6, the subspace T^3 is isotropic. And in so far it contributes in line element just only by the square of the moduli $t = |\mathbf{x}^0|$, $\mathbf{x}^0 \in T^3$, then, the reduction $M_6 \to M_4 = R^3 \oplus T^1$ can be readily achieved if we use $t = |\mathbf{x}^0|$ for conventional *time*.

(2) In case of curved space, the reduction $V_6 \to V_4$ can be achieved if we use the projection $(\breve{e}_0)$ of the temporal component $(\breve{e}_{0\alpha})$ of basis six-vector $\breve{e}(\breve{e}_\alpha, \breve{e}_{0\alpha})$ on the given *universal* direction $(\breve{e}_{0\alpha} \to \breve{e}_0)$. By this we choose the *time* coordinate. Actually, the Lagrangian of physical fields defined on R_6 is a function of scalars such that $A_{(\lambda\alpha)}B^{(\lambda\alpha)} = A_\alpha B^\alpha + A_{0\alpha}B^{0\alpha}$; then upon the reduction of temporal components of six-vectors $A_{0\alpha}B^{0\alpha} = A^{0\alpha}\langle \breve{e}_{0\alpha}, \breve{e}_{0\beta}\rangle B^{0\beta} = A^0 \langle \breve{e}_0, \breve{e}_0 \rangle B^0 = A_0 B^0$ we may fulfill a reduction to V_4.

A distortion of the basis (E.2) comprises the following two steps. We, at first, consider distortion transformations of the ingredient unit vectors O_τ under the distortion gauge field (a):

$$\begin{aligned} \breve{O}_{(+\alpha)}(a) &= \mathbb{Q}^\tau_{(+\alpha)}(a)\, O_\tau = O_+ + æ a_{(+\alpha)} O_-, \\ \breve{O}_{(-\alpha)}(a) &= \mathbb{Q}^\tau_{(-\alpha)}(a)\, O_\tau = O_- + æ a_{(-\alpha)} O_+, \end{aligned} \tag{E.3}$$

where $\mathbb{Q}$ $(=\mathbb{Q}^\tau_{(\lambda\alpha)}(a))$ is an element of the group Q. This induces the distortion transformations of the ingredient unit vectors σ_β, which, in turn, undergo the rotations, $\breve{\sigma}_{(\lambda\alpha)}(\theta) = \mathscr{R}^\beta_{(\lambda\alpha)}(\theta)\sigma_\beta$, where $\mathscr{R}(\theta) \in SO(3)$ is the element of the group of rotations of planes involving two arbitrary axes around the orthogonal third axis in the given ingredient space R^3_λ. In fact, distortion transformations of basis vectors (O) and (σ) are not independent but rather are governed by the spontaneous breaking of the distortion symmetry (for more details see [74]). To avoid a further proliferation of indices, hereafter we will use uppercase Latin (A) in indexing $(\lambda\alpha)$, and so forth. The infinitesimal transformations then read

$$\begin{aligned} \delta \mathbb{Q}^\tau_A(a) &= æ\,\delta a_A X^\tau_\lambda \in Q, \\ \delta \mathscr{R}(\theta) &= -\frac{i}{2} M_{\alpha\beta}\,\delta\omega^{\alpha\beta} \in SO(3), \end{aligned} \tag{E.4}$$

provided by the generators $X^\tau_\lambda = {}^*\delta^\tau_\lambda$ and $I_i = \sigma_i/2$, where σ_i are the Pauli matrices, $M_{\alpha\beta} = \varepsilon_{\alpha\beta\gamma} I_\gamma$, and $\delta\omega^{\alpha\beta} = \varepsilon_{\alpha\beta\gamma}\delta\theta_\gamma$. The transformation matrix $D(a, \theta) = \mathbb{Q}(a) \times \mathscr{R}(\theta)$ is an element of the distortion group $G_D = Q \times SO(3)$:

$$\begin{aligned} D_{(da^A, d\theta^A)} &= I + dD_{(a^A, \theta^A)}, \\ dD_{(a^A, \theta^A)} &= i\left[da^A X_A + d\theta^A I_A\right], \end{aligned} \tag{E.5}$$

where $I_A \equiv I_\alpha$ at given λ. The generators X_A (E.4) of the group Q do not complete the group H to the dynamical group G_D, and therefore they cannot be interpreted as the generators of the quotien space G_D/H, and the distortion fields a_A cannot be identified directly with the Goldstone fields arising in spontaneous breaking of the distortion symmetry G_D. These objections, however, can be circumvented, because, as it is shown by [74], the distortion group $G_D = Q \times SO(3)$ can be mapped in a one-to-one manner onto the group $G_D = SO(3) \times SO(3)$, which is isomorphic to the chiral group $SU(2) \times SU(2)$, in case of which the method of phenomenological Lagrangians is well known. In aftermath, we arrive at the key relation

$$\tan\theta_A = -æ a_A. \tag{E.6}$$

Given the distortion field a_A, the relation (E.6) uniquely determines six angles θ_A of rotations around each of six (A) axes. In pursuing our goal further, we are necessarily led to extending a whole framework of GGP now for the base 12D smooth differentiable manifold:

$$M_{12} = M_6 \oplus \overline{M}_6. \tag{E.7}$$

Here the M_6 is related to the spacetime continuum (E.1), but the $\overline{M}_6$ is displayed as a space of inner degrees of freedom. The

$$e_{(\lambda,\mu,\alpha)} = O_{\lambda,\mu} \otimes \sigma_\alpha \quad (\lambda, \mu = 1,2; \alpha = 1,2,3) \tag{E.8}$$

are basis vectors at the point $p(\zeta)$ of M_{12}:

$$\begin{aligned} &\langle O_{\lambda,\mu}, O_{\tau,\nu} \rangle = {}^*\delta_{\lambda,\tau}\,{}^*\delta_{\mu,\nu}, \qquad O_{\lambda,\mu} = O_\lambda \otimes O_\mu, \\ &O_{\lambda,\mu} \longleftrightarrow {}^*R^4 = {}^*R^2 \otimes {}^*R^2, \qquad \sigma_\alpha \longleftrightarrow R^3, \end{aligned} \tag{E.9}$$

where $\zeta = (\eta, u) \in M_{12}$ ($\eta \in M_6$ and $u \in \overline{M}_6$). So, the decomposition (E.1), together with

$$\begin{aligned} &\overline{M}_6 = \overline{R}^3_+ \oplus \overline{R}^3_- = \overline{T}^3 \oplus \overline{P}^3, \\ &\operatorname{sgn}\left(\overline{T}^3\right) = (+++), \qquad \operatorname{sgn}\left(\overline{P}^3\right) = (---), \end{aligned} \tag{E.10}$$

holds. The 12-dimensional basis (e) transforms under the distortion gauge field $a(\zeta)$ ($\zeta \in M_{12}$):

$$\widetilde{e} = D(a)\,e, \tag{E.11}$$

where the distortion matrix $D(a)$ reads $D(a) = C(a) \otimes R(a)$, provided

$$\widetilde{O} = C(a)\,O, \qquad \widetilde{\sigma} = R(a)\,\sigma. \tag{E.12}$$

The matrices $C(a)$ generate the group of distortion transformations of the bi-pseudo-vectors:

$$C^{\tau,\nu}_{(\lambda\mu\alpha)}(a) = \delta^\tau_\lambda \delta^\nu_\mu + æ a_{(\lambda,\mu,\alpha)}\,{}^*\delta^{\tau}_{\lambda}\,{}^*\delta^\nu_\mu, \tag{E.13}$$

but $R(a) \in SO(3)_{\lambda\mu}$—the group of ordinary rotations of the planes involving two arbitrary bases of the spaces $R^3_{\lambda\mu}$ around the orthogonal third axes. The angles of rotations are determined according to (E.6), but now for the extended indices $A = (\lambda, \mu, \alpha)$ and so forth.

F. Field Equations at Spherical Symmetry

The extended field equations followed at once in terms of Euler-Lagrange variations, respectively, on the spaces M_{12} and $\widetilde{M}_{12}$ [74]. In accordance, the equation of distortion gauge field $a_A = (a_{(\lambda\alpha)}, \overline{a}_{(\tau\beta)})$ reads

$$\begin{aligned} &\partial^B\partial_B a_A - \left(1 - \zeta_0^{-1}\right)\partial_A\partial^B a_B \\ &\quad = J_A = -\frac{1}{2}\sqrt{g}\frac{\partial g^{BC}}{\partial a_A}T_{BC}, \end{aligned} \tag{F.1}$$

where T_{BC} is the energy-momentum tensor and ζ_0 is the gauge fixing parameter. To render our discussion here more transparent, below we clarify the relation between gravitational and coupling constants. To assist in obtaining actual solutions from the field equations, we may consider the weak-field limit and will envisage that the right-hand side of (F.1) should be in the form

$$-\frac{1}{2}\left(4\pi G_N\right)\sqrt{g(x)}\frac{\partial g^{BC}(x)}{\partial x_A}\widetilde{T}_{BC}. \tag{F.2}$$

Hence, we may assign to Newton's gravitational constant G_N the value

$$G_N = \frac{æ^2}{4\pi}. \tag{F.3}$$

The curvature of manifold $M_6 \rightarrow \widetilde{M}_6$ is the familiar distortion induced by the extended field components

$$\begin{aligned} a_{(1,1,\alpha)} &= a_{(2,1,\alpha)} \equiv \frac{1}{\sqrt{2}}a_{(+\alpha)}, \\ a_{(1,2,\alpha)} &= a_{(2,2,\alpha)} \equiv \frac{1}{\sqrt{2}}a_{(-\alpha)}. \end{aligned} \tag{F.4}$$

The other regime of ID presents at

$$\begin{aligned} a_{(1,1,\alpha)} &= -a_{(2,1,\alpha)} \equiv \frac{1}{\sqrt{2}}\overline{a}_{(+\alpha)}, \\ a_{(1,2,\alpha)} &= -a_{(2,2,\alpha)} \equiv \frac{1}{\sqrt{2}}\overline{a}_{(-\alpha)}. \end{aligned} \tag{F.5}$$

To obtain a feeling for this point we may consider physical systems which are static as well as spherically symmetrical. We are interested in the case of a spherical-symmetric gravitational field $a_0(r)$ in presence of one-dimensional space-like ID-field $\overline{a}$:

$$\begin{aligned} &a_{(1,1,3)} = a_{(2,2,3)} = a_{(+3)} = \frac{1}{2}\left(-a_0 + \overline{a}\right), \\ &a_{(1,2,3)} = a_{(2,1,3)} = a_{(-3)} = \frac{1}{2}\left(-a_0 - \overline{a}\right), \\ &a_{(\lambda,\mu,1)} = a_{(\lambda,\mu,2)} = 0, \quad \lambda, \mu = 1, 2. \end{aligned} \tag{F.6}$$

One can then easily determine the basis vectors $(e_{\lambda\alpha}, \overline{e}_{\tau\beta})$, where $\tan\theta_{(\pm 3)} = æ(-a_0 \pm \overline{a})$. Passing back from the $\widetilde{M}_6$ to V_4, the basis vectors read

$$\begin{aligned} &\widetilde{e}_0 = e_0\left(1 - x_0\right) + \overline{e}_3 x, \\ &\widetilde{e}_3 = e_3\left(1 + x_0\right) - \overline{e}_{03} x, \end{aligned}$$

$$\begin{aligned}\tilde{e}_1 = \frac{1}{2}\{&(\cos\theta_{(+3)} + \cos\theta_{(-3)})\, e_1 \\ &+ (\sin\theta_{(+3)} + \sin\theta_{(-3)})\, e_2 \\ &+ (\cos\theta_{(+3)} - \cos\theta_{(-3)})\, \bar{e}_{01} \\ &+ (\sin\theta_{(+3)} - \sin\theta_{(-3)})\, \bar{e}_{02}\}, \\ \tilde{e}_2 = \frac{1}{2}\{&(\cos\theta_{(+3)} + \cos\theta_{(-3)})\, e_2 \\ &- (\sin\theta_{(+3)} + \sin\theta_{(-3)})\, e_1 \\ &+ (\cos\theta_{(+3)} - \cos\theta_{(-3)})\, \bar{e}_{02} \\ &- (\sin\theta_{(+3)} - \sin\theta_{(-3)})\, \bar{e}_{01}\},\end{aligned} \tag{F.7}$$

where $x_0 \equiv æa_0$, $x \equiv æ\bar{a}$.

G. SPC-Configurations

The equations describing the equilibrium SPC include the gravitational and ID field equations (A.2), the hydrostatic equilibrium equation, and the state equation specified for each domain of many layered configurations. The resulting stable SPC is formed, which consists of the protomatter core and the outer layers of ordinary matter. A layering of configurations is a consequence of the onset of different regimes in equation of state. In the density range $\rho < 4.54 \times 10^{12}\,\mathrm{g\,cm^{-3}}$, one uses for both configurations the simple semiempirical formula of state equation given by Harrison and Wheeler, see for example [98]. Above the density $\rho > 4.54 \times 10^{12}\,\mathrm{g\,cm^{-3}}$, for the simplicity, the I-class SPC_I configuration is thought to be composed of regular n-p-e (neutron-proton-electron) gas (in absence of ID) in intermediate density domain $4.54 \times 10^{12}\,\mathrm{g\,cm^{-3}} \le \rho < \rho_d$ and of the n-p-e protomatter in presence of ID at $\rho > \rho_d$. For the II-class SPC_{II} configuration above the density $\rho_{fl} = 4.09 \times 10^{14}\,\mathrm{g\,cm^{-3}}$ one considers an onset of melting down of hadrons when nuclear matter consequently turns to quark matter, found in string flip-flop regime. In domain $\rho_{fl} \le \rho < \rho_d$, to which the distances $0.4\,\mathrm{fm} < r_{NN} \le 1.6\,\mathrm{fm}$ correspond, one has the regular (ID is absent) string flip-flop regime. This is a kind of tunneling effect when the strings joining the quarks stretch themselves violating energy conservation and after touching each other they switch on to the other configuration [71]. The basic technique adopted for calculation of transition matrix element $\widetilde{K}$ is the instanton technique (semiclassical treatment). During the quantum transition from a state ψ_1 of energy $\widetilde{E}_1$ to another one ψ_2 of energy $\widetilde{E}_2$, the lowering of energy of system takes place and the quark matter acquires $\Delta\widetilde{E}$ correction to the classical string energy such that the flip-flop energy lowers the energy of quark matter, consequently by lowering the critical density or critical Fermi momentum. If one, for example, looks for the string flip-flop transition amplitude of simple system of $q\bar{q}q\bar{q}$ described by the Hamiltonian $\widetilde{H}$ and invariant action $\widetilde{S}$, then one has

$$\left\langle \updownarrow\,\updownarrow \right| e^{-\widetilde{H}T} \left| \rightleftarrows \right\rangle = \left\langle \int [d\tilde{\sigma}] e^{-\widetilde{S}} \right\rangle, \tag{G.1}$$

where T is an imaginary time interval and $[d\tilde{\sigma}]$ is the integration over all the possible string motion. The action $\widetilde{S}$ is proportional to the area $\widetilde{A}$ of the surface swept by the strings in the finite region of ID-region of R_4. The strings are initially in the $\rightleftarrows$-configuration and finally in the $\updownarrow\,\updownarrow$-configuration. The maximal contribution to the path integral comes from the surface σ_0 of the minimum surface area "instanton". A computation of the transition amplitude is straightforward by summing over all the small vibrations around σ_0. In domain $\rho_d \le \rho < \rho_{as}$, one has the string flip-flop regime in presence of ID, at distances $0.25\,\mathrm{fm} < r_{NN} \le 0.4\,\mathrm{fm}$. That is, the system is made of quark protomatter in complete β-equilibrium with rearrangement of string connections joining them. In final domain $\rho > \rho_{as}$, the system is made of quarks in one bag in complete β-equilibrium at presence of ID. The quarks are under the weak interactions and gluons, including the effects of QCD-perturbative interactions. The QCD vacuum has a complicated structure, which is intimately connected to the gluon-gluon interaction. In most applications, sufficient accuracy is obtained by assuming that all the quarks are almost massless inside a bag. The latter is regarded as noninteracting Fermi gas found in the ID-region of the space-time continuum, at short distances $r_{NN} \le 0.25\,\mathrm{fm}$. Each configuration is defined by the two free parameters of central values of particle concentration $\tilde{n}(0)$ and dimensionless potential of space-like ID-field $x(0)$. The interior gravitational potential $x_0^{\mathrm{int}}(r)$ matches the exterior one $x_0^{\mathrm{ext}}(r)$ at the surface of the configuration. The central value of the gravitational potential $x_0(0)$ can be found by reiterating integrations when the sewing condition of the interior and exterior potentials holds. The key question of stability of SPC was studied in [72]. In the relativistic case the total mass-energy of SPC is the extremum in equilibrium for all configurations with the same total number of baryons. While the extrema of $\widetilde{M}$ and N occur at the same point in a one-parameter equilibrium sequence, one can look for the extremum of $\widetilde{E} = \widetilde{M}c^2 - \widetilde{m}_B N$ on equal footing. Minimizing the energy will give the equilibrium configuration, and the second derivative of $\widetilde{E}$ will give stability information. Recall that, for spherical configurations of matter, instantaneously at rest, small radial deviations from equilibrium are governed by a Sturm-Liouville linear eigenvalue equation [98], with the imposition of suitable boundary conditions on normal modes with time dependence $\xi^i(\vec{x}, t) = \xi^i(\vec{x})e^{i\omega t}$. A necessary and sufficient condition for stability is that the potential energy be positive defined for all initial data of $\xi^i(\vec{x}, 0)$, namely, in first order approximation when one does not take into account the rotation and magnetic field, if the square of frequency of normal mode of small perturbations is positive. A relativity tends to destabilize configurations. However, numerical integrations of the stability equations of SPC [72] give for the pressure-averaged value of the adiabatic index $\Gamma_1 = (\partial \ln \widetilde{P}/\partial \ln \tilde{\rho})_s$ the following values: $\overline{\Gamma}_1 \approx 2.216$ for the SPC_I and $\overline{\Gamma}_1 \approx 2.4$ for SPC_{II} configurations. This clearly proves the stability of resulting SPC. Note that the SPC is always found inside the event horizon sphere, and therefore it could be observed only in presence of accreting matter.

Conflict of Interests

The author declares that there is no conflict of interests regarding the publication of this paper.

Acknowledgment

The very helpful and knowledgable comments from the anonymous referee which have essentially clarified the paper are much appreciated.

References

[1] J. A. Orosz, "Inventory of black hole binaries," in *Proceedings of the International Astronomical Union Symposium (IAU '03)*, K. van der Hucht, A. Herrero, and C. Esteban, Eds., vol. 212, 2003.

[2] D. Lynden-Bell, "Galactic nuclei as collapsed old quasars," *Nature*, vol. 223, no. 5207, pp. 690–694, 1969.

[3] T. R. Lauer, S. M. Faber, D. Richstone et al., "The masses of nuclear black holes in luminous elliptical galaxies and implications for the space density of the most massive black holes," *Astrophysical Journal Letters*, vol. 662, no. 2 I, pp. 808–834, 2007.

[4] M. Volonteri, "Formation of supermassive black holes," *Astronomy and Astrophysics Review*, vol. 18, no. 3, pp. 279–315, 2010.

[5] J. Woo and C. M. Urry, "Active galactic nucleus black hole masses and bolometric luminosities," *Astrophysical Journal Letters*, vol. 579, no. 2, pp. 530–544, 2002.

[6] E. Treister, P. Natarajan, D. B. Sanders, C. Megan Urry, K. Schawinski, and J. Kartaltepe, "Major galaxy mergers and the growth of supermassive black holes in quasars," *Science*, vol. 328, no. 5978, pp. 600–602, 2010.

[7] A. J. Davis and P. Natarajan, "Angular momentum and clustering properties of early dark matter haloes," *Monthly Notices of the Royal Astronomical Society*, vol. 393, no. 4, pp. 1498–1502, 2009.

[8] M. Vestergaard, "Early growth and efficient accretion of massive black holes at high redshift," *Astrophysical Journal Letters*, vol. 601, no. 2 I, pp. 676–691, 2004.

[9] M. Volonteri, G. Lodato, and P. Natarajan, "The evolution of massive black hole seeds," *Monthly Notices of the Royal Astronomical Society*, vol. 383, no. 3, pp. 1079–1088, 2008.

[10] M. Volonteri and P. Natarajan, "Journey to the MBH-σ relation: the fate of low-mass black holes in the universe," *Monthly Notices of the Royal Astronomical Society*, vol. 400, no. 4, pp. 1911–1918, 2009.

[11] F. Shankar, D. H. Weinberg, and J. Miralda-Escudé, "Self-consistent models of the AGN and black hole populations: duty cycles, accretion rates, and the mean radiative efficiency," *Astrophysical Journal Letters*, vol. 690, no. 1, pp. 20–41, 2009.

[12] B. C. Kelly, M. Vestergaard, X. Fan, P. Hopkins, L. Hernquist, and A. Siemiginowska, "Constraints on black hole growth, quasar lifetimes, and Eddington ratio distributions from the SDSS broad-line quasar black hole mass function," *Astrophysical Journal Letters*, vol. 719, no. 2, pp. 1315–1334, 2010.

[13] P. Natarajan, "The formation and evolution of massive black hole seeds in the early Universe," in *Fluid Flows to Black Holes: A Tribute to S Chandrasekhar on His Birth Centenary*, D. J. Saikia, Ed., pp. 191–206, World Scientific, 2011.

[14] E. Treister and C. M. Urry, "The cosmic history of black hole growth from deep multiwavelength surveys," *Advances in Astronomy*, vol. 2012, Article ID 516193, 21 pages, 2012.

[15] E. Treister, K. Schawinski, M. Volonteri, P. Natarajan, and E. Gawiser, "Black hole growth in the early Universe is self-regulated and largely hidden from view," *Nature*, vol. 474, no. 7351, pp. 356–358, 2011.

[16] C. J. Willott, L. Albert, D. Arzoumanian et al., "Eddington-limited accretion and the black hole mass function at redshift 6," *Astronomical Journal*, vol. 140, no. 2, pp. 546–560, 2010.

[17] V. Bromm and A. Loeb, "Formation of the first supermassive black holes," *Astrophysical Journal Letters*, vol. 596, no. 1, pp. 34–46, 2003.

[18] B. Devecchi and M. Volonteri, "Formation of the first nuclear clusters and massive black holes at high redshift," *Astrophysical Journal Letters*, vol. 694, no. 1, pp. 302–313, 2009.

[19] A. J. Barger, L. L. Cowie, R. F. Mushotzky et al., "The cosmic evolution of hard X-ray-selected active galactic nuclei," *Astronomical Journal*, vol. 129, no. 2, pp. 578–609, 2005.

[20] S. M. Croom, G. T. Richards, T. Shanks et al., "The 2dF-SDSS LRG and QSO survey: the QSO luminosity function at $0.4 < z < 2.6$," *MNRAS*, vol. 399, no. 4, pp. 1755–1772, 2009.

[21] Y. Ueda, M. Akiyama, K. Ohta, and T. Miyaji, "Cosmological evolution of the hard X-ray active galactic nucleus luminosity function and the origin of the hard X-ray background," *The Astrophysical Journal Letters*, vol. 598, no. 2 I, pp. 886–908, 2003.

[22] D. R. Ballantyne, A. R. Draper, K. K. Madsen, J. R. Rigby, and E. Treister, "Lifting the veil on obscured accretion: active galactic nuclei number counts and survey strategies for imaging hard X-ray missions," *Astrophysical Journal*, vol. 736, article 56, no. 1, 2011.

[23] T. Takahashi, K. Mitsuda, R. Kelley et al., "The ASTRO-H mission," in *Space Telescopes and Instrumentation 2010: Ultraviolet to Gamma Ray, 77320Z*, M. Arnaud, S. S. Murray, and T. Takahashi, Eds., vol. 7732 of *Proceedings of SPIE*, San Diego, Calif, USA, June 2010.

[24] L. Yao, E. R. Seaquist, N. Kuno, and L. Dunne, "CO molecular gas in infrared-luminous galaxies," *Astrophysical Journal Letters*, vol. 588, no. 2 I, pp. 771–791, 2003.

[25] A. Castellina and F. Donato, "Astrophysics of galactic charged cosmic rays," in *Planets, Stars and Stellar Systems*, T. D. Oswalt and G. Gilmore, Eds., vol. 5 of *Galactic Structure and Stellar Populations*, pp. 725–788, 2011.

[26] A. Letessier-Selvon and T. Stanev, "Ultrahigh energy cosmic rays," *Reviews of Modern Physics*, vol. 83, no. 3, pp. 907–942, 2011.

[27] G. Sigl, "High energy neutrinos and cosmic rays," http://arxiv.org/abs/1202.0466.

[28] K. Kotera, D. Allard, and A. V. Olinto, "Cosmogenic neutrinos: parameter space and detectabilty from PeV to ZeV," *Journal of Cosmology and Astroparticle Physics*, vol. 2010, article 013, 2010.

[29] D. Semikoz, "High-energy astroparticle physics," CERN Yellow Report CERN-2010-001, 2010.

[30] J. Linsley, "Primary cosmic-rays of energy 1017 to 1020 ev, the energy spectrum and arrival directions," in *Proceedings of the 8th International Cosmic Ray Conference*, vol. 4, p. 77, 1963.

[31] N. Hayashida, K. Honda, M. Honda et al., "Observation of a very energetic cosmic ray well beyond the predicted 2.7 K cutoff in

the primary energy spectrum," *Physical Review Letters*, vol. 73, no. 26, pp. 3491–3494, 1994.

[32] D. J. Bird, S. C. Corbató, H. Y. Dai et al., "Detection of a cosmic ray with measured energy well beyond the expected spectral cutoff due to cosmic microwave radiation," *Astrophysical Journal Letters*, vol. 441, no. 1, pp. 144–150, 1995.

[33] K. Greisen, "End to the cosmic-ray spectrum?" *Physical Review Letters*, vol. 16, no. 17, pp. 748–750, 1966.

[34] G. T. Zatsepin and V. A. Kuz'min, "Upper limit of the spectrum of cosmic rays," *Journal of Experimental and Theoretical Physics Letters*, vol. 4, pp. 78–80, 1966.

[35] R. Abbasi, Y. Abdou, T. Abu-Zayyad et al., "Time-integrated searches for point-like sources of neutrinos with the 40-string IceCube detector," *The Astrophysical Journal*, vol. 732, no. 1, article 18, 2011.

[36] J. Abraham, "Observation of the suppression of the flux of cosmic rays above 4£1019 eV," *Physical Review Letters*, vol. 101, no. 6, Article ID 061101, 2008.

[37] R. J. Protheroe and T. Stanev, "Limits on models of the ultrahigh energy cosmic rays based on topological defects," *Physical Review Letters*, vol. 77, no. 18, pp. 3708–3711, 1996.

[38] D. Fargion, "Ultrahigh energy neutrino scattering onto relic light neutrinos in galactic halo as a possible source of highest extragalactic cosmic rays," in *Proceedings of the 25th International Cosmic Ray Conference (Held July-August, 1997 in Durban, South Africa)*, M. S. Potgieter, C. Raubenheimer, and D. J. van der Walt, Eds., vol. 7, pp. 153–156, Potchefstroom University, Transvaal, South Africa, 1997.

[39] D. Fargion, B. Mele, and A. Salis, "Ultra-high-energy neutrino scattering onto relic light neutrinos in the galactic halo as a possible source of the highest energy extragalactic cosmic rays," *The Astrophysical Journal*, vol. 517, no. 2, pp. 725–733, 1999.

[40] D. Fargion, B. Mele, and A. Salis, "Ultrahigh energy neutrino scattering onto relic light neutrinos in galactic halo as a possible source of highest energy extragalactic cosmic," *Astrophysical Journal*, vol. 517, pp. 725–733, 1999.

[41] T. J. Weiler, "Cosmic-ray neutrino annihilation on relic neutrinos revisited: a mechanism for generating air showers above the Greisen-Zatsepin-Kuzmin cutoff," *Astroparticle Physics*, vol. 11, no. 3, pp. 303–316, 1999.

[42] T. J. Weiler, "Cosmic-ray neutrino annihilation on relic neutrinos revisited: a mechanism for generating air showers above the Greisen-Zatsepin-Kuzmin cutoff," *Astroparticle Physics*, vol. 11, no. 3, pp. 303–316, 1999.

[43] V. K. Dubrovich, D. Fargion, and M. Y. Khlopov, "Primordial bound systems of superheavy particles as the source of ultra-high energy cosmic rays," *Nuclear Physics B—Proceedings Supplements*, vol. 136, no. 1-3, pp. 362–367, 2004.

[44] A. Datta, D. Fargion, and B. Mele, "Supersymmetry—neutrino unbound," *Journal of High Energy Physics*, 0509:007, 2005.

[45] S. Yoshida, G. Sigl, and S. Lee, "Extremely high energy neutrinos, neutrino hot dark matter, and the highest energy cosmic rays," *Physical Review Letters*, vol. 81, no. 25, pp. 5505–5508, 1998.

[46] A. Ringwald, "Possible detection of relic neutrinos and their mass," in *Proceedings of the 27th International Cosmic Ray Conference*, R. Schlickeiser, Ed., Invited, Rapporteur, and Highlight Papers, p. 262, Hamburg, Germany, August 2001.

[47] Z. Fodor, S. D. Katz, and A. Ringwald, "Relic neutrino masses and the highest energy cosmic-rays," *Journal of High Energy Physics*, 2002.

[48] O. E. Kalashev, V. A. Kuzmin, D. V. Semikoz, and G. Sigl, "Ultrahigh-energy neutrino fluxes and their constraints," *Physical Review D*, vol. 66, no. 6, Article ID 063004, 2002.

[49] O. E. Kalashev, V. A. Kuzmin, D. V. Semikoz, and G. Sigl, "Ultra-high energy cosmic rays from neutrino emitting acceleration sources?" *Physical Review D*, vol. 65, no. 10, Article ID 103003, 2002.

[50] A. Y. Neronov and D. V. Semikoz, "Which blazars are neutrino loud?" *Physical Review D*, vol. 66, no. 12, Article ID 123003, 2002.

[51] P. Jain and S. Panda, "Ultra high energy cosmic rays from early decaying primordial black holes," in *Proceedings of the 29th International Cosmic Ray Conference*, B. Sripathi Acharya, Ed., vol. 9, pp. 33–36, Tata Institute of Fundamental Research, Pune, India, 2005.

[52] T. K. Gaisser, F. Halzen, and T. Stanev, "Particle astrophysics with high energy neutrinos," *Physics Report*, vol. 258, no. 3, pp. 173–236, 1995.

[53] J. Alvarez-Muñiz and F. Halzen, "Possible high-energy neutrinos from the cosmic accelerator RX J1713.7-3946," *Astrophysical Journal Letters*, vol. 576, no. 1, pp. L33–L36, 2002.

[54] P. S. Coppi and F. A. Aharonian, "Constraints on the very high energy emissivity of the universe from the diffuse GeV gamma-ray background," *Astrophysical Journal Letters*, vol. 487, no. 1, pp. L9–L12, 1997.

[55] D. Eichler, "High-energy neutrino astronomy—a probe of galactic nuclei," *The Astrophysical Journal*, vol. 232, pp. 106–112, 1979.

[56] F. Halzen and E. Zas, "Neutrino fluxes from active galaxies: a model-independent estimate," *Astrophysical Journal Letters*, vol. 488, no. 2, pp. 669–674, 1997.

[57] S. Adrian-Martinez, I. Al Samarai, A. Albert et al., "Measurement of atmospheric neutrino oscillations with the ANTARES neutrino tele-scope," *Physics Letters B*, vol. 714, no. 2–5, pp. 224–230, 2012.

[58] S. Adrián-Martínez, J. A. Aguilar, I. Al Samarai et al., "First search for point sources of high energy cosmic neutrinos with the ANTARES neutrino telescope," *The Astrophysical Journal Letters*, vol. 743, no. 1, article L14, 2011.

[59] M. G. Aartsen, R. Abbasi, Y. Abdou et al., "Measurement of atmospheric neutrino oscillations with Ice-Cube," *Physical Review Letters*, vol. 111, no. 8, Article ID 081801, 2013.

[60] M. G. Aartsen, R. Abbasi, Y. Abdou et al., "First observation of PeV-energy neutrinos with IceCube," *Physical Review Letters*, vol. 111, Article ID 021103, 2013.

[61] M. G. Aartsen, "Evidence for high-energy extraterrestrial neutrinos at the IceCube detector," *Science*, vol. 342, no. 6161, Article ID 1242856, 2013.

[62] C. Kopper and IceCube Collabotration, "Observation of PeV neutrinos in IceCube," in *Proceedings of the IceCube Particle Astrophysics Symposium (IPA '13)*, Madison, Wis, USA, May 2013, http://wipac.wisc.edu/meetings/home/IPA2013.

[63] N. Kurahashi-Neilson, "Spatial clustering analysis of the very high energy neutrinos in icecube," in *Proceedings of the IceCube Particle Astrophysics Symposium (IPA '13)*, IceCube Collaboration, Madison, Wis, USA, May 2013.

[64] N. Whitehorn and IceCube Collabotration, "Results from IceCube," in *Proceedings of the IceCube Particle Astrophysics Symposium (IPA '13)*, Madison, Wis, USA, May 2013.

[65] A. V. Avrorin, V. M. Aynutdinov, V. A. Balkanov et al., "Search for high-energy neutrinos in the Baikal neutrino experiment," *Astronomy Letters*, vol. 35, no. 10, pp. 651–662, 2009.

[66] T. Abu-Zayyad, R. Aida, M. Allen et al., "The cosmic-ray energy spectrum observed with the surface detector of the tele-scope array experiment," *The Astrophysical Journal Letters*, vol. 768, no. 1, 5 pages, 2013.

[67] P. Abreu, M. Aglietta, M. Ahlers et al., "Large-scale distribution of arrival directions of cosmic rays detected above 10^{18} eV at the pierre auger observatory," *The Astrophysical Journal Supplement Series*, vol. 203, no. 2, p. 20, 2012.

[68] "A search for point sources of EeV neutrons," *The Astrophysical Journal*, vol. 760, no. 2, Article ID 148, 11 pages, 2012.

[69] T. Ebisuzaki, Y. Takahashi, F. Kajino et al., "The JEM-EUSO mission to explore the extreme Universe," in *Proceedings of the 7th Tours Symposium on Nuclear Physics and Astrophysics*, vol. 1238, pp. 369–376, Kobe, Japan, November 2009.

[70] K. Murase, M. Ahlers, and B. C. Lacki, "Testing the hadronuclear origin of PeV neutrinos observed with IceCube," *Physical Review D*, vol. 88, Article ID 121301(R), 2013.

[71] G. Ter-Kazarian, "Protomatter and EHE C.R," *Journal of the Physical Society of Japan B*, vol. 70, pp. 84–98, 2001.

[72] G. Ter-Kazarian, S. Shidhani, and L. Sargsyan, "Neutrino radiation of the AGN black holes," *Astrophysics and Space Science*, vol. 310, no. 1-2, pp. 93–110, 2007.

[73] G. Ter-Kazarian and L. Sargsyan, "Signature of plausible accreting supermassive black holes in Mrk 261/262 and Mrk 266," *Advances in Astronomy*, vol. 2013, Article ID 710906, 12 pages, 2013.

[74] G. T. Ter-Kazarian, "Gravitation and inertia; a rearrangement of vacuum in gravity," *Astrophysics and Space Science*, vol. 327, no. 1, pp. 91–109, 2010.

[75] G. Ter-Kazarian, "Ultra-high energy neutrino fluxes from supermassive AGN black holes," *Astrophysics & Space Science*, vol. 349, pp. 919–938, 2014.

[76] G. T. Ter-Kazarian, "Gravitation gauge group," *Il Nuovo Cimento*, vol. 112, no. 6, pp. 825–838, 1997.

[77] A. Neronov, D. Semikoz, F. Aharonian, and O. Kalashev, "Large-scale extragalactic jets powered by very-high-energy gamma rays," *Physical Review Letters*, vol. 89, no. 5, pp. 1–4, 2002.

[78] T. Eguchi, P. B. Gilkey, and A. J. Hanson, "Gravitation, gauge theories and differential geometry," *Physics Reports C*, vol. 66, no. 6, pp. 213–393, 1980.

[79] S. Kobayashi and K. Nomizu, *Foundations of Differential Geometry*, Interscience Publishers, New York, NY, USA, 1963.

[80] J. Plebanski, "Forms and riemannian geometry," in *Proceedings of the International School of Cosmology and Gravitation*, Erice, Italy, 1972.

[81] A. Trautman, *Differential Geometry for Physicists*, vol. 2 of *Monographs and Textbooks in Physical Science*, Bibliopolis, Naples, Fla, USA, 1984.

[82] G. Ter-Kazarian, "Two-step spacetime deformation induced dynamical torsion," *Classical and Quantum Gravity*, vol. 28, no. 5, Article ID 055003, 2011.

[83] L. S. Pontryagin, *Continous Groups*, Nauka, Moscow, Russia, 1984.

[84] B. A. Dubrovin, *Contemporary Geometry*, Nauka, Moscow, Russia, 1986.

[85] S. Coleman, J. Wess, and B. Zumino, "Structure of phenomenological lagrangians. I," *Physical Review*, vol. 177, no. 5, pp. 2239–2247, 1969.

[86] C. G. Callan, S. Coleman, J. Wess, and B. Zumino, "Structure of phenomenological lagrangians. II," *Physical Review*, vol. 177, no. 5, pp. 2247–2250, 1969.

[87] S. Weinberg, *Brandeis Lectures*, MIT Press, Cambridge, Mass, USA, 1970.

[88] A. Salam and J. Strathdee, "Nonlinear realizations: II. Conformal symmetry," *Physical Review*, vol. 184, pp. 1760–1768, 1969.

[89] C. J. Isham, A. Salam, and J. Strathdee, "Nonlinear realizations of space-time symmetries. Scalar and tensor gravity," *Annals of Physics*, vol. 62, pp. 98–119, 1971.

[90] D. V. Volkov, "Phenomenological lagrangians," *Soviet Journal of Particles and Nuclei*, vol. 4, pp. 1–17, 1973.

[91] V. I. Ogievetsky, "Nonlinear realizations of internal and space-time symmetries," in *Proceedings of 10th Winter School of Theoretical Physics in Karpacz*, vol. 1, p. 117, Wroclaw, Poland, 1974.

[92] V. de Sabbata and M. Gasperini, "Introduction to gravitation," in *Unified Field Theories of more than Four Dimensions*, V. de Sabbata and E. Schmutzer, Eds., p. 152, World Scientific, Singapore, 1985.

[93] V. de Sabbata and M. Gasperini, "On the Maxwell equations in a Riemann-Cartan space," *Physics Letters A*, vol. 77, no. 5, pp. 300–302, 1980.

[94] V. de Sabbata and C. Sivaram, *Spin and Torsion in Gravitation*, World Scientific, Singapore, 1994.

[95] N. J. Poplawski, "Spacetime and fields," http://arxiv.org/abs/0911.0334.

[96] A. Trautman, "On the structure of the Einstein-Cartan equations," in *Differential Geometry*, vol. 12 of *Symposia Mathematica*, pp. 139–162, Academic Press, London, UK, 1973.

[97] J.-P. Francoise, G. L. Naber, and S. T. Tsou, Eds., *Encyclopedia of Mathematical Physics*, Elsevier, Oxford, UK, 2006.

[98] S. L. Shapiro and S. A. Teukolsky, *Black Holes, White Dwarfs, and Neutron Stars*, A Wiley-Intercience Publication, Wiley-Intercience, New York, NY, USA, 1983.

5

Energetic Electron Enhancements below the Radiation Belt and X-Ray Contamination at Low-Orbiting Satellites

Alla V. Suvorova,[1,2] Alexei V. Dmitriev,[1,2] and Chien-Ming Huang[1]

[1] *Institute of Space Science, National Central University, No 300 Jungda Road, Jhongli, Taoyuan 32001, Taiwan*
[2] *Skobeltsyn Institute of Nuclear Physics, Lomonosov Moscow State University, Moscow 119234, Russia*

Correspondence should be addressed to Alla V. Suvorova; suvorova_alla@yahoo.com

Academic Editor: Athina Meli

The work concerns a problem of electron-induced contaminant at relatively low latitudes to high-energy astrophysical measurements on board the low-orbiting satellites. We show the results of a statistical analysis of the energetic electron enhancements in energy range 30–300 keV observed by a fleet of NOAA/POES low-orbiting satellites over the time period from 1999 to 2012. We demonstrate geographical distributions of great and moderate long-lasting enhancements caused by different type of the solar wind drivers.

1. Introduction

Instrumental measurements in high-energy astrophysics require knowledge of contaminating background radiation of local (magnetospheric) origin [1]. Either by direct penetration or by secondary radiations produced in payload materials, photon detectors may at times give spurious responses, particularly if the "background" radiations are nonsteady [2]. It was recently pointed out that the most important effect limiting the accuracy of the cosmic X-ray background measurements is related to the intrinsic background variation in detectors [3, 4]. This problem was comprehensively discussed at the Workshop on Electron Contamination in X-Ray Astronomy Experiments in 1974 [5]. It was shown that detectors of X- and gamma-ray on sounding rockets, on balloons, and on board the low-orbiting satellites are subject to in-orbit enhanced background noise caused by the magnetospheric electrons, especially in the Earth's auroral zone and zone of trapped radiation (radiation belt), that is, at high latitudes, and also in South Atlantic Anomaly (SAA) at low latitudes (see also [6, 7]). To minimize the contamination, most cosmic and galactic X- and gamma-ray measurements made from within the magnetosphere are conducted at equatorial or low-to-middle magnetic latitudes, where the influence of auroral and radiation belt effects is expected to be small. It is considered that data below the radiation belt outside the region of SAA are particularly valuable to satellite missions in high-energy astrophysics. However, even then during the early 1970s, the X-ray astronomers unexpectedly revealed the electron-induced contaminant at relatively low latitudes as well, which was a few times higher than the cosmic X-ray background [2, 8]. They found several events when flux intensity of electrons with energy of tens of keV was as large as $\sim 10^3$ el/(cm^2 s sr)$^{-1}$, exceeding quiet level by 2 orders of magnitude. Note that it is still much lower than in the radiation belt and auroral zones (see [9]). In addition to astrophysical measurements, ionospheric and atmospheric studies [10–17] and satellite data failures studies (e.g., [18, 19]) also found several effects suggesting that electron impact is important factor at low and middle latitudes. That is, more importantly, the occasional electron flux increases below the radiation belt were discovered even earlier in direct satellite-borne measurements [20, 21] and then corroborated in several studies [22–25]. They reported about sporadic fluxes of very high intensity which was comparable with the auroral precipitation. However, the direct observations of sporadic events caused strong argument due to a doubt about validity of measured high intensity (see review by Paulikas [26]).

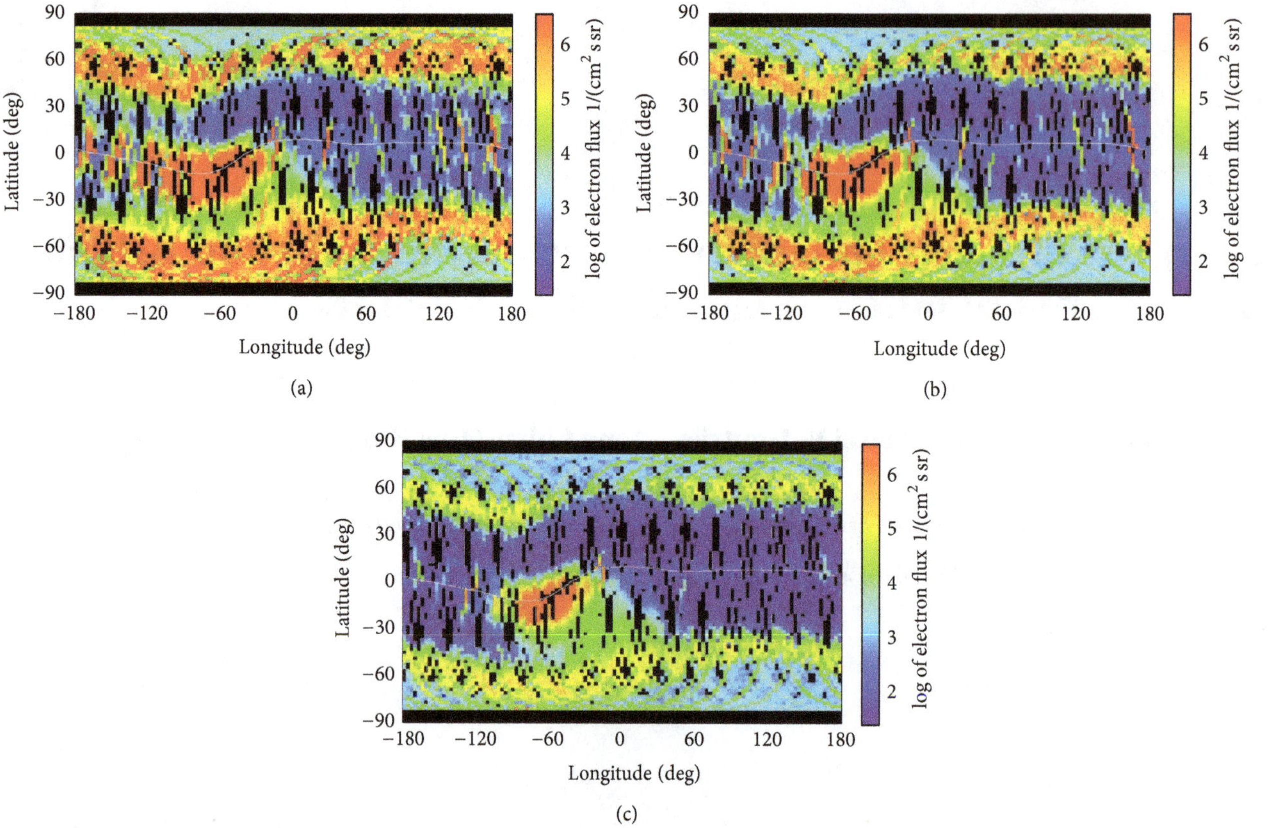

FIGURE 1: Energetic electron enhancement during the major geomagnetic storm on 15-16 May 2005. Global distribution of the electron fluxes in energy range: (a) >30 keV; (b) >100 keV; (c) >300 keV measured by the NOAA/POES-15, -16, -17 satellites at altitude of ~850 km. The maps are composed of data retrieved from two orthogonally directed detectors (see details in the text). The white curve indicates the geomagnetic equator. Intensity of energetic electron fluxes extremely and globally enhances at equator-to-low latitudes (IRB and below it) even exceeding one at high latitudes (ORB and auroral zone).

As a result of this, despite importance of low-latitude measurement of electron fluxes recognized earlier [5], further investigation of the enhanced electrons phenomenon was not carried out.

Until recently, sporadic enhancement of energetic electrons below the inner radiation belt (IRB) was a poor-studied phenomenon [15, 27]. Comprehensive studies based on large statistics collected for more than ten years [28–31] have showed that fluxes of quasitrapped electrons within the energy range 10–300 keV can increase dramatically by a few orders of magnitude relative to the quiet level at very low L shells ($L < 1.1$), in a region called a forbidden zone. The most extreme intensity of forbidden zone fluxes of the order of auroral precipitation, $\sim 10^6$–10^7 (cm^2 s sr)$^{-1}$, was observed during some major storms driven by a coronal mass ejection (CME). Nevertheless, CME-type or major storms themselves are not a necessary condition for electron enhancements in the forbidden zone. Another important solar wind driver resulting in significant flux enhancements is the extremely strong solar wind dynamic pressure, as it occurred on 21 January 2005 [32]. It can be easily understood that large enhancements occur much less frequently than moderate ones. The moderate fluxes are smaller by one-two orders of magnitudes. They are mostly associated with low-to-moderate level of geomagnetic activity and minor storms (major storms can also contribute, though). Minor storms, as known, are mainly driven by corotating interaction regions (CIR) and high speed solar wind streams (HSS) [33].

This paper describes the results of a statistical analysis of the energetic electron enhancements observed by a fleet of NOAA/POES low-orbiting satellites over the time period from 1999 to 2012. We demonstrate geographical distributions of great and moderate enhancements caused by different type of the solar wind drivers.

2. Data from NOAA Satellites

We used time profiles of 30–300 keV electrons fluxes measured on board the polar orbiting NOAA/POES satellites [34]. The POES satellites have Sun-synchronous orbits at altitudes of ~800–850 km (with ~100 minute periods of revolution). It is well known that the electron measurements can be distorted by proton contamination and nonideal detector efficiency. According to a comprehensive study [35] the 30 keV electron fluxes should be, on an average, more than two times larger than uncorrected fluxes. The 100 keV

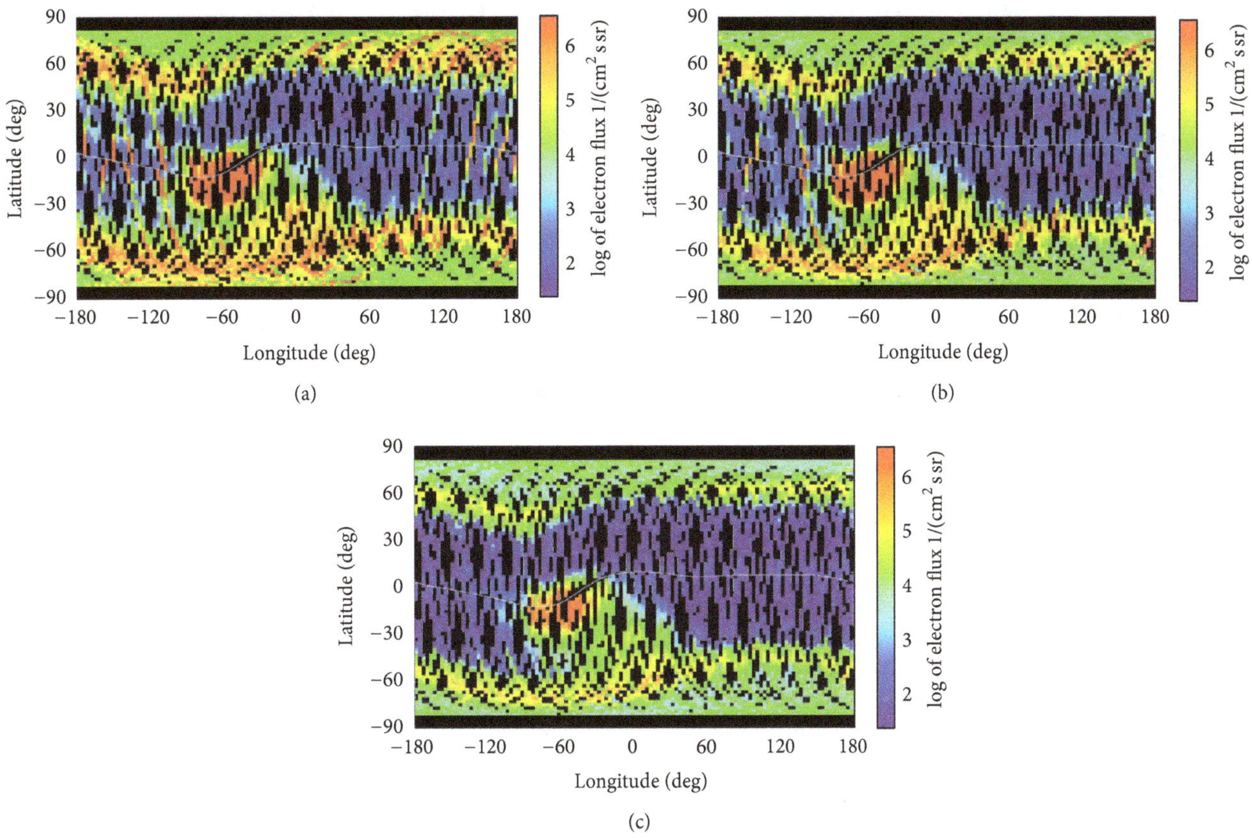

FIGURE 2: The same as Figure 1 but for extremely strong magnetospheric compression and moderate geomagnetic storm on 21 January 2005.

electron fluxes practically do not change, while the 300 keV electron fluxes should be decreased by about twenty percent. Because this factor is not crucial for the current study, we present uncorrected fluxes.

3. Enhancements of the Quasitrapped Energetic Electrons

Figure 1 presents geographical distributions of energetic electron fluxes in three energy ranges: >30 keV, >100 keV, and >300 keV. The data was compiled from measurements by two orthogonally oriented detectors (0°-detector and 90°-detector) of three POES satellites (NOAA-15, NOAA-16, NOAA-17) during the major (CME-driven) geomagnetic storm on 15-16 May 2005. In each spatial bin, the maximal value of flux instead of the averaged one was used. The flux intensities below IRB in all three energy bands exceeded the quiet level by ~5-6 orders of magnitude. The lower energy electrons outside the SAA region achieved an extremely large value of $3 \cdot 10^7$ (cm^2 s sr)$^{-1}$, as much as in the IRB (including SAA) and auroral zone with outer radiation belt. The enhancements in >30 keV and >100 keV were long-lasting. The most prolonged equatorial enhancement in >30 keV occupied the forbidden zone at L shell of 1.05–1.15 for more than 20 hours.

Figure 2 presents geographical distributions of energetic electron fluxes during the prolonged compression of the magnetosphere by extremely high solar wind dynamic pressure of more than 150 nPa [32]. The magnetic storm was of moderate strength. Due to the compression, the Earth's magnetopause shrunk to about ~3-4 Re in the subsolar region, radiation belt, and ring current moved closely to the Earth. The enhancement of >30 keV electrons was observed during 6 hours. Fluxes of electrons with higher energies were also increased, but due to fast azimuthal drift they disappeared in one or two hours.

Figure 3 demonstrates that >30 keV electrons can appear even during weak geomagnetic storms. The global map was compiled from measurements by five NOAA/POES satellites for one year 2008. During this year of solar activity minimum, there were only minor (CIR/HSS-driven) storms of intensities less than 50 nT. In course of the year there were 60 days when the electron fluxes below IRB increased to ~10^4 (cm^2 s sr)$^{-1}$. However, even moderate electron enhancements can significantly contaminate to the X-ray background.

4. Summary

In this paper we are concerned with a very important problem of the electron contamination to high-energy astrophysical measurements. The study is based on long-term statistics of

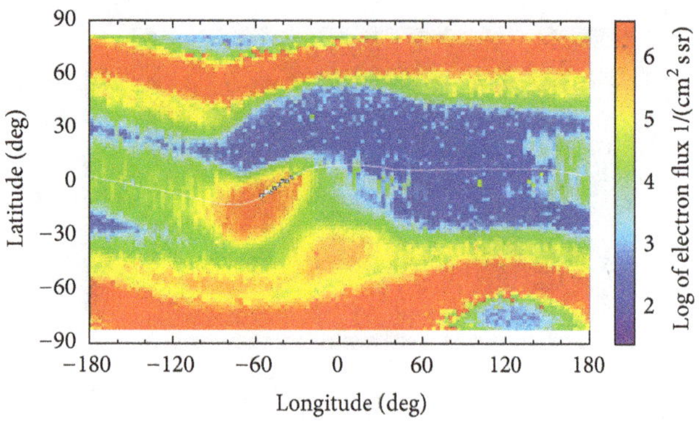

FIGURE 3: Global distribution of the electron fluxes in energy range >30 keV measured by the NOAA/POES-15, -16, -17, -18, and METOP-02 satellites during the whole year 2008 of the solar activity minimum. Low-latitude electron enhancements were observed during only 60 days in the course of the year.

the energetic electron observations by low-orbiting satellites. We have demonstrated three cases of electron fluxes that significantly exceeded a quite level: a major geomagnetic storm, a strong compression of the magnetosphere, and one-year period of the solar activity minimum leading to a weak geomagnetic activity.

The phenomenon of "forbidden zone electron" relates to the magnetospheric electric fields driven by external parameters, the solar wind, and interplanetary electric field [31]. A notable feature of the "forbidden-zone" 30 keV electrons is their long persistence for about several hours. It is important that the significant and longtime electron enhancements at equatorial latitudes occur quite often during moderate CIR/HSS-storms.

Conflict of Interests

The authors declare that there is no conflict of interests regarding the publication of this paper.

Acknowledgments

The authors thank a team of NOAA's Polar Orbiting Environmental Satellites for providing experimental data about energetic particles. The work of Alla V. Suvorova was supported by Grant NSC-102-2811-M-008-045 from the National Science Council of Taiwan. Alla V. Suvorova and Alexei V. Dmitriev gratefully acknowledge the support of part of this work by Grant NSC103-2923-M-006-002-MY3/14-05-92002HHC_a from Taiwan-Russia Research Cooperation.

References

[1] A. J. Dean, F. Lei, and P. J. Knight, "Background in space-borne low-energy γ-ray telescopes," *Space Science Reviews*, vol. 57, no. 1-2, pp. 109–186, 1991.

[2] F. D. Seward, R. J. Grader, A. Toor, G. A. Burginyon, and R. W. Hill, "Electrons at low altitudes: a difficult background problem for soft X-ray astronomy," in *Proceedings of the Workshop on Electron Contamination in X-Ray Astronomy Experiments*, S. S. Holt, Ed., pp. 661–74, 1974, NASA GSFC Rep. X-661-74-130.

[3] D. E. Gruber, J. L. Matteson, L. E. Peterson, and G. V. Jung, "The spectrum of diffuse cosmic hard X-rays measured with HEAO 1," *Astrophysical Journal Letters*, vol. 520, no. 1, pp. 124–129, 1999.

[4] G. S. Bisnovatyi-Kogan and A. S. Pozanenko, "About the measurements of the hard X-ray background," *Astrophysics and Space Science*, vol. 332, no. 1, pp. 57–63, 2011.

[5] S. S. Holt, "Proceedings of the workshop on electron contamination in X-ray astronomy experiments , Washington, 26 April 1974," NASA GSFC Report X-661-74-130, 1974.

[6] R. Bučík, K. Kudela, A. V. Dmitriev, S. N. Kuznetsov, I. N. Myagkova, and S. P. Ryumin, "Spatial distribution of low energy gamma-rays associated with trapped particles," *Advances in Space Research*, vol. 30, no. 12, pp. 2843–2848, 2002.

[7] A. Gusev, I. Martin, and G. Pugacheva, "The soft X-ray emission of nocturnal atmosphere during the descending phase of the 23rd solar cycle," *Sun and Geosphere*, vol. 7, no. 2, pp. 127–131, 2012.

[8] J. E. Neighbours and G. W. Clark, "A survey of trapped low energy electrons near the inner boundary of the inner radiation zone from the OSO-7," in *Proceedings of the Workshop on Electron Contamination in X-Ray Astronomy Experiments*, S. S. Holt, Ed., 1974, NASA GSFC Rep. X-661-74-130.

[9] R. Bučík, K. Kudela, A. V. Dmitriev, S. N. Kuznetsov, I. N. Myagkova, and S. P. Ryumin, "Review of electron fluxes within the local drift loss cone: measurements on CORONAS-I," *Advances in Space Research*, vol. 36, no. 10, pp. 1979–1983, 2005.

[10] L. A. Antonova and G. S. Ivanov-Kholodny, "Corpuscular hypothesis for the ionization of the night ionosphere," *Geomagnetizm I Aeronomiya*, vol. 1, no. 2, pp. 164–173, 1961.

[11] L. A. Antonova and T. V. Kazarchevskaya, "Measurements of soft electron streams in the upper atmosphere at altitude to 500 km," in *Space Research, Transactions of the All-Union Conference on Space Physics*, G. A. Skuridin, Y. L. Al'pert, V. I. Krasovskiy, and V. V. Shvarev, Eds., Science Publishing House, Moscow, Russia, June 1965.

[12] W. C. Knudsen and G. W. Sharp, "F2-region electron concentration enhancement from inner radiation belt particles," *Journal of Geophysical Research*, vol. 73, no. 19, pp. 6275–6283, 1968.

[13] R. H. Doherty, "Observations suggesting particle precipitation at latitudes below 40N," *Radio Science*, vol. 6, no. 6, pp. 639–646, 1971.

[14] T. A. Potemra and T. J. Rosenberg, "VLF propagation disturbances and electron precipitation at mid-latitudes," *Journal of Geophysical Research*, vol. 78, no. 10, pp. 1572–1580, 1973.

[15] A. V. Dmitriev and H.-C. Yeh, "Storm-time ionization enhancements at the topside low-latitude ionosphere," *Annales Geophysicae*, vol. 26, no. 4, pp. 867–876, 2008.

[16] A. V. Dmitriev, H.-C. Yeh, M. I. Panasyuk et al., "Latitudinal profile of UV nightglow and electron precipitations," *Planetary and Space Science*, vol. 59, no. 8, pp. 733–740, 2011.

[17] P. Bobik, M. Putis, M. Bertaina et al., "UV night background estimation in South Atlantic anomaly," in *Proceedings of the 33rd International Cosmic Ray Conference (ICRC '13)*, Rio de Janeiro, Brazil, July 2013.

[18] A. V. Dmitriev, I. I. Guilfanov, and M. I. Panasyuk, "Data failures in the "Riabina-2" experiment on MIR orbital station," *Radiation Measurements*, vol. 35, no. 5, pp. 499–504, 2002.

[19] J. F. Fennell, J. L. Roeder, and H. C. Koons, "Substorms and magnetic storms from the satellite charging perspective," *COSPAR Colloquia Series*, vol. 12, pp. 163–173, 2002.

[20] V. I. Krasovskii, Y. M. Kushner, G. A. Bordovskii, G. F. Zakharov, and E. M. Svetlitskii, "The observation of corpuscles by means of the third artificial earth satellite," *Planetary and Space Science*, vol. 5, no. 3, pp. 248–249, 1961.

[21] I. A. Savenko, P. I. Shavrin, and N. F. Pisarenko, "Soft particle radiation at an altitude of 320 km in the latitudes near the equator," *Iskusstvennye Sputniki Zemli*, vol. 11, pp. 75–80, 1962 (Russian), Planetary and Space Science, vol. 11, pp. 431–436, 1963.

[22] W. J. Heikkila, "Soft particle fluxes near the equator," *Journal of Geophysical Research*, vol. 76, no. 4, pp. 1076–1078, 1971.

[23] J. D. Winningham, ""Low energy (10 eV to 10 keV) equatorial particle fluxes", Proceedings of the Workshop on Electron Contamination in X-Ray Astronomy Experiments, edited by S.S. Holt," NASA GSFC Report X-661-74-130, 1974.

[24] R. A. Goldberg, "Rocket observation of soft energetic particles at the magnetic equator," *Journal of Geophysical Research*, vol. 79, pp. 5299–5303, 1974.

[25] R. Lieu, J. Watermann, K. Wilhelm, J. J. Quenby, and W. I. Axford, "Observations of low-latitude electron precipitation," *Journal of Geophysical Research*, vol. 93, pp. 4131–4133, 1988.

[26] G. A. Paulikas, "Precipitation of particles at low and middle latitudes," *Reviews of Geophysics and Space Physics*, vol. 13, no. 5, pp. 709–734, 1975.

[27] A. V. Suvorova, L.-C. Tsai, and A. V. Dmitriev, "On relation between mid-latitude ionospheric ionization and quasi-trapped energetic electrons during 15 December 2006 magnetic storm," *Planetary and Space Science*, vol. 60, no. 1, pp. 363–369, 2012.

[28] A. V. Suvorova, L.-C. Tsai, and A. V. Dmitriev, "On magnetospheric source for positive ionospheric storms," *Sun and Geosphere*, vol. 7, no. 2, pp. 91–96, 2012.

[29] A. V. Suvorova, L.-C. Tsai, and A. V. Dmitriev, "TEC enhancement due to energetic electrons above Taiwan and the West Pacific," *Terrestrial, Atmospheric and Oceanic Sciences*, vol. 24, no. 2, pp. 213–224, 2013.

[30] A. V. Suvorova, A. V. Dmitriev, and L.-C. Tsai, "Evidence for near-equatorial deposition by energetic electrons in the ionospheric F-layer," in *Proceedings of the International Conference "Modern Engineering and Technologies of the Future"*, A. Khnykin, Ed., pp. 68–82, Krasnoyarsk, Russia, February 2013.

[31] A. V. Suvorova, A. V. Dmitriev, L.-C. Tsai et al., "TEC evidence for near-equatorial energy deposition by 30 keV electrons in the topside ionosphere," *Journal of Geophysical Research A: Space Physics*, vol. 118, no. 7, pp. 4672–4695, 2013.

[32] A. V. Dmitriev, A. V. Suvorova, J.-K. Chao et al., "Anomalous dynamics of the extremely compressed magnetosphere during 21 January 2005 magnetic storm," *Journal of Geophysical Research*, vol. 119, pp. 877–896, 2005.

[33] B. T. Tsurutani, W. D. Gonzalez, A. L. C. Gonzalez et al., "Corotating solar wind streams and recurrent geomagnetic activity: a review," *Journal of Geophysical Research B: Solid Earth*, vol. 111, no. 7, Article ID A07S01, 2006.

[34] D. S. Evans and M. S. Greer, "Polar orbiting environmental satellite space environment monitor: 2. Instrument descriptions and archive data documentation," Technical Memo version 1.4, NOAA Space Environ. Lab, Boulder, Colo, USA, 2004.

[35] T. Asikainen and K. Mursula, "Correcting the NOAA/MEPED energetic electron fluxes for detector efficiency and proton contamination," *Journal of Geophysical Research A: Space Physics*, vol. 118, no. 10, pp. 6500–6510, 2013.

6

Planar and Nonplanar Solitary Waves in a Four-Component Relativistic Degenerate Dense Plasma

M. R. Hossen, L. Nahar, and A. A. Mamun

Department of Physics, Jahangirnagar University, Savar, Dhaka 1342, Bangladesh

Correspondence should be addressed to M. R. Hossen; rasel.plasma@gmail.com

Academic Editor: Milan S. Dimitrijevic

The nonlinear propagation of electrostatic perturbation modes in an unmagnetized, collisionless, relativistic, degenerate plasma (containing both nonrelativistic and ultrarelativistic degenerate electrons, nonrelativistic degenerate ions, and arbitrarily charged static heavy ions) has been investigated theoretically. The Korteweg-de Vries (K-dV) equation has been derived by employing the reductive perturbation method. Their solitary wave solution is obtained and numerically analyzed in case of both planar and nonplanar (cylindrical and spherical) geometry. It has been observed that the ion-acoustic (IA) and modified ion-acoustic (mIA) solitary waves have been significantly changed due to the effects of degenerate plasma pressure and number densities of the arbitrarily charged heavy ions. It has been also found that properties of planar K-dV solitons are quite different from those of nonplanar K-dV solitons. There are numerous variations in case of mIA solitary waves due to the polarity of heavy ions. The basic features and the underlying physics of IA and mIA solitary waves, which are relevant to some astrophysical compact objects, are briefly discussed.

1. Introduction

A large fraction of matter in the universe is in the plasma state. Significant attention has been devoted to the study of ion-acoustic (IA) waves in plasmas not only from an academic point of view, but also from the view of its vital role in understanding the nonlinear features of localized electrostatic disturbances in laboratory and space environments [1–6]. The basic features of solitary waves associated with IA waves, in which electron thermal pressure gives rise to a restoring force and ion mass provides the inertia, were first theoretically predicted by Washimi and Tanuiti by assuming an ideal plasma containing cold ions and isothermal electrons. These basic features [7] were verified by a novel laboratory experiment of Ikezi et al. [6].

Presently, relativistic degeneracy of plasmas has received great attention because of its vital role in different astrophysical environments [8, 9], where particle velocities become comparable to the speed of light. Astrophysical compact objects such as white dwarfs, neutron stars, quasars, black holes, and pulsars are examples where relativistic degenerate plasmas are dominant and interesting new phenomena are investigated by several nonlinear effects in such plasmas. The basic constituents of white dwarfs are mainly oxygen, carbon, and helium with an envelope of hydrogen gas. In some relatively massive white dwarfs, one can think of the presence of heavier element like iron within the stars. The existence of heavy elements is found to form in a prestellar stage of the evolution of the universe, when all matter was compressed to extremely high densities and possessed correspondingly high temperatures [10]. In case of such a compact object the degenerate electron number density is so high (in white dwarfs it can be of the order of 10^{30} cm^{-3}, even more [8]).

Chandrasekhar [11, 12] presented a general expression for the relativistic ion and electron pressures in his classical papers. The pressure for ion fluid can be given by the following equation:

$$P_i = K_i n_i^{\alpha}, \tag{1}$$

where

$$\alpha = \frac{5}{3}; \qquad K_i = \frac{3}{5}\left(\frac{\pi}{3}\right)^{1/3}\frac{\pi\hbar^2}{m} \simeq \frac{3}{5}\Lambda_c \hbar c, \tag{2}$$

for the nonrelativistic limit (where $\Lambda_c = \pi\hbar/mc = 1.2 \times 10^{-10}$ cm and $\hbar$ is the Planck constant divided by 2π), while for the electron fluid,

$$P_e = K_e n_e^{\gamma}, \tag{3}$$

where

$$\gamma = \alpha; \qquad K_e = K_i \quad \text{for nonrelativistic limit,}$$
$$\gamma = \frac{4}{3}; \qquad K_e = \frac{3}{4}\left(\frac{\pi^2}{9}\right)^{1/3} \hbar c \simeq \frac{3}{4}\hbar c, \tag{4}$$

in the ultrarelativistic limit [11–14].

We note that we have considered both electron and ion degeneracy, but light electrons are assumed to be relativistic (nonrelativistic and ultrarelativistic) and ions (which are heavier than electrons) are assumed to be nonrelativistic. Equations (1) and (3) represent the equations of state of ion and electron pressure in high density regime.

Recently, several authors have used the pressure laws to observe the linear and nonlinear properties of electrostatic [15–19] and electromagnetic [20, 21] waves by using only nonrelativistic or both nonrelativistic and ultrarelativistic quantum hydrodynamics (QHD) [22] and quantum-magnetohydrodynamics (QMHD) models [23, 24]. The solution of the Dirac equation for electrostatic and electromagnetic waves in a relativistic quantum plasmas has been discussed by Mendonca and Serbeto [25]. Masood and Eliasson considered an unmagnetized quantum plasma with relativistically degenerate electrons and cold fluid ions and they studied the basic features of the solitary structures [26]. Later on, Shukla et al. [27] theoretically investigated the nonlinear propagation of electrostatic waves in degenerate quantum plasma. They considered strongly coupled nondegenerate ions and degenerate electron fluids in an unmagnetized dense plasma and studied the basic properties of solitary and shock structures. Zeba et al. [28] considered a warm collisionless electron-positron-ion plasma with ultrarelativistic degenerate electrons and positrons and investigated theoretically the existence regions for ion solitary pulses. Since the dense astrophysical quantum plasmas can be confined by stationary heavy ions, therefore the effect of the heavy ions has to be taken into account, especially for astrophysical observations (such as white dwarfs, neutron stars, and black holes) where the degenerate plasma pressure and heavy ions play an important role in the formation and stability of the existing waves. Recently, Zobaer et al. [9, 15, 16, 29] also considered nonrelativistic ion fluids and both nonrelativistic and ultrarelativistic electron fluids and theoretically investigated the basic features of solitary, shock, and double layer structures. All of these investigations are mainly based on planar geometry and they did not consider the effect of heavy ions, which may not be a realistic situation in space environments. Since the waves observed in astrophysical compact objects are certainly not infinite in one dimension (1D) and there is a great possibility of having both positively and negatively charged heavy ions, there are several cases of practical importance where planar geometry does not work and one would have to consider a nonplanar geometry. The notable examples where nonplanar geometry plays a vital role are white dwarfs, neutron stars, black holes, circumstellar disks, dark molecular clouds, cometary tails, and so forth. Till now, no theoretical investigation has been made to study the extreme conditions of matter for both nonrelativistic and ultrarelativistic limits and arbitrarily (both positively and negatively) charged heavy ions on a planar and nonplanar (cylindrical and spherical) geometry. Therefore, in our present work, we attempt to study the basic features of planar and nonplanar IA and mIA solitary waves by deriving the Korteweg-de Vries equation in a dense plasma containing degenerate electron and ion fluids and arbitrarily charged static heavy ions.

The paper is organized as follows. The governing equations are provided in Section 2. The K-dV equation is derived in Section 3. The profiles of K-dV solitons are presented in Section 4 and numerically analyzed in order. Finally, a brief discussion is given in Section 5.

2. Governing Equations

We consider a nonplanar (cylindrical and spherical) geometry and nonlinear propagation of mIA waves (consisting of arbitrarily charged static heavy ions, nonrelativistic degenerate cold ions, and both nonrelativistic and ultrarelativistic degenerate electron fluids) in an unmagnetized, collisionless dense plasma. Hence, at equilibrium, we have $Z_i n_{i0} + jZ_h n_{h0} = n_{e0}$, where n_{s0} is the particle number density of the species s with $s = e(i)h$ for electrons (ions) heavy ions, Z_h is the number of ions residing onto the heavy ion surface, and e is the magnitude of charge of an electron and $j = +1$ ($j = -1$) for positively (negatively) charged heavy ions. The nonlinear dynamics of the electrostatic waves propagating in such a degenerate dense plasma system is governed by the following normalized equations:

$$\begin{aligned}
&\frac{\partial n_i}{\partial t} + \frac{1}{r^{\nu}}\frac{\partial}{\partial r}\left(r^{\nu} n_i u_i\right) = 0,\\
&\frac{\partial u_i}{\partial t} + u_i\frac{\partial u_i}{\partial r} + \frac{\partial \phi}{\partial r} + \frac{K_1}{n_i}\frac{\partial n_i^{\alpha}}{\partial r} = 0,\\
&n_e\frac{\partial \phi}{\partial r} - K_2\frac{\partial n_e^{\gamma}}{\partial r} = 0,\\
&\frac{1}{r^{\nu}}\frac{\partial}{\partial r}\left(r^{\nu}\frac{\partial \phi}{\partial r}\right) = -\rho,\\
&\rho = n_i - (1 + j\mu)\, n_e + j\mu,
\end{aligned} \tag{5}$$

where $\nu = 0$ for one-dimensional planar geometry and $\nu = 1(2)$ for nonplanar cylindrical (spherical) geometry; n_i (n_e) is the ion (electron) number densities normalized by its equilibrium value n_{i0} (n_{e0}), u_i is the ion fluid speed normalized by $C_i = (m_e c^2/m_i)^{1/2}$ with m_e (m_i) being the electron (ion) rest mass, c is the speed of light in vacuum, and ϕ is the electrostatic wave potential normalized by $m_e c^2/e$ with e being the magnitude of the charge of an electron. Here $\mu(= Z_h n_{h0}/Z_p n_{i0})$ is the ratio of the number density of heavy ions and positive ions multiplied by their charge per ion (Z_h/Z_p).

When $\mu = 0$ the situation acts as normal electron-ion plasma system. The polarity of the heavy ions depends on the parameter j. When $j = +1(-1)$ the heavy ions act as positively (negatively) charged in this plasma system. The time variable (t) is normalized by $\omega_{pi} = (4\pi n_0 e^2/m_i)^{1/2}$, and the space variable (r) is normalized by $\lambda_s = (m_e c^2/4\pi n_0 e^2)^{1/2}$. The relativistic constants are $K_1 = n_{i_0}^{\alpha-1} K_i/m_e c^2$, $K_2 = n_{e_0}^{\gamma-1} K_e/m_e c^2$, $K_1' = \alpha K_1$, and $K_2' = \gamma K_2$.

3. Derivation of K-dV Equation

Now we derive the K-dV equation by employing the reductive perturbation technique to examine the electrostatic perturbations propagating in this dense plasma system and introduce the stretched coordinates [30] as follows:

$$\xi = -\epsilon^{1/2}\left(r + V_p t\right), \qquad \tau = \epsilon^{3/2} t, \tag{6}$$

where V_p is the wave phase speed (ω/k with ω being the angular frequency and k being the wave number of the perturbation mode) and ϵ is a smallness parameter measuring the weakness of the dispersion ($0 < \epsilon < 1$). We then expand the parameters n_i, n_e, u_i, ϕ, and ρ in power series of ϵ:

$$\begin{aligned} n_i &= 1 + \epsilon n_i^{(1)} + \epsilon^2 n_i^{(2)} + \cdots, \\ n_e &= 1 + \epsilon n_e^{(1)} + \epsilon^2 n_e^{(2)} + \cdots, \\ u_i &= \epsilon u_i^{(1)} + \epsilon^2 u_i^{(2)} + \cdots, \\ \phi &= \epsilon \phi^{(1)} + \epsilon^2 \phi^{(2)} + \cdots, \\ \rho &= \epsilon \rho^{(1)} + \epsilon^2 \rho^{(2)} + \cdots. \end{aligned} \tag{7}$$

Now, expressing (5) (using (6)) in terms of ξ and τ and substituting (7) into (5), one can easily develop different sets of equations in various powers of ϵ. To the lowest order in ϵ, we have $u_i^{(1)} = -V_p \phi^{(1)}/(V_p^2 - K_1')$, $n_i^{(1)} = \phi^{(1)}/(V_p^2 - K_1')$, $n_e^{(1)} = \phi^{(1)}/K_2'$, and $V_p = \sqrt{(K_2'/(1 + j\mu)) + K_1'}$, where $K_1' = \alpha K_1$ and $K_2' = \gamma K_2$. The relation $V_p = \sqrt{(K_2'/(1 + j\mu)) + K_1'}$ represents the dispersion relation as well as the phase velocity for the mIA-type electrostatic waves in this degenerate plasma under consideration.

To the next higher order in ϵ, we obtain a set of equations:

$$\begin{aligned} &\frac{\partial n_i^{(1)}}{\partial \tau} - V_p \frac{\partial n_i^{(2)}}{\partial \xi} - \frac{\partial}{\partial \xi}\left[u_i^{(2)} + n_i^{(1)} u_i^{(1)}\right] - \frac{\nu u_i^{(1)}}{V_p \tau} = 0, \\ &\frac{\partial u_i^{(1)}}{\partial \tau} - V_p \frac{\partial u_i^{(2)}}{\partial \xi} - u_i^{(1)} \frac{\partial u_i^{(1)}}{\partial \xi} - \frac{\partial \phi^{(2)}}{\partial \xi} \\ &\quad - K_1' \frac{\partial}{\partial \xi}\left[n_i^{(2)} + \frac{(\alpha - 2)}{2}\left(n_i^{(1)}\right)^2\right] = 0, \\ &\frac{\partial \phi^{(2)}}{\partial \xi} - K_2' \frac{\partial}{\partial \xi}\left[n_e^{(2)} + \frac{(\gamma - 2)}{2}\left(n_e^{(1)}\right)^2\right] = 0, \\ &\frac{\partial^2 \phi^{(1)}}{\partial \xi^2} = -\rho^{(1)}, \\ &\rho = n_i - (1 + j\mu) n_e + j\mu. \end{aligned} \tag{8}$$

Now, combining (8) we deduce a Korteweg-de Vries (K-dV) equation:

$$\frac{\partial \phi^{(1)}}{\partial \tau} + A\phi^{(1)} \frac{\partial \phi^{(1)}}{\partial \xi} + B \frac{\partial^3 \phi^{(1)}}{\partial \xi^3} + \frac{\nu \phi^{(1)}}{2\tau} = 0, \tag{9}$$

where

$$\begin{aligned} A &= \frac{\left(V_p^2 - K_1'\right)^2}{2V_p}\left[\frac{3V_p^2 + K_1'(\alpha - 2)}{\left(V_p^2 - K_1'\right)^3} + \frac{(\gamma - 2)(1 + j\mu)}{K_2'^2}\right], \\ B &= \frac{\left(V_p^2 - K_1'\right)^2}{2V_p}. \end{aligned} \tag{10}$$

4. Numerical Analysis and Results

Generally, IA waves are low frequency electrostatic waves in which inertia is provided by ion mass and restoring force comes from electron thermal pressure. The modified ion-acoustic (mIA) waves are nothing but IA-type waves in the presence of static heavy ions. It is important to note that if we neglect the heavy ions then we get the usual IA waves. In order to analyze the IA and mIA solitary waves, we turn to (9) with the term $\nu\phi^{(1)}/2\tau$, which is due to the effects of the nonplanar (cylindrical and spherical) geometry. An exact analytic solution of (9) is not possible. However, for clear understanding, we first briefly discuss the stationary solitary wave solution for (9) with $\nu = 0$, though the solution is similar for both IA and mIA waves. We should note that for large value of τ, the term $\nu\phi^{(1)}/2\tau$ is negligible. So, in our numerical analysis, we start with a large value of τ (namely, $\tau = -14$), and at this large (negative) value of τ, we choose the stationary solitary wave solution of (9) (without the term $\nu\phi^{(1)}/2\tau$) as our initial pulse. For a moving frame moving with a speed u_0, the stationary solitary wave solution of (9) is given by

$$\phi^{(1)}_{(\nu=0)} = \phi_m \operatorname{sech}^2\left(\frac{\xi}{\Delta}\right), \tag{11}$$

where the maximum amplitude $\phi_m = 3u_0/A$, and the width

$$\Delta = \left(\frac{4B}{u_0}\right)^{1/2}. \tag{12}$$

The profiles of solitary waves caused by the balance between nonlinearity and dispersion are shown in Figures 3 to 14, which show how the effects of cylindrical ($\nu = 1$) and spherical ($\nu = 2$) geometry modify the time-dependent IA and mIA solitary structures and how it differs from planar ($\nu = 0$) ones.

The conditions for the existence of cylindrical and spherical solitary structures and their basic features are found to be significantly modified in the presence of nonrelativistic ions, both nonrelativistic and ultrarelativistic electrons and arbitrarily charged static heavy ions. The IA and mIA waves are modified when ion is nonrelativistic degenerate ($\alpha = 5/3$) and electron is ultrarelativistic degenerate ($\gamma = 4/3$) than both electron and ion being nonrelativistic degenerate ($\alpha = \gamma = 5/3$). The ranges ($u_0 = 0.01$–1 and $\mu = 0.2$–0.6) [13–15] of plasma parameters used in this numerical analysis are very wide and correspond to space and laboratory plasma situations.

The nonlinearity has an important role as the solitary waves are caused by the balance between nonlinearity and dispersion. This is a unique phenomenon that the degree of nonlinearity is proportional to the potential of the plasma system as a highly nonlinear medium causes high electrostatic potential. We have found that the phase speed with negatively (positively) charged heavy ions for ultrarelativistic case is higher (lower) than that for nonrelativistic case (it is expected from the expression of V_p). We have also found that as time decreases the amplitude of the solitary waves in cylindrical and spherical geometry increases. It is also examined that, in spherical geometries, the amplitude is always distinctly higher than cylindrical geometries for K-dV solitons, which indicates that the density compression can be more effectively obtained in a spherical geometry. It is observed that the amplitude of the K-dV solitons is always distinctly higher for nonrelativistic case than for ultrarelativistic one. It is due to the effects of relativistic factors; that is, $\alpha > \gamma$ for ultrarelativistic case and $\alpha = \gamma$ for nonrelativistic case. We also found that the amplitude of the K-dV solitons is lower for positively charged heavy ions than for negatively charged heavy ions. Actually, this happens due to the reason that it decreases the value of the nonlinearity coefficient A. It is obvious from (12) that the height of the amplitude of the solitary structures is directly proportional to the solitary speed moving with speed u_0 and inversely proportional to A. On the other hand, (12) also implies that the width of these solitary structures is directly proportional to the square root of the constant B and inversely proportional to the solitary speed moving with speed u_0.

We have first graphically represented the effects of μ on the phase speed of mIA waves as shown in Figures 1 and 2. From the figures, for negatively charged static heavy ions ($j = -1$), we observe that the upper (dashed orange) line represents the ultrarelativistic case and the lower (dashed blue) line represents the nonrelativistic case. On the other hand, for positively charged static heavy ions ($j = +1$), the upper (dashed red) line represents the nonrelativistic case and the lower (dashed green) line represents the ultrarelativistic case. The profiles of IA and mIA waves for 1D planar ($\nu = 0$) system comparing the nonrelativistic and ultrarelativistic limits are shown in Figures 3 and 4.

The effects of degenerate electron and ion fluids and arbitrarily charged static heavy ions significantly modify

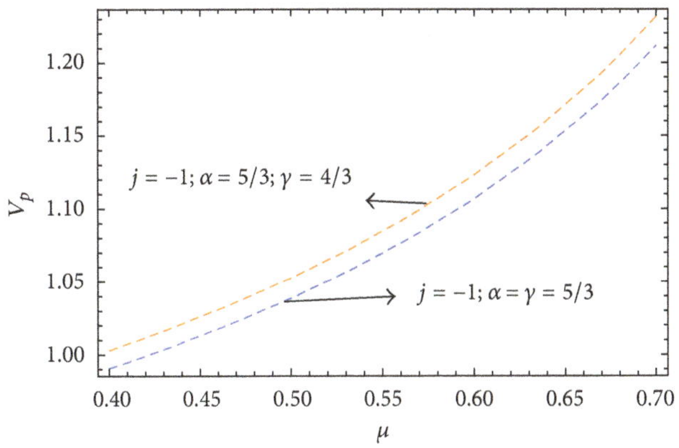

FIGURE 1: Showing the effects of μ on the phase speed of mIA K-dV solitons for negatively charged static heavy ions. The upper (dashed orange) line represents the ultrarelativistic case and the lower (dashed blue) line the nonrelativistic case.

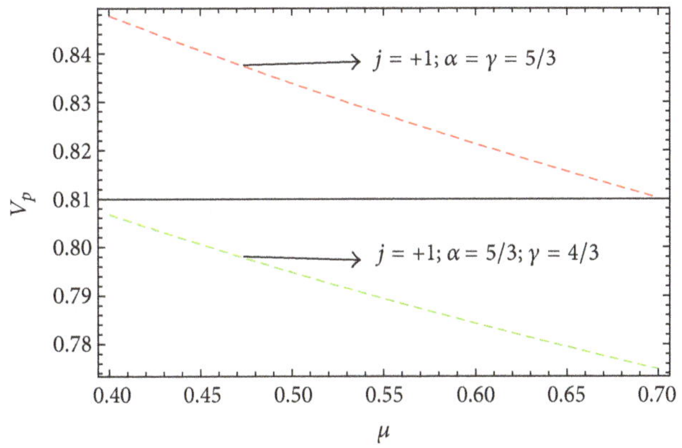

FIGURE 2: Showing the effects of μ on the phase speed of mIA K-dV solitons for positively charged static heavy ions. The upper (dashed red) line represents the nonrelativistic case and the lower (dashed green) line the ultrarelativistic case.

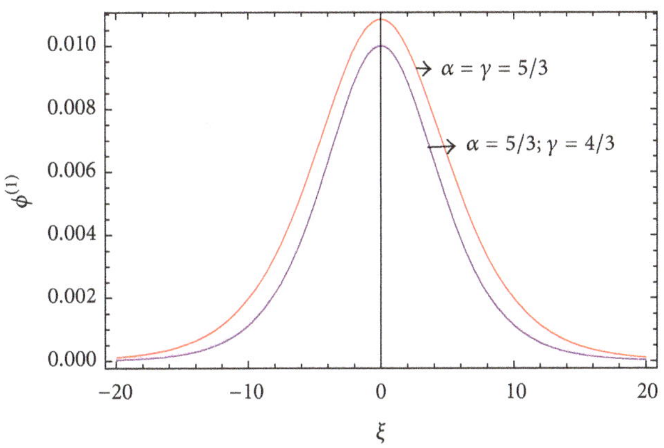

FIGURE 3: Showing the variation of amplitude of IA K-dV solitons in case of planar ($\nu = 0$) geometry. The lower (purple) line represents ion being nonrelativistic degenerate and electron being ultrarelativistic degenerate and the upper (red) one represents both electron and ion being nonrelativistic degenerate.

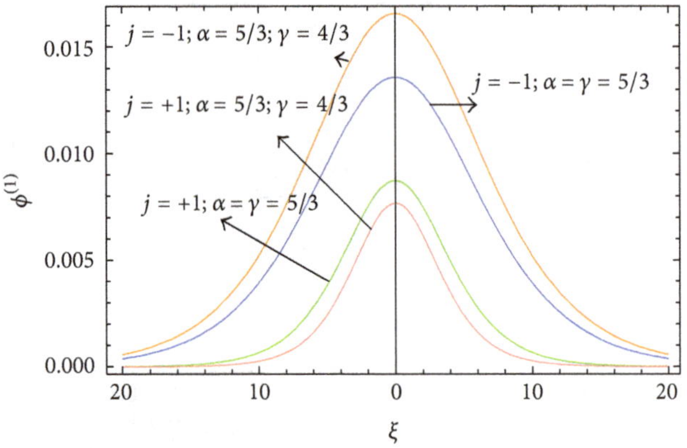

FIGURE 4: Showing the variation of amplitude of mIA K-dV solitons in the presence of arbitrarily charged static heavy ions for planar ($\nu = 0$) geometry. The upper orange and blue lines represent the relativistic cases for negatively charged static heavy ions and the lower green and pink lines represent the relativistic cases for positively charged static heavy ions.

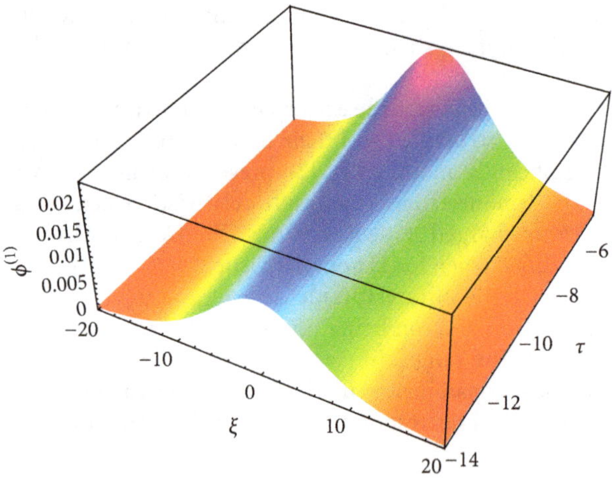

FIGURE 6: Effects of cylindrical geometry on mIA K-dV solitons in the presence of negatively charged static heavy ions when both electron and ion are nonrelativistic degenerate ($\nu = 1$; $j = -1$; $\mu = 0.5$; $n_{e0} = 1.1 \times 10^{30}$; $u_0 = 0.01$; $\alpha = \gamma = 5/3$).

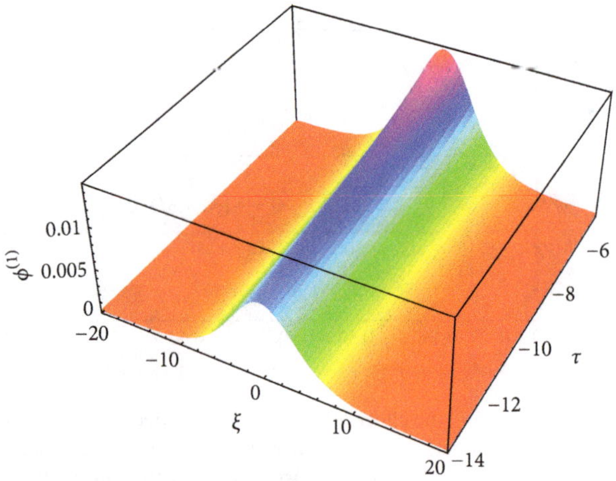

FIGURE 5: Effects of cylindrical geometry on mIA K-dV solitons in the presence of positively charged static heavy ions when both electron and ion are nonrelativistic degenerate ($\nu = 1$; $j = +1$; $\mu = 0.5$; $n_{e0} = 1.1 \times 10^{30}$; $u_0 = 0.01$; $\alpha = \gamma = 5/3$).

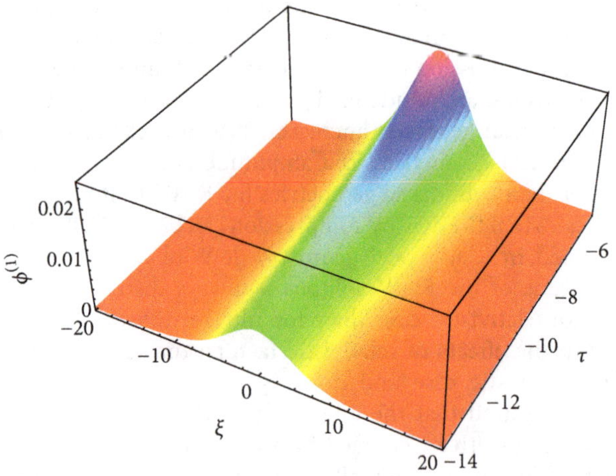

FIGURE 7: Effects of spherical geometry on mIA K-dV solitons in the presence of positively charged static heavy ions when both electron and ion are nonrelativistic degenerate ($\nu = 2$; $j = +1$; $\mu = 0.5$; $n_{e0} = 1.1 \times 10^{30}$; $u_0 = 0.01$; $\alpha = \gamma = 5/3$).

the basic properties (speed, amplitude, and width) of the IA and mIA K-dV solitons. When we compare the $\nu = 0$ case for IA wave (Figure 3), we observe that the red line represents the nonrelativistic case and the purple line represents the ultrarelativistic case. But in case of mIA wave, for negatively charged heavy ions ($j = -1$), we found that the upper orange line represents the ultrarelativistic case and the lower blue line represents the nonrelativistic case (Figure 4). On the other hand, for positively charged heavy ions ($j = +1$), the upper green line represents the nonrelativistic case and the lower pink line represents the ultrarelativistic case (Figure 4). After that, we observe the $\nu = 1$ and $\nu = 2$ case for both IA (Figures 13-14) and mIA waves (Figures 5–12) comparing the nonrelativistic and ultrarelativistic limits. Finally, the results that we have found in this investigation can be summarized as follows.

(1) The plasma system under consideration supports finite amplitude K-dV solitons, whose basic properties (polarity, amplitude, width, etc.) depend on the degenerate electrons and the arbitrarily charged stationary heavy ions number densities.

(2) For negatively (positively) charged heavy ions, the phase speed increases (decreases) with the increase of μ (Figures 1 and 2). The phase speed is always higher (lower) for negatively (positively) charged static heavy ions for ultrarelativistic case than for nonrelativistic case.

(3) We can compare the IA and mIA solitary structures in case of planar geometry as shown in Figures 3 and 4. In these profiles, we observe that the amplitude is higher (lower) for negatively (positively) charged

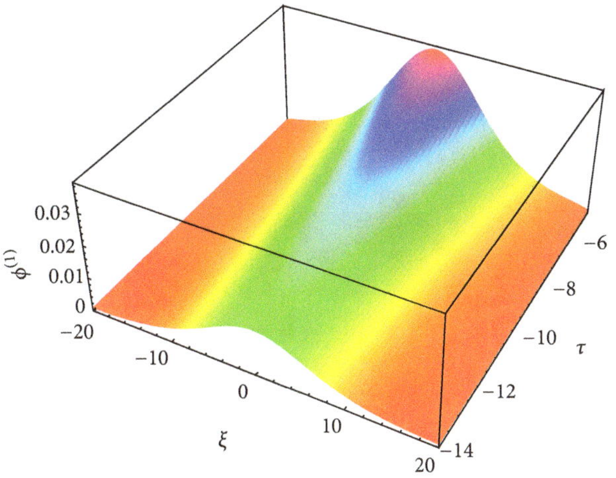

FIGURE 8: Effects of spherical geometry on mIA K-dV solitons in the presence of negatively charged static heavy ions when both electron and ion are nonrelativistic degenerate ($\nu = 2$; $j = -1$; $\mu = 0.5$; $n_{e0} = 1.1 \times 10^{30}$; $u_0 = 0.01$; $\alpha = \gamma = 5/3$).

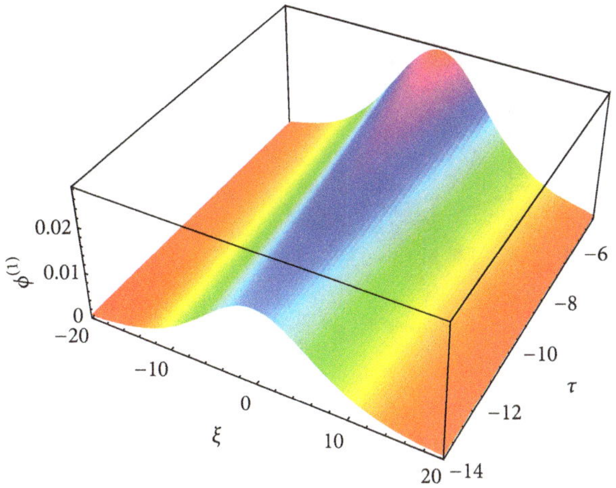

FIGURE 10: Effects of cylindrical geometry on mIA K-dV solitons in the presence of negatively charged static heavy ions when ion is nonrelativistic degenerate and electron is ultrarelativistic degenerate ($\nu = 1$; $j = -1$; $\mu = 0.5$; $n_{e0} = 1.1\times10^{30}$; $u_0 = 0.01$; $\alpha = 5/3$; $\gamma = 4/3$).

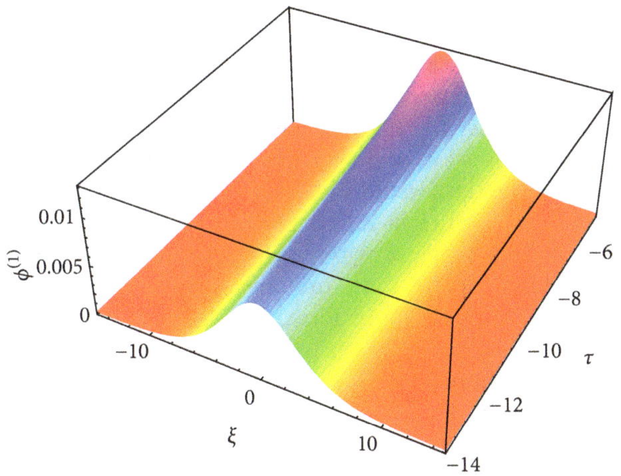

FIGURE 9: Effects of cylindrical geometry on mIA K-dV solitons in the presence of positively charged static heavy ions when ion is nonrelativistic degenerate and electron is ultrarelativistic degenerate ($\nu = 1$; $j = +1$; $\mu = 0.5$; $n_{e0} = 1.1\times10^{30}$; $u_0 = 0.01$; $\alpha = 5/3$; $\gamma = 4/3$).

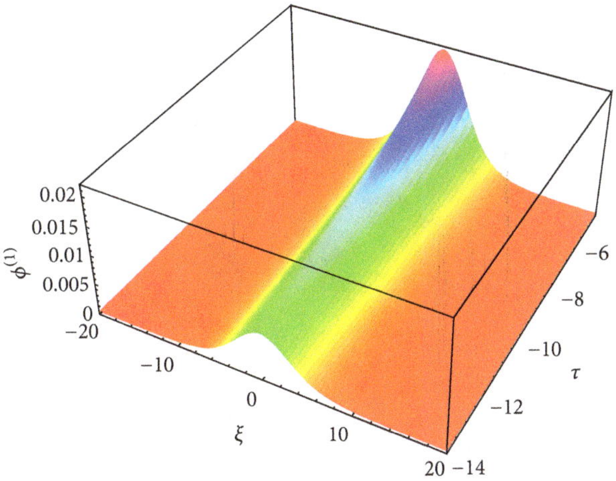

FIGURE 11: Effects of spherical geometry on mIA K-dV solitons in the presence of positively charged static heavy ions when ion is nonrelativistic degenerate and electron is ultrarelativistic degenerate ($\nu = 2$; $j = +1$; $u_0 = 0.01$; $n_{e0} = 1.1 \times 10^{30}$; $\alpha = 5/3$; $\gamma = 4/3$).

mIA waves than for the IA ones. It is also valid for cylindrical and spherical IA (Figures 13-14) and mIA (Figures 5–12) solitary structures.

(4) The large value of τ kills the possibility of formation of nonplanar solitons. It has been found that as the value of τ decreases the amplitude of these localized pulses increases for both IA (Figures 13-14) and mIA (Figures 5–12) waves. From the observation we can say that the amplitude of cylindrical (Figures 13 and 14 for IA and Figures 5, 6, 9, and 10 for mIA wave) K-dV solitons is larger than 1D planar (Figure 3 for IA and Figure 4 for mIA wave) ones but smaller than that of the spherical ones (Figures 7, 8, 11, and 12 for mIA wave).

(5) From the observation of Figure 3, we can clearly say that the amplitude is lower for ultrarelativistic case than for nonrelativistic case and from Figure 4, the amplitude is higher for ultrarelativistic case than for nonrelativistic case. It is also valid for all other cylindrical and spherical graphs of IA (Figures 13-14) and mIA (Figures 5–12) waves.

(6) The amplitude of the mIA K-dV solitons significantly differ with the polarity of heavy ions and relativistic parameters. It is found from Figure 4 that the amplitude of mIA wave in planar case is always higher (lower) for negatively (positively) charged heavy ions for both nonrelativistic and ultrarelativistic limits which is also valid in nonplanar case (Figures 5–12).

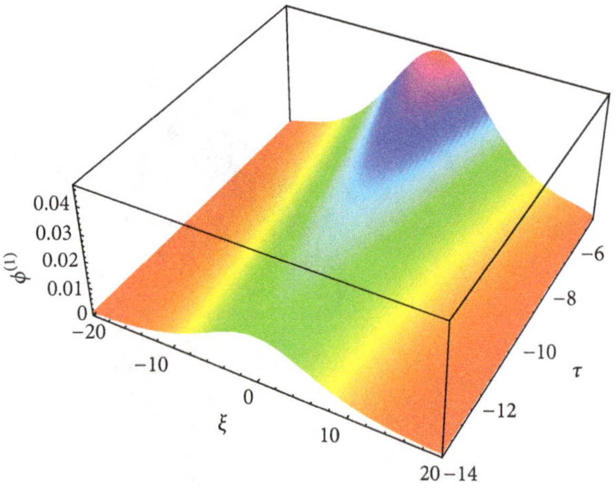

FIGURE 12: Effects of spherical geometry on mIA K-dV solitons in the presence of negatively charged static heavy ions when both electron and ion are nonrelativistic degenerate ($\nu = 2$; $j = -1$; $\mu = 0.5$; $n_{e0} = 1.1 \times 10^{30}$; $u_0 = 0.01$; $\alpha = 5/3$; $\gamma = 4/3$).

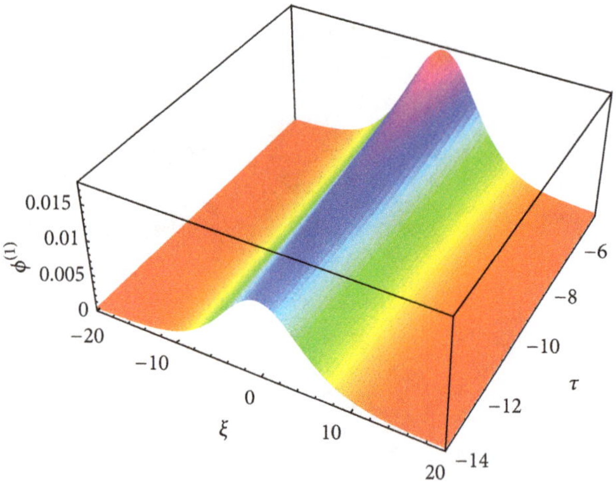

FIGURE 14: Effects of cylindrical geometry on IA K-dV solitons when ion is nonrelativistic degenerate and electron is ultrarelativistic degenerate ($\nu = 1$; $u_0 = 0.01$; $n_{e0} = 1.1 \times 10^{30}$; $\alpha = 5/3$; $\gamma = 4/3$).

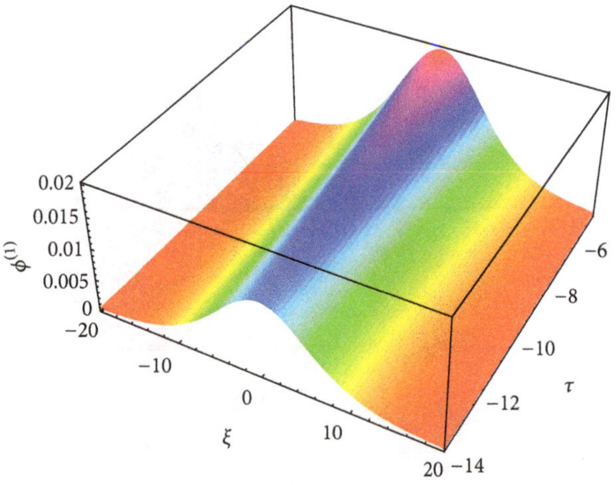

FIGURE 13: Effects of cylindrical geometry on IA K-dV solitons when both electron and ion are nonrelativistic degenerate ($\nu = 1$; $u_0 = 0.01$; $n_{e0} = 1.1 \times 10^{30}$; $\alpha = \gamma = 5/3$).

5. Discussion

We have studied the nonlinear propagation of IA and mIA solitary waves in an unmagnetized, collisionless dense plasma (containing both nonrelativistic and ultrarelativistic degenerate electrons, nonrelativistic degenerate ions, and arbitrarily charged static heavy ions). The degenerate electron and ion number densities and arbitrarily charged static heavy ions significantly modify the basic properties of the solitary waves. It is also found that the amplitude of the solitary waves has been modified by the term μ. In our numerical analysis, we have tried to give the idea of the variation among nonrelativistic and ultrarelativistic degenerate plasma pressure, IA and mIA solitary structures in case of planar and nonplanar geometry which makes our present work significant to understand the localized electrostatic disturbances in many space and astrophysical plasma environments (namely, white dwarfs, neutron stars, compact planets like massive Jupiter, other exotic dense stars, and black holes). The relativistic effects of the compact objects such as white dwarf stars make them gravitationally unstable for masses larger than about 1.4 solar masses [31] and formation of neutron stars begins. This investigation would be useful to study the effects of degenerate pressure in interstellar and space plasmas [32], particularly in stellar polytropes [33], hadronic matter and quark-gluon plasma [34], protoneutron stars [35], dark-matter halos [36], and so forth. Our present investigation is a theoretical work which is applicable for matter under extreme conditions, for example, IA and mIA solitary waves propagation in the interior of compact objects [1, 2] where the planar and nonplanar geometry, arbitrarily charged static heavy ions, inertial ions, and degenerate plasma pressure are taken into account. This theory could be rigorously important for global nonlinear models of astrophysical compact objects.

Conflict of Interests

The authors declare that there is no conflict of interests regarding the publication of this paper.

References

[1] S. L. Shapiro and S. A. Teukolsky, *Black Holes, White Dwarfs and Neutron Stars: The Physics of Compact Objects*, John Wiley & Sons, New York, NY, USA, 1983.

[2] A. K. Harding and D. Lai, "Physics of strongly magnetized neutron stars," *Reports on Progress in Physics*, vol. 69, no. 9, p. 2631, 2006.

[3] M. Temerin, K. Cerny, W. Lotko, and F. S. Mozer, "Observations of double layers and solitary waves in the auroral plasma," *Physical Review Letters*, vol. 48, no. 17, pp. 1175–1179, 1982.

[4] R. Boström, G. Gustafsson, B. Holback, G. Holmgren, H. Koskinen, and P. Kintner, "Characteristics of solitary waves and

weak double layers in the magnetospheric plasma," *Physical Review Letters*, vol. 61, p. 82, 1988.

[5] P. O. Dovner, A. I. Eriksson, R. Bostrom, and B. Holback, "Freja multiprobe observations of electrostatic solitary structures," *Geophysical Research Letters*, vol. 21, no. 17, pp. 1827–1830, 1994.

[6] H. Ikezi, R. J. Tailor, and D. R. Baker, "Formation and interaction of ion-acoustic solitions," *Physical Review Letters*, vol. 25, article 11, 1970.

[7] H. Washimi and T. Taniuti, "Propagation of ion-acoustic solitary waves of small amplitude," *Physical Review Letters*, vol. 17, no. 19, pp. 996–998, 1966.

[8] N. Roy, S. Tasnim, and A. A. Mamun, "Solitary waves and double layers in an ultra-relativistic degenerate dusty electron-positron-ion plasma," *Physics of Plasmas*, vol. 19, no. 3, Article ID 033705, 2012.

[9] L. Nahar, M. S. Zobaer, N. Roy, and A. A. Mamun, "Ion-acoustic K-dV and mK-dV solitons in a degenerate electron-ion dense plasma," *Physics of Plasmas*, vol. 20, Article ID 022304, 2013.

[10] G. B. van Albada, "On the origin of the heavy elements," *Astrophysical Journal*, vol. 105, p. 393, 1947.

[11] S. Chandrasekhar, "The density of white Dwarf stars," *Philosophical Magazine*, vol. 11, pp. 592–596, 1931.

[12] S. Chandrasekhar, "Stellar configurations with degenerate cores," *Monthly Notices of the Royal Astronomical Society*, vol. 95, p. 676, 1935.

[13] M. R. Hossen, L. Nahar, and A. A. Mamun, "Roles of arbitrarily charged heavy ions and degenerate plasma pressure in cylindrical and spherical IA shock waves," *Physica Scripta*, vol. 89, no. 10, Article ID 105603, 2014.

[14] A. A. Mamun and P. K. Shukla, "Solitary waves in an ultrarelativistic degenerate dense plasma," *Physics of Plasmas*, vol. 17, no. 10, Article ID 104504, 2010.

[15] M. S. Zobaer, N. Roy, and A. A. Mamun, "DIA solitary and shock waves in dusty multi-ion dense plasma with arbitrary charged dust," *Journal of Modern Physics*, vol. 3, no. 8, pp. 755–761, 2012.

[16] M. S. Zobaer, N. Roy, and A. A. Mamun, "Nonlinear propagation of dust ion-acoustic waves in dusty multi-ion dense plasma," *Astrophysics and Space Science*, vol. 343, no. 2, pp. 675–681, 2013.

[17] M. M. Hossain, A. A. Mamun, and K. S. Ashrafi, "Cylindrical and spherical dust ion-acoustic Gardner solitons in a quantum plasma," *Physics of Plasmas*, vol. 18, no. 10, Article ID 103704, 2011.

[18] S. Chandra, S. N. Paul, and B. Ghosh, "Electron-acoustic solitary waves in a relativistically degenerate quantum plasma with two-temperature electrons," *Astrophysics and Space Science*, vol. 343, no. 1, pp. 213–219, 2013.

[19] M. Akbari-Moghanjoughi, "Nonlinear ion waves in Fermi-Dirac pair plasmas," *Physics of Plasmas*, vol. 18, Article ID 012701, 2011.

[20] S. Chandra and B. Ghosh, "Modulational instability of electron-acoustic waves in relativistically degenerate quantum plasma," *Astrophysics and Space Science*, vol. 342, no. 2, pp. 417–424, 2012.

[21] M. Akbari-Moghanjoughi, "Comment on "the effects of Bohm potential on ion-acoustic solitary waves interaction in a nonplanar quantum plasma" [Phys. Plasmas 17, 082307 (2010)]," *Physics of Plasmas*, vol. 17, no. 11, Article ID 114701, 2010.

[22] G. Manfredi, "How to model quantum plasmas," *Fields Institute Communications*, vol. 46, pp. 263–287, 2005.

[23] G. Brodin and M. Marklund, "Spin magnetohydrodynamics," *New Journal of Physics*, vol. 9, article 277, 2007.

[24] W. F. El-Taibany, A. A. Mamun, and K. H. El-Shorbagy, "Nonlinear electromagnetic perturbations in a degenerate electron-positron plasma," *Advances in Space Research*, vol. 50, no. 1, pp. 101–107, 2012.

[25] J. T. Mendonca and A. Serbeto, "Volkov solutions for relativistic quantum plasmas," *Physical Review E*, vol. 83, Article ID 026406, 2011.

[26] W. Masood and B. Eliasson, "Electrostatic solitary waves in a quantum plasma with relativistically degenerate electrons," *Physics of Plasmas*, vol. 18, no. 3, Article ID 034503, 2011.

[27] P. K. Shukla, A. A. Mamun, and D. A. Mendis, "Nonlinear ion modes in a dense plasma with strongly coupled ions and degenerate electron fluids," *Physical Review E*, vol. 84, Article ID 026405, 2011.

[28] I. Zeba, W. M. Moslem, and P. K. Shukla, "Ion solitary pulses in warm plasmas with ultrarelativistic degenerate electrons and positrons," *Astrophysical Journal*, vol. 750, no. 1, article 72, 2012.

[29] M. S. Zobaer, N. Roy, and A. A. Mamun, "Ion-acoustic shock waves in a degenerate dense plasma," *Journal of Plasma Physics*, vol. 79, no. 1, pp. 65–68, 2013.

[30] S. Maxon and J. Viecelli, "Spherical Solitons," *Physical Review Letters*, vol. 32, no. 1, pp. 4–6, 1974.

[31] B. Eliasson and P. K. Shukla, "The formation of electrostatic shocks in quantum plasmas with relativistically degenerate electrons," *Europhysics letters*, vol. 97, p. 15001, 2011.

[32] F. Ferro, A. Lavagno, and P. Quarati, "Non-extensive resonant reaction rates in astrophysical plasmas," *European Physical Journal A*, vol. 21, no. 3, pp. 529–534, 2004.

[33] A. R. Plastino and A. Plastino, "Stellar polytropes and Tsallis' entropy," *Physics Letters A*, vol. 174, no. 5-6, pp. 384–386, 1993.

[34] G. Gervino, A. Lavagno, and D. Pigato, "Nonextensive statistical effects in the quark-gluon plasma formation at relativistic heavy-ion collisions energies," *Central European Journal of Physics*, vol. 10, no. 3, pp. 594–601, 2012.

[35] A. Lavagno and D. Pigato, "Nonextensive statistical effects in protoneutron stars," *European Physical Journal A*, vol. 47, no. 4, article 52, 2011.

[36] C. Féron and J. Hjorth, "Simulated dark-matter halos as a test of nonextensive statistical mechanics," *Physical Review E*, vol. 77, Article ID 022106, 2008.

7

Field Independent Cosmic Evolution

Nayem Sk[1] and Abhik Kumar Sanyal[2]

[1] *Department of Physics, University of Kalyani, Nadia 741235, India*
[2] *Department of Physics, Jangipur College, Murshidabad 742213, India*

Correspondence should be addressed to Abhik Kumar Sanyal; sanyal_ak@yahoo.com

Academic Editors: M. Biesiada, K. Bolejko, G. Chincarini, M. S. Dimitrijevic, D. Mota, and E. Saridakis

It has been shown earlier that Noether symmetry does not admit a form of $F(R)$ corresponding to an action in which $F(R)$ is coupled to scalar-tensor theory of gravity or even for pure $F(R)$ theory of gravity taking anisotropic model into account. Here, we prove that $F(R)$ theory of gravity does not admit Noether symmetry even if it is coupled to tachyonic field and considering a gauge in addition. To handle such a theory, a general conserved current has been constructed under a condition which decouples higher-order curvature part from the field part. This condition, in principle, solves for the scale-factor independently. Thus, cosmological evolution remains independent of the form of the chosen field, whether it is a scalar or a tachyon.

1. Introduction

Interest in $F(R)$ theory of gravity has increased predominantly in recent years, since it appears to explain most of the presently available cosmological data unifying early inflation with late time cosmic acceleration (see [1, 2] for recent reviews and also references therein). However, most of these interesting results are the outcome of scalar-tensor equivalence under some arbitrary choice of the form of $F(R)$. It is, therefore, important to test if the same results are obtainable from $F(R)$ theory of gravity without invoking scalar-tensor equivalence. But then how to choose a form of $F(R)$ out of indefinitely large number of curvature invariant terms and how to find exact solutions are big questions. From physical ground, namely, to obtain a renormalizable theory of gravity, a form of $F(R) = \alpha R + \beta R^2 + \gamma R_{\mu\nu}R^{\mu\nu}$ had been found in the context of early universe, which contains ghosts [3]. A ghost-free action has also been presented in recent years [4, 5]. Likewise, the only physically meaningful technique to obtain a form of $F(R)$ to explain late time cosmological evolution is to invoke Noether symmetry as a selection rule. This requires canonical formulation, and for a general $F(R)$ theory of gravity, it is only possible treating R as an auxiliary variable, provided $F''(R) \neq 0$ (here, prime represents derivative with respect to R). In the process, it is possible to construct a point Lagrangian, and one can demand Noether symmetry to find a suitable form of $F(R)$. Following this technique, several authors [6–12] have found $F(R) \propto R^{3/2}$ in the Robertson-Walker metric both in vacuum ($\rho = p = 0$) and pressureless dust ($p = 0$). Although such a form of $F(R)$ shows accelerating expansion in the matter dominated era ($p = 0$), nevertheless, early decelerating phase tracks as $a \propto t^{1/2}$ in the matter dominated era instead of usual $t^{2/3}$ and $a \propto t^{4/3}$ in the radiation dominated era ($p = \rho/3$) instead of usual $t^{1/2}$, creating problem in explaining structure formation [13]. Thus, $R^{3/2}$ alone, in the absence of a linear term in the action, is not worth explaining presently available cosmological data [13]. It has been also noticed [13] that instead of the scale-factor a, if one would have started with the basic variable $h_{ij} = z = a^2$, then z becomes cyclic both in vacuum and matter dominated era, and therefore, such symmetry is independent of the choice of configuration space variables. Thus, the Noether symmetry obtained in the process is inbuilt and trivial. The situation could have been improved if the Noether symmetry would allow linear term R in the action, but it does not. To obtain a better form of $F(R)$, scalar field has been incorporated both minimally and nonminimally, but even then Noether symmetry remains absent [14].

Despite the fact that Noether symmetry yields nothing other than $F(R) \propto R^{3/2}$ in vacuum or in matter dominated era and that too only in isotropic space-time, some authors have recently claimed [15] to have obtained arbitrary form of $F(R)$ taking into account a gauge term in the Noether theory. Note that the theory of gravity is very special in the context that it admits reparametrization invariance leading to the $\binom{0}{0}$ equation of Einstein or the so-called Hamiltonian constraint equation $H = 0$. This reparametrization invariance is not reflected in the Noether equations. Thus, solution obtained from the Noether equations, namely, the conserved current, must satisfy the Hamilton constraint equation. It has been shown that the result [15] is not true, since neither all the Noether equations nor the Hamilton constraint equation is satisfied for such symmetry [16]. More recently, there is yet another claim [17] that the gauge Noether symmetry yields $F(R) \propto R^2$ taking tachyon field into account. It is important to review this fact since in our earlier works [13, 14] tachyon field had not been accounted for.

Rolling tachyon condensates originated from string theories and have interesting cosmological consequences [18–25] particularly because its equation of state parameter ($w = p_T/\rho_T = -[1-\dot{\phi}^2]$, $\rho_T = V(\phi)/\sqrt{1-\dot{\phi}^2}$, $p_T = -V(\phi)\sqrt{1-\dot{\phi}^2}$) smoothly interpolates between 0 and −1 [26]. As a result, it behaves like pressureless dust and cosmological constant in the limits. Such a nice feature, initiated to construct viable cosmological models treating tachyon as the inflaton field by coupling it minimally to the gravitational field taking different self-interacting potential densities, in the form of power law, exponential, and hyperbolic functions of the tachyon field [27–40]. Nevertheless, such models have been found to suffer from serious disease associated with density perturbations and reheating [41]. Tachyon fields with such kinds of potential densities were also used in order to describe the present accelerating period of the universe, where it behaves as dark energy [42–52]. Further, it has been observed that a tachyon field with an exponential potential plays the role of inflaton in the early universe and dark energy at the late [53]. Thus, it appears to be a good candidate that might explain late time cosmological evolution. Nonetheless, present observations suggest that the state parameter might even cross the phantom divide line $w < -1$, which is not realizable under minimal coupling. Such crossing may be possible if the tachyon field is nonminimally coupled to the gravitational field, which already exists in the literature [54, 55]. Another possibility appears if the tachyon field is coupled to $F(R)$ theory of gravity, which is our present concern. We therefore, in following sections, proceed to check if $F(R)$ theory of gravity being coupled to tachyon field admits Noether symmetry. The result we find is null, since the conserved current, thus obtained, does not satisfy the Hamilton constraint equation. This completes our "Tour de Noether symmetry" of $F(R)$ theory of gravity, which now states that $F(R)$ theory of gravity only admits the trivial symmetry $F(R) \propto R^{3/2}$ in vacuum or matter dominated era and that too in isotropic space-time only.

In view of the above discussion, we organize the present work in the following manner. In Section 2, we review the gauge Noether symmetry of $F(R)$ theory of gravity being coupled to tachyonic field to show that indeed such symmetry is not admissible. In Section 3, we take up a more general action and attempt Noether symmetry once again but in vain. At this end, we would like to mention that on the contrary, starting from $F(R) = R^2$, *a priori*, the Noether symmetry is obtainable for a scalar-tensor theory of gravity [56] which we have not been able to recover. This is important, since R^2 is special in the sense that the corresponding action is scale invariant which is also somehow related to Noether symmetry [57]. In Section 4, we briefly enunciate our view in this regard. Section 5 is devoted to construct a general conserved current and to explain its utility in extracting solutions choosing arbitrary form of $F(R)$. In view of such conserved current, we show that the form of the field in no way dictates the cosmological evolution, but rather it depends only on the form of $F(R)$ chosen. Finally, in Section 6, we find the constraint to recover the Newtonian limit for such an action under weak field approximation. Section 7 concludes the present work.

2. Action, Field Equations, and Noether Equation

Among all the dynamical symmetries, transformations that map solutions of the equations of motion into solutions, one can single out the Noether symmetries as the continuous transformations that leave the action invariant—except for boundary terms. In formal language, the Noether symmetry states that for any regular system if there exists a vector field $X^{(1)}$, such that

$$\left(\pounds_{X^{(1)}} + \frac{d\eta}{dt}\right) L = \left(X^{(1)} + \frac{d\eta}{dt}\right) L = \frac{dB}{dt}, \quad (1)$$

in the presence of a gauge function $B(q_i, t)$, where $X^{(1)}$ is the first prolongation of the vector field X given by

$$X^{(1)} = X + \sum_i \left[(\dot{\alpha}_i - \dot{\eta}\dot{q}_i) \frac{\partial}{\partial \dot{q}_i} \right],$$
$$X = \eta \frac{\partial}{\partial t} + \sum_i \alpha_i \frac{\partial}{\partial q_i}, \quad (2)$$

with $\alpha_i = \alpha_i(q_i, t)$, $\eta = \eta(q_i, t)$, then there exists a conserved current

$$I = \sum_i (\alpha_i - \eta \dot{q}_i) \frac{\partial L(q_i, \dot{q}_i, t)}{\partial \dot{q}_i} + \eta L(q_i, \dot{q}_i, t) - B(q_i, t)$$
$$= \sum_i \alpha_i p_i - B(q_i, t) - \eta(q_i, t) H(q_i, p_i, t). \quad (3)$$

In some earlier works [16, 58] it has been shown that under infinitesimal coordinate and temporal transformations ($q_i' = q_i + \epsilon\alpha_i(q_i, t)$ and $t' = t + \epsilon\eta(q_i, t)$), time dependence may be introduced in a Lagrangian which does not contain time explicitly, through a time dependent gauge function. Invariance of Hamilton's principal function finally yields

$$\int L(q_i, \dot{q}_i, t)\, dt = \int L(q_i, \dot{q}_i)\, dt - \epsilon \frac{d}{dt} \int \Big[\eta \Big(L(q_i, \dot{q}_i, t) - \sum \dot{q}_i \frac{\partial L(q_i, \dot{q}_i, t)}{\partial \dot{q}_i} \Big) + \sum \alpha_i \frac{\partial L(q_i, \dot{q}_i, t)}{\partial \dot{q}_i} - B(q_i, t) \Big] dt = \int L(q_i, \dot{q}_i)\, dt - \epsilon \frac{d}{dt} \int \Big[-\eta H + \sum \alpha_i p_i - B(q_i, t) \Big] dt, \quad (4)$$

retaining only up to first-order term in the Taylor expansion. It is apparent that the conserved current (3) may be obtained from (4) only if the Lagrangian in the left-hand side and that in the right-hand side of (4) get cancelled. This is possible provided that the Lagrangian appearing in the left-hand side is time independent. Remember that time-dependence has been generated in the Lagrangian only through a time-dependent gauge function $B(q_i, t)$ [16]. Thus for a time-independent Lagrangian (as in the case of gravity under consideration), the gauge function has to be time independent as in the case of harmonic oscillator. Further, it is known that the Noether integral is the Hamiltonian for the trivial Noether point symmetry $\partial/\partial t$. Other way round, the Hamiltonian is the generator of time translation, and so conservation of the Hamiltonian requires η to be constant. Further, in gravity, the Hamiltonian is not only conserved, but it is constrained to vanish. As a result, η does not play any significant role in the Noether symmetry as is clearly observed from (4) and may even be set equal to zero. Time independence of the Lagrangian, the gauge term, and η automatically enforces time independence of α_i. In the process, the integral of motion (3) reduces simply to

$$I = \sum_i \alpha_i (q_i)\, p_i - B(q_i). \quad (5)$$

Essentially, if gauge turns out to be zero while solving the Noether equations, the integral of motion remains the same for the Noether symmetry without gauge, and no new result is expected. Although it is clear that time translation is unnecessary, still we keep it explicitly in the following, to keep track with the earlier work [17] performed in this context.

Now, considering Robertson-Walker line element

$$ds^2 = -dt^2 + a^2 \left[\frac{dr^2}{1 - kr^2} + r^2 d\theta^2 + r^2 \sin^2\theta\, d\varphi^2 \right], \quad (6)$$

the following Born-Infeld effective 4-dimensional action for a rolling tachyon field ϕ with Lagrangian density $\mathscr{L}(\phi) = -V(\phi)\sqrt{1 + \zeta\phi_{,\mu}\phi^{,\mu}}$ being minimally coupled to $F(R)$

$$A = \int d^4x \sqrt{-g} \left[F(R) - V(\phi) \sqrt{1 + \zeta\phi_{,\mu}\phi^{,\mu}} \right] \quad (7)$$

leads to a point Lagrangian:

$$L = 6a\dot{a}^2 F' + 6a^2 \dot{a}\dot{R} F'' + a^3 (F'R - F) + a^3 V(\phi) \sqrt{1 - \zeta\dot{\phi}^2}, \quad (8)$$

for the spatially flat ($k = 0$) case, treating

$$R = 6\left(\frac{\ddot{a}}{a} + \frac{\dot{a}^2}{a^2} \right) \quad (9)$$

as a constraint of the theory and effectively spanning the Lagrangian by a set of configuration space variables $(a, R, \phi, \dot{a}, \dot{R}, \dot{\phi})$. In the above action ζ is treated as the coupling constant required to make the kinetic part of the action dimensionless. For the above point Lagrangian the Noether equation (1) reads

$$\begin{aligned} &\alpha \left[6\dot{a}^2 F' + 12a\dot{a}\dot{R}F'' + 3a^2 (F'R - F) + 3a^2 V \sqrt{1 - \zeta\dot{\phi}^2} \right] \\ &+ \beta \left[6a\dot{a}^2 F'' + 6a^2 \dot{a}\dot{R} F''' + a^3 R F'' \right] \\ &+ \gamma \left[a^3 V_{,\phi} \sqrt{1 - \zeta\dot{\phi}^2} \right] \\ &+ \left[(\alpha_{,t} - \dot{a}\eta_{,t}) + \alpha_{,a}\dot{a} + \alpha' \dot{R} + \alpha_{,\phi}\dot{\phi} - \eta_{,a}\dot{a}^2 - \eta' \dot{a}\dot{R} - \eta_{,\phi}\dot{a}\dot{\phi} \right] \left[12a\dot{a}F' + 6a^2 \dot{R} F'' \right] \\ &+ \left[(\beta_{,t} - \dot{R}\eta_{,t}) + \beta_{,a}\dot{a} + \beta' \dot{R} + \beta_{,\phi}\dot{\phi} - \eta_{,a}\dot{a}\dot{R} - \eta' \dot{R}^2 - \eta_{,\phi}\dot{R}\dot{\phi} \right] \left[6a^2 \dot{a} F'' \right] \\ &+ \left[(\gamma_{,t} - \dot{\phi}\eta_{,t}) + \gamma_{,a}\dot{a} + \gamma' \dot{R} + \gamma_{,\phi}\dot{\phi} - \eta_{,a}\dot{a}\dot{\phi} - \eta' \dot{R}\dot{\phi} - \eta_{,\phi}\dot{\phi}^2 \right] \left[-\frac{\zeta a^3 V \dot{\phi}}{\sqrt{1 - \zeta\dot{\phi}^2}} \right] \\ &+ \left[\eta_{,t} + \eta_{,a}\dot{a} + \eta' \dot{R} + \eta_{,\phi}\dot{\phi} \right] \times \left[6a\dot{a}^2 F' + 6a^2 \dot{a}\dot{R} F'' + a^3 (F'R - F) + a^3 V \sqrt{1 - \zeta\dot{\phi}^2} \right] \\ &= B_{,a}\dot{a} + B' \dot{R} + B_{,\phi}\dot{\phi} + B_{,t}, \end{aligned} \quad (10)$$

where prime ($'$) denotes derivative with respect to R. Note that we have kept both time translation and a time-dependent gauge function to keep track with the work performed by Jamil et al. [17]. Now, equating coefficients as usual, we obtain a large overdeterminant set of the Noether equations for $F''(R) \neq 0$. These are

$$\eta_{,a} = \eta' = \eta_{,\phi} = 0, \quad (11)$$

$$\alpha' = \alpha_{,\phi} = 0, \quad (12)$$

$$\beta_{,\phi} = 0, \quad (13)$$

$$\gamma_{,t} = \gamma' = \gamma_{,a} = 0, \quad (14)$$

$$B_{,\phi} = 0, \quad (15)$$

$$\gamma_{,\phi} - \eta_{,t} = 0, \quad (16)$$

$$6a^2 F'' \alpha_{,t} = B', \quad (17)$$

$$12aF'\alpha_{,t} + 6a^2 F'' \beta_{,t} = B_{,a}, \quad (18)$$

$$a^2 (3\alpha + a\eta_{,t}) (F'R - F) + a^3 \beta R F'' = B_{,t}, \quad (19)$$

$$\alpha F' + a\beta F'' + 2a\alpha_{,a} F' + a^2 \beta_{,a} F'' - aF'\eta_{,t} = 0, \quad (20)$$

$$2a\alpha F'' + a^2 \beta F''' + a^2 F'' (\alpha_{,a} + \beta' - \eta_{,t}) = 0, \quad (21)$$

$$3\alpha V + a\gamma V_{,\phi} + aV\eta_{,t} = 0. \quad (22)$$

Before we proceed let us mention that if $V(\phi) = 0$, the action (7) corresponds to pure $F(R)$ theory of gravity, for which nothing other than $F(R) \propto R^{3/2}$ is possible [13]. In what follows, we will assume $V(\phi) \neq 0$ in (22), and so it will never be possible to recover pure $F(R)$ case. Now, in view of (11) through (15), it is apparent that $\eta = \eta(t)$, $\alpha = \alpha(t, a)$, $\beta = \beta(t, a, R)$, $\gamma = \gamma(\phi)$, and $B = B(t, a, R)$. Therefore, (16) dictates that η may be at most linear in t. Thus, (22) clearly states that α has to be time independent (as γ is), and it should be linear in the scale-factor a. Thus, taking $\alpha = c_4 a$, where c_4 is a constant, it is clear that β has to be independent of t and a in view of (20) and (21) and the gauge term B must be independent of R as is apparent from (17). Further, since both α and β are time independent, therefore, the gauge term further becomes independent of the scale-factor a in view of (18). Finally, (19) is satisfied provided $B \neq B(t)$. In the process, the gauge term B turns out to be a constant and hence plays no role in the Noether symmetry. Thus, one can set $B = 0$, without loss of generality. In view of the previous analysis, we now have

$$\eta = c_1 t + c_3, \qquad \alpha = c_4 a, \qquad \beta = \beta(R), \qquad \gamma = c_1\phi + c_2, \qquad B = 0, \quad (23)$$

where c_1, c_2, c_3, and c_4 are constants. Having obtained explicit forms of η, α, and γ, we are now left to find the explicit forms of $\beta(R)$, $F(R)$, and the potential $V(\phi)$ from the set of overdeterminant equations (19) through (22) which in view of (23) now get reduced to, following four equations

$$(3c_4 + c_1)(F'R - F) + F''R\beta = 0, \quad (24)$$

$$(3c_4 - c_1) F' + \beta F'' = 0, \quad (25)$$

$$\beta F''' + F'' (3c_4 + \beta' - c_1) = 0, \quad (26)$$

$$(3c_4 + c_1) V + (c_1\phi + c_2) V_{,\phi} = 0. \quad (27)$$

2.1. Reviewing Jamil et al.'s Work. Equation (27) may now be solved for the following form of the potential:

$$V = V_{10}(c_1\phi + c_2)^{-((3c_4+c_1)/c_1)} = c_1 V_{10}\left(\phi + \frac{c_2}{c_1}\right)^{-((3c_4+c_1)/c_1)}, \quad (28)$$

where V_{10} is a constant. However, the authors [17] claimed the following form of the potential:

$$V = V_0 (\phi + \phi_0)^{-4}, \quad (29)$$

which is possible only under the choices $V_0 = c_1 V_{10}$, $\phi_0 = c_2/c_1$, and in particular $c_1 = c_4$, which, as we will show shortly, create severe problem. One can easily check that under the choice $c_1 = c_4$, (26) upon integration yields (25), provided the constant of integration is zero. Thus, we are now left with two independent equations (24) and (25) to find the form of $F(R)$ and $\beta(R)$. However, (25) now reads

$$\beta F'' = -2c_1 F'. \quad (30)$$

Now eliminating β between (24) and (30) one ends up with

$$F(R) \propto R^2 \quad (31)$$

in a straightforward manner, and β gets solved as $\beta = -2c_1 R$. At this end the authors [17] obtained two conserved currents I_1 and I_2. I_1 is the invariance under time translation, as stated correctly by the authors (here we point out a typographical error in (31) of [17]; I_1 should be $I_1 = \tau[L - (\dot{a}(\partial L/\partial \dot{a}) + \dot{R}(\partial L/\partial \dot{R}) + \dot{\phi}(\partial L/\partial \dot{\phi}))]$ instead, where τ used in [17] stands for η here). Nevertheless, the third bracketed term is the Hamiltonian, which is constrained to vanish. Thus, if $I_1 = 0$ is substituted in I_2 (as already mentioned), the conserved current is simply

$$I_2 = \sum \alpha_i p_i = \left[12F_0\dot{R} + \frac{\zeta V_0 \dot{\phi}}{(\phi + \phi_0)^3 \sqrt{1 - \zeta\dot{\phi}^2}} \right] a^3. \quad (32)$$

Conserved current is not an independent equation, but rather it is the first integral of certain combination of the field equations. Thus, it is essential to check if the previous conserved current obtained for $F(R) = F_0 R^2$ and $V = V_0(\phi + \phi_0)^{-4}$ satisfies the field equations, which was not performed by the authors [17]. Now the corresponding field equations are

$$F_0\left[2\frac{\ddot{a}}{a}+\frac{\dot{a}^2}{a^2}+\frac{\ddot{R}}{R}+2\frac{\dot{a}\dot{R}}{aR}-\frac{R}{4}\right]+\frac{V}{4R}\sqrt{1-\zeta\dot{\phi}^2}=0,$$

$$\ddot{\phi}+3\frac{\dot{a}}{a}\dot{\phi}-3\zeta\frac{\dot{a}}{a}\dot{\phi}^3+\frac{V_{,\phi}}{\zeta V}\left(1-\zeta\dot{\phi}^2\right)=0,$$

$$H=F_0\left[12a\dot{a}^2R+12a^2\dot{a}\dot{R}-a^3R^2\right]+\frac{a^3V(\phi)}{\sqrt{1-\zeta\dot{\phi}^2}}=0. \tag{33}$$

It is not difficult to check that the previous conserved current (I_2) satisfies the field equations only under the trivial condition $R=0$ together with the condition $V_0=0$, that is, for vanishing potential, which leads to inconsistency, since it has been restricted at the beginning. Thus, $F(R)\propto R^2$ is not a symmetry of the action under consideration, and the work performed by Jamil et al. [17] is completely wrong. One can even look at the consequence in a straightforward manner. It is well-known that for the Noether point symmetry $\partial/\partial t$, the Noether integral is the Hamiltonian; that is, I_1 is the Hamiltonian H. But here, $I_1=\eta H=(c_1t+c_3)H$. Thus, unless $\eta=1$, implying $c_1=0$ and $c_3=1$, Hamiltonian is not obtained as the integral of motion. However, since $c_1=c_4$, it cannot be set equal to zero at this stage, as it makes $\alpha=\beta=0$ and $\gamma=$ constant, while the form of the potential (28) becomes undefined. Thus, the Hamiltonian is never recovered as the Noether integral. Therefore, to recover the Hamiltonian constraint equation as an outcome of Noether point symmetry $\partial/\partial t$, one should start with $c_1=0$, *a priori*, which we consider in the following subsection.

2.2. No Need to Consider Time Translation, so $\eta=c_3=1$. It must be clear by this time why earlier authors did not consider time translation. Likewise, if one starts with $\eta=1$, which implies $c_1=0$ and $c_3=1$, γ turns out to be a constant and may be set as $\gamma=c_2=1$, without loss of generality. Thus, (24) through (27) take the following forms:

$$3c_4\left(F'R-F\right)+RF''\beta=0, \tag{34}$$

$$3c_4F'+F''\beta=0, \tag{35}$$

$$\beta F'''+F''\left(3c_4+\beta'\right)=0, \tag{36}$$

$$3c_4V+V_{,\phi}=0. \tag{37}$$

Now, just multiplying (35) by R and comparing it with (34), one can observe that $F(R)=0$. Thus, search for a form of $F(R)$ by imposing the Noether symmetry in the action (7) went in vain.

3. In Search of Noether Symmetry for a More General Tachyonic Action

Having obtained null result in connection with the Noether symmetry for Born-Infeld action being coupled to $F(R)$, let us now turn our attention to a more general action containing a linear curvature invariant term being nonminimally coupled to Born-Infeld-$F(R)$ action (7). However, as we have already mentioned that the gravitational Hamiltonian is not only conserved but is also constrained to vanish and therefore is a part and parcel of Einstein's equation, namely, the $\binom{0}{0}$ component, so time translation is overall unnecessary. Further, in all our earlier analysis, we have shown that a gauge term does not contribute to the Noether equations, since it becomes constant and therefore may be set equal to zero. Therefore, in the following analysis neither do we consider time translation nor a gauge term. In the Robertson-Walker line element (6) the following action

$$A=\int d^4x\sqrt{-g}\left[h(\phi)R+\mathcal{B}F(R)-V(\phi)\sqrt{1+\zeta\phi_{,\mu}\phi^{,\mu}}\right] \tag{38}$$

leads to the point Lagrangian:

$$\begin{aligned}L&=6a\dot{a}^2h+6a^2\dot{a}\dot{\phi}h_{,\phi}-6kah\\&+\mathcal{B}\left[6a\dot{a}^2F'+6a^2\dot{a}\dot{R}F''+a^3\left(F-F'R\right)-6kaF'\right]\\&+a^3V(\phi)\sqrt{1-\zeta\dot{\phi}^2},\end{aligned} \tag{39}$$

following the same technique as before. For the previous point Lagrangian the Noether equation (1) reads

$$\begin{aligned}&\alpha\Big[6\dot{a}^2h+12a\dot{a}\dot{\phi}h_{,\phi}-6kh\\&+\mathcal{B}\left[6\dot{a}^2F'+12a\dot{a}\dot{R}F''+3a^2\left(F-F'R\right)-6kF'\right]\\&+3a^2V\sqrt{1-\zeta\dot{\phi}^2}\Big]\\&+\mathcal{B}\beta\left[6a\dot{a}^2F''+6a^2\dot{a}\dot{R}F'''-a^3RF''-6kaF''\right]\\&+\gamma\Big[6a\dot{a}^2h_{,\phi}+6a^2\dot{a}\dot{\phi}h_{,\phi\phi}-6kah_{,\phi}\\&+a^3V_{,\phi}\sqrt{1-\zeta\dot{\phi}^2}\Big]+\left[\alpha_{,a}\dot{a}+\alpha'\dot{R}+\alpha_{,\phi}\dot{\phi}\right]\\&\times\left[12a\dot{a}h+6a^2h_{,\phi}\dot{\phi}+\mathcal{B}\left[12a\dot{a}F'+6a^2\dot{R}F''\right]\right]\\&+\left[\beta_{,a}\dot{a}+\beta'\dot{R}+\beta_{,\phi}\dot{\phi}\right]\left[6\mathcal{B}a^2\dot{a}F''\right]\\&+\left[\gamma_{,a}\dot{a}+\gamma'\dot{R}+\gamma_{,\phi}\dot{\phi}\right]\left[6a^2\dot{a}h_{,\phi}-\frac{\zeta a^3V\dot{\phi}}{\sqrt{1-\zeta\dot{\phi}^2}}\right]=0.\end{aligned} \tag{40}$$

Now, equating coefficients as usual, we obtain the following overdeterminant set of Noether equations:

$$\alpha' = \alpha_{,\phi} = 0, \tag{41}$$

$$\begin{aligned} 2\alpha h_{,\phi} + a\gamma h_{,\phi\phi} + ah_{,\phi}\alpha_{,a} \\ + ah_{,\phi}\gamma_{,\phi} + \mathscr{B}aF''\beta_{,\phi} = 0, \end{aligned} \tag{42}$$

$$\gamma' = \gamma_{,a} = 0, \tag{43}$$

$$\gamma_{,\phi} = 0, \tag{44}$$

$$\begin{aligned} \mathscr{B}\left[3a^2\alpha\left(F - F'R\right) - a^3\beta RF'' \right. \\ \left. -6\alpha kF' - 6\beta kaF''\right] - 6\alpha kh = 0, \end{aligned} \tag{45}$$

$$\begin{aligned} \alpha h + a\gamma h_{,\phi} + 2ah\alpha_{,a} \\ + \mathscr{B}\left[\alpha F' + a\beta F'' + 2a\alpha_{,a}F' + a^2\beta_{,a}F''\right] = 0, \end{aligned} \tag{46}$$

$$2a\alpha F'' + a^2\beta F''' + a^2F''\left(\alpha_{,a} + \beta'\right) = 0, \tag{47}$$

$$3\alpha V + a\gamma V_{,\phi} = 0. \tag{48}$$

3.1. Solutions. Equations (41) through (48) imply $\alpha = \alpha(a)$, $\beta = \beta(a, R, \phi)$, and $\gamma = \gamma_0$, where γ_0 is a constant. Equation (42) then determines α as a linear function of the scale-factor, namely,

$$\alpha = \alpha_0 a, \tag{49}$$

and (42) gets reduced to

$$3\alpha_0 h_{,\phi} + \gamma_0 h_{,\phi\phi} + \mathscr{B}F''\beta_{,\phi} = 0. \tag{50}$$

Equation (45) is then satisfied only for spatially flat $k = 0$ case and thus reduces to

$$3\alpha_0\left(F - F'R\right) = \beta RF'', \tag{51}$$

forcing β to be a function of R only; that is, $\beta = \beta(R)$. Equation (46) then in view of (45) makes $F(R)$ a linear function of R as $F(R) = F_0R$; F_0 is a constant. Thus, search for the Noether symmetry for $F''(R) \neq 0$ again went in vain. Nevertheless, the coupling parameter $h(\phi)$ gets solved as

$$h(\phi) = F_0 e^{-\phi/\phi_0}, \tag{52}$$

where $\phi_0 = \gamma_0/3\alpha_0$ is a constant. Equation (47) is then trivially satisfied while (48) yields the following form of the potential:

$$V(\phi) = V_0 e^{-\phi/\phi_0}. \tag{53}$$

The previous forms of the potential $V(\phi)$ and the coupling parameter $h(\phi)$ along with the conserved current had been found earlier by de Souza and Kremer [55] as a consequence of the Noether symmetry for linear gravity, and further, they had explicitly studied the cosmological evolution. Hence, we leave our discussion here. Nonetheless, we observe that $F(R)$ theory of gravity does not admit the Noether symmetry even for a general Born-Infeld action and so our earlier conclusion that "it is not possible to find a form of $F(R)$ other than the trivial and very special one, namely, $R^{3/2}$, by imposing Noether symmetry" stands.

4. On the Absence of Noether Symmetry and Exploring the Special Feature of $R^{3/2}$

We have mentioned that starting from $F(R) = R^2$, *a priori*, the Noether symmetry is obtainable for a scalar-tensor theory of gravity [56]. Further, it has been proved that $F(R) = R^2$ leads to a scale-invariant action [57]. So, it is expected that starting from arbitrary form of $F(R)$ if the Noether symmetry is claimed, it should end up with $F(R) \propto R^2$. But we have not been able to recover this result. This contradiction puts up doubt in treating R as an auxiliary variable for canonical formulation of $F(R)$ theory of gravity, since an auxiliary variable Q different from R has been considered for canonical formulation of R^2 theory of gravity. Let us briefly describe the issue.

In the quantum domain observable depends on the choice of momentum, while momentum is different for different choice of auxiliary variable. The canonical formulation of R^2 gravity in view of the auxiliary variable $Q = \partial A/\partial \ddot{h}_{ij}$ (A being the action) leads to a Schrödinger-like quantum dynamics, with a hermitian effective Hamiltonian leading to the straightforward probability interpretation [59–64], provided that the total derivative terms in the action are taken care of *a priori*. From metric variation principle, it is known that R^2 theory of gravity must be supplemented by a boundary term $\Sigma = 4\beta \int ({}^4R)K\sqrt{h}\,d^3x$, where symbols have their usual meaning. It was shown [63, 64] that to obtain Schrödinger-like quantum equation as mentioned, it is required first to express the action in terms of the first fundamental form h_{ij} and then to split the above boundary term into $\Sigma = \sigma_1 + \sigma_2$, where $\sigma_1 = 4\beta \int ({}^3R)K\sqrt{h}\,d^3x$ and $\sigma_2 = 4\beta \int ({}^4R - {}^3R)K\sqrt{h}\,d^3x$. Canonical programme then follows by eliminating the available total derivative term from the action, which gets cancelled with the boundary term σ_1, and then introducing the auxiliary variable, $Q = \partial A/\partial \ddot{h}_{ij}$, as suggested by Horowitz [65] thereafter. Thus, Q is different from R in the R^2 theory of gravity. It is not possible to follow such technique for an action containing a general $F(R)$ theory of gravity. It is also important to mention that classical field equations require derivative of momentum, and an arbitrary choice of auxiliary variable reproduces correct classical field equations. However, as for quantization one requires momentum (rather than its derivative, which is the reason for obtaining different quantum dynamics with different auxiliary variables); likewise, for Noether symmetry one again requires momentum instead of its derivative, and this may be the reason why it could not reproduce Noether symmetry for $F(R) = R^2$ theory of gravity already available in the literature [56]. All the beauty of $F(R)$ theory of gravity observed in the context of cosmological data fitting, are outcome of scalar-tensor equivalence, since otherwise it is almost impossible to find exact solutions. Scalar-tensor equivalence is a mathematical artifact, and it gives totally different quantum description [66] in comparison to one discussed earlier [62, 63]. The difference at the classical level has also been established to some extent [67, 68]. Thus, the prospect of $F(R)$ theory of gravity appears to be at stake, unless one can find a better way out to handle the theory. Indeed a better technique has been expatiated earlier by finding a general (non-Noether) conserved current for $F(R)$ theory of gravity being minimally coupled to scalar-tensor theory of gravity [69]. The technique was also found useful to extract exact solutions of the theory [14, 70].

The obvious question is how then $F(R) \propto R^{3/2}$ was found in vacuum and in radiation dominated era? Let us brief the beauty of $F(R) \propto R^{3/2}$ in isotropic space-time [13]. Remember that in metric variation technique, the field equations are obtained by varying the action with respect to $g_{\mu\nu}$, while the scale-factor ($a = \sqrt{h_{ij}}$) is taken as the basic variable to express the canonical action. Instead, if we take $h_{ij} = a^2 = z$ to be the basic variable, then $R = 6(\ddot{z}/2z + k/z)$ and so the action

$$A = \int \sqrt{-g}\, d^4x F(R) + 2\int_{\Sigma} \sqrt{h} F_{,R} K\, d^3x \tag{54}$$

for $F(R) = R^{3/2}$ reads

$$A = 3\sqrt{3}\int (\ddot{z} + 2k)^{3/2} dt - 2\int_{\sigma} \frac{3}{2} z^{3/2} \sqrt{R} K\, d^3x. \tag{55}$$

Introducing an auxiliary variable

$$Q = \frac{\partial A}{\partial \ddot{z}} = \frac{9\sqrt{3}}{2} (\ddot{z} + 2k)^{1/2} \tag{56}$$

(which is clearly different from R) in the action and removing appropriate surface terms, the canonical form of the action is

$$A = \int \left[-\dot{Q}\dot{z} + 2kQ - \frac{4Q^3}{729B^2}\right] dt. \tag{57}$$

Clearly z is cyclic and a Noether conserved current

$$\frac{d}{dt}\left(a\sqrt{R}\right) = \text{constant} \tag{58}$$

is apparent, which may be solved trivially to yield

$$a = \left[a_4 t^4 + a_3 t^3 + a_2 t^2 + a_1 t + a_0\right]^{1/2}, \tag{59}$$

while Q variation equation only reproduces the definition of Q, given previously. Thus, $F(R) = R^{3/2}$ leads to a trivial Noether current when viewed in terms of the basic variable h_{ij}. The above solution obtained by several authors clearly leads to power law inflation. It is also admissible in the matter dominated era, leading to present acceleration, but early deceleration remains absent.

5. Handling $F(R)$ Theory of Gravity in View of a General Conserved Current

In the absence of Noether symmetry of $F(R)$ theory of gravity coupled to tachyon field, even if a suitable form of $F(R)$ is chosen by hand, it is extremely difficult, if not impossible to find exact solution. However, in the literature there exists a technique to find a conserved current for nonminimally coupled scalar-tensor theory of gravity [71, 72] and also for $F(R)$ being minimally coupled to a non-minimal scalar-tensor theory of gravity [14, 69, 70]. This conserved current has been found useful to extract solutions. Here, we explore the same corresponding to the action (38). We first briefly review the issue of conserved current already explored [14, 69, 70]. The field equations corresponding to the following action containing scalar-tensor theory of gravity in the presence of $F(R)$

$$A = \int \left[h(\phi) R + BF(R) - \frac{\omega(\phi)}{\phi}\phi_{,\mu}\phi^{,\mu} - V(\phi) - \kappa L_m\right] \times \sqrt{-g}\, d^4x \tag{60}$$

are

$$\begin{aligned} &h\left(R_{\mu\nu} - \frac{1}{2}g_{\mu\nu}R\right) + h^{;\alpha}_{;\alpha}g_{\mu\nu} - h_{;\mu;\nu} - \frac{\omega}{\phi}\phi_{,\mu}\phi_{,\nu} \\ &\quad + \frac{1}{2}g_{\mu\nu}\left(\frac{\omega}{\phi}\phi_{,\alpha}\phi^{,\alpha} + V(\phi)\right) \\ &\quad + B\left[F'R_{\mu\nu} - \frac{1}{2}Fg_{\mu\nu} + \left(F'\right)^{;\alpha}_{;\alpha}g_{\mu\nu} - \left(F'\right)_{;\mu;\nu}\right] \\ &= \frac{\kappa}{2}T_{\mu\nu}, \end{aligned} \tag{61}$$

$$Rh_{,\phi} + 2\frac{\omega}{\phi}\phi^{;\mu}_{;\mu} + \left(\frac{\omega_{,\phi}}{\phi} - \frac{\omega}{\phi^2}\right)\phi^{,\mu}\phi_{,\mu} - V_{,\phi}(\phi) = 0. \tag{62}$$

The trace of (61) is the following:

$$\begin{aligned} &Rh - 3h^{;\mu}_{;\mu} - \frac{\omega}{\phi}\phi^{,\mu}\phi_{,\mu} - 2V \\ &\quad - B\left[RF' + 3\Box\left(F'\right) - 2F\right] \\ &= -\frac{\kappa}{2}T^{\mu}_{\mu}. \end{aligned} \tag{63}$$

Now eliminating the first term between (62) and (63), then substituting $\Box h = h_{,\phi}\phi^{;\mu}_{;\mu} + h_{,\phi\phi}\phi_{,\alpha}\phi^{,\alpha}$ and following a little algebra, one can arrive at the following equation:

$$\begin{aligned} &\left[\left(3h_{,\phi}^2 + 2h\frac{\omega}{\phi}\right)^{1/2}\phi^{;\mu}\right]_{;\mu} + \left(3h_{,\phi}^2 + 2h\frac{\omega}{\phi}\right)^{-1/2} \\ &\quad \times \left[Bh_{,\phi}\left[RF' + 3\Box\left(F'\right) - 2F\right]\right. \\ &\quad \left. - \frac{\kappa}{2}h_{,\phi}T^{\mu}_{\mu} - h^3\left(\frac{V}{h^2}\right)_{,\phi}\right] = 0. \end{aligned} \tag{64}$$

In view of the above equation one can conclude that under the following condition:

$$B\left[RF' + 3\Box\left(F'\right) - 2F\right] = \frac{\kappa}{2}T^{\mu}_{\mu} + \frac{h^3}{h_{,\phi}}\left(\frac{V}{h^2}\right)_{,\phi} \quad (65)$$

there exists a conserved current J^{μ}, where

$$J^{\mu}_{;\mu} = \left[\left(3h^2_{,\phi} + 2h\frac{\omega}{\phi}\right)^{1/2}\phi^{;\mu}\right]_{;\mu} = 0. \quad (66)$$

Further, assuming $V \propto h^2$, one finds that the condition (65) for the existence of the conserved current does not depend on the scalar field and the choice of potential. Thus, the cosmological evolution only depends on the form of $F(R)$, which may be chosen by hand. Note that T^{μ}_{μ} vanishes in vacuum and radiation dominated era, and so in these era, the fluid distribution does not play any role. In fact if one takes, say, for example, $F(R) \propto R^2$, power law inflation is realized in vacuum era. Inflation makes the space-time flat, and so assuming $k = 0$ in the isotropic and homogeneous metric, the cosmological evolution of the scale-factor in the radiation era behaves like Friedmann solution ($a \propto \sqrt{t}$). Thus, higher-order curvature invariant term does not affect baryogenesis, nucleosynthesis, and structure formation (see [70]). Let us now turn our attention to find similar conserved current corresponding to the action (38) containing tachyonic field. The field equations corresponding to the action (38) containing a matter part in addition are the following:

$$\begin{aligned} &h\left(R_{\mu\nu} - \frac{1}{2}g_{\mu\nu}R\right) + \Box h g_{\mu\nu} - h_{;\mu;\nu} \\ &\quad + \mathscr{B}\left[F'R_{\mu\nu} - \frac{1}{2}Fg_{\mu\nu} + \left(\Box F'\right)g_{\mu\nu} - \left(F'\right)_{;\mu;\nu}\right] \\ &= \frac{1}{2}\left(\mathscr{T}_{\mu\nu} + T_{\mu\nu}\right), \end{aligned} \quad (67)$$

$$Rh_{,\phi} + \left(\frac{\zeta V\phi^{;\alpha}}{\sqrt{1+\zeta\phi_{,\mu}\phi^{,\mu}}}\right)_{,\alpha} - V_{,\phi}\sqrt{1+\zeta\phi_{,\mu}\phi^{,\mu}} = 0, \quad (68)$$

where $\mathscr{T}_{\mu\nu}$ is the energy-momentum tensor for the tachyon field given by

$$\mathscr{T}_{\alpha\beta} = g_{\alpha\beta}V(\phi)\sqrt{1+\zeta\phi_{,\mu}\phi^{,\mu}} - \frac{\zeta V\phi_{\alpha}\phi_{\beta}}{\sqrt{1+\zeta\phi_{,\mu}\phi^{,\mu}}}, \quad (69)$$

while $T_{\mu\nu}$ is that for the usual matter field. Trace of (67) is

$$Rh - 3\Box h - \mathscr{B}\left[RF' + 3\Box F' - 2F\right] = -\frac{1}{2}\left(\mathscr{T}^{\mu}_{\mu} + T^{\mu}_{\mu}\right). \quad (70)$$

Now eliminating the first terms between (68) and (70), we obtain

$$\begin{aligned} &\left(3h^2_{,\phi} + \frac{\zeta Vh}{\sqrt{1+\zeta\phi_{,\mu}\phi^{,\mu}}}\right)\phi^{;\alpha}_{;\alpha} + 3h_{,\phi}h_{,\phi\phi}\phi_{,\alpha}\phi^{,\alpha} \\ &\quad + \frac{\zeta hV_{,\phi}\phi_{,\alpha}\phi^{\alpha}}{\sqrt{1+\zeta\phi_{,\mu}\phi^{,\mu}}} + \frac{\zeta Vh\phi^{,\alpha}}{\left(\sqrt{1+\zeta\phi_{,\mu}\phi^{,\mu}}\right)_{,\alpha}} \\ &\quad + \frac{\zeta Vh_{,\phi}\phi_{,\alpha}\phi^{,\alpha}}{2\sqrt{1+\zeta\phi_{,\mu}\phi^{,\mu}}} - 2Vh_{,\phi}\sqrt{1+\zeta\phi_{,\mu}\phi^{,\mu}} \\ &\quad - V_{,\phi}h\sqrt{1+\zeta\phi_{,\mu}\phi^{,\mu}} \\ &\quad + h_{,\phi}\left[\mathscr{B}\left(RF' + 3\Box F' - 2F\right) - \frac{T^{\mu}_{\mu}}{2}\right] = 0, \end{aligned} \quad (71)$$

where we have substituted $\Box h = h_{,\phi}\phi^{;\mu}_{;\mu} + h_{,\phi\phi}\phi_{,\alpha}\phi^{,\alpha}$ and the trace $\mathscr{T}^{\mu}_{\mu}$ of the energy-momentum tensor corresponding to the Tachyon field. The previous equation can be rearranged as,

$$\begin{aligned} &\left[\left(\sqrt{3h^2_{,\phi} + \frac{\zeta Vh}{\sqrt{1+\zeta\phi_{,\mu}\phi^{,\mu}}}}\right)\phi^{;\alpha}\right]_{;\alpha} \\ &\quad + \left(3h^2_{,\phi} + \frac{\zeta Vh}{\sqrt{1+\zeta\phi_{,\mu}\phi^{,\mu}}}\right)^{-1/2} \\ &\quad \times \left[h_{,\phi}\left(\mathscr{B}\left(RF' + 3\Box F' - 2F\right) - \frac{T^{\mu}_{\mu}}{2}\right)\right. \\ &\quad + \frac{\zeta h}{2}\left(\frac{V}{\sqrt{1+\zeta\phi_{,\mu}\phi^{,\mu}}}\right)_{;\alpha} \\ &\quad \left. - \sqrt{1+\zeta\phi_{,\mu}\phi^{,\mu}}h^3\left(\frac{V}{h^2}\right)_{,\phi}\right] = 0. \end{aligned} \quad (72)$$

One can now conclude that there exists a conserved current:

$$\begin{aligned} J^{\alpha}_{;\alpha} &= \left[\left(\sqrt{3h^2_{,\phi} + \frac{\zeta Vh}{\sqrt{1+\zeta\phi_{,\mu}\phi^{,\mu}}}}\right)\phi^{;\alpha}\right]_{;\alpha} \\ &= \frac{1}{\sqrt{-g}}\left[\left(\sqrt{3h^2_{,\phi} + \frac{\zeta Vh}{\sqrt{1+\zeta\phi_{,\mu}\phi^{,\mu}}}}\right)\phi^{;\alpha}\sqrt{-g}\right]_{,\alpha} = 0, \end{aligned} \quad (73)$$

provided

$$\left[h_{,\phi}\left(\mathscr{B}\left(RF' + 3\Box F' - 2F\right) - \frac{T^{\mu}_{\mu}}{2}\right)\right.$$
$$\left. + \frac{\zeta h}{2}\left(\frac{V}{\sqrt{1+\zeta\phi_{,\mu}\phi^{,\mu}}}\right)_{;\alpha} - \sqrt{1+\zeta\phi_{,\mu}\phi^{,\mu}}h^3\left(\frac{V}{h^2}\right)_{,\phi}\right] = 0. \tag{74}$$

Note that, treating the scalar curvature R as a single variable, we had only a couple of equations, (67) and (68), to solve for the five variables, namely, R, $F(R)$, ϕ, $h(\phi)$, and $V(\phi)$. So for exact solution, we require altogether three physically reasonable assumptions. Above conserved current has been constructed out of the two field equations. So taking one of the field equations (say (68)), the conserved current (73) and the condition (74) altogether give us three equations. We therefore need two more assumptions to extract exact solution for the system. One of these may be set by separating the condition (74), so that cosmic evolution, that is, the solution of R, becomes field independent and depends only on the form of $F(R)$ chosen by hand, as before. Thus, we separate the condition (74) for the existence of the conserved current as

$$\mathscr{B}\left(RF' + 3\Box F' - 2F\right) = \frac{T^{\mu}_{\mu}}{2}, \tag{75}$$

$$\frac{\zeta h}{2}\left(\frac{V}{\sqrt{1+\zeta\phi_{,\mu}\phi^{,\mu}}}\right)_{;\alpha} - \sqrt{1+\zeta\phi_{,\mu}\phi^{,\mu}}h^3\left(\frac{V}{h^2}\right)_{,\phi} = 0, \tag{76}$$

since $h_{,\phi} \neq 0$. Clearly the condition (65) (under the choice $V \propto h^2$) is the same as obtained here in (75). Therefore, the cosmological evolution is independent of choice of the field, even if it is tachyon. Thus, we have completed our goal to find a general conserved current for tachyonic field such that the cosmological evolution does not depend on the field and the choice of potential at all. Finally, as soon as a form of $F(R)$ would be chosen by hand-one can, in principle, have explicit solution of the field equations under consideration. It is important to note that the cosmological evolution depends strictly on the form of $F(R)$, and it does not in any way depend on the linear curvature invariant term. In the following, we will take up Robertson-Walker line element (6) to explicitly demonstrate the applicability of the treatment developed. Equation (72) states that there exists a conserved current, namely,

$$a^3\dot{\phi}\sqrt{3h^2_{,\phi} + \frac{\zeta V h}{\sqrt{1-\zeta\dot{\phi}^2}}} = C, \tag{77}$$

C being a constant, provided

$$\zeta\dot{\phi}\ddot{\phi} + \left(1-\zeta\dot{\phi}^2\right)\left[\dot{\phi} - \frac{2}{\zeta}\left(1-\zeta\dot{\phi}^2\right)\right]\frac{V_{,\phi}}{V} + \frac{4}{\zeta}\left(1-\zeta\dot{\phi}^2\right)^2\frac{h_{,\phi}}{h} = 0, \tag{78}$$

$$\mathscr{B}\left[RF' + 3\Box F' - 2F\right] = \frac{1}{2}T^{\mu}_{\mu}. \tag{79}$$

Note that if matter part ($T_{\mu\nu}$) is ignored for the time being, still we have only a couple of independent field equations to solve for $F(R), h(\phi), V(\phi), \phi(t)$, and the scale-factor $a(t)$. Thus, it is required to make at least three physically reasonable assumptions to obtain explicit solutions of the system under consideration. Here, to obtain the conserved current, we have essentially made two assumptions, namely, (78) and (79). Therefore, if a form of $F(R)$ is assumed, one can solve (79) for the scale-factor $a(t)$, and the remaining three, namely, $h(\phi), V(\phi)$ and $\phi(t)$, may be solved in view of the conserved current (77), (78), and one of the field equations (say, (68)). Interestingly enough, in the process, curvature part (79) is decoupled from the field under consideration and the solution of the scale-factor which emerges from (79). Therefore, cosmic evolution remains independent of the form of the field chosen. For example, in the flat case $k = 0$, it has been shown [70] that for $F(R) = R^2$, (79) leads to power law inflationary solution in the vacuum dominated era, and $a \propto \sqrt{t}$ in the radiation dominated era ($T^{\mu}_{\nu} = 0$), which is the result of the standard Friedmann model, despite strong coupling with tachyon field being incorporated and $a \propto t^{4/3}$ in the matter dominated era ($p = 0$). On the other hand, if $F(R) = R^{3/2}$ is chosen [14], (79) leads to $a \propto \sqrt{t}$ in the radiation dominated era, and $a \propto t$ emerges as a particular solution in matter dominated era.

6. Weak Energy Limit

Indeed, it is true that solar system puts up severe constraints on alternative theories of gravity [73, 74]. For an action

$$A = \int \sqrt{-g}\, d^4x R^n, \tag{80}$$

the gravitational potential [75] in the weak field limit [76] is found as

$$\Phi(r) = -\frac{Gm}{2r}\left[1 + \left(\frac{r}{r_c}\right)^{\beta}\right], \tag{81}$$

where r_c is an arbitrary parameter varying within the range $(1 - 10^4)$ AU taking into account the velocity of the earth to be 30 Km s^{-1} [73], while β is related to n as

$$\beta = \frac{12n^2 - 7n - 1 - \sqrt{36n^4 + 12n^3 - 83n^2 + 50n + 1}}{6n^2 - 4n + 2}. \tag{82}$$

Clearly, for $n = 1$, $\beta = 0$ and Newtonian gravitational field is recovered. For $n = 3/2$, $\beta \sim 0.5$ Newtonian limit is not realized, and such value of β is ruled out from light bending data in the sun limb and planetary periods [73].

Here, we provide additional constraint required to fit the Newtonian limit for the more general action (38) being supplemented by ordinary matter action. In weak field approximation $g_{\mu\nu} = \eta_{\mu\nu} + h_{\mu\nu}$, where $|h_{\mu\nu}| \ll 1$. Retaining only linear terms in $h_{\mu\nu}$, we have

$$R_{\mu\nu} \simeq \frac{1}{2}\Box h_{\mu\nu}, \qquad R \simeq \frac{1}{2}\Box h, \quad \text{where } h = h^{\mu}_{\mu}, \tag{83}$$

along with the time-time component of the field equation (61), taking $F(R) = R^{3/2}$ as

$$\begin{aligned}&\left(R_{00} - \frac{1}{2}g_{00}R\right) - \mathscr{B}R^{1/2}\left(3R_{00} - Rg_{00}\right) - \frac{3\mathscr{B}}{2}R^{-3/2}\\ &\qquad\times\left[\left(R\Box R - \frac{1}{2}R_{;\lambda}R^{;\lambda}\right)g_{00} + RR_{;0;0} - \frac{1}{2}R_{;0}R_{;0}\right]\\ &\quad = \mathscr{T}_{00} + T_{00},\end{aligned} \tag{84}$$

provided $\phi \to$ constant with the expansion, so that $h(\phi) \to 1/16\pi G$ with the cosmic evolution. Under such constraint, (84) in static-background spacetime, retaining only linear term in $h_{\mu\nu}$ yields

$$\nabla^2 h_{00} \simeq \rho_e, \tag{85}$$

where $\rho_e = \rho + \rho_{\mathscr{T}}$ is the effective matter density, ρ and $\rho_{\mathscr{T}}$ being the matter densities corresponding to the pressureless dust and tachyonic field, respectively. Considering next higher-order term in $h_{\mu\nu}$, (84) gives

$$\nabla^2 h_{00} - 3\mathscr{B}\sqrt{\frac{1}{2}\nabla^2 h}\left(\nabla^2 h_{00} - \frac{1}{6}\nabla^2 h\right) \simeq \rho_e. \tag{86}$$

Since at low energy limit Poisson equation is obtained, the Newtonian gravity is valid at weak energy limit. Nevertheless, to pass solar test, an additional constraint $h(\phi) \to 1/16\pi G$ under weak field approximation is required.

7. Concluding Remarks

Earlier attempts to find the Noether symmetry for $F(R)$ theory of gravity in vacuum and matter dominated era lead to $F(R) \propto R^{3/2}$ [6–13]. Such a form is not suitable to explain presently available cosmological data [13]. Search of a better form of $F(R)$ taking a minimally or nonminimally coupled scalar field into account failed to produce symmetry [14]. Apparently, it is not a problem, since not all actions admit symmetry. But then, such a symmetry for $F(R) \propto R^2$ already exists in the literature, where an auxiliary variable Q different from R was introduced for the purpose of canonization [56]. Further, $F(R) \propto R^2$ leads to a scale-invariant action [57]. Hence, the result should have been reproduced from $F(R)$ theory of gravity. Canonization of the action, treating R as the auxiliary variable, creates problem in quantum domain, while canonical formulation of a general $F(R)$ theory of gravity is possible only by treating R as an auxiliary variable, and so we presume that it might be the root of trouble.

Earlier, some authors [15] have claimed that $F(R)$ in vacuum admits the Noether symmetry for arbitrary power of R, if a gauge term is introduced. Such a claim is not true since neither all of the Noether equations are satisfied nor the conserved current obtained satisfies the Hamilton constraint equation [16]. There is yet another more recent claim that Noether symmetry exists for $F(R) \propto R^2$ in the presence of tachyon field [17]. Here, we show that not only the conserved current does not satisfy the field equations, but also the Hamiltonian is not constrained to vanish, which is fundamental of the theory of gravitation. We have also taken up a more general action in Section 3 and found that the Noether symmetry is not allowed for $F(R)'' \neq 0$. Since all attempts to find the Noether symmetry for $F(R)$ theory of gravity (other than the trivial one, viz., $F(R) \propto R^{3/2}$) so far have failed, we conclude that it is not possible to find a form of $F(R)$ by demanding the Noether symmetry.

To handle $F(R)$ theory of gravity in the presence of tachyonic field, we have presented a general conserved current in Section 5, which is different from one obtained earlier for $F(R)$ being coupled to scalar-tensor theory of gravity [14, 69, 70]. However, one of the conditions required for such conserved current, containing higher-order curvature invariant terms, remains unaltered. This part is decoupled from the matter field and may be solved in principle for the scale-factor $a(t)$ independently, choosing a suitable form of $F(R)$ by hand. Thus, cosmic evolution remains independent of the choice of the field (Tachyon or scalar field). The conserved current obtained in the process might be an outcome of some higher symmetry, which is not known at present.

References

[1] S. Nojiri and S. D. Odintsov, "Unified cosmic history in modified gravity: from $F(R)$ theory to Lorentz non-invariant models," *Physics Reports*, vol. 505, no. 2–4, pp. 59–144, 2011.

[2] S. Capozziello and M. de Laurentis, "Extended theories of gravity," *Physics Reports*, vol. 509, no. 4-5, pp. 167–321, 2011.

[3] K. S. Stelle, "Renormalization of higher-derivative quantum gravity," *Physical Review D*, vol. 16, no. 4, pp. 953–969, 1977.

[4] T. Biswas, E. Gerwick, T. Koivisto, and A. Mazumdar, "Towards singularity- and ghost-free theories of gravity," *Physical Review Letters*, vol. 108, no. 3, Article ID 031101, 4 pages, 2012.

[5] T. Biswas, T. Koivisto, and A. Mazumdar, "Nonlocal theories of gravity:the flat space propagator," Proceedings of the Barcelona Postgrad Encounters on Fundamental Physics, 2013.

[6] S. Capozziello and G. Lambiase, "Selection rules in minisuper space quantum cosmology," *General Relativity and Gravitation*, vol. 32, no. 4, pp. 673–696, 2000.

[7] S. Capozziello, A. Stabile, and A. Troisi, "Spherically symmetric solutions in $F(R)$ gravity via the Noether symmetry approach," *Classical and Quantum Gravity*, vol. 24, no. 8, article 013, pp. 2153–2166, 2007.

[8] S. Capozziello and A. De Felice, "$F(R)$ cosmology by Noether's symmetry," *Journal of Cosmology and Astroparticle Physics*, vol. 2008, 22 pages, 2008.

[9] B. Vakili, "Noether symmetry in $f(R)$ cosmology," *Physics Letters B*, vol. 664, pp. 16–20, 2008.

[10] B. Vakili, "Noether symmetric $f(R)$ quantum cosmology and its classical correlations," *Physics Letters B*, vol. 669, no. 3-4, pp. 206–211, 2008.

[11] S. Capozziello, P. Martin-Moruno, and C. Rubano, "Dark energy and dust matter phases from an exact $F(R)$-cosmology model," *Physics Letters B*, vol. 664, no. 1-2, pp. 12–15, 2008.

[12] S. Capozziello, P. Martin-Moruno, and C. Rubano, "Exact $f(R)$-cosmological model coming from the request of the existence of a Noether symmetry," *AIP Conference Proceedings*, vol. 1122, pp. 213–216, 2009.

[13] K. Sarkar, S. K. Nayem, S. Ruz, S. Debnath, and A. K. Sanyal, "Why Noether symmetry of $F(R)$ theory yields three-half power law?" *International Journal of Theoretical Physics*, vol. 52, no. 5, pp. 1515–1531, 2013.

[14] K. Sarkar, S. K. Nayem, S. Debnath, and A. K. Sanyal, "Viability of Noether symmetry of $F(R)$ theory of gravity," *International Journal of Theoretical Physics*, vol. 52, no. 4, pp. 1194–1213, 2012.

[15] I. Hussain, M. Jamil, and F. M. Mahomed, "Noether gauge symmetry approach in $F(R)$ gravity," *Astrophysics and Space Science*, vol. 337, no. 1, pp. 373–377, 2012.

[16] N. Sk and A. K. Sanyal, "Revisiting Noether gauge symmetry for $F(R)$ theory of gravity," *Astrophysics and Space Science*, vol. 342, pp. 549–555, 2012.

[17] M. Jamil, F. M. Mahomed, and D. Momeni, "Noether symmetry approach in $F(R)$-tachyon model," *Physics Letters B*, vol. 702, no. 5, pp. 315–319, 2011.

[18] A. Sen, "Space—like branes," *Journal of High Energy Physics*, vol. 0204, 23 pages, 2002.

[19] A. Sen, "Tachyon matter," *Journal of High Energy Physics*, vol. 0207, 14 pages, 2002.

[20] A. Sen, "Supersymmetric world volume action for nonBPS D-branes," *Journal of High Energy Physics*, vol. 9910, 16 pages, 1999.

[21] M. R. Garousi, "Tachyon couplings on non-BPS D-branes and dirac-born-infeld action," *Nuclear Physics B*, vol. 584, no. 1-2, pp. 284–299, 2000.

[22] M. R. Garousi, "On shell S matrix and tachyonic effective actions," *Nuclear Physics B*, vol. 647, pp. 117–130, 2002.

[23] M. R. Garousi, "Slowly varying tachyon and tachyon potential," *Journal Of High Energy Physics*, vol. 0305, 11 pages, 2003.

[24] E. A. Bergshoeff, M. de Roo, T. C. de Wit, E. Eyras, and S. Panda, "T duality and actions for non BPS D-branes," *Journal of High Energy Physics*, vol. 0005, 10 pages, 2000.

[25] J. Kluson, "Proposal for non—BPS D—brane action," *Physical Review D*, vol. 62, Article ID 126003, 7 pages, 2000.

[26] G. W. Gibbons, "Cosmological evolution of the rolling tachyon," *Physics Letters B*, vol. 537, no. 1-2, pp. 1–4, 2002.

[27] A. Mazumdar, S. Panda, and A. Pérez-Lorenzana, "Assisted inflation via tachyon condensation," *Nuclear Physics B*, vol. 614, no. 1-2, pp. 101–116, 2001.

[28] M. Fairbairn and M. H. G. Tytgat, "Inflation from a tachyon fluid?" *Physics Letters B*, vol. 546, no. 1-2, pp. 1–7, 2002.

[29] A. Feinstein, "Power-law inflation from the rolling tachyon," *Physical Review D*, vol. 66, no. 6, Article ID 063511, 3 pages, 2002.

[30] M. Sami, P. Chingangbam, and T. Qureshi, "Aspects of tachyonic inflation with an exponential potential," *Physical Review D*, vol. 66, no. 4, Article ID 043530, 5 pages, 2002.

[31] M. Sami, "Implementing power law inflation with tachyon rolling on the brane," *Modern Physics Letters A*, vol. 18, no. 10, pp. 691–697, 2003.

[32] Y. S. Piao, R. G. Cai, X. m. Zhang, and Y. Z. Zhang, "Assisted tachyonic inflation," *Physical Review D*, vol. 66, no. 12, Article ID 121301, 3 pages, 2002.

[33] L. R. W. Abramo and F. Finelli, "Cosmological dynamics of the tachyon with an inverse power-law potential," *Physics Letters B*, vol. 575, no. 3-4, pp. 165–171, 2003.

[34] D.-J. Liu and X.-Z. Li, "Cosmological perturbations and non-commutative tachyon inflation," *Physical Review D*, vol. 70, no. 12, Article ID 123504, 10 pages, 2004.

[35] G. M. Kremer and D. S. M. Alves, "Acceleration field of a universe modeled as a mixture of scalar and matter fields," *General Relativity and Gravitation*, vol. 36, no. 9, pp. 2039–2051, 2004.

[36] D. A. Steer and F. Vernizzi, "Tachyon inflation: tests and comparison with single scalar field inflation," *Physical Review D*, vol. 70, no. 4, Article ID 43527, 15 pages, 2004.

[37] C. Campuzano, S. Del Campo, and R. Herrera, "Curvaton reheating in tachyonic inflationary models," *Physics Letters B*, vol. 633, no. 2-3, pp. 149–154, 2006.

[38] R. Herrera, S. Del Campo, and C. Campuzano, "Tachyon warm inflationary universe models," *Journal of Cosmology and Astroparticle Physics*, no. 10, article 009, 2006.

[39] H. Xiong and J. Zhu, "Tachyon field in loop quantum cosmology: inflation and evolution picture ," *Physical Review D*, vol. 75, no. 8, Article ID 084023, 8 pages, 2007.

[40] L. Balart, S. del Campo, R. Herrera, P. Labraña, and J. Saavedra, "Tachyonic open inflationary universes," *Physics Letters B*, vol. 647, no. 5-6, pp. 313–319, 2007.

[41] L. Kofman and A. Linde, "Problems with tachyon inflation," *Journal of High Energy Physics*, vol. 2002, no. 7, article 4, 2002.

[42] T. Padmanabhan, "Accelerated expansion of the universe driven by tachyonic matter," *Physical Review D*, vol. 66, no. 2, Article ID 021301, 4 pages, 2002.

[43] J. S. Bagla, H. K. Jassal, and T. Padmanabhan, "Cosmology with tachyon field as dark energy," *Physical Review D*, vol. 67, no. 6, Article ID 063504, 11 pages, 2003.

[44] J. Hao and X. Li, "Reconstructing the equation of state of the tachyon," *Physical Review D*, vol. 66, no. 8, Article ID 087301, 4 pages, 2002.

[45] L. R. W. Abramo and F. Finelli, "Cosmological dynamics of the tachyon with an inverse power-law potential," *Physics Letters B*, vol. 575, no. 3-4, pp. 165–171, 2003.

[46] J. M. Aguirregabiria and R. Lazkoz, "A note on the structural stability of tachyonic inflation," *Modern Physics Letters A*, vol. 19, no. 12, pp. 927–930, 2004.

[47] Z. K. Guo and Y. Z. Zhang, "Cosmological scaling solutions of multiple tachyon fields with inverse square potentials ," *Journal of Cosmology and Astroparticle Physics*, vol. 2004, no. 8, article 10, 2004.

[48] E. J. Copeland, M. R. Garousi, M. Sami, and S. Tsujikawa, "What is needed of a tachyon if it is to be the dark energy?" *Physical Review D*, vol. 71, no. 4, pp. 43003–43011, 2005.

[49] S. Del Campo, R. Herrera, and D. Pavon, "Soft coincidence in late acceleration," *Physical Review D*, vol. 71, no. 12, Article ID 123529, 11 pages, 2005.

[50] A. Das, S. Gupta, T. D. Saini, and S. Kar, "Cosmology with decaying tachyon matter," *Physical Review D*, vol. 72, no. 4, Article ID 043528, 6 pages, 2005.

[51] G. Panotopoulos, "Cosmological constraints on a dark energy-model with a non-linear scalar field," http://arxiv.org/abs/astro-ph/0606249v2.

[52] J. Ren and X. Meng, "Tachyon field-inspired dark energy and supernovae constraints," *International Journal of Modern Physics D*, vol. 17, no. 12, pp. 2325–2335, 2008.

[53] V. H. Cardenas, "Tachyonic quintessential inflation," *Physical Review D*, vol. 73, no. 10, Article ID 103512, 5 pages, 2006.

[54] S. K. Srivastava, "Tachyon as a dark energy source," *High Energy Physics*, Article ID 040907, 2005.

[55] C. de Souza and G. M. Kremer, "Constraining non-minimally coupled tachyon fields by the Noether symmetry," *Classical and Quantum Gravity*, vol. 26, no. 13, Article ID 135008, 2009.

[56] A. K. Sanyal, B. Modak, C. Rubano, and E. Piedipalumbo, "Noether symmetry in the higher order gravity theory," *General Relativity and Gravitation*, vol. 37, no. 2, pp. 407–417, 2005.

[57] H.-J. Schmidt and D. Singleton, "Isotropic universe with almost scale-invariant fourth-order gravity," *Journal of Mathematical Physics*, vol. 54, Article ID 062502, 2013.

[58] N. Sk and A. K. Sanyal, "Revisiting conserved currents in $F(R)$ theory of gravity via Noether symmetry," *Chinese Physics Letters*, vol. 30, no. 2, Article ID 020401, 2013.

[59] A. K. Sanyal and B. Modak, "Quantum cosmology with a curvature squared action," *Physical Review D*, vol. 63, no. 6, Article ID 064021, 6 pages, 2001.

[60] A. K. Sanyal and B. Modak, "Quantum cosmology with $R + R^2$ gravity," *Classical and Quantum Gravity*, vol. 19, no. 3, pp. 515–525, 2002.

[61] A. K. Sanyal, "Quantum mechanical probability interpretation in the mini-superspace model of higher order gravity theory," *Physics Letters B*, vol. 542, no. 1-2, pp. 147–159, 2002.

[62] A. K. Sanyal, "Quantum mechanical formulation of quantum cosmology for brane world effective action," in *Focus on Astrophysics Research*, L. V. Ross, Ed., pp. 109–117, Nova Science Publishers, 2003.

[63] A. K. Sanyal, "Hamiltonian formulation of curvature squared action," *General Relativity and Gravitation*, vol. 37, no. 12, pp. 1957–1983, 2005.

[64] A. K. Sanyal, S. Debnath, and S. Ruz, "Canonical formulation of the curvature-squared action in the presence of a lapse function," *Classical and Quantum Gravity*, vol. 29, no. 21, Article ID 215007, 2012.

[65] G. T. Horowitz, "Quantum cosmology with a positive-definite action," *Physical Review D*, vol. 31, no. 6, pp. 1169–1177, 1985.

[66] E. Dyer and K. Hinterbichler, "Boundary terms, variational principles, and higher derivative modified gravity," *Physical Review D*, vol. 79, no. 2, Article ID 024028, 20 pages, 2009.

[67] A. Bhadra, K. Sarkar, D. P. Datta, and K. K. Nandi, "Brans-dicke theory: jordan versus Einstein frame," *Modern Physics Letters A*, vol. 22, no. 5, pp. 367–375, 2007.

[68] S. Capozziello, P. Martin-Moruno, and C. Rubano, "Physical non-equivalence of the Jordan and Einstein frames," *Physics Letters B*, vol. 689, no. 4-5, pp. 117–121, 2010.

[69] A. K. Sanyal, "Scalar-tensor theory of gravity carrying a conserved current," *Physics Letters B*, vol. 624, no. 1-2, pp. 81–92, 2005.

[70] A. K. Sanyal, "Study of symmetry in $F(R)$ theory of gravity," *Modern Physics Letters A*, vol. 25, no. 31, pp. 2667–2676, 2010.

[71] A. K. Sanyal and B. Modak, "Is the Nöther symmetric approach consistent with the dynamical equation in non-minimal scalar-tensor theories?" *Classical and Quantum Gravity*, vol. 18, no. 17, pp. 3767–3774, 2001.

[72] A. K. Sanyal, "Noether and some other dynamical symmetries in Kantowski-Sachs model," *Physics Letters B*, vol. 524, no. 1-2, pp. 177–184, 2002.

[73] A. F. Zakharov, A. A. Nucita, F. De Paolis, and G. Ingrosso, "Solar system constraints on R^n gravity," *Physical Review D*, vol. 74, no. 10, Article ID 107101, 4 pages, 2006.

[74] A. F. Zakharov, S. Capozziello, F. De Paolis, G. Ingrosso, and A. A. Nucita, "The role of dark matter and dark energy in cosmological models: theoretical overview," *Space Science Reviews*, vol. 148, no. 1–4, pp. 301–313, 2009.

[75] S. Capozziello, V. F. Cardone, and A. Troisi, "Gravitational lensing in fourth order gravity," *Physical Review D*, vol. 73, no. 10, Article ID 104019, 10 pages, 2006.

[76] T. Clifton and J. D. Barrow, "The power of general relativity," *Physical Review D*, vol. 72, Article ID 103005, 21 pages, 2005.

8

Constraint on Heavy Element Production in Inhomogeneous Big-Bang Nucleosynthesis from the Light Element Observations

Riou Nakamura,[1] Masa-aki Hashimoto,[1] Shin-ichiro Fujimoto,[2] and Katsuhiko Sato[3,4]

[1] *Department of Physics, Graduate School of Sciences, Kyushu University, 6-10-1 Hakozaki, Higashi-ku, Fukuoka 812-8581, Japan*
[2] *Department of Control and Information Systems Engineering, Kumamoto National College of Technology, 2659-2 Suya, Koshi, Kumamoto 861-1102, Japan*
[3] *Institute for the Physics and Mathematics of the Universe, University of Tokyo, Kashiwa, Chiba 277-8568, Japan*
[4] *National Institutes of Natural Sciences, Kamiyacho Central Place 2F, 4-3-13 Toranomon, Minato-ku, Tokyo 104-0001, Japan*

Correspondence should be addressed to Riou Nakamura; riou@astrog.phys.kyushu-u.ac.jp

Academic Editors: G. Chincarini, X. Dai, A. De Rosa, M. S. Dimitrijevic, and G. Wegner

We investigate the observational constraints on the inhomogeneous big-bang nucleosynthesis that Matsuura et al. (2005) suggested that states the possibility of the heavy element production beyond ^{7}Li in the early universe. From the observational constraints on light elements of ^{4}He and D, possible regions are found on the plane of the volume fraction of the high-density region against the ratio between high- and low-density regions. In these allowed regions, we have confirmed that the heavy elements beyond Ni can be produced appreciably, where *p*- and/or *r*-process elements are produced well simultaneously.

1. Introduction

Big-bang nucleosynthesis (BBN) has been investigated to explain the origin of the light elements, such as ^{4}He, D, ^{3}He, and ^{7}Li, during the first few minutes [1–4]. Standard model of BBN (SBBN) can succeed in explaining the observation of those elements, ^{4}He [5–9], D [10–13], and ^{3}He [14, 15], except for ^{7}Li. The study of SBBN has been done under the assumption of the homogeneous universe, where the model has only one parameter, the baryon-to-photon ratio η. If the present value of η is determined, SBBN can be calculated from the thermodynamical history with the use of the nuclear reaction network. We can obtain the reasonable value of η by comparing the calculated abundances with observations. In the meanwhile, the value of η is obtained as $\eta = (5.1 - 6.5) \times 10^{-10}$ [1] from the observations of ^{4}He and D. These values agree well with the observation of the cosmic microwave background: $\eta = (6.19 \pm 0.14) \times 10^{-10}$ [16].

On the other hand, BBN with the inhomogeneous baryon distribution also has been investigated. The model is called as inhomogeneous BBN (IBBN). IBBN relies on the inhomogeneity of baryon concentrations that could be induced by baryogenesis (e.g., [17]) or phase transitions such as QCD or electro-weak phase transition [18–21] during the expansion of the universe. Although a large-scale inhomogeneity is inhibited by many observations [16, 22–24], a small scale one has been advocated within the present accuracy of the observations. Therefore, it remains a possibility for IBBN to occur in some degree during the early era. In IBBN, the heavy element nucleosynthesis beyond the mass number A = 8 has been proposed [17, 18, 25–35]. In addition, peculiar observations of abundances for heavy elements and/or ^{4}He could be understood in the way of IBBN. For example, the quasar metallicity of C, N, and Si could have been explained from IBBN [36]. Furthermore, from recent observations of globular clusters, a possibility of inhomogeneous helium distribution is pointed out [37], where some separate groups of different main sequences in blue band of low mass stars are assumed due to high primordial helium abundances compared to the standard value [38, 39]. Although baryogenesis could be the origin of the inhomogeneity, the mechanism of it has not been clarified due to unknown properties of the supersymmetric Grand Unified Theory [40].

Despite a negative opinion against IBBN due to insufficient consideration of the scale of inhomogeneity [41], Matsuura et al. have found that the heavy element synthesis for both p- and r-processes is possible if $\eta > 10^{-4}$ [42], where they have also shown that the high η regions are compatible with the observations of the light elements, ^{4}He and D [43]. However, their analysis is only limited to a parameter of a specific baryon number concentration. In this paper, we extend the investigations of Matsuura et al. [42, 43] to check the validity of their conclusion from a wide parameter space of the IBBN model.

In Section 2, we review and give the adopted model of IBBN which is the same one as that of Matsuura et al. [43]. Constraints on the critical parameters of IBBN due to light element observations are shown in Section 3, and the possible heavy elements of nucleosynthesis are presented in Section 4. Finally, Section 5 is devoted to the summary and discussion.

2. Model

In this section, we introduce the model of IBBN. We adopt the two-zone model for the inhomogeneous BBN. In the IBBN model, we assume the existence of spherical high-density region inside the horizon. For simplicity, we ignore in the present study the diffusion effects before (10^{10} K $< T <$ 10^{11} K) and during the primordial nucleosynthesis (10^7 K $<$ $T < 10^{10}$ K), because the timescale of the neutron diffusion is longer than that of the cosmic expansion [25, 37].

To find the parameters compatible with the observations, we consider the average abundances between the high- and low-density regions. We get at least the parameters for the extreme case by averaging the abundances in two regions. Let us define the notations, $n_{\rm ave}$, $n_{\rm high}$, and $n_{\rm low}$ as average, high, and low baryon number densities. f_v is the volume fraction of the high baryon density region. $X_i^{\rm ave}$, $X_i^{\rm high}$, and $X_i^{\rm low}$ are mass fractions of each element i in average, and high- and low-density regions, respectively. Then, basic relations are written as follows [43]:

$$n_{\rm ave} = f_v n_{\rm high} + (1 - f_v)\, n_{\rm low},$$
$$n_{\rm ave} X_i^{\rm ave} = f_v n_{\rm high} X_i^{\rm high} + (1 - f_v)\, n_{\rm low} X_i^{\rm low}. \tag{1}$$

Here, we assume the baryon fluctuation to be isothermal as was done in previous studies (e.g., [18, 19, 30]). Under that assumption, since the baryon-to-photon ratio is defined by the number density of photon in standard BBN, (1) is rewritten as follows:

$$\eta_{\rm ave} = f_v \eta_{\rm high} + (1 - f_v)\, \eta_{\rm low}, \tag{2}$$

$$\eta_{\rm ave} X_i^{\rm ave} = f_v X_i^{\rm high} \eta_{\rm high} + (1 - f_v)\, X_i^{\rm low} \eta_{\rm low}, \tag{3}$$

where ηs with subscripts are the baryon-to-photon ratios in each region. In the present paper, we fix $\eta_{\rm ave} = 6.19 \times 10^{-10}$ from the cosmic microwave background observation [16]. The values of $\eta_{\rm high}$ and $\eta_{\rm low}$ are obtained from both f_v and the density ratio between high- and low-density regions: $R \equiv n_{\rm high}/n_{\rm low} = \eta_{\rm high}/\eta_{\rm low}$.

To calculate the evolution of the universe, we solve the following Friedmann equation:

$$\left(\frac{\dot{x}}{x}\right)^2 = \frac{8\pi G}{3}\rho, \tag{4}$$

where x is the cosmic scale factor and G is the gravitational constant. The total energy density ρ in (4) is the sum of decomposed parts:

$$\rho = \rho_\gamma + \rho_\nu + \rho_{e^\pm} + \rho_b. \tag{5}$$

Here, the subscripts γ, ν, and $e^\pm$ indicate photons, neutrino, and electrons/positrons, respectively. The final term is the baryon density obtained as $\rho_b \simeq m_p n_{\rm ave}$.

We should note the energy density of baryon. To get the time evolution of the baryon density in both regions, the energy conservation law is used as follows:

$$\frac{d}{dt}\left(\rho x^3\right) + p\frac{d}{dt}\left(x^3\right) = 0, \tag{6}$$

where p is the pressure of the fluid. When we solve (6), initial values in both regions are obtained from (2) with f_v and R fixed. For $\eta_{\rm high} \geq 2 \times 10^{-4}$, the baryon density in the high-density region, $\rho_{\rm high}$, is larger than the radiation component at $T > 10^9$ K. However, we note that the contribution to (5) is not $\rho_{\rm high}$, but $f_v \rho_{\rm high}$. In our research, the ratio of $f_v \rho_{\rm high}$ to ρ_γ is about 10^{-7} at BBN epoch. Therefore, we can neglect the final term of (5) in the same way as it has been done in SBBN during the calculation of (4).

3. Constraints from Light Element Observations

In this section, we calculate the nucleosynthesis in high- and low-density regions with the use of the BBN code [44] which includes 24 nuclei from neutron to ^{16}O. We adopt the reaction rates of Descouvemont et al. [45], the neutron lifetime $\tau_N = 885.7$ sec [1], and consider three massless neutrinos.

Let us consider the range of f_v. For $f_v \ll 0.1$, the heavier elements can be synthesized in the high-density regions as discussed in [33]. For $f_v > 0.1$, contribution of the low-density region to $\eta_{\rm ave}$ can be neglected, and therefore to be consistent with the observations of light elements, we need to impose the condition of $f_v < 0.1$.

Figure 1 illustrates the light element synthesis in the high- and low-density regions with $f_v = 10^{-6}$ and $R = 10^6$ that corresponds to $\eta_{\rm high} = 3.05 \times 10^{-4}$ and $\eta_{\rm low} = 3.05 \times 10^{-10}$. Light elements synthesized in these calculations are shown in Table 1. In the low-density region, the evolution of the elements is almost the same as the case of SBBN. In the high-density region, while ^{4}He is more abundant than that in the low-density region, ^{7}Li (or ^{7}Be) is much less produced.

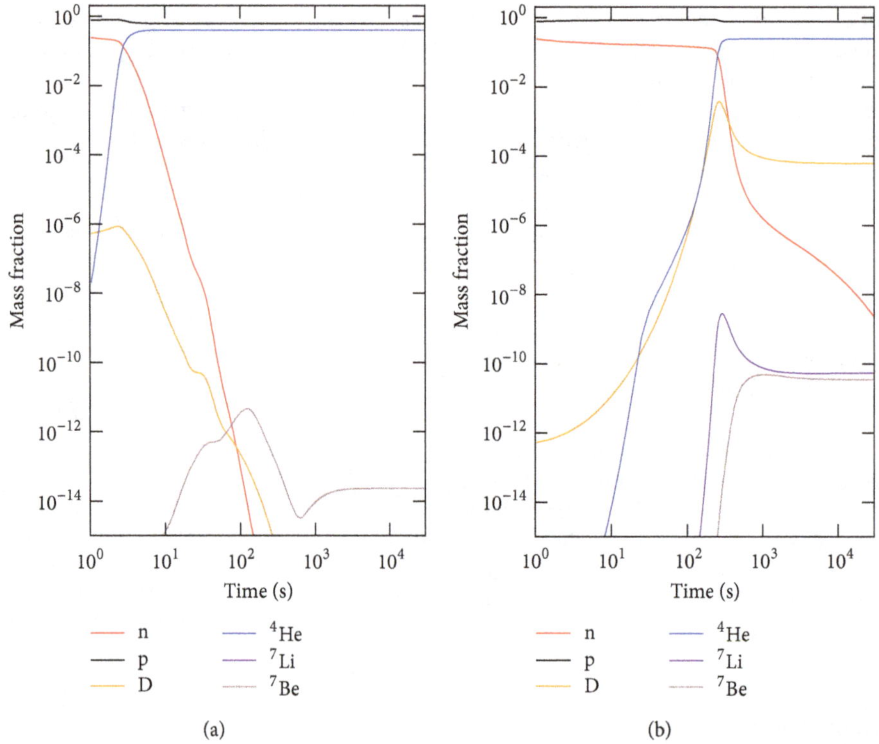

FIGURE 1: Illustration of the nucleosynthesis in the two-zone IBBN model with $f_\nu = 10^{-6}$ and $R = 10^6$. The baryon-to-photon ratios in the high (a) and low (b) density regions are $\eta_{\text{high}} = 3.05 \times 10^{-4}$ and $\eta_{\text{low}} = 3.05 \times 10^{-10}$, respectively.

TABLE 1: The numerical abundances of light elements synthesized as shown in Figure 1.

Elements	X_i^{high}	X_i^{low}	X_i
p	0.608	0.759	0.684
D	3.07×10^{-18}	1.19×10^{-4}	5.95×10^{-5}
$T + {}^3He$	1.15×10^{-13}	3.41×10^{-5}	1.71×10^{-5}
^4He	0.392	0.241	0.316
$^7Li + {}^7Be$	8.2×10^{-13}	6.29×10^{-10}	3.14×10^{-10}

In this case, we can see that average values such as ^{4}He and D are overproduced as shown in Table 1. However, this overproduction can be saved by choosing the parameters carefully. We need to find the reasonable parameter ranges for both f_ν and R by comparing with the observation of the light elements.

Now, we put constraints on f_ν and R by comparing the average values of ^{4}He and D obtained from (3) with the following observational values. First, we consider the primordial ^{4}He abundance reported in [8]:

$$Y_p = 0.2565 \pm 0.0010 \pm 0.0050, \tag{7}$$

and [9]:

$$Y_p = 0.2534 \pm 0.0083. \tag{8}$$

We adopt ^{4}He abundances as follows:

$$0.2415 < Y_p < 0.2617. \tag{9}$$

Next, we take the primordial abundance from the D/H observation reported in [12]:

$$\frac{\text{D}}{\text{H}} = (2.84 \pm 0.26) \times 10^{-5}, \tag{10}$$

and [13]:

$$\frac{\text{D}}{\text{H}} = (2.535 \pm 0.05) \times 10^{-5}, \qquad \frac{\text{D}}{\text{H}} = (2.48 \pm 0.12) \times 10^{-5}. \tag{11}$$

Considering those observations with errors, we adopt the primordial D/H abundance as follows:

$$2.36 < \frac{\text{D}}{\text{H}} \times 10^5 < 3.02. \tag{12}$$

Figure 2 illustrates the constraints on the $f_\nu - R$ plane from the above light element observations with contours of constant η_{high}. The solid and dashed lines indicate the upper limits from (9) and (12), respectively. From the results, we can obtain approximately the following relations between f_ν and R:

$$R \leq \begin{cases} 10^4 \times f_\nu^{-0.3} & \text{for } f_\nu > 7.4 \times 10^{-6}, \\ 0.13 \times f_\nu^{-0.98} & \text{for } f_\nu \leq 7.4 \times 10^{-6}. \end{cases} \tag{13}$$

The ^{4}He observation (9) gives the upper bound for $f_\nu < 7.4 \times 10^{-6}$, and the limit for $f_\nu > 7.4 \times 10^{-6}$ is obtained from

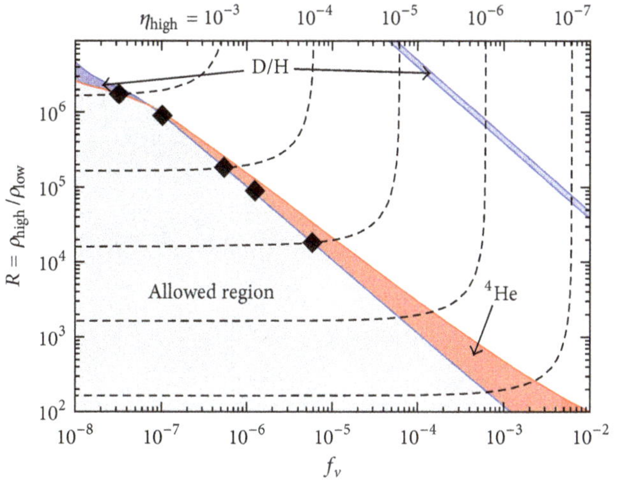

FIGURE 2: Constraints on the $f_v - R$ plane from the observations of light element abundances. The region below the red line is an allowed one obtained from ^{4}He observation (9). Constraints from the D/H observation (12) are shown by the region below the blue line. The gray region corresponds to the allowed parameters determined from the two observations of ^{4}He and D/H. There is another region which is still consistent with only D/H in the upper right direction. This is the contribution of the low-density region with $\eta_{low} \sim 10^{-12}$. The D abundance tends to decrease against the baryon density for $\eta > 10^{-12}$. The dotted lines show the contours of the baryon-to-photon ratio in the high-density region. Filled squares indicate the parameters for heavy element nucleosynthesis adopted in Section 4.

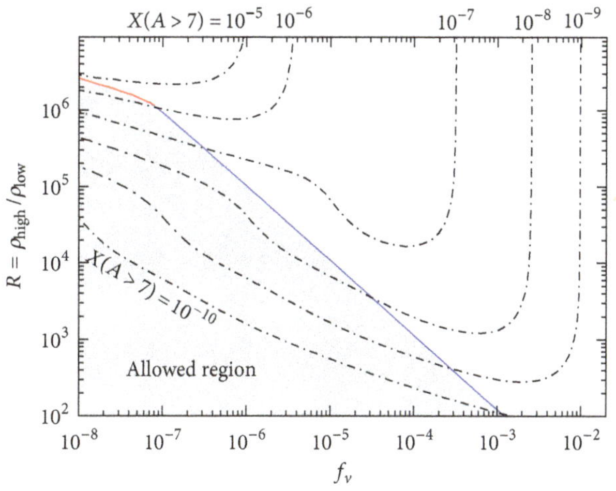

FIGURE 3: Contours of the average total mass fractions which are the sum of the nuclei heavier than ^{7}Li, where we find a consistent region of the produced elements with ^{4}He and D observations.

D observation (12). As shown in Figure 2, we can find the allowed regions which include the very high-density region such as $\eta_{high} = 10^{-3}$.

We should note that η_{high} takes a larger value, nuclei which are heavier than ^{7}Li are synthesized more and more. Then we can estimate the amount of total CNO elements in the allowed region. Figure 3 illustrates the contours of the summation of the average values of the heavier nuclei ($A > 7$), which correspond to Figure 2 and are drawn using the constraint from ^{4}He and D/H observations. As a consequence, we get the upper limit of total mass fractions for heavier nuclei as follows:

$$X(A > 7) \leq 10^{-5}. \quad (14)$$

4. Heavy Element Production

In the previous section, we have obtained the amount of CNO elements produced in the two-zone IBBN model. However, it is not enough to examine the nuclear production beyond $A > 8$ because the baryon density in the high-density region becomes so high that elements beyond CNO isotopes can be produced [17, 31, 32, 34, 42]. In this section, we investigate the heavy element nucleosynthesis in the high-density region considering the constraints shown in Figure 2. Abundance change is calculated with a large nuclear reaction network, which includes 4463 nuclei from neutron (n) and proton (p) to Americium ($Z = 95$ and $A = 292$). Nuclear data, such as reaction rates, nuclear masses, and partition functions, are the same as the ones used in [46–49] except for the neutron-proton interaction. We use the weak interaction of Kawano code [50], which is adequate for the high-temperature epoch of $T > 10^{10}$ K.

As seen in Figure 3, heavy elements of $X(A > 7) > 10^{-9}$ are produced nearly along the upper limit of R. Therefore, to examine the efficiency of the heavy element production, we select five models with the following parameters: $\eta_{high} = 10^{-3}$, 5.1×10^{-4}, 10^{-4}, 5.0×10^{-5}, and 10^{-5} corresponded to $(f_v, R) = (3.24 \times 10^{-8}, 1.74 \times 10^6)$, $(1.03 \times 10^{-8}, 9.00 \times 10^5)$, $(5.41 \times 10^{-7}, 1.84 \times 10^5)$, $(1.50 \times 10^{-6}, 9.20 \times 10^4)$, and $(5.87 \times 10^{-6}, 1.82 \times 10^4)$. Adopted parameters are indicated by filled squares in Figure 2.

First, we evaluate the validity of the nucleosynthesis code with 4463 nuclei. Table 2 shows the results of the light elements, p, D, ^{4}He, ^{3}He, and ^{7}Li. The results of the high-density region are calculated by the extended nucleosynthesis code, and the abundances in the low-density region are obtained by BBN code. The average abundances are obtained by (3). Since the average values of ^{4}He and D are consistent with the observations, there is no difference between BBN code and the extended nucleosynthesis code in regard to the average abundances of light elements.

Figure 4 shows the results of nucleosynthesis in the high-density regions with $\eta_{high} \simeq 10^{-4}$ and 10^{-3}. In Figure 4(a), we see the time evolution of the abundances of Gd and Eu for the mass number 159. First, ^{159}Tb (stable r-element) is synthesized and later ^{159}Gd and ^{159}Eu are synthesized through the neutron captures. After $t = 10^3$ sec, ^{159}Eu decays to nuclei by way of ^{159}Eu $\rightarrow$ ^{159}Gd $\rightarrow$ ^{159}Tb, where the half-lifes of ^{159}Eu and ^{159}Gd are 26.1 min and 18.479 h, respectively.

For $\eta_{high} \simeq 10^{-3}$, the result is seen in Figure 4(b). ^{108}Sn, which is a proton-rich nuclei is synthesized. After that, stable nuclei ^{108}Cd is synthesized by way of ^{108}Sn $\rightarrow$ ^{108}In $\rightarrow$

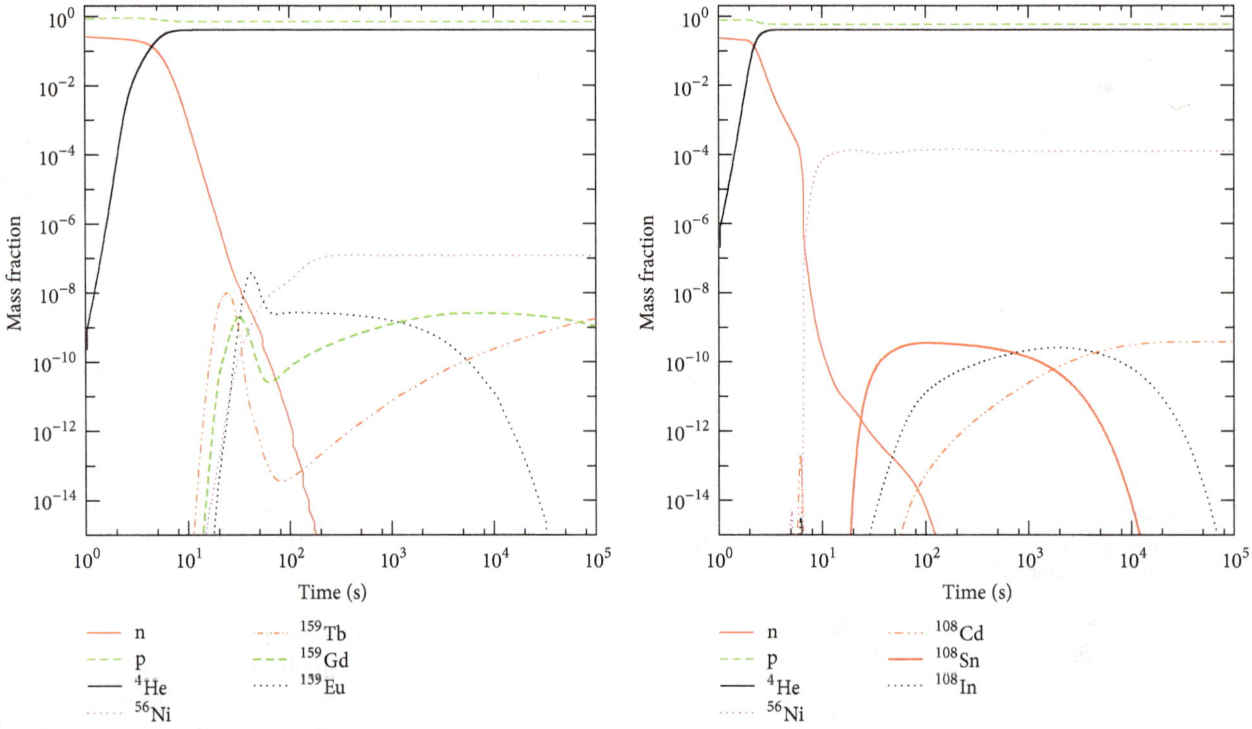

(a) $\eta_{\text{high}} = 1.02 \times 10^{-4}$ (b) $\eta_{\text{high}} = 1.06 \times 10^{-3}$

FIGURE 4: Time evolution of the mass fractions in high-density regions of (a) $\eta_{\text{high}} = 1.02 \times 10^{-4}$ and (b) $\eta_{\text{high}} = 1.06 \times 10^{-3}$.

TABLE 2: Mass fractions of light elements for the four cases: $\eta_{\text{high}} \simeq 10^{-3}$, $\eta_{\text{high}} = 5 \times 10^{-4}$, $\eta_{\text{high}} \simeq 10^{-4}$, and $\eta_{\text{high}} = 10^{-5}$. t_{fin} and T_{fin} are the time and temperature at the final stage of the calculations.

(a) For cases of $\eta_{\text{high}} = 10^{-3}$ and $\eta_{\text{high}} = 5 \times 10^{-4}$

f_ν, R (η_{high}, η_{low})	3.23×10^{-8}, 1.74×10^{6} (1.02×10^{-3}, 5.86×10^{-10})			1.03×10^{-7}, 9.00×10^{5} (5.10×10^{-4}, 5.67×10^{-10})		
(t_{fin}, T_{fin})	1.0×10^{5} sec, 4.2×10^{7} K			1.1×10^{5} sec, 4.9×10^{7} K		
Elements	High	Low	Average	High	Low	Average
p	0.586	0.753	0.744	0.600	0.753	0.740
D	1.76×10^{-21}	4.50×10^{-5}	4.26×10^{-5}	3.43×10^{-21}	4.75×10^{-5}	4.34×10^{-5}
^{3}He + T	2.91×10^{-14}	2.18×10^{-5}	2.07×10^{-5}	2.77×10^{-14}	2.23×10^{-5}	2.04×10^{-5}
^{4}He	0.413	0.247	0.256	0.400	0.247	0.260
^{7}Li + ^{7}Be	1.63×10^{-13}	1.78×10^{-9}	1.68×10^{-9}	6.80×10^{-14}	1.65×10^{-9}	1.52×10^{-9}

(b) For cases of $\eta_{\text{high}} = 10^{-4}$ and $\eta_{\text{high}} = 10^{-5}$

f_ν, R (η_{high}, η_{low})	5.41×10^{-7}, 1.84×10^{5} (1.04×10^{-4}, 5.62×10^{-10})			5.87×10^{-6}, 1.82×10^{4} (1.02×10^{-5}, 5.59×10^{-10})		
(t_{fin}, T_{fin})	1.2×10^{5} sec, 4.3×10^{7} K			1.2×10^{5} sec, 4.5×10^{7} K		
Elements	High	Low	Average	High	Low	Average
p	0.638	0.753	0.742	0.670	0.753	0.745
D	6.84×10^{-22}	4.79×10^{-5}	4.36×10^{-5}	1.12×10^{-22}	4.48×10^{-5}	4.37×10^{-5}
^{3}He + T	1.63×10^{-13}	2.23×10^{-5}	2.04×10^{-5}	1.49×10^{-9}	2.25×10^{-5}	2.03×10^{-5}
^{4}He	0.362	0.247	0.258	0.330	0.247	0.254
^{7}Li + ^{7}Be	7.42×10^{-13}	1.64×10^{-9}	1.49×10^{-9}	6.73×10^{-8}	1.62×10^{-9}	7.96×10^{-9}

TABLE 3: Mass fractions of heavy elements ($A > 7$) for three cases of $\eta_{high} \simeq 10^{-3}$, $\eta_{high} = 5.33 \times 10^{-4}$, and $\eta_{high} \simeq 10^{-4}$.

$f_v = 3.23 \times 10^{-8}$, $R = 1.74 \times 10^6$ ($\eta_{high} = 1.06 \times 10^{-3}$)			$f_v = 1.03 \times 10^{-7}$, $R = 9.00 \times 10^5$ ($\eta_{high} = 5.33 \times 10^{-4}$)			$f_v = 5.41 \times 10^{-7}$, $R = 1.84 \times 10^5$ ($\eta_{high} = 1.02 \times 10^{-4}$)		
Element	High	Average	Element	High	Average	Element	High	Average
Ni56	1.247×10^{-4}	6.658×10^{-6}	Nd142	2.051×10^{-5}	1.738×10^{-6}	Nd145	3.692×10^{-7}	3.342×10^{-8}
Co57	1.590×10^{-5}	8.487×10^{-7}	Ni56	1.270×10^{-5}	1.077×10^{-6}	Ca40	2.706×10^{-7}	2.450×10^{-8}
Sr86	1.061×10^{-5}	5.662×10^{-7}	Sm148	1.059×10^{-5}	8.976×10^{-7}	Mn52	2.417×10^{-7}	2.188×10^{-8}
Sr87	9.772×10^{-6}	5.214×10^{-7}	Pm147	6.996×10^{-6}	5.930×10^{-7}	Eu155	2.374×10^{-7}	2.149×10^{-8}
Se74	9.745×10^{-6}	5.200×10^{-7}	Pm145	6.559×10^{-6}	5.559×10^{-7}	Ce140	1.931×10^{-7}	1.748×10^{-8}
Sr84	9.172×10^{-6}	4.894×10^{-7}	Sm146	6.539×10^{-6}	5.542×10^{-7}	Cr51	1.546×10^{-7}	1.400×10^{-8}
Kr82	8.910×10^{-6}	4.754×10^{-7}	Nd143	4.146×10^{-6}	3.514×10^{-7}	Ce142	1.114×10^{-7}	1.008×10^{-8}
Kr81	7.797×10^{-6}	4.160×10^{-7}	Pr141	3.957×10^{-6}	3.354×10^{-7}	Ni56	1.100×10^{-7}	9.964×10^{-9}
Ge72	7.674×10^{-6}	4.095×10^{-7}	Nd144	3.952×10^{-6}	3.350×10^{-7}	Nd146	1.049×10^{-7}	9.501×10^{-9}
Kr78	7.602×10^{-6}	4.057×10^{-7}	Sm147	3.752×10^{-6}	3.180×10^{-7}	Eu156	9.436×10^{-8}	8.542×10^{-9}
Kr80	7.063×10^{-6}	3.769×10^{-7}	Sm149	3.322×10^{-6}	2.815×10^{-7}	Nd148	9.361×10^{-8}	8.474×10^{-9}
Kr83	6.252×10^{-6}	3.336×10^{-7}	Pm146	2.629×10^{-6}	2.228×10^{-7}	Fe52	8.974×10^{-8}	8.124×10^{-9}
Ge73	6.144×10^{-6}	3.278×10^{-7}	Sm144	2.207×10^{-6}	1.870×10^{-7}	Tb161	8.956×10^{-8}	8.108×10^{-9}
Se76	5.929×10^{-6}	3.164×10^{-7}	Sm150	1.683×10^{-6}	1.426×10^{-7}	La139	8.804×10^{-8}	7.971×10^{-9}
Br79	5.904×10^{-6}	3.150×10^{-7}	Pm144	1.581×10^{-6}	1.340×10^{-7}	N14	8.736×10^{-8}	7.909×10^{-9}
Se77	5.345×10^{-6}	2.852×10^{-7}	Pm143	1.575×10^{-6}	1.335×10^{-7}	Cr48	8.561×10^{-8}	7.750×10^{-9}
Y89	4.759×10^{-6}	2.539×10^{-7}	Sm145	1.010×10^{-6}	8.568×10^{-8}	Ba138	7.955×10^{-8}	7.202×10^{-9}
Zr90	4.412×10^{-6}	2.354×10^{-7}	Co57	8.643×10^{-7}	7.326×10^{-8}	C12	7.672×10^{-8}	6.945×10^{-9}
Rb85	4.324×10^{-6}	2.307×10^{-7}	Eu153	5.563×10^{-7}	4.715×10^{-8}	Dy162	6.835×10^{-8}	6.188×10^{-9}
Rb83	4.082×10^{-6}	2.178×10^{-7}	Ce140	4.944×10^{-7}	4.191×10^{-8}	C13	6.428×10^{-8}	5.819×10^{-9}
Y88	3.845×10^{-6}	2.052×10^{-7}	Nd145	4.376×10^{-7}	3.709×10^{-8}	O16	6.301×10^{-8}	5.704×10^{-9}
Zr88	3.546×10^{-6}	1.892×10^{-7}	Eu155	4.224×10^{-7}	3.581×10^{-8}	Gd158	5.845×10^{-8}	5.292×10^{-9}
As73	3.519×10^{-6}	1.878×10^{-7}	Eu151	4.106×10^{-7}	3.480×10^{-8}	Cs137	5.559×10^{-8}	5.033×10^{-9}
Ga71	3.388×10^{-6}	1.808×10^{-7}	Cr52	4.071×10^{-7}	3.450×10^{-8}	Nd147	3.962×10^{-8}	3.587×10^{-9}
Se75	2.933×10^{-6}	1.565×10^{-7}	Cd108	3.596×10^{-7}	3.048×10^{-8}	Ho165	3.770×10^{-8}	3.413×10^{-9}
Nb91	2.896×10^{-6}	1.545×10^{-7}	Gd156	3.368×10^{-7}	2.854×10^{-8}	Pr143	3.111×10^{-8}	2.817×10^{-9}
As75	2.856×10^{-6}	1.524×10^{-7}	Cd110	3.103×10^{-7}	2.630×10^{-8}	Ce141	2.998×10^{-8}	2.714×10^{-9}
Mo92	2.442×10^{-6}	1.303×10^{-7}	Eu152	2.809×10^{-7}	2.381×10^{-8}	Gd160	2.950×10^{-8}	2.670×10^{-9}
Ge70	2.318×10^{-6}	1.237×10^{-7}	Sm151	2.795×10^{-7}	2.369×10^{-8}	Xe136	2.771×10^{-8}	2.509×10^{-9}
Sr88	2.021×10^{-6}	1.078×10^{-7}	Eu154	2.759×10^{-7}	2.339×10^{-8}	Xe134	2.238×10^{-8}	2.026×10^{-9}
$\sum_{A>7} X(A)$	3.010×10^{-4}	1.652×10^{-5}	$\sum_{A>7} X(A)$	1.062×10^{-4}	9.006×10^{-6}	$\sum_{A>7} X(A)$	3.850×10^{-6}	3.485×10^{-7}

^{108}Cd, where the half-lifes of ^{108}Sn and ^{108}In are 10.3 min and 58.0 min, respectively. These results are qualitatively the same as Matsuura et al. [42].

In addition, we notice the production of radioactive nuclei of ^{56}Ni and ^{57}Co, where ^{56}Ni is produced at early times, just after the formation of ^{4}He. Usually, nuclei such as ^{56}Ni and ^{57}Co are produced in supernova explosions, which are assumed to be the events after the first star formation (e.g., [51]). In IBBN model, however, this production can be found to occur at an extremely high-density region of $\eta_{high} \geq 10^{-3}$ as the primary elements without supernova events in the early universe.

Final results ($T = 4 \times 10^7$ K) of nucleosynthesis calculations are shown in Table 3. When we calculate the average values, we set the abundances of $A > 16$ to be zero for low-density side. For $\eta_{high} \simeq 10^{-4}$, a lot of nuclei of $A > 7$ are synthesized whose amounts are comparable to that of ^{7}Li. Produced elements in this case include both s-element (i.e., ^{138}Ba) and r-elements (for instance, ^{142}Ce and ^{148}Nd). For $\eta_{high} \simeq 10^{-3}$, there are few r-elements while both s-elements (i.e., ^{82}Kr and ^{89}Y) and p-elements (i.e., ^{74}Se and ^{78}Kr) are synthesized such as the case of supernova explosions. For $\eta_{high} = 10^{-3}$, the heavy elements are produced slightly more than the total mass fraction (shown in Figure 3) derived from the BBN code calculations. This is because our BBN code used in Section 3 includes the elements up to $A = 16$ and the actual abundance flow proceeds to much heavier elements.

Figure 5 shows the average abundances between high- and low-density regions using (3) in comparing with the solar system abundances [52]. For $\eta_{high} \simeq 10^{-4}$, abundance productions of $120 < A < 180$ are comparable to the solar values. For $\eta_{high} \simeq 10^{-3}$, those of $50 < A < 100$ have been synthesized well. In the case of $\eta_{high} = 5 \times 10^{-4}$, there are two outstanding peaks: one is around $A = 56$ ($N = 28$) and

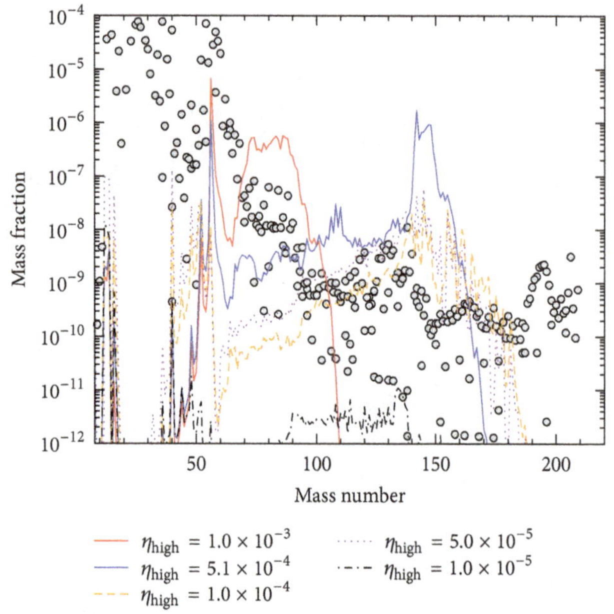

FIGURE 5: Comparison of the average mass fractions in the two-zone model with the solar system abundances [52] (indicated by dots).

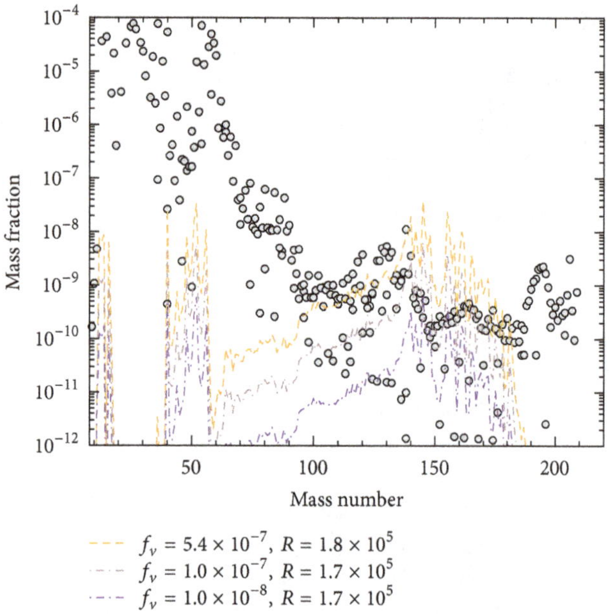

FIGURE 6: Same as Figure 5, but η_{high} is fixed as 10^{-4}.

the other can be found around $A = 140$. Abundance patterns are very different from that of the solar system ones, because IBBN occurs under the condition of a significant amount of abundances of both neutrons and protons.

5. Summary and Discussion

We extend the previous studies of Matsuura et al. [42, 43] and investigate the consistency between the light element abundances in the IBBN model and the observation of ^{4}He and D/H.

First, we have done the nucleosynthesis calculation using the BBN code with 24 nuclei for both regions. The time evolution of the light elements at the high-density region differs significantly from that at the low-density region. The nucleosynthesis begins faster and ^{4}He is more abundant than that in the low-density region. By comparing the average abundances with the ^{4}He and D/H observations, we can get the allowed parameters of the two-zone model: the volume fraction f_v of the high-density region and the density ratio R between the two regions.

Second, we calculate the nucleosynthesis that includes 4463 nuclei in the high-density regions. Qualitatively, results of nucleosynthesis are the same as those in [42]. In the present results, we showed that p- and r-elements are synthesized simultaneously at high-density region with $\eta_{high} \simeq 10^{-4}$.

We find that the average mass fractions in IBBN amount to as much as the solar system abundances. As seen from Figure 5, there are overproduced elements around $A = 150$ (for $\eta_{high} = 10^{-4}$) and $A = 80$ (for $\eta_{high} = 10^{-3}$). Although it seems to conflict with the chemical evolution in the universe, this problem could be solved by the careful choice of f_v and/or R. Figure 6 illustrates the mass fractions with $\eta_{high} = 1.0 \times 10^{-4}$ for three sets of $f_v - R$. It is shown that the abundances can become lower than the solar system abundances. If we put a constraint on the f_v–R plane from the heavy element observations [53–56], the parameters in IBBN model should be tightly determined.

In the meanwhile, we would like to touch on the consistency against the primordial ^{7}Li. We have obtained interesting results about ^{7}Li abundances in our model. For the recent study of ^{7}Li, the lithium problem arises from the discrepancy among the ^{7}Li abundance predicted by SBBN theory, the baryon density of WMAP, and abundance inferred from the observations of metal-poor stars (see [57, 58]). As seen in Table 2, ^{7}Li is clearly overproduced such as $^7\mathrm{Li}/\mathrm{H}|_{ave} = 1.52 \times 10^{-9}$ for $\eta_{high} = 10^{-5}$, although we adopt the highest observational value $^7\mathrm{Li/H} = (2.75 - 4.17) \times 10^{-10}$ [59]. However, for cases of $\eta_{high} = 10^{-3}$, 5×10^{-4} and 10^{-4}, the values of $^7\mathrm{Li}/\mathrm{H}|_{ave}$ agree with the observation. Usually, the consistency with BBN has been checked using observations of ^{4}He, D/H, and ^{7}Li/H. Then, the parameters such as $\eta_{high} = 10^{-5}$ ought to be excluded. However, the abundance of $^7\mathrm{Li}/\mathrm{H}|_{ave}$ is sensitive to the values of both η_{high} and η_{low}. As for the future work of IBBN, we will study in detail the ^{7}Li production. In addition, recent ^{4}He observation could suggest the need for a nonstandard BBN model [8]. IBBN may also give a clue to the problems.

Acknowledgments

This work has been supported in part by a Grant-in-Aid for Scientific Research (no. 24540278) of the Ministry of Education, Culture, Sports, Science, and Technology of Japan, and in part by a grant for Basic Science Research Projects from the Sumitomo Foundation (no. 080933).

References

[1] J. Beringer, J. F. Arguin, R. M. Barnett et al., "Review of particle physics," *Physical Review D*, vol. 86, no. 1, Article ID 010001, 1528 pages, 2012.

[2] G. Steigman, "Primordial nucleosynthesis in the precision cosmology era," *Annual Review of Nuclear and Particle Science*, vol. 57, pp. 463–491, 2007.

[3] F. Iocco, G. Mangano, G. Miele, O. Pisanti, and P. D. Serpico, "Primordial nucleosynthesis: from precision cosmology to fundamental physics," *Physics Reports*, vol. 472, no. 1-6, pp. 1–76, 2009.

[4] A. Coc, S. Goriely, Y. Xu, M. Saimpert, and E. Vangioni, "Standard big bang nucleosynthesis up to CNO with an improved extended nuclear network," *The Astrophysical Journal*, vol. 744, no. 2, article 158, 2012.

[5] V. Luridiana, A. Peimbert, M. Peimbert, and M. Cerviño, "The effect of collisional enhancement of Balmer lines on the determination of the primordial helium abundance," *The Astrophysical Journal Letters*, vol. 592, no. 2, pp. 846–865, 2003.

[6] K. A. Olive and E. D. Skillman, "A realistic determination of the error on the primordial helium abundance: steps toward nonparametric nebular helium abundances," *The Astrophysical Journal*, vol. 617, no. 1, pp. 29–40, 2004.

[7] Y. I. Izotov, T. X. Thuan, and G. Stasińska, "The primordial abundance of ^{4}He: a self-consistent empirical analysis of systematic effects in a large sample of low-metallicity H II regions," *The Astrophysical Journal*, vol. 662, no. 1, article 15, 2007.

[8] Y. I. Izotov and T. X. Thuan, "The primordial abundance of ^{4}He: evidence for non-standard big bang nucleosynthesis," *The Astrophysical Journal Letters*, vol. 710, no. 1, pp. L67–L71, 2010.

[9] E. Aver, K. A. Olive, and E. D. Skillman, "An MCMC determination of the primordial helium abundance," *Journal of Cosmology and Astroparticle Physics*, vol. 2012, no. 4, article 4, 2012.

[10] D. Kirkman, D. Tytler, N. Suzuki, J. M. O'Meara, and D. Lubin, "The cosmological baryon density from the deuterium-to-hydrogen ratio in QSO absorption systems: D/H toward Q1243+3047," *The Astrophysical Journal Supplement Series*, vol. 149, no. 1, article 1, 2003.

[11] J. M. O'Meara, S. Burles, J. X. Prochaska, G. E. Prochter, R. A. Bernstein, and K. M. Burgess, "The deuterium-to-hydrogen abundance ratio toward the QSO SDSS J155810.16-003120.0," *The Astrophysical Journal Letters*, vol. 649, no. 2, article L61, 2006.

[12] M. Pettini, B. J. Zych, M. T. Murphy, A. Lewis, and C. C. Steidel, "Deuterium abundance in the most metal-poor damped Lyman alpha system: converging on Ωb,0h2," *Monthly Notices of the Royal Astronomical Society*, vol. 391, no. 4, pp. 1499–1510, 2008.

[13] M. Pettini and R. Cooke, "A new, precise measurement of the primordial abundance of deuterium," *Monthly Notices of the Royal Astronomical Society*, vol. 425, no. 4, pp. 2477–2486, 2012.

[14] T. M. Bania, R. T. Rood, and D. S. Balser, "The cosmological density of baryons from observations of $^3He^+$ in the Milky Way," *Nature*, vol. 415, no. 6867, pp. 54–57, 2002.

[15] E. Vangioni-Flam, K. A. Olive, B. D. Fields, and M. Cassé, "On the baryometric status of ^{3}He," *The Astrophysical Journal*, vol. 585, no. 2, article 611, 2003.

[16] C. L. Bennett, D. Larson, J. L. Weiland et al., "Nine-year Wilkinson microwave anisotropy probe (WMAP) observations: final maps and results," *Astrophysical Journal Supplement Series*, http://arxiv.org/abs/1212.5225.

[17] S. Matsuura, A. D. Dolgov, S. Nagataki, and K. Sato, "Affleck-dine baryogenesis and heavy element production from inhomogeneous big bang nucleosynthesis," *Progress of Theoretical Physics*, vol. 112, no. 6, pp. 971–981, 2004.

[18] C. Alcock, G. M. Fuller, and G. J. Mathews, "The quark-hadron phase transition and primordial nucleosynthesis," *The Astrophysical Journal*, vol. 320, pp. 439–447, 1987.

[19] G. M. Fuller, G. J. Mathews, and C. R. Alcock, "Quark-hadron phase transition in the early Universe: Isothermal baryon-number fluctuations and primordial nucleosynthesis," *Physical Review D*, vol. 37, no. 6, pp. 1380–1400, 1988.

[20] H. Kurki-Suonio and R. A. Matzner, "Effect of small-scale baryon inhomogeneity on cosmic nucleosynthesis," *Physical Review D*, vol. 39, no. 4, pp. 1046–1053, 1989.

[21] H. Kurki-Suonio and R. A. Matzner, "Overproduction of ^{4}He in strongly inhomogeneous Ωb=1 models of primordial nucleosynthesis," *Physical Review D*, vol. 42, no. 4, pp. 1047–1056, 1990.

[22] C. L. Bennett, M. Halpern, G. Hinshaw et al., "First-year Wilkinson microwave anisotropy probe (WMAP) observations: preliminary maps and basic results," *The Astrophysical Journal Supplement Series*, vol. 148, no. 1, article 1, 2003.

[23] D. N. Spergel, R. Bean, O. Doré et al., "Three-year Wilkinson microwave anisotropy probe (WMAP) observations: implications for cosmology," *The Astrophysical Journal Supplement Series*, vol. 170, no. 2, article 377, 2007.

[24] J. Dunkley, E. Komatsu, M. R. Nolta et al., "Five-year Wilkinson microwave anisotropy probe observations: likelihoods and parameters from the WMAP data," *The Astrophysical Journal Supplement Series*, vol. 180, no. 2, article 306, 2009.

[25] J. H. Applegate, C. J. Hogan, and R. J. Scherrer, "Cosmological baryon diffusion and nucleosynthesis," *Physical Review D*, vol. 35, no. 4, pp. 1151–1160, 1987.

[26] R. M. Malaney and W. A. Fowler, "Late-time neutron diffusion and nucleosynthesis in a post-QCD inhomogeneous Ω(b) = 1 universe," *The Astrophysical Journal*, vol. 333, pp. 14–20, 1988.

[27] J. H. Applegate, C. J. Hogan, and R. J. Scherrer, "Cosmological quantum chromodynamics, neutron diffusion, and the production of primordial heavy elements," *The Astrophysical Journal*, vol. 329, pp. 572–579, 1988.

[28] N. Terasawa and K. Sato, "Production of Be-9 and heavy elements in the inhomogeneous universe," *The Astrophysical Journal*, vol. 362, pp. L47–L49, 1990.

[29] D. Thomas, D. N. Schramm, K. A. Olive, G. J. Mathews, B. S. Meyer, and B. D. Fields, "Production of lithium, beryllium, and boron from baryon inhomogeneous primordial nucleosynthesis," *The Astrophysical Journal*, vol. 430, no. 1, pp. 291–299, 1994.

[30] N. Terasawa and K. Sato, "Neutron diffusion and nucleosynthesis in the Universe with isothermal fluctuations produced by quark-hadron phase transition," *Physical Review D*, vol. 39, no. 10, pp. 2893–2900, 1989.

[31] K. Jedamzik and J. B. Rehm, "Inhomogeneous big bang nucleosynthesis: upper limit on Ωb and production of lithium, beryllium, and boron," *Physical Review D*, vol. 64, no. 2, Article ID 023510, 8 pages, 2001.

[32] T. Rauscher, H. Applegate, J. Cowan, F. Thielmann, and M. Wiescher, "Production of heavy elements in inhomogeneous

cosmologies," *The Astrophysical Journal*, vol. 429, no. 2, pp. 499–530, 1994.

[33] K. Jedamzik, G. M. Fuller, G. J. Mathews, and T. Kajino, "Enhanced heavy-element formation in baryon-inhomogeneous big bang models," *The Astrophysical Journal Letters*, vol. 422, no. 2, pp. 423–429, 1994.

[34] R. V. Wagoner, W. A. Fowler, and F. Hoyle, "On the synthesis of elements at very high temperatures," *The Astrophysical Journal*, vol. 148, article 3, 1967.

[35] R. V. Wagoner, "Big bang nucleosynthesis revisited," *The Astrophysical Journal*, vol. 179, pp. 343–360, 1973.

[36] Y. Juarez, R. Maiolino, R. Mujica et al., "The metallicity of the most distant quasars," *Astronomy and Astrophysics*, vol. 494, no. 2, pp. L25–L28, 2009.

[37] T. Moriya and T. Shigeyama, "Multiple main sequence of globular clusters as a result of inhomogeneous big bang nucleosynthesis," *Physical Review D*, vol. 81, no. 4, Article ID 043004, 7 pages, 2010.

[38] L. R. Bedin, G. Piotto, J. Anderson et al., "ω centauri: the population puzzle goes deeper," *The Astrophysical Journal Letters*, vol. 605, no. 2, article L125, 2004.

[39] G. Piotto, L. R. Bedin, J. Anderson et al., "A triple main sequence in the globular cluster NGC 2808," *The Astrophysical Journal Letters*, vol. 661, no. 1, article L53, 2007.

[40] I. Affleck and M. Dine, "A new mechanism for baryogenesis," *Nuclear Physics B*, vol. 249, no. 2, pp. 361–380, 1985.

[41] T. Rauscher, "Comment on 'heavy element production in inhomogeneous big bang nucleosynthesis'," *Physical Review D*, vol. 75, no. 6, Article ID 068301, 2 pages, 2007.

[42] S. Matsuura, S. I. Fujimoto, S. Nishimura, M. A. Hashimoto, and K. Sato, "Heavy element production in inhomogeneous big bang nucleosynthesis," *Physical Review D*, vol. 72, no. 12, Article ID 123505, 6 pages, 2005.

[43] S. Matsuura, S. I. Fujimoto, M. A. Hashimoto, and K. Sato, "Reply to 'Comment on heavy element production in inhomogeneous big bang nucleosynthesis'," *Physical Review D*, vol. 75, no. 6, Article ID 068302, 5 pages, 2007.

[44] M. Hashimoto and K. Arai, "The nuclear reaction network," *Physics Reports of Kumamoto University*, vol. 7, no. 2, pp. 47–65, 1985.

[45] P. Descouvemont, A. Adahchour, C. Angulo, A. Coc, and E. Vangioni-Flam, "Compilation and R-matrix analysis of big bang nuclear reaction rates," *Atomic Data and Nuclear Data Tables*, vol. 88, no. 1, pp. 203–236, 2004.

[46] S. Fujimoto, M. Hashimoto, O. Koike, K. Arai, and R. Matsuba, "p-process nucleosynthesis inside supernova-driven supercritical accretion disks," *The Astrophysical Journal*, vol. 585, no. 1, article 418, 2003.

[47] O. Koike, M. Hashimoto, R. Kuromizu, and S. Fujimoto, "Final products of the rp-process on accreting neutron stars," *The Astrophysical Journal*, vol. 603, no. 1, article 592, 2004.

[48] S. Fujimoto, M. Hashimoto, K. Arai, and R. Matsuba, "Nucleosynthesis inside an accretion disk and disk winds related to gamma-ray bursts," *The Astrophysical Journal*, vol. 614, no. 2, article 847, 2004.

[49] S. Nishimura, K. Kotake, M. Hashimoto et al., "r-process nucleosynthesis in magnetohydrodynamic jet explosions of core-collapse supernovae," *The Astrophysical Journal*, vol. 642, no. 1, article 410, 2006.

[50] K. Kawano, "Let's go: early universe 2. Primordial nucleosynthesis. The computer way," FERMILAB-Pub-92/04-A, 58 pages, 1992.

[51] M. Hashimoto, "Supernova nucleosynthesis in massive stars," *Progress of Theoretical Physics*, vol. 94, no. 5, pp. 663–736, 1995.

[52] E. Anders and N. Grevesse, "Abundances of the elements: meteoritic and solar," *Geochimica et Cosmochimica Acta*, vol. 53, no. 1, pp. 197–214, 1989.

[53] A. Frebel, N. Christlieb, J. E. Norris, C. Thom, T. C. Beers, and J. Rhee, "Discovery of HE 1523-0901, a strongly r-process-enhanced metal-poor star with detected uranium," *The Astrophysical Journal Letters*, vol. 660, no. 2, pp. L117–L120, 2007.

[54] A. Frebel, J. E. Norris, W. Aoki et al., "Chemical abundance analysis of the extremely metal-poor star HE 1300+0157," *The Astrophysical Journal*, vol. 658, no. 1, article 534, 2007.

[55] C. Siqueira Mello, M. Spite, B. Barbuy et al., "First stars: XVI. HST/STIS abundances of heavy elements in the uranium-rich metal-poor star CS 31082-001," *Astronomy and Astrophysics*, vol. 550, article A122, 17 pages, 2013.

[56] C. C. Worley, V. Hill, J. Sobeck, and E. Carretta, "Ba and Eu abundances in M 15 giant stars," *Astronomy and Astrophysics*, vol. 553, article A47, 20 pages, 2013.

[57] A. Coc, E. Vangioni-Flam, P. Descouvemont, A. Adahchour, and C. Angulo, "Updated big bang nucleosynthesis compared with Wilkinson microwave anisotropy probe observations and the abundance of light elements," *The Astrophysical Journal Letters*, vol. 600, no. 2, pp. 544–552, 2004.

[58] R. H. Cyburt, B. D. Fields, K. A. Olive, and JCAP, "An update on the big bang nucleosynthesis prediction for ^{7}Li: the problem worsens," *Journal of Cosmology and Astroparticle Physics*, vol. 2008, no. 11, article 12, 2008.

[59] A. J. Korn, F. Grundahl, O. Richard et al., "A probable stellar solution to the cosmological lithium discrepancy," *Nature*, vol. 442, pp. 657–659, 2006.

9

Modulation Instability of Ion-Acoustic Waves in Plasma with Nonthermal Electrons

Basudev Ghosh[1] and Sreyasi Banerjee[2]

[1] *Department of Physics, Jadavpur University, Kolkata 700 032, India*
[2] *Department of Electronics, Vidyasagar College, Kolkata 700 006, India*

Correspondence should be addressed to Basudev Ghosh; bsdvghosh@gmail.com

Academic Editor: Milan S. Dimitrijevic

Modulational instability of ion-acoustic waves has been theoretically investigated in an unmagnetized collisionless plasma with nonthermal electrons, Boltzmann positrons, and warm positive ions. To describe the nonlinear evolution of the wave amplitude a nonlinear Schrödinger (NLS) equation has been derived by using multiple scale perturbation technique. The nonthermal parameter, positron concentration, and ion temperature are shown to play significant role in the modulational instability of ion-acoustic waves and the formation of envelope solitons.

1. Introduction

Electron-positron-ion (*e-p-i*) plasmas occur in many astrophysical environments such as active galactic nuclei [1], pulsar magnetospheres [2], polar regions of neutron stars [3], centres of our galaxy [4], the early universe [5, 6], and the solar atmosphere [7]. For this, over the last two decades there has been a great deal of interest in the study of nonlinear wave phenomena in *e-p-i* plasmas [8–12]. Positrons are produced by pair production in high energy processes occurring in many astrophysical environments. Popel et al. [9] have reported decrease in soliton amplitude in the presence of positrons. Jehan et al. [13] have shown that solitons become narrower as the concentration of positron increases. The presence of non-Maxwellian electron is common in space and astrophysical plasmas including the magnetosphere [12] and auroral zones [14]. The presence of such non-Maxwellian electrons gives rise to many interesting characteristics in the nonlinear propagation of waves including the ion-acoustic solitons [15, 16]. The solitary structures with density depression in the magnetosphere observed by the Freja satellites [17, 18] have been explained by Cairns et al. [19] by assuming electron distribution to be nonthermal. Nonlinear ion-acoustic solitary waves in *e-p-i* plasma have been considered by some authors [9, 20, 21] assuming ions to be cold. In practice ions have finite temperature and the ionic temperature can significantly affect the characteristics of nonlinear ion-acoustic structures [10, 22, 23]. Chawla et al. [24] have considered ion-acoustic waves in *e-p-i* plasma with warm adiabatic ions and isothermal electrons. Baluku and Hellberg [25] have considered ion-acoustic solitary waves in *e-p-i* plasma with cold ions and nonthermal electrons. Hence it is interesting to study the nonlinear ion-acoustic waves in *e-p-i* plasma assuming simultaneous presence of nonthermal electrons, warm negative ions, and the positrons. Recently Pakzad [11] has shown that the presence of warm ions and nonthermal electrons can modify parametric regions of existence of ion-acoustic solitary waves. A nonlinear theory of ion-acoustic waves in *e-p-i* plasma has been developed by Dubinov and Sazonkin [26] considering polytropic laws of compression and rarefaction for all plasma components. Survey of the past literatures shows that a large number of works on KdV type and large amplitude solitary structure formation in *e-p-i* plasmas have been reported. Nonlinear propagation of waves in a dispersive medium is generically subject to amplitude modulation due to carrier wave self-interaction or intrinsic nonlinearity of the medium. Modulational instability is an important phenomenon in connection with stable wave propagation. However, only a few works have been reported in recent years on the modulational instability and formation of envelope soliton in *e-p-i* plasmas [20, 21, 24]. It has been shown that the presence of positrons shifts the critical wave number separating the stability and instability regions to higher values and for fixed amplitude, width of envelope solitons decreases with the increase of positron concentration. Mahmood et al. [27] have studied modulational instability of ion-acoustic waves in *e-p-i* plasma with warm ions and isothermal electrons and positrons at the same temperature. Chawla et al. [24] have studied the effects of ion temperature, positron concentration, and positron temperature on the modulational instability of ion-acoustic waves in *e-p-i* plasma with isothermal electrons and positrons at different temperatures. Bains et al. [28] have considered modulational instability of ion-acoustic waves in *e-p-i* plasma with dust particles. Eslami et al. [29] have considered modulational instability of ion-acoustic waves in *e-p-i* plasma with electrons and positrons following q-nonextensive distribution. Gill et al. [21] have studied modulational instability of ion-acoustic waves in *e-p-i* plasma with superthermal electrons and isothermal positrons. Zhang et al. [30] have investigated modulational instability of ion-acoustic waves in *e-p-i* plasma with nonthermally distributed electrons and cold ions. Modulational instability and excitation of ion-acoustic envelope solitons in *e-p-i* plasma with nonthermal electrons have been investigated by Gill et al. [31] including ion temperature. The purpose of the present paper is to make a detailed study of modulational instability of ion-acoustic waves in *e-p-i* plasma including simultaneously both the effects of nonthermality of electrons and ion-temperature.

2. Basic Formulation

We consider an unmagnetized collisionless plasma consisting of warm positive ions, Boltzmann positrons, and nonthermal electrons. The normalized basic equations governing ion dynamics for one-dimensional propagation in such plasma in dimensionless form are as follows [28]:

$$\frac{\partial n_i}{\partial t} + \frac{\partial}{\partial x}(n_i v_i) = 0,$$

$$\frac{\partial v_i}{\partial t} + v_i \frac{\partial v_i}{\partial x} + \frac{3\sigma_i}{(1-\chi)^2} n_i \frac{\partial n_i}{\partial x} = -\frac{\partial \phi}{\partial x}, \quad (1)$$

$$\frac{\partial^2 \phi}{\partial x^2} = n_e - n_p - n_i.$$

In aforementioned equations, the parameters n_i, v_i are, respectively, the concentration and velocity of the positive ions; n_e and n_p are, respectively, the concentration of electrons and positrons; ϕ denotes the electrostatic potential; other parameters have their usual meaning. Different quantities are normalized as follows: the velocities by ion-acoustic speed $C_s = \sqrt{k_B T_e / m_i}$, the densities by equilibrium electron density n_{e0}, all the length x by the electron Debye length $\lambda_{De} = \sqrt{k_B T_e / 4e^2 n_{e0}}$, time by λ_{De}/C_s, ion temperature T_i by T_e ($\sigma_i = T_i/T_e$) and the potential ϕ by $k_B T_e/e$, where k_B is the Boltzmann's constant. The nonthermal electron density is given by [19]

$$n_e = \left(1 - \beta\phi + \beta\phi^2\right)\exp(\phi), \quad (2)$$

where $\beta = 4\delta/(1 + 3\delta)$ measures the deviation from the thermalized state and δ determines the presence of nonthermal electrons inside the plasma. The density of Boltzmann positrons is given by

$$n_p = \chi \exp\left(-\sigma_p \phi\right), \quad (3)$$

where $\chi = n_{p0}/n_{e0}$ is the ratio between the unperturbed positron and electron number densities and $\sigma_p = T_e/T_p$ is the ratio between electron and positron temperatures. The equilibrium charge neutrality condition in normalized form is given by

$$\chi + n_{i0} = 1, \quad (4)$$

in which n_{i0} is the equilibrium ion density normalized by the equilibrium electron density.

Using (2) and (3), Poisson's equation in (1) is rewritten as

$$\frac{\partial^2 \phi}{\partial x^2} = \left(1 - \beta + \beta\phi^2\right)\exp(\phi) - \chi \exp\left(-\sigma_p \phi\right) - n_i. \quad (5)$$

3. Derivation of the Evolution Equation

Following the usual procedure we make the following Fourier expansions for the field quantities [28, 32–34]:

$$F = \varepsilon^2 F_0' + \sum_{s=1}^{\infty} \varepsilon_s \left\{F_s \exp(is\psi) + F_s^* \exp(-is\psi)\right\}, \quad (6)$$

where F stands for the field quantities n_i, v_i, and ϕ; F_0' and F_s are assumed to vary slowly with space and time; that is, they are supposed to be functions of $\xi = \varepsilon(x - C_g t)$ and $\tau = \varepsilon^2 \tau$, with ε being a small parameter and C_g the group velocity; $\psi = kx - \omega t$ (ω, k being two constants satisfying linear dispersion relation). Substituting the expansion (6) in (1) and (5) and then equating from both sides the coefficients of $\exp(i\psi)$, $\exp(2i\psi)$, and terms independent of ψ we obtain three sets of equations which we call, respectively, I, II, and III. To solve these equations we make the following perturbation expansion for the field quantities, F_0' and F_s, which we denote by X:

$$X = X^{(1)} + \varepsilon X^{(2)} + \varepsilon X^{(3)} + \cdots. \quad (7)$$

Solving the lowest order equations obtained from the set of equations I after substituting the expansion (7) we get the following solutions for the first harmonic quantities in the lowest order:

$$n_{i1}^{(1)} = \left(1 - \beta + \chi\sigma_p + k^2\right) \cdot \alpha,$$

$$v_{i1}^{(1)} = \frac{\omega \cdot \left(1 - \beta + \chi\sigma_p + k^2\right)}{(1-\chi)} \cdot \alpha, \quad (8)$$

where

$$\alpha = \phi_1^{(1)}. \tag{9}$$

The linear dispersion relation is obtained as

$$\omega^2 = k^2 \left[\frac{(1-\chi)}{\left(1-\beta+\chi\sigma_p+k^2\right)} + \frac{3\sigma_i}{(1-\chi)} \right]. \tag{10}$$

The wave frequency is found to increase with the increase in the nonthermal parameter β and the ion temperature. On the other hand, increase in positron concentration decreases the wave frequency. In this connection it is pertinent to mention that Pakzad [35] reported an incorrect result and it was pointed out and corrected by Baluku and Hellberg [25]. If we put $\beta = 0$, $\chi = 0$, and $\sigma_i = 0$, we get the linear dispersion relation for ion-acoustic waves in *e-i* plasma as obtained by Kakutani and Sugimoto [36]. In the limit $k \to 0$ (10) leads to the normalized ion-acoustic speed (V_s) modified by the presence of positrons, ion-temperature, and non-Maxwellian electron distribution:

$$V_s^2 = \frac{(1-\chi)}{\left(1-\beta+\chi\sigma_p\right)} + \frac{3\sigma_i}{(1-\chi)}. \tag{11}$$

It agrees with the results obtained by Baluku and Hellberg [25] for the case of cold ions ($\sigma_i = 0$). Equation (11) shows that, for the case of cold ions, increase in positron concentration decreases the phase speed [15], increase in the nonthermal parameter (β) leads to an increase in phase speed, and also increase in ion temperature increases the phase speed.

First harmonic quantities in the second order are obtained from the solutions (8) by replacing $-i\omega$ by $-i\omega - \varepsilon C_g(\partial/\partial\xi) + \varepsilon^2(\partial/\partial\tau)$ and ik by $ik + \varepsilon(\partial/\partial\xi)$ and then picking out order ε terms. These are as follows:

$$\begin{aligned}
\phi_1^{(2)} &= 0,\\
n_{i_1}^{(2)} &= -i2k\frac{\partial\alpha}{\partial\xi},\\
v_{i_1}^{(2)} &= \left[\left\{\left(\frac{\omega}{k^2} - \frac{C_g}{k}\right)\left(\frac{1-\beta+\chi\sigma_p}{1-\chi}\right)\right\} - \frac{2\omega k}{1-\chi} - \frac{k^2 C_g}{1-\chi}\right]\frac{\partial\alpha}{\partial\xi}.
\end{aligned} \tag{12}$$

The second harmonic quantities in the lowest order obtained from the set of equations II after substituting the expansion (7) are as follows:

$$\begin{aligned}
\phi_2^{(1)} &= A_1 \cdot \alpha^2,\\
n_{i_2}^{(1)} &= \left[\left\{A_1\left(1-\beta+\chi\sigma_p+4k^2\right)^2\right\} + \left\{\frac{\chi\sigma_p^2}{2} - \beta\right\}\right] \cdot \alpha^2,\\
v_{i_2}^{(1)} &= \frac{\omega}{k(1-\chi)}\left[\left\{A_1\left(1-\beta+\chi\sigma_p+4k^2\right)\right\} - \frac{\left(1-\beta+\chi\sigma_p+k^2\right)^2}{(1-\chi)} + \left(\frac{\chi\sigma_p^2}{2} - \beta\right)\right] \cdot \alpha^2,
\end{aligned} \tag{13}$$

where

$$\begin{aligned}
A_1 = {} & \left[\left(\frac{2\omega^2}{k(1-\chi)} - \frac{6\sigma_i}{(1-\chi)}\right)\left(\frac{\chi\sigma_p^2}{2} - \beta\right)\right] - \left[\left(\frac{3\omega^2}{k(1-\chi)^2} - \frac{3\sigma_i}{(1-\chi)^2}\right)\left(1-\beta+\chi\sigma_p+k^2\right)^2\right]\\
& \times \left(\left[\frac{6\sigma_i k}{(1-\chi)}\left(1-\beta+\chi\sigma_p+4k^2\right)\right] + 2k - \frac{\left[2\omega^2\left(1-\beta+\chi\sigma_p+4k^2\right)\right]}{k(1-\chi)}\right)^{-1}.
\end{aligned} \tag{14}$$

The zeroth harmonic components generated through nonlinear self-interaction of the finite amplitude wave are obtained from the set of equations III after substituting the expansion (7):

$$\begin{aligned}
\phi_0^{(1)} &= B_1 \cdot \alpha\alpha^*,\\
n_{i_0}^{(1)} &= \left[\left\{B_1\left(1+\chi\sigma_p\right)\right\} - 2\beta\right] \cdot \alpha\alpha^*,\\
v_{i_0}^{(1)} &= \left[\left\{B_1\frac{C_g\left(1+\chi\sigma_p\right)}{1-\chi}\right\} - \frac{2\beta C_g}{1-\chi} - \frac{\left\{2\omega\left(1-\beta+\chi\sigma_p+k^2\right)^2\right\}}{k(1-\chi)^2}\right]\alpha\alpha^*,
\end{aligned} \tag{15}$$

where

$$\begin{aligned}
B_1 = {} & \frac{6\beta\sigma_i}{1-\chi} - \frac{2\beta C_g^2}{1-\chi} - \left[\frac{\left(1-\beta+\chi\sigma_p+k^2\right)^2}{(1-\chi)^2}\left(3\sigma_i + 2C_g\frac{\omega}{k} + \frac{\omega^2}{k^2}\right)\right]\\
& \times \left(1 - \frac{C_g^2\left(1+\chi\sigma_p\right)}{1-\chi} + \left[\frac{3\sigma_i}{1-\chi}\left(1+\chi\sigma_p\right)\right]\right)^{-1}.
\end{aligned} \tag{16}$$

Now in order to derive the NLS equation, we need to consider first harmonic quantities in the third order. Collecting coefficients of ε^3 from both sides of the set of equations I after substituting perturbation expansion (7), we get a set of equations for first harmonic quantities in the third order from which after proper elimination we obtain the following desired NLS equation:

$$i\frac{\partial \alpha}{\partial \tau} + P \cdot \frac{\partial^2 \alpha}{\partial \xi^2} = Q \cdot \alpha\alpha^*, \tag{17}$$

$$P = \frac{k(1-\chi)}{2\omega\left(1-\beta+\chi\sigma_p+k^2\right)} \times \left[\frac{2\omega C_g}{1-\chi} - \frac{\omega^2}{k(1-\chi)} - \frac{2\omega^2}{(1-\chi)^2} - \frac{kC_g\omega}{(1-\chi)^2} - C_g \right.$$
$$\times \left[\left\{\left(\frac{\omega}{k^2} - \frac{C_g}{k}\right)\left(\frac{1-\beta+\chi\sigma_p}{1-\chi}\right)\right\} - \frac{2\omega k}{1-\chi} - \frac{k^2 C_g}{1-\chi}\right]$$
$$\left. - \frac{3\sigma_i k}{(1-\chi)^2} + \frac{\omega}{k}\left\{\left(\frac{\omega}{k^2} - \frac{C_g}{k}\right)\left(\frac{1-\beta+\chi\sigma_p}{1-\chi}\right)\right\}\right], \tag{18}$$

$$Q = \frac{k(1-\chi)}{2\omega\left(1-\beta+\chi\sigma_p+k^2\right)} \left[F_2 k - \frac{\omega^2 F_3}{k(1-\chi)} + \frac{\omega F_1}{(1-\chi)}\right], \tag{19}$$

where

$$F_1 = \left[\left(1+\chi\sigma_p\right)B_1 - 2\beta\right]\frac{\omega\left(1-\beta+\chi\sigma_p+k^2\right)}{k(1-\chi)}$$
$$+ \left(1-\beta+\chi\sigma_p+k^2\right)$$
$$\times \left[\frac{C_g\left(1+\chi\sigma_p\right)}{1-\chi} - \frac{2\beta C_g}{1-\chi} - \frac{2\omega\left(1-\beta+\chi\sigma_p+k^2\right)^2}{k(1-\chi)^2}\right]$$
$$+ \frac{\omega\left(1-\beta+\chi\sigma_p+k^2\right)}{k(1-\chi)}$$
$$\times \left[\frac{C_g\left(1+\chi\sigma_p\right)}{1-\chi} - \frac{2\beta C_g}{1-\chi} - \frac{\left(1-\beta+\chi\sigma_p+k^2\right)^2}{1-\chi}\right]$$
$$+ \frac{\omega\left(1-\beta+\chi\sigma_p+k^2\right)}{k(1-\chi)}$$
$$\times \left[\left(1-\beta+\chi\sigma_p+4k^2\right)A_1 + \left(\frac{\chi\sigma_p^2}{2} - \beta\right)\right],$$

$$F_2 = \left[\frac{C_g\left(1+\chi\sigma_p\right)}{1-\chi}B_1 - \frac{2\beta C_g}{1-\chi} - \frac{2\omega\left(1-\beta+\chi\sigma_p+k^2\right)^2}{k(1-\chi)^2}\right]$$
$$\times \frac{\omega\left(1-\beta+\chi\sigma_p+k^2\right)}{k(1-\chi)} \frac{\omega^2\left(1-\beta+\chi\sigma_p+k^2\right)}{k^2(1-\chi)^2}$$
$$\times \left[\left(1-\beta+\chi\sigma_p+4k^2\right) + \left(\frac{\chi\sigma_p^2}{2} - \beta\right)\right.$$
$$\left. - \frac{\left(1-\beta+\chi\sigma_p+k^2\right)^2}{(1-\chi)}\right]$$
$$+ \frac{3\sigma_i}{(1-\chi)^2}\left(1-\beta+\chi\sigma_p+k^2\right)$$
$$\times \left[\left(1+\chi\sigma_p\right)B_1 - 2\beta + \left(1-\beta+\chi\sigma_p+4k^2\right)A_1\right.$$
$$\left. + \left(\frac{\chi\sigma_p^2}{2} - \beta\right)\right],$$
$$F_3 = B_1\left(2\beta + \chi\sigma_p^2\right) + \beta + \beta A_1. \tag{20}$$

4. Modulational Instability and Envelope Solitons

NLS equation (17) describes the nonlinear evolution of the amplitude of IAWs in *e-p-i* plasma with warm ions, nonthermal electrons, and Boltzmann positrons. NLS equation (17) has been studied extensively in connection with the nonlinear propagation of different wave modes. It is well known that a uniform wave train may be modulationally stable or unstable depending on the sign of the product of the group dispersive and the nonlinearity coefficient, that is, PQ. As the coefficients depend on the plasma parameters such as nonthermal parameter β, ion temperature σ_i, and positron concentration χ, the product of PQ can have both positive and negative values over different parametric regions. The wave is modulationally unstable if $PQ < 0$ and the growth rate of instability has a maximum value g_m given by

$$g_m = |Q|\,\alpha_0^2, \tag{21}$$

where α_0 is the constant real amplitude of the carrier wave. For $PQ > 0$, the IAW is modulationally stable. As the product can have both positive and negative signs for different values of β, σ_i, and χ, there are accordingly two types of localized solitary wave solutions of the NLS equation (17). To obtain the soliton profile we let

$$\alpha = \rho \exp(i\theta), \tag{22}$$

where ρ and θ are two real variables. Solving the resulting equations for ρ and θ with $PQ < 0$ we get the following bright envelope soliton solution:

$$\rho = \frac{\sqrt{2|P/Q|}}{L}\operatorname{sech}\left(\frac{\xi - U\tau}{L}\right), \tag{23}$$

where U is the envelope speed and L is the spatial width of the pulse. It encloses high frequency carrier oscillations and vanishes at infinity. On the other hand, if $PQ > 0$, a stable gray or dark soliton (a potential hole or a localized region of deceased amplitude) is obtained:

$$\rho = \frac{\sqrt{2P/Q}}{Ld}\sqrt{1 - d^2\operatorname{sech}^2\left(\frac{\xi - U\tau}{L}\right)}, \tag{24}$$

where the parameter d determines the depth of the modulation. For $d = 1$ we get a dark soliton:

$$\rho = \frac{\sqrt{2P/Q}}{Ld}\tanh\left(\frac{\xi - U\tau}{L}\right). \tag{25}$$

Thus the sign of the product PQ determines the stability/instability profile of IAWs as well as the type of soliton structure. The soliton width is determined by the ratio $|P/Q|$.

We have numerically examined different parametric regions where some of the above excitations may occur. As the coefficients P and Q depend on nonthermal parameter β, ion-to-electron temperature ratio σ_i, and positron-to-electron concentration ratio χ, these parameters would definitely determine the modulational instability and the formation of envelope solitons. Numerical plots in Figures 1–3 show P/Q as a function of k for different values of β, σ_i, and χ. It shows that the IAWs remain modulationally stable for k less than certain critical value k_c and for $k > k_c$ the wave is modulationally unstable.

In Figure 1 the variation of P/Q with wave number has been plotted for different values of nonthermal parameter (β), keeping positron concentration (χ) and ion temperature (σ_i) fixed. It shows that as β increases the value of critical wave number separating stable and unstable regions decreases. It is also noticed that as β increases the width of the dark solitons increases, but that of the bright solitons decreases.

In Figure 2 P/Q is plotted as function of k for different values of ion temperature (σ_i) taking other plasma parameters such as positron concentration (χ) and nonthermal parameter (β) as constant. It is seen that as σ_i increases critical wave number decreases; the width of dark solitons increases but that of bright solitons decreases.

Figure 3 is a P/Q versus wave number plot for different values of positron concentration (χ), keeping the values of nonthermal parameter (β) and ion temperature (σ_i) constant. It shows that as the value of χ increases the critical wave number increases. The width of dark solitons decreases and that of bright solitons increases as χ increases.

Qualitatively these results agree with those obtained by Gill et al. [31] but quantitatively there are differences. We find that the critical wave number is more sensitive to the variation in β, σ_i, and χ than that predicted by Gill et al. [31].

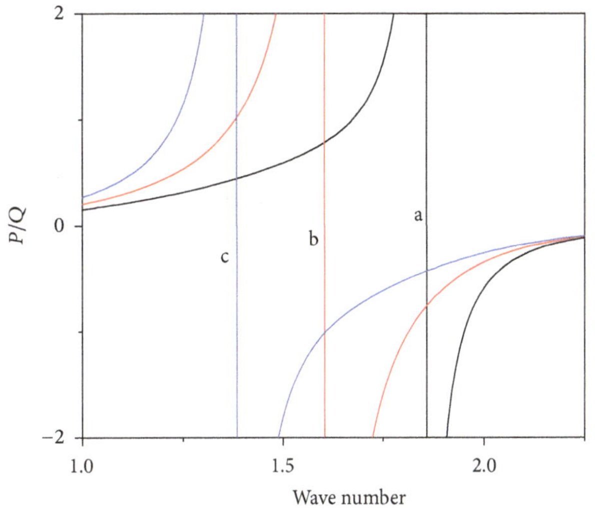

FIGURE 1: Plot of P/Q versus wave number k for different values of nonthermal parameter (β). Curves labelled a, b, and c correspond to $\beta = 0$, 0.055, and 0.11, respectively. $\chi = 0.22$, $\sigma_p = 0.01$, and $\sigma_i = 0.02$.

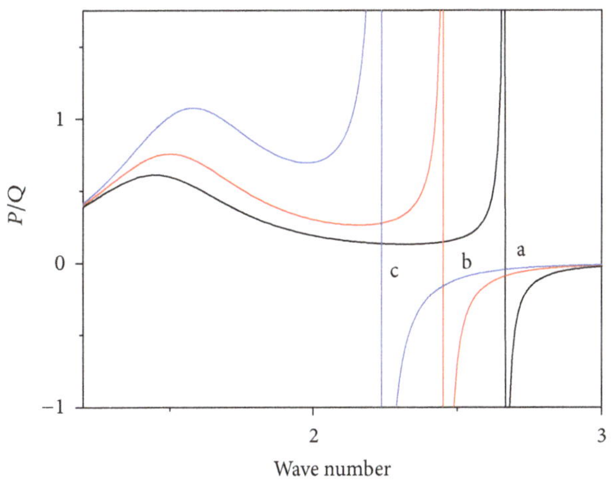

FIGURE 2: Plot of P/Q versus wave number k for different values of ion temperature (σ_i). Curves labelled a, b, and c correspond to σ_i = 0.0145, 0.0155, and 0.0165, respectively. $\chi = 0.2$, $\sigma_p = 0.015$, and $\beta = 0.022$.

In addition, we have numerically studied the dependence of growth rate of instability on all the plasma parameters β, σ_i, and χ. The results are shown in Figures 4, 5, and 6. It is shown that the growth rate of instability increases with increase in the nonthermality of electrons and ion temperature but the increase of positron concentration reduces instability growth rate.

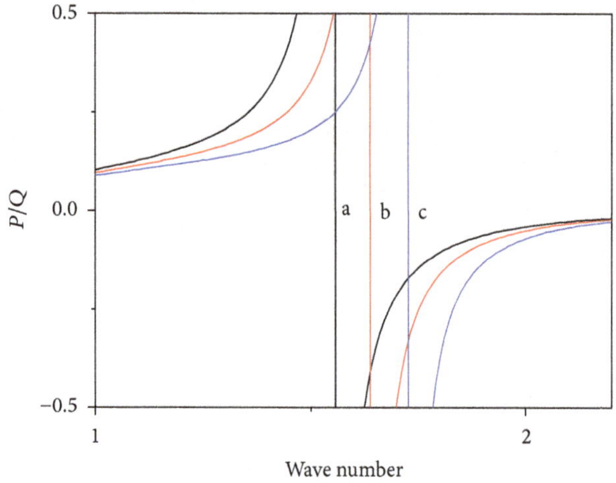

FIGURE 3: Plot of P/Q versus wave number k for different values of positron concentration (χ). Curves labelled a, b, and c correspond to $\chi = 0.25, 0.26$, and 0.27, respectively. $\beta = 0.022$, $\sigma_p = 0.01$, and $\sigma_i = 0.052$.

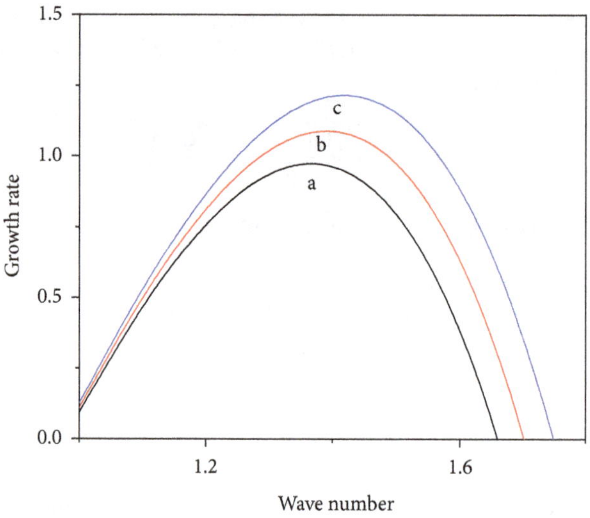

FIGURE 5: Plot of growth rate versus wave number k for different values of ion temperature (σ_i). Curves labelled a, b, and c correspond to $\sigma_i = 0.0012, 0.0024$, and 0.0036, respectively. $\chi = 0.001$, $\sigma_p = 0.01$, and $\beta = 0.001$.

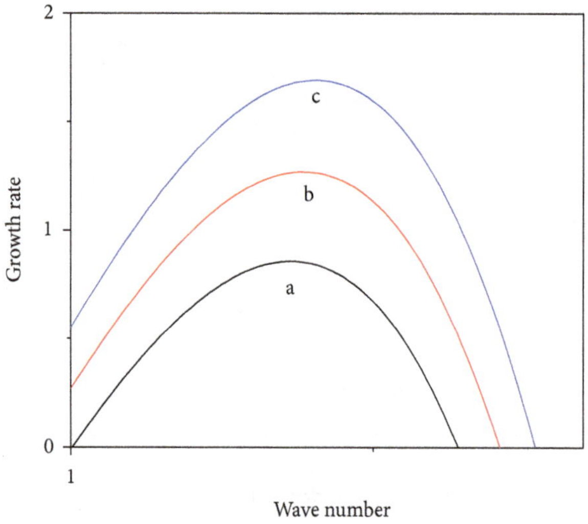

FIGURE 4: Plot of growth rate versus wave number k for different values of nonthermal parameter (β). Curves labelled a, b, and c correspond to $\beta = 0, 0.055$, and 0.11, respectively. $\chi = 0.02$, $\sigma_p = 0.01$, and $\sigma_i = 0.002$.

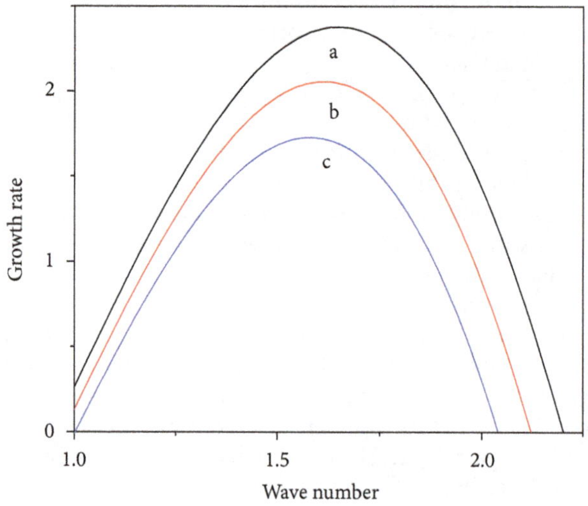

FIGURE 6: Plot of growth rate versus wave number k for different values of positron concentration (χ). Curves labelled a, b, and c correspond to $\chi = 0, 0.02$, and 0.04, respectively. $\beta = 0.01$, $\sigma_p = 0.01$, and $\sigma_i = 0.01$.

5. Conclusions

In the present work, we have investigated modulational instability and envelope excitations of IAWs in the *e-p-i* plasma in detail including simultaneously the effects of nonthermality of electrons and temperatures of ions. Our main findings are summarized below.

(i) The wave frequency increases with increase in nonthermality of electrons and the temperature of ions whereas the increase in positron concentration decreases the wave frequency.

(ii) There exists a critical wave number k_c below which the wave is modulationally stable and above which the wave is modulationally unstable.

(iii) The value of the critical wave number and the characteristics of bright/dark envelope solitons depend significantly on the nonthermal parameter (β), ion temperature (σ_i), and positron concentration (χ).

Finally we would like to mention that the results presented in this paper may be useful to explain modulational instability and envelope soliton excitations of IAWs in some astrophysical and space environments where e-p-i plasmas with nonthermal electrons are present.

Conflict of Interests

The authors declare that there is no conflict of interests regarding the publication of this paper.

Acknowledgment

The authors would like to thank the reviewers for various suggestions and helpful comments in bringing the paper to the present form.

References

[1] H. R. Miller and P. J. Witta, *Active Galactic Nuclei*, Springer, Berlin, Germany, 1978.

[2] F. C. Michel, "Theory of pulsar magnetospheres," *Reviews of Modern Physics*, vol. 54, no. 1, pp. 1–66, 1982.

[3] F. C. Michel, *Theory of Neutron Star Magnetosphere*, Chicago University Press, Chicago, Ill, USA, 1991.

[4] M. I. Barns, *Positron Electron Pairs in Astrophysics*, American Institute of Physics, New York, NY, USA, 1983.

[5] W. K. Misner, S. Thorne, and J. A . Wheeler, *Gravitation*, Freeman, San Francisco, Calif, USA, 1973.

[6] M. J. Rees, G. W. Gibbons, S. W. Hawking, and S. Siklaseds, *The Early Universe*, Cambridge University Press, Cambridge, UK, 1983.

[7] E. Tandberg-Hanssen and A. Gordon Emslie, *The Physics of Solar Flares*, Cambridge University Press, Cambridge, UK, 1988.

[8] A. Cairns, R. Bingham, R. O. Dendy, C. M. C. Nairn, P. K. Shukla, and A. A. Mamun, "Ion sound solitary waves with density depressions," *Journal of Physics IV. France*, vol. 5, no. C6, pp. 43–48, 1995.

[9] S. I. Popel, S. V. Vladimirov, and P. K. Shukla, "Ion-acoustic solitons in electron-positron-ion plasmas," *Physics of Plasmas*, vol. 2, no. 3, pp. 716–719, 1995.

[10] Y. N. Nejoh, "The effect of the ion temperature on large amplitude ion-acoustic waves in an electron-positron-ion plasma," *Physics of Plasmas*, vol. 3, no. 4, pp. 1447–1451, 1996.

[11] H. R. Pakzad, "Ion acoustic solitary waves in plasma with nonthermal electron, positron and warm ion," *Astrophysics and Space Science*, vol. 323, no. 4, pp. 345–350, 2009.

[12] S. Ghosh and R. Bharuthram, "Ion acoustic solitons and double layers in electron-positron-ion plasmas with dust particulates," *Astrophysics and Space Science*, vol. 314, no. 1-3, pp. 121–127, 2008.

[13] N. Jehan, W. Masood, and A. M. Mirza, "Planar and nonplanar dust acoustic solitary waves in electronpositron-ion- dust plasmas," *Physica Scripta*, vol. 80, no. 3, Article ID 035506, 2009.

[14] R. A. Cairns, A. A. Mamun, R. Bingham, and P. K. Shukla, "Ion acoustic solitons in a magnetised plasma with nonthermal electrons," *Physica Scripta*, vol. 63, pp. 80–86, 1996.

[15] B. Ghosh, S. Banerjee, and S. N. Paul, "Effect of non-thermal electrons and warm negative ions on ion-acoustic solitary waves in multi-component drifting plasma," *Indian Journal of Pure and Applied Physics*, vol. 51, no. 7, pp. 488–493, 2013.

[16] B. Ghosh, S. N. Paul, C. Das, I. Paul, and S. Banerjee, "Electrostatic double layers in a multicomponent drifting plasma having nonthermal electrons," *Brazilian Journal of Physics*, vol. 43, no. 1-2, pp. 28–33, 2013.

[17] P. O. Dovner, A. I. Eriksson, R. Bostrom, and B. Holback, "Freja multiprobe observations of electrostatic solitary structures," *Geophysical Research Letters*, vol. 21, no. 17, pp. 1827–1830, 1994.

[18] R. Boström, G. Gustafsson, B. Holback, G. Holmgren, H. Koskinen, and P. Kintner, "Characteristics of solitary waves and weak double layers in the magnetospheric plasma," *Physical Review Letters*, vol. 61, no. 1, pp. 82–85, 1988.

[19] R. A. Cairns, A. A. Mamun, R. Bingham et al., "Electrostatic solitary structures in non-thermal plasmas," *Geophysical Research Letters*, vol. 22, no. 20, pp. 2709–2712, 1995.

[20] M. Salahuddin, H. Saleem, and M. Saddiq, "Ion-acoustic envelope solitons in electron-positron-ion plasmas," *Physical Review E*, vol. 66, no. 3, Article ID 036407, 2002.

[21] T. S. Gill, C. Bedi, and A. S. Bains, "Envelope excitations of ion acoustic solitary waves in a plasma with superthermal electrons and positrons," *Physica Scripta*, vol. 81, no. 5, Article ID 055503, 2010.

[22] G. Murtaza and M. Salahuddin, "Modulational instability of ion acoustic waves in a magnetised plasma," *Plasma Physics*, vol. 24, no. 5, pp. 451–456, 1982.

[23] Yashvir, T. N. Bhatnagar, and S. R. Sharma, "Nonlinear ion-acoustic waves and solitons in warm-ion magnetized plasma," *Plasma Physics and Controlled Fusion*, vol. 26, no. 11, article 004, pp. 1303–1310, 1984.

[24] J. K. Chawla, M. K. Mishra, and R. S. Tiwari, "Modulational instability of ion-acoustic waves in electron-positron-ion plasmas," *Astrophysics and Space Science*, vol. 347, pp. 283–292, 2013.

[25] T. K. Baluku and M. A. Hellberg, "Ion acoustic solitary waves in an electron-positron-ion plasma with non-thermal electrons," *Plasma Physics and Controlled Fusion*, vol. 53, no. 9, Article ID 095007, 2011.

[26] A. E. Dubinov and M. A. Sazonkin, "Nonlinear theory of ion-acoustic waves in an electron-positron-ion plasma," *Plasma Physics Reports*, vol. 35, no. 1, pp. 14–24, 2009.

[27] S. Mahmood, S. Siddiqui, and N. Jehan, "Modulational instability of ion acoustic wave with warm ions in electron-positron-ion plasmas," *Physics of Plasmas*, vol. 18, no. 5, Article ID 052309, 2011.

[28] A. S. Bains, N. S. Saini, and T. S. Gill, "Modulational instability of ion-acoustic soliton in electron-positron-ion plasma with dust particulates," *Astrophysics and Space Science*, vol. 343, no. 1, pp. 293–299, 2013.

[29] P. Eslami, M. Mottaghizadeh, and H. R. Pakzad, "Modulational instability of ion acoustic waves in e-p-i plasmas with electrons and positrons following a q-nonextensive distribution," *Physics of Plasmas*, vol. 18, no. 10, Article ID 102313, 2011.

[30] J. Zhang, Y. Wang, and L. Wu, "Modulation instability of ion acoustic waves, solitons, and their interactions in nonthermal electron-positron-ion plasmas," *Physics of Plasmas*, vol. 16, no. 6, Article ID 062102, 2009.

[31] T. S. Gill, A. S. Bains, N. S. Saini, and C. Bedi, "Ion-acoustic envelope excitations in electron-positron-ion plasma with non-thermal electrons," *Physics Letters A*, vol. 374, no. 31-32, pp. 3210–3215, 2010.

[32] B. Ghosh, S. N. Paul, C. Das, and I. Paul, "Modulational instability of high frequency surface waves on warm plasma half-space," *Canadian Journal of Physics*, vol. 90, no. 3, pp. 291–297, 2012.

[33] B. Ghosh and K. P. Das, "Modulational instability of electron plasma waves in a cylindrical wave guide," *Plasma Physics and Controlled Fusion*, vol. 27, no. 9, pp. 969–982, 1985.

[34] B. Ghosh, S. Chandra, and S. N. Paul, "Amplitude modulation of electron plasma waves in a quantum plasma," *Physics of Plasmas*, vol. 18, no. 1, Article ID 012106, 2011.

[35] H. R. Pakzad, "Ion acoustic solitary waves in plasma with nonthermal electron and positron," *Physics Letters A: General, Atomic and Solid State Physics*, vol. 373, no. 8-9, pp. 847–850, 2009.

[36] T. Kakutani and N. Sugimoto, "Krylov-Bogoliubov-Mitropolsky method for nonlinear wave modulation," *The Physics of Fluids*, vol. 17, pp. 1617–1625, 1974.

10

Analytical Results Connecting Stellar Structure Parameters and Extended Reaction Rates

Hans J. Haubold[1,2] **and Dilip Kumar**[2]

[1] *Office for Outer Space Affairs, United Nations, Vienna International Centre, P.O. Box 500, 1400 Vienna, Austria*
[2] *Centre for Mathematical Sciences Pala Campus, Arunapuram P.O., Palai, Kerala 686 574, India*

Correspondence should be addressed to Hans J. Haubold; hans.haubold@gmail.com

Academic Editor: Milan S. Dimitrijevic

Possible modification in the velocity distribution in the nonresonant reaction rates leads to an extended reaction rate probability integral. The closed form representation for these thermonuclear functions is used to obtain the stellar luminosity and neutrino emission rates. The composite parameter $\mathscr{C}$ that determines the standard nuclear reaction rate through the Maxwell-Boltzmann energy distribution is extended to $\mathscr{C}^*$ by the extended reaction rates through a more general distribution than the Maxwell-Boltzmann distribution. The new distribution is obtained by the pathway model introduced by Mathai (2005). Simple analytic models considered by various authors are utilized for evaluating stellar luminosity and neutrino emission rates and are obtained in generalized special functions such as Meijer's G-function and Fox's H-function. The standard and extended nonresonant thermonuclear functions are compared by plotting them. Behaviour of the new energy distribution, which is more general than the Maxwell-Boltzmann, is also studied.

1. Introduction

The mystery behind the distant universe is explored so far by the understanding of the sun, the star near to us. It is the only star whose mass, radius, and luminosity are fairly accurately known. The structural change in the sun is due to the central thermonuclear reactor in it. Solar nuclear energy generation and solar neutrino emission are governed by chains of nuclear reactions in the gravitationally stabilized solar fusion reactor [1, 2]. Qualitative calculations of specific reaction rates require a large amount of experimental inputs and theoretical assumptions. By using the theories from nuclear physics and kinetic theory of gases one can determine the reaction rate for low-energy nonresonant thermonuclear reactions in nondegenerate plasma [3]. The formalization of the calculation of the reaction rate of interacting articles under cosmological or stellar conditions was presented by many authors [4, 5]. For the most common case, a nuclear reaction in which a particle of type 1 strikes a particle of type 2 producing a nucleus 3 and a new particle 4 is symbolically represented as

$$1 + 2 \longrightarrow 3 + 4 + E_{12}, \tag{1}$$

where E_{12} is the energy release given by $E_{12} = (m_1 + m_2 - m_3 - m_4)c^2$, where m_i, $i = 1, 2, 3, 4$ denote the masses of the particles and c denotes the velocity of light. The reaction rate r_{12} of the interacting particles 1 and 2 is obtained by averaging the reaction cross section over the normalized density function of the relative velocity of the particles [5–7]. Let n_1 and n_2 denote the number densities of the particles 1 and 2, respectively, and let $\sigma(v)$ be the reaction cross section where v is the relative velocity of the particles and $f(v)$ is the normalized velocity density; then the reaction rate r_{12} is given by

$$\begin{aligned} r_{12} &= \left(1 - \frac{1}{2}\delta_{12}\right) n_1 n_2 \langle \sigma v \rangle_{12} \\ &= \left(1 - \frac{1}{2}\delta_{12}\right) n_1 n_2 \int_0^\infty v\sigma(v) f(v)\,\mathrm{d}v \\ &= \left(1 - \frac{1}{2}\delta_{12}\right) n_1 n_2 \int_0^\infty \sigma(E) \left(\frac{2E}{\mu}\right)^{1/2} f(E)\,\mathrm{d}E, \end{aligned} \tag{2}$$

where δ_{12} is the Kronecker delta which is introduced to avoid double counting in the reaction if particles 1 and 2 are identical. $\langle\sigma v\rangle_{12}$ is the thermally averaged product which is in fact the probability per unit time that two particles 1 and 2 confined to a unit volume will react with each other. μ is the reduced mass of the particles given by $\mu = (m_1 m_2)/(m_1 + m_2)$. $E = \mu v^2/2$ is the kinetic energy of the particles in the centre of mass system. From literature [4, 5, 7] it may be noted that all the analytic expressions for astrophysically relevant nuclear reaction rates underline the hypothesis that the distribution of the relative velocities of the reacting particles always remains Maxwell-Boltzmann for a nonrelativistic nondegenerate plasma of nuclei in thermodynamic equilibrium. The Maxwell-Boltzmann relative kinetic energy distribution can be written as

$$f_{\text{MBD}}(E)\,\mathrm{d}E = 2\pi\left(\frac{1}{\pi kT}\right)^{3/2}\exp\left(-\frac{E}{kT}\right)\sqrt{E}\,\mathrm{d}E, \quad (3)$$

where k is the Boltzmann constant and T is the temperature. Substituting (3) in (2) we get

$$r_{12} = \left(1-\frac{1}{2}\delta_{12}\right)n_1 n_2\left(\frac{8}{\pi\mu}\right)^{1/2}\left(\frac{1}{kT}\right)^{3/2} \times \int_0^\infty E\sigma(E)\exp\left(-\frac{E}{kT}\right)\mathrm{d}E. \quad (4)$$

The thermonuclear fusion depends on three physical variables, the temperature T, the Gamow energy E_G, and the nuclear fusion factor $S(E)$. If two nuclei of charges $Z_1 e$ and $Z_2 e$ collide at low energies below the Coulomb barrier, then the Gamow energy E_G is given by [8, 9]

$$E_G = 2\mu(\pi\alpha Z_1 Z_2 c)^2, \quad (5)$$

where α is the electromagnetic fine structure constant given by

$$\alpha = \frac{e^2}{\hbar c}, \quad (6)$$

where e is the quantum of electric charge, $\hbar$ is Planck's quantum of action, and α is approximately $1/137$ [9] for our universe. Thus the Gamow factor, which is determined by the electromagnetic force, and the nuclear fusion factor $S(E)$ set the nuclear reaction cross section at low energies for nonresonant charged particles as [7, 10]

$$\sigma(E) = \frac{S(E)}{E}\exp\left[-\left(\frac{E_G}{E}\right)^{1/2}\right], \quad (7)$$

and $S(E)$ is the cross section factor which is often found to be constant or a slowly varying function of energy over a limited range of energy given by [4, 5]

$$S(E) \approx S(0) + \frac{\mathrm{d}S(0)}{\mathrm{d}E}E + \frac{1}{2}\frac{\mathrm{d}^2S(0)}{\mathrm{d}E^2}E^2 = \sum_{\nu=0}^{2}\frac{S^{(\nu)}(0)}{\nu!}E^\nu. \quad (8)$$

Substituting (7) and (8) in (4) we obtain

$$r_{12} = \left(1-\frac{1}{2}\delta_{12}\right)n_1 n_2\left(\frac{8}{\pi\mu}\right)^{1/2} \times \left(\frac{1}{kT}\right)^{3/2}\sum_{\nu=0}^{2}\frac{S^{(\nu)}(0)}{\nu!} \times \int_0^\infty E^\nu \exp\left[-\frac{E}{kT}-\left(\frac{E_G}{E}\right)^{1/2}\right]\mathrm{d}E. \quad (9)$$

This is the nonresonant reaction rate probability integral in the Maxwell-Boltzmann case. The closed form evaluation of this integral can be seen in a series of papers by Mathai and Haubold; see, for example, Haubold and Mathai [6, 11], Mathai and Haubold [5], and so forth. The main aim of the present work is to extend the reaction rate probability integral given in (9) by replacing the Maxwell-Boltzmann energy distribution by a more general energy distribution called the pathway energy distribution obtained by using the pathway model of Mathai introduced in 2005.

The paper is organized as follows. In the next section we discuss a more general energy distribution than the Maxwell-Boltzmann distribution and obtain the extended reaction rate probability integral in the nonresonant case. We take advantage of the closed form representation of the extended thermonuclear reaction rate for finding the luminosity and the neutrino emission rate of the nonlinear stellar model under consideration in Section 3. Section 4 is devoted to finding the desired connection between stellar structure parameters and the neutrino emission of the stellar model by using the closed form analytic representation of the extended reaction rates. A comparison of the Maxwell-Boltzmann energy distribution with the pathway energy distribution is done with the help of graphs in Section 5. Also we try to discriminate the standard and extended reaction rates. Concluding remarks are included in Section 6.

2. Extended Nonresonant Thermonuclear Reaction Rate and Its Closed Forms

In recent years, possible deviations of the velocity distribution of the plasma particles from the Maxwell-Boltzmann in connection with the production of neutrinos in the gravitationally stabilized solar fusion reactor have been pointed out [2, 10, 12–15]. It was initiated by Tsallis, the originator of nonextensive statistical mechanics [16–18], who has used q-exponential function as the fundamental distribution instead of the Maxwell-Boltzmann distribution. An initial attempt to extend the standard theories of reaction rates to Tsallis statistics was done by many authors; see Mathai and Haubold [19] and Saxena et al. [20]. In 2005, Mathai introduced the pathway model by which even more general distributions can be incorporated in the theory of reaction rates [19, 21]. Initially, pathway model was introduced for the matrix variate case to cover many of the matrix variate statistical densities. The scalar case is a particular one there. Later, Mathai, his coworkers, and others found connection of pathway model with the information theory, the fractional calculus, the Mittag-Leffler functions, and so forth. The pathway model can be effectively used in any situation in which we need to switch between three different functional forms, namely, generalized type-1 beta form, generalized type-2 beta form, and generalized gamma form, using the pathway parameter q. In practical purpose of fitting experimental data, pathway model can be utilized to switch between different parametric families with thicker or thinner tail. The pathway model for the real scalar case can be explained as follows:

$$f_1(x) = c_1 x^{\gamma-1}\left[1 - a(1-q)x^{\delta}\right]^{1/(1-q)}, \quad a>0,\ \delta>0,\ 1-a(1-q)x^{\delta}>0,\ \gamma>0,\ q<1, \tag{10}$$

is the generalized type-1 beta form of the pathway model. This is a model with right tail cut-off for $q < 1$. The Tsallis statistics for $q < 1$ can be obtained from this model by putting $\gamma = 1$ [16–18]. Other cases available are the regular type-1 beta density, the Pareto density, the power function, and the triangular and related models [22]. The generalized type-2 beta form of the pathway model is given by

$$f_2(x) = c_2 x^{\gamma-1}\left[1 + a(q-1)x^{\delta}\right]^{-1/(q-1)}, \quad 0<x<\infty,\ q>1,\ a>0,\ \gamma>0,\ \delta>0. \tag{11}$$

Here also for $\gamma = 1$ we get the Tsallis statistics for $q > 1$ [16–18]. Other standard distributions coming from this model are regular type-2 beta density, F-distribution, the Lévy model, and related models [22]. When $q \to 1$, $f_1(x)$ and $f_2(x)$ will reduce to the generalized gamma form of the pathway model given by

$$f_3(x) = c_3 x^{\gamma-1} e^{-ax^{\delta}}, \quad x>0. \tag{12}$$

This model covers generalized gamma, gamma, exponential, chi-square, the Weibull, the Maxwell-Boltzmann, the Rayleigh, and related densities. c_1, c_2, and c_3 defined in (10), (11), and (12), respectively, are the normalizing constants if we consider statistical densities.

By a suitable modification of the Maxwell-Boltzmann distribution given in (3) through the pathway model, we get a more general energy distribution called the pathway energy distribution given by the density

$$f_{\mathrm{PD}}(E)\,dE = \frac{2\pi(q-1)^{3/2}}{(\pi kT)^{3/2}} \frac{\Gamma(1/(q-1))}{\Gamma(1/(q-1)-3/2)} \times \sqrt{E}\left[1+(q-1)\frac{E}{kT}\right]^{-1/(q-1)} dE, \tag{13}$$

for $q > 1$, $1/(q-1)-3/2 > 0$. The Maxwell-Boltzmann energy distribution can be retrieved from (13) by taking $q \to 1$. Thus the reaction rate probability integral given in (4) can be modified by using (13) and we get the extended reaction rate as

$$\tilde{r}_{12} = \left(1-\frac{1}{2}\delta_{12}\right) n_1 n_2 \left(\frac{8}{\pi\mu}\right)^{1/2} \left(\frac{q-1}{kT}\right)^{3/2} \times \frac{\Gamma(1/(q-1))}{\Gamma(1/(q-1)-3/2)} \sum_{\nu=0}^{2} \frac{S^{(\nu)}(0)}{\nu!} \times \int_0^{\infty} E^{\nu}\left[1+(q-1)\frac{E}{kT}\right]^{-1/(q-1)} \times \exp\left[-\left(\frac{E_G}{E}\right)^{1/2}\right] dE \tag{14}$$

for $q > 1$, $1/(q-1) - 3/2 > 0$. Substituting $y = E/kT$ and $x = (E_G/kT)^{1/2}$ we obtain the above integral in a more convenient form as follows:

$$\tilde{r}_{12} = \left(1-\frac{1}{2}\delta_{12}\right) n_1 n_2 \left(\frac{8}{\pi\mu}\right)^{1/2} (q-1)^{3/2} \times \frac{\Gamma(1/(q-1))}{\Gamma(1/(q-1)-3/2)} \sum_{\nu=0}^{2} \left(\frac{1}{kT}\right)^{-\nu+(1/2)} \frac{S^{(\nu)}(0)}{\nu!} \times \int_0^{\infty} y^{\nu}\left[1+(q-1)y\right]^{-1/(q-1)} e^{-xy^{-1/2}} dy, \tag{15}$$

for $q > 1$, $1/(q-1) - 3/2 > 0$. Here we consider the integral to be evaluated as

$$I_{1q} = \int_0^{\infty} y^{\nu}\left[1+(q-1)y\right]^{-1/(q-1)} \times e^{-xy^{-1/2}} dy. \tag{16}$$

The integral can be evaluated by the techniques in applied analysis and can be obtained in closed form via Meijer's G-function as [2, 23, 24]

$$I_{1q} = \frac{(\pi)^{-1/2}}{(q-1)^{\nu+1}\Gamma(1/(q-1))} \times G_{1,3}^{3,1}\left(\frac{(q-1)x^2}{4}\,\middle|\,{}^{2-1/(q-1)+\nu}_{0,1/2,\nu+1}\right) \tag{17}$$

which yields the nonresonant reaction rate probability integral in the extended case as

$$\tilde{r}_{12} = \left(1-\frac{1}{2}\delta_{12}\right) n_1 n_2 \left(\frac{8}{\mu}\right)^{1/2} \frac{\pi^{-1}}{\Gamma(1/(q-1)-3/2)} \times \sum_{\nu=0}^{2} \left(\frac{q-1}{kT}\right)^{-\nu+(1/2)} \frac{S^{(\nu)}(0)}{\nu!} \times G_{1,3}^{3,1}\left[\frac{(q-1)E_G}{4kT}\,\middle|\,{}^{2-1/(q-1)+\nu}_{0,1/2,\nu+1}\right]. \tag{18}$$

Meijer's G-function and its properties can be seen in Mathai and Saxena [25] and Mathai [26]. We can obtain series expansions of the G-function given in (18) by combining the theories of residue calculus and generalized special functions; see Kumar and Haubold [24] for series expansions for all possible values of ν. In many cases the nuclear factor $S^{(\nu)}(0)$ is approximately constant across the fusion window. Taking $S^{(\nu)}(0) = 0$ for $\nu = 1$ and $\nu = 2$ and taking $S^{0}(0) = S(0)$, we obtain the extended reaction rate probability integral as

$$\tilde{r}_{12} = \left(1 - \frac{1}{2}\delta_{12}\right) n_1 n_2 \left[\frac{8(q-1)}{\mu kT}\right]^{1/2} \frac{\pi^{-1}}{\Gamma(1/(q-1) - 3/2)} \times S(0)\, G^{3,1}_{1,3}\left[\frac{(q-1)E_G}{4kT}\bigg|^{2-1/(q-1)}_{0,1/2,1}\right]. \tag{19}$$

The series representation for (19) can be obtained as

$$\begin{aligned}\tilde{r}_{12} &= \left(1 - \frac{1}{2}\delta_{12}\right) n_1 n_2 \left[\frac{8(q-1)}{\mu kT}\right]^{1/2} \frac{\pi^{-1}}{\Gamma(1/(q-1) - 3/2)} \\ &\times S(0)\Bigg\{\sqrt{\pi}\,\Gamma(1/(q-1) - 1) \\ &- 2\pi\Gamma\left(\frac{1}{q-1} - \frac{1}{2}\right)\left[\frac{(q-1)E_G}{4kT}\right]^{1/2} \\ &\times {}_1F_2\left(\frac{1}{q-1} - \frac{1}{2}; \frac{3}{2}, \frac{1}{2}; -\frac{(q-1)E_G}{4kT}\right) \\ &+ \left(\frac{2\sqrt{\pi}(q-1)E_G}{4kT}\right) \\ &\times \sum_{r=0}^{\infty}\left(\frac{(q-1)E_G}{4kT}\right)^r \\ &\times \left[A_r - \ln\left(\frac{(q-1)E_G}{4kT}\right)\right] B_r\Bigg\},\end{aligned} \tag{20}$$

where

$$\begin{aligned}A_r &= \Psi\left(-\frac{1}{2} - r\right) + \Psi\left(\frac{1}{q-1} + r\right) \\ &+ \Psi(1+r) + \Psi(2+r),\end{aligned} \tag{21}$$

$$B_r = \frac{(-1)^r\Gamma(1/(q-1) + r)}{(3/2)_r r!\,(1+r)!}.$$

See the Appendix for detailed evaluation. As $q \to 1$ in (19), then, by using Stirling's formula for gamma functions given by

$$\Gamma(z+a) \approx (2\pi)^{1/2} z^{z+a-1/2} e^{-z}, \quad |z| \to \infty, \quad a \text{ is bounded}, \tag{22}$$

we get the reaction rate probability integral in the Maxwell-Boltzmann case as

$$r_{12} = \left(1 - \frac{1}{2}\delta_{12}\right) n_1 n_2 \left(\frac{8}{\pi\mu}\right)^{1/2} \left(\frac{1}{kT}\right)^{3/2} S(0) \times \int_0^\infty \exp\left[-\frac{E}{kT} - \left(\frac{E_G}{E}\right)^{1/2}\right] dE \tag{23}$$

$$= \left(1 - \frac{1}{2}\delta_{12}\right) n_1 n_2 \left(\frac{8}{\mu kT}\right)^{1/2} S(0)\,\pi^{-1} \times G^{3,0}_{0,3}\left[\frac{E_G}{4kT}\bigg|^{-}_{0,1/2,1}\right] \tag{24}$$

which is obtained in a series of papers by Mathai and Haubold; see, for example, Mathai and Haubold [5]. The integral in (23) is dominated by the minimum value of $E/kT + (E_G/E)^{1/2} = g(E)$ (say). The minimum value of the function $g(E)$, say E_0, can be determined as

$$\frac{d}{dE}\left[\frac{E}{kT} + \left(\frac{E_G}{E}\right)^{1/2}\right]_{E=E_0} = \frac{1}{kT} - \frac{1}{2}E_G^{1/2}E_0^{-3/2} = 0 \Longrightarrow E_0 = E_G^{1/2}\left(\frac{kT}{2}\right)^{3/2} \tag{25}$$

and the function

$$g(E_0) = 3\left(\frac{E_G}{4kT}\right)^{1/3} = 3\Theta, \tag{26}$$

where $\Theta = (E_G/4kT)^{1/3}$. Now by using the Laplace method [27, 28] we can obtain an approximate value for (23) as

$$r_{12} \approx \left(1 - \frac{1}{2}\delta_{12}\right) n_1 n_2 \frac{8S(0)\,\Theta^2 \exp(-3\Theta)}{\sqrt{3}\pi\mu\alpha Z_1 Z_2 c}. \tag{27}$$

In the next section we will obtain the mass, pressure, and temperature for the case of analytic stellar models characterized by density distribution and corresponding temperature distribution suggested by Haubold and Mathai [11].

3. Closed Forms of the Integral over the Stellar Nuclear Energy Generation Rate

Let us consider the density distribution $\varrho(r)$ considered by Haubold and Mathai [6, 11] and Mathai and Haubold [5] in the form

$$\varrho(r) = \varrho_c\left[1 - \left(\frac{r}{R}\right)^\delta\right], \quad \delta > 0, \tag{28}$$

where ϱ_c is the central density of the star, r is an arbitrary distance from the center, and R is the solar radius. This density function is capable of producing different density distributions by choosing the free parameter δ. Now we determine the quantities $M(r)$, $P(r)$, and $T(r)$, the mass, the pressure, and the temperature at r.

By the equation of the mass conservation

$$\frac{dM(r)}{dr} = 4\pi r^2 \varrho(r) \tag{29}$$

we get

$$\begin{aligned} M(r) &= 4\pi\varrho_c \int_0^r t^2 \left[1 - \left(\frac{t}{R}\right)^\delta\right] dt \\ &= \frac{4\pi}{3}\varrho_c r^3 \left[1 - \frac{3}{\delta+3}\left(\frac{r}{R}\right)^\delta\right]. \end{aligned} \tag{30}$$

From (30), we get the central density as

$$\varrho_c = \frac{3(\delta+3)}{4\pi\delta}\frac{M(R)}{R^3}. \tag{31}$$

If an element of a matter at a distance r from the center of a spherical system is in hydrostatic equilibrium, then setting the sum of the radial forces acting on it to zero we obtain

$$\frac{dP(r)}{dr} = -\frac{G\varrho(r)M(r)}{r^2}, \tag{32}$$

where G is the gravitational constant. Assuming that the pressure at the center of the sun is P_c and at the surface is zero, we get

$$\begin{aligned} P(r) &= P_c - G\int_0^r \frac{M(t)\varrho(t)}{t^2}dt \\ &= \frac{4\pi}{3}G\varrho_c^2 R^2 \left[\xi - \frac{1}{2}\left(\frac{r}{R}\right)^2 + \frac{\delta+6}{(\delta+2)(\delta+3)}\left(\frac{r}{R}\right)^{\delta+2} \right. \\ &\quad \left. - \frac{3}{2(\delta+1)(\delta+3)}\left(\frac{r}{R}\right)^{2\delta+2}\right]. \end{aligned} \tag{33}$$

Using the boundary conditions $P(R) = 0$, we get $P_c = (4\pi/3)G\xi\varrho_c^2 R^2$, where

$$\xi = \frac{1}{2} - \frac{\delta+6}{(\delta+2)(\delta+3)} + \frac{3}{2(\delta+1)(\delta+3)}. \tag{34}$$

By the kinetic theory of gases, for a perfect gas, the pressure is given by

$$P(r) = \frac{kN_A}{\mu}\varrho(r)T(r). \tag{35}$$

For the temperature of interest for stellar models, we neglect the negligible radiation pressure from the total pressure and obtain from (35) the following:

$$\begin{aligned} T(r) &= \frac{\mu}{kN_A}\frac{P(r)}{\varrho(r)} \\ &= \frac{4\pi}{3kN_A}\frac{G\mu\varrho_c R^2}{\left[1-(r/R)^\delta\right]} \\ &\quad \times \left[\xi - \frac{1}{2}\left(\frac{r}{R}\right)^2 + \frac{\delta+6}{(\delta+2)(\delta+3)}\left(\frac{r}{R}\right)^{\delta+2} \right. \\ &\quad \left. - \frac{3}{2(\delta+1)(\delta+3)}\left(\frac{r}{R}\right)^{2\delta+2}\right]. \end{aligned} \tag{36}$$

The central temperature $T_c = (4/kN_A)G\mu\xi(M(R)/R)$, where ξ is as defined in (34).

Thus we have obtained the mass, the pressure, and the temperature throughout the nonlinear stellar model with the density distribution defined in (28). Next our aim is to obtain analytical results for stellar luminosity and neutrino emission rates for various stellar models.

4. Stellar Luminosity and Neutrino Emission Rate

The energy conservation equation states that the net increase in the rate of energy flux coming out of a spherical shell from the inside is the same as the energy produced within the shell [29]. If we denote $L_r = L(r)$ as the energy flux through the sphere of radius r, then we have

$$\frac{dL_r}{dr} = 4\pi r^2 \varrho(r)\varepsilon(r), \tag{37}$$

where $\varepsilon(r)$ is the energy produced per second by nuclear reactions in each gram of stellar matter. $\varepsilon(r)$ depends on the chemical composition in each gram of stellar matter. Here usually L_r is a constant but will be equal to L at the surface of the star. We assume here that the star is chemically homogeneous (that is a star where chemical composition throughout is a constant). Also we assume the energy generation rate $\varepsilon(r)$ for one particular nuclear reaction. Now if we denote $\tilde{r}_{12}(\varrho(r), T(r))$ as the extended nonresonant thermonuclear reaction rate for the particles 1 and 2 defined by (19), then we will consider the energy generation rate $\varepsilon_{12}(r)$ and it can be written in terms of the extended reaction rates via

$$\varepsilon_{12}(r) = \frac{1}{\varrho(r)}\mathscr{C}^*\varrho^2 G_{1,3}^{3,1}\left[\frac{(q-1)E_G}{4kT}\bigg|_{0,1/2,1}^{2-1/(q-1)}\right], \tag{38}$$

where

$$\mathscr{C}^* = \frac{E_{12}\tilde{r}_{12}(\varrho(r), T(r))}{\varrho^2 G_{1,3}^{3,1}\left[(q-1)E_G/4kT\big|_{0,1/2,1}^{2-1/(q-1)}\right]} \tag{39}$$

in which E_{12} is the amount of energy given off in a single reaction. It is to be noted that, by using the asymptotic behaviour of $G_{1,3}^{3,1}((q-1)E_G/4kT)$ [26] and as $q \to 1$, $\mathscr{C}^* \to \mathscr{C}$, the composite parameter considered by [9], which is defined as

$$\mathscr{C} = \frac{E_{12}r_{12}}{\varrho^2\Theta^2}\exp(3\Theta) \tag{40}$$

for our universe $\mathscr{C} \approx 2\times 10^4$ for proton-proton fusion under typical stellar conditions [9]. Then from (37) we have the total luminosity of the star by integration as follows:

$$L(R) = \int_0^R 4\pi r^2\varrho(r)\varepsilon(r)\,dr. \tag{41}$$

If we are considering only one specific reaction defined as in (1), then we have

$$L_{12}(R) = \int_0^R 4\pi r^2\varrho(r)\varepsilon_{12}(r)\,dr, \tag{42}$$

where the energy generation rate is defined in (38) and $\varrho(r)$ is a suitable density distribution explaining the sun. Writing (42) in terms of $\tilde{r}_{12}(\varrho(r), T(r))$ we get

$$L_{12}(R) = \int_0^R 4\pi r^2 \mathscr{C}^* \varrho^2 G_{1,3}^{3,1}\left[\frac{(q-1)E_G}{4kT}\Bigg|_{0,1/2,1}^{2-1/(q-1)}\right] dr = \int_0^R 4\pi r^2 E_{12}\tilde{r}_{12}(\varrho(r), T(r))\, dr. \tag{43}$$

The number density n_i of a particle i, for a gas of mean density $\varrho(r)$, can be expressed as

$$n_i(r) = \varrho(r) N_A \frac{X_i}{A_i}, \tag{44}$$

where N_A stands for Avagadro's constant, A_i is the atomic mass of particle i in atomic mass units, and X_i is the mass fraction of particle i such that $\sum_i X_i = 1$. Substituting $\tilde{r}_{12}(\varrho(r), T(r))$ from (19) and using (44) we have

$$\begin{aligned} L_{12}(R) &= \int_0^R 4\pi r^2 E_{12}\left(1-\frac{1}{2}\delta_{12}\right) N_A^2 \varrho^2(r) \frac{X_1X_2}{A_1A_2}\left[\frac{8(q-1)}{\mu k T(r)}\right]^{1/2} \\ &\times \frac{\pi^{-1}}{\Gamma(1/(q-1)-3/2)} S(0) G_{1,3}^{3,1} \\ &\times \left[\frac{(q-1)\pi^2\mu}{2kT(r)}\left(\frac{Z_1Z_2e^2}{\hbar}\right)^2\Bigg|_{0,1/2,1}^{2-1/(q-1)}\right] dr. \end{aligned} \tag{45}$$

If we divide the "internal luminosity" $L_{12}(R_\odot)$ by the amount of energy E_{12}, then we get the total number of particles per second N_{12} liberated in the reaction given by (1) as follows:

$$\begin{aligned} N_{12} &= \frac{L_{12}(R)}{E_{12}} \\ &= 4\left(1-\frac{1}{2}\delta_{12}\right) N_A^2 \frac{X_1X_2}{A_1A_2}\left[\frac{8(q-1)}{\mu k}\right]^{1/2} \\ &\times \frac{1}{\Gamma(1/(q-1)-3/2)} S(0) \\ &\times \int_0^R \frac{r^2\varrho^2(r)}{[T(r)]^{1/2}} G_{1,3}^{3,1}\left[\frac{(q-1)\pi^2\mu}{2kT(r)}\right. \\ &\times \left.\left(\frac{Z_1Z_2e^2}{\hbar}\right)^2\Bigg|_{0,1/2,1}^{2-1/(q-1)}\right] dr \\ &= 4\left(1-\frac{1}{2}\delta_{12}\right) N_A^2 \frac{X_1X_2}{A_1A_2}\left[\frac{8(q-1)}{\mu k}\right]^{1/2} \\ &\times \frac{1}{\Gamma(1/(q-1)-3/2)} S(0)\frac{1}{2\pi i} \\ &\times \int_L \Gamma(s)\Gamma\left(\frac{1}{2}+s\right)\Gamma(1+s) \\ &\times \Gamma\left(\frac{1}{q-1}-1-s\right)\left[\frac{(q-1)\pi^2\mu}{2k}\left(\frac{Z_1Z_2e^2}{\hbar}\right)^2\right]^{-s} \\ &\times \int_0^R r^2\varrho^2(r)[T(r)]^{-1/2+s} dr\, ds. \end{aligned} \tag{46}$$

For the density distribution defined in (28) introduced by Haubold and Mathai [5, 6] and the corresponding temperature distribution (36), we get

$$\begin{aligned} &\int_0^R r^2\varrho^2(r)[T(r)]^{-1/2+s} dr \\ &= \left[\frac{4\pi G\mu}{3kN_A}\varrho_c R^2\right]^{s-1/2} \varrho_c^2 \\ &\times \int_0^R r^2\left[1-\left(\frac{r}{R}\right)^\delta\right]^{5/2-s} \\ &\times \left[-\frac{1}{2}\left(\frac{r}{R}\right)^2 + \frac{\delta+6}{(\delta+2)(\delta+3)}\left(\frac{r}{R}\right)^{\delta+2}\right. \\ &\left. -\frac{3}{2(\delta+1)(\delta+3)}\left(\frac{r}{R}\right)^{2\delta+2}\right]^{s-1/2} dr, \end{aligned} \tag{47}$$

where ξ is as defined in (34). If we put a substitution $r = xR$, then we get

$$\begin{aligned} &\int_0^R r^2\varrho^2(r)[T(r)]^{-1/2+s} dr \\ &= \left[\frac{4\pi G\mu}{3kN_A}\varrho_c R^2\right]^{s-1/2} \varrho_c^2 R^3 \\ &\times \int_0^1 x^2\left[1-x^\delta\right]^{5/2-s} \\ &\times \left[\xi-\frac{1}{2}x^2 + \frac{\delta+6}{(\delta+2)(\delta+3)}x^{\delta+2}\right. \\ &\left. -\frac{3}{2(\delta+1)(\delta+3)}x^{2\delta+2}\right]^{s-1/2} dx. \end{aligned} \tag{48}$$

Putting $x^\delta = y$ and simplifying we obtain

$$\int_0^R r^2 \varrho^2(r)[T(r)]^{-1/2+s} dr = \left[\frac{4\pi G\mu\xi}{3kN_A}\varrho_c R^2\right]^{s-1/2} \frac{\varrho_c^2 R^3}{\delta} \times \int_0^1 y^{3/\delta-1}[1-y]^{5/2-s}[1-u(y)]^{s-1/2} dy, \quad (49)$$

where $u(y)$ is defined as

$$u(y) = \frac{y^{2/\delta}}{\xi}\left[\frac{1}{2} - \frac{\delta+6}{(\delta+2)(\delta+3)}y + \frac{3}{2(\delta+1)(\delta+3)}y^2\right]. \quad (50)$$

As $y \to 0, u(y) \to 0$, and $y \to 1, u(y) \to 1$. If we take

$$v(y) = \frac{1}{2} - \frac{\delta+6}{(\delta+2)(\delta+3)}y + \frac{3}{2(\delta+1)(\delta+3)}y^2, \quad (51)$$

we have $v(0) = 1/2$. The minimum value of $v(y)$ is at $y = (\delta+1)(\delta+6)/3(\delta+2)$ and the value is $1/2 - 1/6((\delta+6)^2(\delta+1)/(\delta+3)(\delta+2)^2)$. Thus the minimum value is nonnegative since $(\delta+6)^2(\delta+1)/(\delta+3)(\delta+2)^2$ decreases steadily from 3 to 1 for all $\delta > 0$. Therefore $v(y) \le 0$. Since $\xi > 0$, for all $\delta > 0$, $u(y) \le 0$ for all $\delta > 0, \xi > 0$. Thus $[1-u(y)]^{s-1/2} \le 0$. Hence $0 < u(y) < 1$ for $0 < y < 1$ and for $\delta > 0$. Thus by using the binomial expansion we obtain

$$[1-u(y)]^{s-1/2} = \sum_{m=0}^{\infty}\frac{(1/2-s)_m}{m!}[u(y)]^m, \quad (52)$$

where $(1/2-s)_m$ is the Pochhammer symbol defined for $a \in \mathbb{C}$ by

$$(a)_0 = 1,$$
$$(a)_m = \begin{cases} a(a+1)\cdots(a+m-1), \\ \dfrac{\Gamma(a+m)}{\Gamma(a)}, \end{cases} \quad m = 1, 2, \ldots, a \neq 0 \quad (53)$$

whenever $\Gamma(a)$ exists. Taking $\delta = 2$ we get $\xi = 1/5$ and

$$[1-u(y)]^{s-1/2} = (1-y)^{2s-1}\left(1-\frac{1}{2}y\right)^{s-1/2}. \quad (54)$$

Then from (49) we obtain

$$\begin{aligned}\int_0^R r^2\varrho^2(r)[T(r)]^{-1/2+s}dr &= \left[\frac{4\pi G\mu}{15kN_A}\varrho_c R^2\right]^{s-1/2}\frac{\varrho_c^2R^3}{2} \times \int_0^1 y^{3/2-1}[1-y]^{5/2+s-1}\left(1-\frac{1}{2}y\right)^{s-1/2}dy \\ &= \left[\frac{4\pi G\mu}{15kN_A}\varrho_c R^2\right]^{s-1/2}\frac{\varrho_c^2R^3}{2} \times \sum_{m=0}^{\infty}\frac{(1/2-s)_m}{m!}\frac{1}{2^m} \times \int_0^1 y^{3/2+m-1}[1-y]^{5/2+s-1}dy.\end{aligned} \quad (55)$$

By using beta integral and using (53) we obtain

$$\int_0^R r^2\varrho^2(r)[T(r)]^{-1/2+s}dr = \left[\frac{4\pi G\mu}{15kN_A}\varrho_c R^2\right]^{s-1/2}\frac{\varrho_c^2R^3}{2} \times \sum_{m=0}^{\infty}\frac{\Gamma(1/2-s+m)}{m!\Gamma(1/2-s)}\frac{1}{2^m}\frac{\Gamma(3/2+m)\Gamma(5/2+s)}{\Gamma(4+s+m)}. \quad (56)$$

Now from (46) we obtain the total number of particles per second liberated in the reaction (1) as follows:

$$\begin{aligned}N_{12} = {} & \frac{2N_A^2\varrho_c^2R^2}{\mu}\left(1-\frac{1}{2}\delta_{12}\right)\frac{X_1X_2}{A_1A_2}\left[\frac{30(q-1)N_A}{\pi G\varrho_c}\right]^{1/2} \\ & \times S(0)\frac{1}{\Gamma(1/(q-1)-3/2)} \times \sum_{m=0}^{\infty}\frac{1}{2^m}\frac{\Gamma(3/2+m)}{m!}\frac{1}{2\pi i} \\ & \times \int_L \Gamma(s)\Gamma\left(\frac{1}{2}+s\right)\Gamma(1+s)\Gamma\left(\frac{5}{2}+s\right) \\ & \times \frac{\Gamma(1/(q-1)-1-s)\Gamma(1/2+m-s)}{\Gamma(1/2-s)\Gamma(4+m+s)} \\ & \times \left[\frac{15\pi(q-1)N_A}{8G\varrho_cR^2}\left(\frac{Z_1Z_2e^2}{\hbar}\right)^2\right]^{-s} ds\end{aligned}$$

$$N_{12} = \frac{2N_A^2 \varrho_c^2 R^2}{\mu} \left(1 - \frac{1}{2}\delta_{12}\right) \frac{X_1 X_2}{A_1 A_2} \left[\frac{30(q-1) N_A}{\pi G \varrho_c}\right]^{1/2}$$
$$\times S(0) \frac{1}{\Gamma(1/(q-1) - 3/2)}$$
$$\times \sum_{m=0}^{\infty} \frac{1}{2^m} \frac{\Gamma(3/2+m)}{m!}$$
$$\times G_{3,5}^{4,2} \left[\frac{15\pi(q-1) N_A}{8 G \varrho_c R^2} \right.$$
$$\left. \times \left(\frac{Z_1 Z_2 e^2}{\hbar}\right)^2 \Bigg|_{0,1/2,1,5/2,1/2}^{2-1/(q-1),1/2-m,4+m} \right]. \quad (57)$$

For more details on G-function and its properties see [25, 26]. Thus we have obtained the the total number of particles per second liberated in the reaction (1) in terms of the density distribution considered by Haubold and Mathai [6, 29].

5. Comparison of Pathway Energy Density and the Maxwell-Boltzmann Energy Density

In Figure 1 it can be obtserved that, for the nuclei to react at energy E, they have to borrow an energy E from the thermal environment. The probability of such an energy is proportional to the Maxwell-Boltzmann energy $\exp[-E/kT]$. The fusion will take place when the nuclei penerate the Coulomb barrier keeping them apart. The probability of penetration is given by the factor $\exp[-(E_G/E)^{1/2}]$. The product of these two factors illustrates that fusion mostly occurs in the energy window given in the figure.

In Figure 2 the pathway energy density is plotted for $q = 0.7, 0.9, 1, 1.2, 1.4$, respectively. For different values of q we get different energy densities (curves (a), (b), (c), (d), and (e)). The nonresonant cross section is also plotted. The product of the pathway energy density and the nonresonant cross section for different values of q, namely, $q = 0.7, 0.9, 1, 1.2, 1.4$, is also plotted. It is to be noted that as $q \to 1$ the pathway energy density coincides with the Maxwell-Boltzmann energy density and also the fusion window for the Maxwell-Boltzmann case in Figure 1.

The curves in the figure represent pathway density $[1 + (q-1)(E/kT)]^{-1/(q-1)}$ for (a) $q = 0.7$, (b) $q = 0.9$, (c) $q = 1$, (d) $q = 1.2$, and (e) $q = 1.4$. (f) represents the nonresonant cross section $\exp[-(E_G/E)^{1/2}]$. The product $[1 + (q-1)(E/kT)]^{-1/(q-1)} \exp[-(E_G/E)^{1/2}]$ is also represented for (g) $q = 0.7$, (h) $q = 0.9$, (i) $q = 1$, (j) $q = 1.2$, and (k) $q = 1.4$.

Figure 3(a) shows the pathway energy density defined in (13) for $q = 1, 1.2, 1.3, 1.4$ and Figure 3(b) shows the Maxwell-Boltzmann energy density defined in (3). As $q \to 1$ the pathway energy density reduces to the Maxwell-Boltzmann energy density. The pathway energy density covers many stable and unstable situations as the value of q varies. If the Maxwell-Boltzmann density is the equilibrium situation, many other nonequilibrium situations are covered by the pathway energy density.

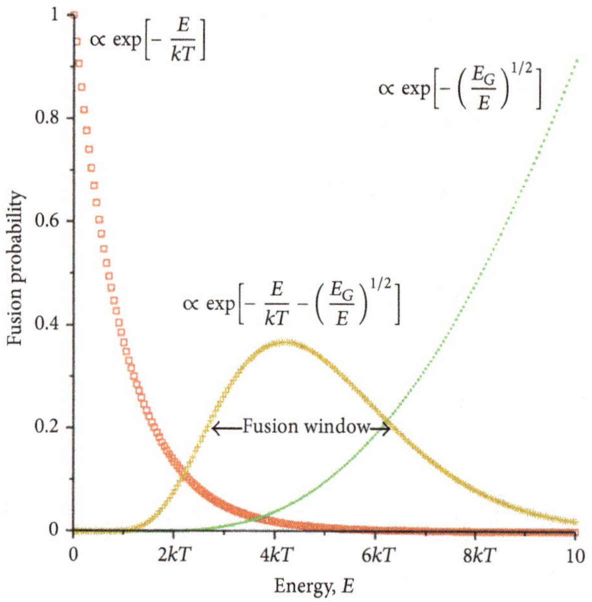

FIGURE 1: Schematic plot of the enrgy-dependent factors for the reaction rate probability integral: the Maxwell-Boltzmann energy density, nonresonant nuclear cross section, and the product of the Maxwell-Boltzmann density and the nonresonant cross section.

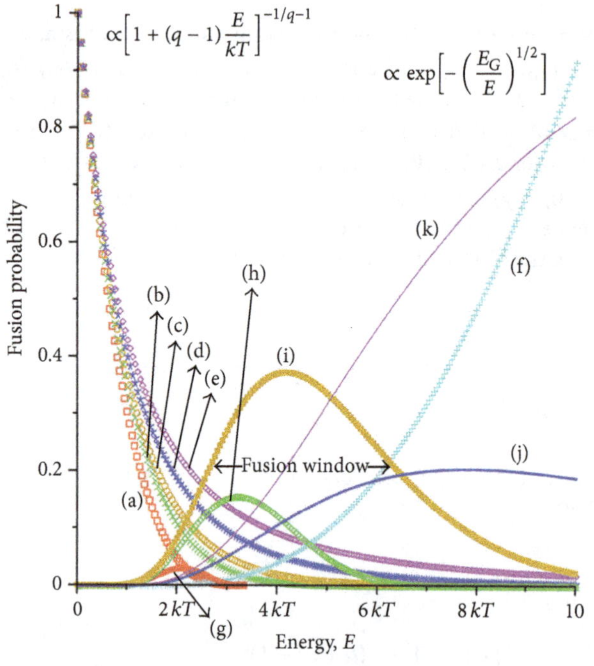

FIGURE 2: Schematic plot of the energy-dependent factors for the extended reaction rate probability integral: pathway energy density, nonresonant nuclear cross section, and the product of the pathway density and the nonresonant cross section.

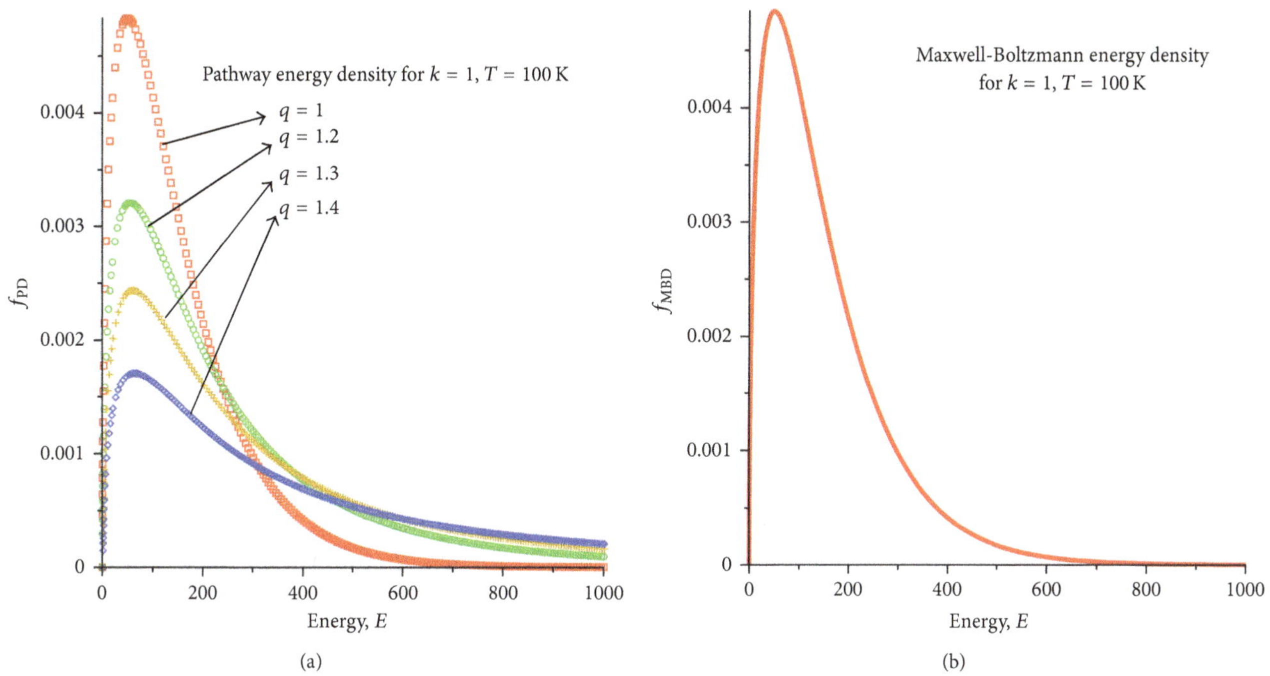

FIGURE 3: (a) Pathway energy density for $q = 1, 1.2, 1.3, 1.4$ and for $k = 1, T = 100\,K$. (b) The Maxwell-Boltzmann energy density for $k = 1, T = 100\,K$.

6. Concluding Remarks

In this paper we have modified the energy distribution for a nonresonant reaction rate probability integral. The composition parameter $\mathscr{C}$ considered by [9] is extended to $\mathscr{C}^*$ by the pathway energy density. Considering the analytic density distributions developed by Haubold and Mathai [6, 29], they are used to obtain the stellar luminosity and the neutrino emission rates and are obtained in generalized special functions such as Meijer's G-function. The pathway energy density considered here covers many density functions and hence the extended reaction rate integral covers a wider class of integral. Pathway energy density helps us to obtain various fusion windows by giving different values to q, the pathway parameter, which in turn leads to a new opening in the fusion research. The graphs plotted here are by using Maple 14 in Windows XP platform.

Appendix

Series Representation

The series representation for the right-hand side of (19) can be obtained through the following procedure. Here we apply residue calculus on the G-function given in (19). Consider the G-function as follows:

$$G_{1,3}^{3,1}\left(\frac{(q-1)E_G}{4kT}\Big|_{0,1/2,1}^{2-1/(q-1)}\right)$$

$$= \frac{1}{2\pi i}\int_{c-i\infty}^{c+i\infty}\Gamma(s)\,\Gamma\left(\frac{1}{2}+s\right)\Gamma(1+s) \times\Gamma\left(\frac{1}{q-1}-1-s\right)\left(\frac{(q-1)E_G}{4kT}\right)^{-s}\mathrm{d}s. \tag{A.1}$$

The right-hand side is the sum of the residues of the integrand. The poles of the gammas in the integral representation in (19) are as follows:

$$\text{poles of } \Gamma(s) : s = 0, -1, -2, \ldots;$$

$$\text{poles of } \Gamma(1/2+s) : s = -1/2, -3/2, -5/2, \ldots;$$

$$\text{poles of } \Gamma(1+s) : s = -1, -2, -3, \ldots.$$

Here the poles of $\Gamma(s)$ and $\Gamma(1+s)$ will coincide with each other at all points except at $s = 0$. Note that the pole $s = 0$ is a pole of order 1, $s = -1/2, -3/2, -5/2, \ldots$ are each of order 1, and $s = -1, -2, -3, \ldots$ are each of order 2. We know that

$$\lim_{s\to -r}(s+r)\,\Gamma(s) = \frac{(-1)^r}{r!},$$
$$\Gamma(a-r) = \frac{(-1)^r\,\Gamma(a)}{(1-a)_r}, \tag{A.2}$$
$$\Gamma(a+m) = \Gamma(a)\,(a)_m$$

when $\Gamma(a)$ is defined, $r = 0, 1, 2, \ldots; \Gamma(1/2) = \pi^{1/2}$,

$$(a)_r = \begin{cases} a(a+1)\cdots(a+r-1) & \text{if } r \geq 1,\ a \neq 0 \\ 1 & \text{if } r = 0. \end{cases} \quad \text{(A.3)}$$

The sum of the residues corresponding to the poles $s = 0$ is given by

$$R_1 = \sqrt{\pi}\Gamma\left(\frac{1}{q-1} - 1\right). \quad \text{(A.4)}$$

The sum of the residues corresponding to the poles $s = -1/2, -3/2, -5/2, \ldots$ is

$$\begin{aligned} R_2 &= \sum_{r=0}^{\infty} \frac{(-1)^r}{r!}\Gamma\left(-\frac{1}{2} - r\right)\Gamma\left(\frac{1}{2} - r\right) \\ &\quad \times \Gamma\left(\frac{1}{q-1} - \frac{1}{2} + r\right)\left[\frac{(q-1)E_G}{4kT}\right]^{-1/2+r} \\ &= -2\pi\Gamma\left(\frac{1}{q-1} - \frac{1}{2}\right)\left[\frac{(q-1)E_G}{4kT}\right]^{1/2} \\ &\quad \times {}_1F_2\left(\frac{1}{q-1} - \frac{1}{2}; \frac{3}{2}, \frac{1}{2}; -\frac{(q-1)E_G}{4kT}\right), \end{aligned} \quad \text{(A.5)}$$

where ${}_1F_2$ is the hypergeometric function defined by

$${}_1F_2(a; b, c; x) = \sum_{r=0}^{\infty} \frac{(a)_r}{(b)_r (c)_r} \frac{x^r}{r!}. \quad \text{(A.6)}$$

To obtain the sum of the residues corresponding to poles $s = -1, -2, -3, \ldots$ of order 2, we proceed as follows:

$$\begin{aligned} R_3 & \\ &= \sum_{r=0}^{\infty} \lim_{s \to -1-r} \frac{\partial}{\partial s}\left[(s+1+r)^2\Gamma(1+s)\Gamma(s)\Gamma\left(\frac{1}{2} + s\right)\right. \\ &\quad \left. \times \Gamma\left(\frac{1}{q-1} - 1 - s\right)\left(\frac{(q-1)E_G}{4kT}\right)^{-s}\right] \\ &= \sum_{r=0}^{\infty} \lim_{s \to -1-r} \frac{\partial}{\partial s}\left[\left(\Gamma^2(2+s+r)\Gamma(1/2+s)\right.\right. \\ &\quad \times \Gamma(1/(q-1) - 1 - s)) \\ &\quad \times \left((s+r)^2(s+r-1)^2\cdots(s+1)^2 s\right)^{-1} \\ &\quad \left. \times \left(\frac{(q-1)E_G}{4kT}\right)^{-s}\right] \\ &= \sum_{r=0}^{\infty} \lim_{s \to -1-r} \frac{\partial}{\partial s}\Phi(s), \end{aligned} \quad \text{(A.7)}$$

where

$$\Phi(s) = \frac{\Gamma^2(2+s+r)\Gamma(1/2+s)\Gamma(1/(q-1) - 1 - s)}{(s+r)^2(s+r-1)^2\cdots(s+1)^2 s} \times \left(\frac{(q-1)E_G}{4kT}\right)^{-s}. \quad \text{(A.8)}$$

We have

$$\frac{\partial}{\partial s}\Phi(s) = \Phi(s)\frac{\partial}{\partial s}[\ln[\Phi(s)]]$$

$$\begin{aligned} \ln\Phi(s) &= 2\ln[\Gamma(2+s+r)] + \ln\left[\Gamma\left(\frac{1}{2} + s\right)\right] \\ &\quad + \ln\left[\Gamma\left(\frac{1}{q-1} - 1 - s\right)\right] - s\ln\left(\frac{(q-1)E_G}{4kT}\right) \\ &\quad - 2\ln(s+r) - 2\ln(s+r-1) - \cdots \\ &\quad - 2\ln(s+1) - \ln(s) \end{aligned}$$

$$\begin{aligned} \frac{\partial}{\partial s}[\ln[\Phi(s)]] &= 2\Psi(2+s+r) + \Psi\left(\frac{1}{2} + s\right) \\ &\quad + \Psi\left(\frac{1}{q-1} - 1 - s\right) \\ &\quad - \ln\left(\frac{(q-1)E_G}{4kT}\right) - \frac{2}{s+r} \\ &\quad - \frac{2}{s+r-1} - \cdots - \frac{2}{s+1} - \frac{1}{s} \end{aligned}$$

$$\begin{aligned} \lim_{s \to -1-r}\left\{\frac{\partial}{\partial s}\ln[\Phi(s)]\right\} &= \Psi\left(-\frac{1}{2} - r\right) + \Psi\left(\frac{1}{q-1} + r\right) \\ &\quad + \Psi(1+r) + \Psi(2+r) \\ &\quad - \ln\left(\frac{(q-1)E_G}{4kT}\right), \end{aligned} \quad \text{(A.9)}$$

where $\Psi(z)$ is a Psi function or digamma function (see Mathai [26]) and $\Psi(1) = -\gamma$, $\gamma = 0.5772156649\ldots$ is Euler's constant. Now

$$\lim_{s \to -1-r}\Phi(s) = \frac{(-1)^{1+r} 2\sqrt{\pi}\Gamma(1/(q-1) + r)}{(3/2)_r r!(1+r)!} \times \left(\frac{(q-1)E_G}{4kT}\right)^{1+r}. \quad \text{(A.10)}$$

Then by using (A.7), (A.9), and (A.10) we get

$$\begin{aligned} R_3 &= \sum_{r=0}^{\infty} \frac{(-1)^{1+r} 2\sqrt{\pi}\Gamma(1/(q-1)+r)}{(3/2)_r r!(1+r)!} \left(\frac{(q-1)E_G}{4kT}\right)^{1+r} \\ &\quad \times \left[\Psi\left(-\frac{1}{2}-r\right)+\Psi\left(\frac{1}{q-1}+r\right)+\Psi(1+r)\right. \\ &\quad \left. +\Psi(2+r)-\ln\left(\frac{(q-1)E_G}{4kT}\right)\right] \\ &= \left(\frac{2\sqrt{\pi}(q-1)E_G}{4kT}\right) \\ &\quad \times \sum_{r=0}^{\infty}\left(\frac{(q-1)E_G}{4kT}\right)^r \left[A_r - \ln\left(\frac{(q-1)E_G}{4kT}\right)\right] B_r, \end{aligned} \tag{A.11}$$

where

$$\begin{aligned} A_r &= \Psi\left(-\frac{1}{2}-r\right)+\Psi\left(\frac{1}{q-1}+r\right) \\ &\quad +\Psi(1+r)+\Psi(2+r), \end{aligned} \tag{A.12}$$

$$B_r = \frac{(-1)^r\Gamma(1/(q-1)+r)}{(3/2)_r r!(1+r)!}.$$

Thus from (A.4), (A.5), and (A.11) we get (20).

Conflict of Interests

The authors declare that there is no conflict of interests regarding the publication of this paper.

Acknowledgments

The authors would like to thank the Department of Science and Technology, Government of India, New Delhi, for the financial assistance for this work under Project no. SR/S4/MS:287/05 and the Centre for Mathematical Sciences for providing all facilities.

References

[1] R. Davis Jr., "Nobel Lecture: a half-century with solar neutrinos," *Reviews of Modern Physics*, vol. 75, no. 3, pp. 985–994, 2003.

[2] H. J. Haubold and D. Kumar, "Extension of thermonuclear functions through the pathway model including Maxwell-Boltzmann and Tsallis distributions," *Astroparticle Physics*, vol. 29, no. 1, pp. 70–76, 2008.

[3] H. J. Haubold and R. W. John, "On the evaluation of an integral connected with the thermonuclear reaction rate in closed-form," *Astronomische Nachrichten*, vol. 299, no. 5, pp. 225–232, 1978.

[4] W. A. Fowler, G. R. Caughlan, and B. A. Zimmerman, "Thermonuclear rection rates," *Annual Review of Astronomy and Astrophysics*, vol. 5, pp. 525–570, 1967.

[5] A. M. Mathai and H. J. Haubold, *Modern Problems in Nuclear and Neutrino Astrophysics*, Akademie, Berlin, Germany, 1988.

[6] H. J. Haubold and A. M. Mathai, "Analytic representations of modified non-resonant thermonuclear reaction rates," *Journal of Applied Mathematics and Physics*, vol. 37, no. 5, pp. 685–695, 1986.

[7] W. A. Fowler, "Experimental and theoretical nuclear astrophysics: the quest for the origin of the elements," *Reviews of Modern Physics*, vol. 56, no. 2, pp. 149–179, 1984.

[8] A. C. Phillips, *The Physics of Stars*, John Wiley & Sons, Chichester, UK, 2nd edition, 1999.

[9] F. C. Adams, "Stars in other universes: stellar structure with different fundamental constants," *Journal of Cosmology and Astroparticle Physics*, vol. 2008, no. 8, article 10, 2008.

[10] M. Coraddu, G. Kaniadakis, A. Lavagno, M. Lissia, G. Mezzorani, and P. Quarati, "Thermal distributions in stellar plasmas, nuclear reactions and solar neutrinos," *Brazilian Journal of Physics*, vol. 29, no. 1, pp. 153–168, 1999.

[11] H. J. Haubold and A. M. Mathai, "On nuclear reaction rate theory," *Annalen der Physik*, vol. 41, pp. 380–396, 1984.

[12] M. Coraddu, M. Lissia, G. Mezzorani, and P. Quarati, "Super-Kamiokande hep neutrino best fit: a possible signal of non-Maxwellian solar plasma," *Physica A: Statistical Mechanics and Its Applications*, vol. 326, no. 3-4, pp. 473–481, 2003.

[13] A. Lavagno and P. Quarati, "Classical and quantum non-extensive statistics effects in nuclear many-body problems," *Chaos, Solitons and Fractals*, vol. 13, no. 3, pp. 569–580, 2002.

[14] A. Lavagno and P. Quarati, "Metastability of electron-nuclear astrophysical plasmas: motivations, signals and conditions," *Astrophysics and Space Science*, vol. 305, no. 3, pp. 253–259, 2006.

[15] M. Lissia and P. Quarati, "Nuclear astrophysical plasmas: ion distribution functions and fusion rates," *Europhysics News*, vol. 36, no. 6, pp. 211–214, 2005.

[16] C. Tsallis, "Possible generalization of Boltzmann-Gibbs statistics," *Journal of Statistical Physics*, vol. 52, no. 1-2, pp. 479–487, 1988.

[17] C. Tsallis, *Introduction to Non-Extensive Statistical Mechanics*, Springer, New York, NY, USA, 2009.

[18] M. Gell-Mann and C. Tsallis, Eds., *Nonextensive Entropy: Interdisciplinary Applications*, Oxford University Press, New York, NY, USA, 2004.

[19] A. M. Mathai and H. J. Haubold, "Pathway model, superstatistics, Tsallis statistics, and a generalized measure of entropy," *Physica A: Statistical Mechanics and Its Applications*, vol. 375, no. 1, pp. 110–122, 2007.

[20] R. K. Saxena, A. M. Mathai, and H. J. Haubold, "Astrophysical thermonuclear functions for Boltzmann-Gibbs statistics and Tsallis statistics," *Physica A: Statistical Mechanics and Its Applications*, vol. 344, no. 3-4, pp. 649–656, 2004.

[21] A. M. Mathai, "A pathway to matrix-variate gamma and normal densities," *Linear Algebra and Its Applications*, vol. 396, no. 1–3, pp. 317–328, 2005.

[22] A. M. Mathai and H. J. Haubold, "On generalized distributions and pathways," *Physics Letters A: General, Atomic and Solid State Physics*, vol. 372, no. 12, pp. 2109–2113, 2008.

[23] H. J. Haubold and D. Kumar, "Fusion yield: Guderley model and Tsallis statistics," *Journal of Plasma Physics*, vol. 77, no. 1, pp. 1–14, 2011.

[24] D. Kumar and H. J. Haubold, "On extended thermonuclear functions through pathway model," *Advances in Space Research*, vol. 45, no. 5, pp. 698–708, 2010.

[25] A. M. Mathai and R. K. Saxena, *Generalized Hypergeometric Functions with Applications in Statistics and Physical Sciences*, vol. 348 of *Lecture Notes in Mathematics*, Springer, New York, NY, USA, 1973.

[26] A. M. Mathai, *A Handbook of Generalized Special Functions for Statistics and Physical Sciences*, Clarendo Press, Oxford, UK, 1993.

[27] A. Erdélyi, *Asymptotic Expansions*, Dover, New York, NY, USA, 1956.

[28] F. W. J. Olver, *Asymptotics and Spcecial Functions*, Academic Press, New York, NY, USA, 1974.

[29] H. J. Haubold and A. M. Mathai, "Analytical results connecting stellar structure parameters and neutrino uxes," *Annalen der Physik*, vol. 44, no. 2, pp. 103–116, 1987.

11

New Classes of Charged Spheroidal Models

S. Thirukkanesh

Department of Mathematics, Eastern University, 30350 Chenkalady, Sri Lanka

Correspondence should be addressed to S. Thirukkanesh; thirukkanesh@yahoo.co.uk

Academic Editors: M. S. Dimitrijevic, A. Mesinger, L. Nicastro, A. S. Pozanenko, and A. Pradhan

New classes of exact solutions to the Einstein-Maxwell system is found in closed form by assuming that the hypersurface t = constant is spheroidal. This is achieved by choosing a particular form for the electric field intensity. A class of solution is found for all positive spheroidal parameter K for a specific form of electric field intensity. In general, the condition of pressure isotropy reduces to a difference equation with variable, rational coefficients that can be solved. Consequently, an explicit solution in series form is found. By placing restrictions on the parameters, it is shown that the series terminates and there exist two classes of solutions in terms of elementary functions. These solutions contain the models found previously in the limit of vanishing charge. Solutions found are directly relating the spheroidal parameter and electric field intensity. Masses obtained are consistent with the previously reported experimental and theoretical studies describing strange stars. A physical analysis indicates that these models may be used to describe a charged sphere.

1. Introduction

In recent years, there have been several investigations into the Einstein-Maxwell system of equations for static spherically symmetric gravitational fields with isotropic pressures in the presence of the electromagnetic field. In such study, regular interior spacetime is matched smoothly at the pressure free interface to the Reissner-Nordstrom exterior model. The models generated are useful to describe charged relativistic bodies with strong gravitational fields such as neutron stars. Gravitational collapse of a spherically symmetric distribution of matter to a point singularity may be avoided in the presence of electromagnetic field. In this situation, the gravitational attraction is counterbalanced by the repulsive Columbian force with the pressure gradient and, hence, charged fluids have a tendency to resist the gravitational collapse. This property persuades the researchers to work on charged perfect fluid distribution. Bonnor [1] has shown that charged dust solutions are expected to form a point like model of electron when its radius shrinks to zero. The presence of electromagnetic field affects the value of redshifts, luminosities, and maximum mass of a compact relativistic object (Ivanov [2], Sharma et al. [3]). Many exact solutions which satisfy the conditions for a physically acceptable charged relativistic sphere have been given by Ivanov [2], Thirukkanesh and Maharaj [4], and Gupta and Maurya [5], among others. Detailed studies of Sharma et al. [6] in cold compact objects, Sharma and Mukherjee [7] analysis of strange matter and binary pulsars and Sharma and Mukherjee [8] analysis of qark-diquark mixtures in equilibrium are of interest physically. Thomas et al. [9], Tikekar and Thomas [10], and Paul and Tikekar [11] have shown that charged relativistic matter is relevant in modeling core-envelope stellar system in which the stellar core is an isotropic fluid surrounded by a layer of anisotropic fluid.

Vaidya and Tikekar [12] proposed the geometry of the spacelike hypersurfaces generated by t = constant are of 3-spheroid to generate exact solutions since it provides a clear geometrical interpretation: the models with spheroidal geometries directly related to the physical situations. Tikekar [13] found an exact solution for a particular spheroidal geometry which could be used to model superdense neutron stars of densities in the range of 10^{14} gcm^{-3}; this solution has been generalized by Maharaj and Leach [14]. There have been extensive studies on charged spheroidal stars by considering a particular form for the electric field in recent years [15–23]. These charged spheroidal models contain uncharged neutron stars in the relevant limit and are consequently relevant in the description of dense astrophysical objects. Therefore, the study of charged fluid spheres in static spherically symmetric spacetimes is important in relativistic astrophysics.

The objective of this paper is to generate new classes of charged spheroidal solutions in terms of elementary function, which may be used to describe the interior of a relativistic compact sphere. In Section 2, the Einstein-Maxwell system of equations is expressed for static spherically symmetric spacetime. In Section 3, particular forms for one of the gravitational potentials with spheroidal parameter and the electric field intensity are chosen, which reduces the condition of pressure isotropy to a second order linear differential equation in the remaining gravitational potential. In Section 4, a class of solutions for a particular parameter value is first obtained. In general, the solution is obtained in series form using the method of Frobenius, and then two categories of solutions in terms of elementarily functions are derived by placing restrictions on the parameters. The physical features are illustrated graphically, and numerical values of some physical quantities are calculated for a particular example in Section 5.

2. Field Equations

The gravitational field should be static and spherically symmetric to describe the internal structure of a charged dense compact relativistic sphere. Therefore, the generic form of the line element for describing such configuration is given by

$$ds^2 = -e^{2\nu(r)}dt^2 + e^{2\lambda(r)}dr^2 + r^2\left(d\theta^2 + \sin^2\theta d\phi^2\right) \quad (1)$$

in Schwarzschild coordinates $(x^a) = (t, r, \theta, \phi)$, where $\nu(r)$ and $\lambda(r)$ are arbitrary function of radial coordinate r. The Einstein-Maxwell system of field equations, for the metric (1), can be written in the form

$$\frac{1}{r^2}\left(1 - e^{-2\lambda}\right) + \frac{2\lambda'}{r}e^{-2\lambda} = \rho + \frac{1}{2}E^2, \quad (2a)$$

$$\frac{-1}{r^2}\left(1 - e^{-2\lambda}\right) + \frac{2\nu'}{r}e^{-2\lambda} = p - \frac{1}{2}E^2, \quad (2b)$$

$$e^{-2\lambda}\left(\nu'' + \nu'^2 + \frac{\nu'}{r} - \nu'\lambda' - \frac{\lambda'}{r}\right) = p + \frac{1}{2}E^2, \quad (2c)$$

$$\sigma = \frac{1}{r^2}e^{-\lambda}\left(r^2E\right)'. \quad (2d)$$

The energy density ρ and the pressure p are measured relative to the commoving fluid 4-velocity $u^a = e^{-\nu}\delta_0^a$ and primes denote differentiation with respect to the radial coordinate r. The quantities E and σ denote the electric field intensity and the proper charge density, respectively. In the system (2a)–(2d), the units used are such that the coupling constant $8\pi G/c^4 = 1$ and the speed of light $c = 1$. This system of equations determines the behaviour of the gravitational field for a charged perfect fluid source. When $E = 0$ the Einstein-Maxwell system (2a)–(2d) reduces to the uncharged Einstein system.

3. Choosing Gravitational Potential and Electric Field Intensity

The aim is to seek solutions to the Einstein-Maxwell system (2a)–(2d) by making explicit choices for the gravitational potential $e^{2\lambda(r)}$ and electric field intensity E on physical grounds. The system (2a)–(2d) comprises four equation with six unknowns λ, ν, ρ, p, E, and σ so that it is necessary to choose two of the variables to integrate the system. In this treatment, λ and E are specified. A particular choice for λ is made such that

$$e^{2\lambda(r)} = \frac{1 - Kr^2/R^2}{1 - r^2/R^2}, \quad (3)$$

where K and R are real constants. The above form has been used previously by Tikekar [13] and Maharaj and Leach [14] to study the behaviour of uncharged superdense stars. Note that the choice (3) for the gravitational potential λ restricts the geometry of the 3-dimensional hypersurfaces t = constant to be spheroidal for $K \neq 0$ and spherical for $K = 0$.

Eliminating p from (2b) and (2c), for the particular form (3), one obtain the condition of pressure isotropy:

$$\left(1 - \frac{Kr^2}{R^2}\right)^2 E^2 = \left(1 - \frac{Kr^2}{R^2}\right)\left(1 - \frac{r^2}{R^2}\right)\left(\nu'' + \nu'^2 - \frac{\nu'}{r}\right) - (1 - K)\left(\frac{r}{R^2}\right)\left(\nu' + \frac{1}{r}\right) + \frac{1 - K}{R^2}\left(1 - \frac{Kr^2}{R^2}\right). \quad (4)$$

The transformation

$$\psi(x) = e^{\nu(r)}, \qquad x^2 = 1 - \frac{r^2}{R^2} \quad (5)$$

reduces the condition (4) for pressure isotropy to a convenient form

$$\left(1 - K + Kx^2\right)\ddot{\psi} - Kx\dot{\psi} + \left(\frac{\left(1 - K + Kx^2\right)^2 R^2E^2}{x^2 - 1} + K(K - 1)\right)\psi = 0 \quad (6)$$

in terms of the new variables ψ and x, where dots denote differentiation with respect to x.

In terms of new variable x, the Einstein-Maxwell system (2a)–(2d) becomes

$$\rho = \frac{1 - K}{R^2}\frac{\left(3 - K + Kx^2\right)}{\left(1 - K + Kx^2\right)^2} - \frac{1}{2}E^2, \quad (7a)$$

$$p = \frac{1}{R^2\left(1 - K + Kx^2\right)}\left(-2x\frac{\dot{\psi}}{\psi} + K - 1\right) + \frac{1}{2}E^2, \quad (7b)$$

$$\sigma^2 = \frac{\left[2xE - \left(1 - x^2\right)\dot{E}\right]^2}{R^2\left(1 - x^2\right)\left(1 - K + Kx^2\right)} \quad (7c)$$

for the choice (3). Note that in (7a)–(7c), ρ, p, and σ are defined in terms of E. Equation (6) may be integrable if a particular choice of the electric field intensity E is made. For mathematical convenient one may take

$$E^2 = \frac{2\beta Kx\left(1-x^2\right)}{\left(1-K+Kx^2\right)^2 R^2}\frac{\dot{\psi}}{\psi}, \quad \text{that is, } E^2 = \frac{\beta Kr}{\left(r^2-R^2\right)}\frac{\nu'}{e^{4\lambda}}, \tag{8}$$

where β is a constant which is different from the choice of Komathiraj and Maharaj [15]. On substituting (8) into (6), we obtain a second order linear differential equation

$$\left(1-K+Kx^2\right)\ddot{\psi} - K\left(1+2\beta\right)x\dot{\psi} + K\left(K-1\right)\psi = 0 \tag{9}$$

in ψ. It is expected that investigation of (9) will produce physically reasonable models of charged stars since $\beta = 0$ yields models found previously by Maharaj and Leach [14] for neutron stars which contain Tikekar [13] superdense stars as special case.

4. Solution

It is clear that the solution of the Einstein-Maxwell system depends on the integrability of (9). One may consider the following two cases.

4.1. Particular Case. When $\beta = -1$, (9) becomes

$$\left(1-K+Kx^2\right)\ddot{\psi} + Kx\dot{\psi} + K\left(K-1\right)\psi = 0. \tag{10}$$

If we utilize the transformation

$$Z = \frac{1}{\sqrt{K}}\ln\left[Kx + \sqrt{K^2x^2 + K\left(1-K\right)}\right] \tag{11}$$

equation (10) reduces to the equation of free oscillation

$$\frac{d^2\psi}{dZ^2} + K\left(K-1\right)\psi = 0, \tag{12}$$

for $K > 0$. The solutions of (12) become

$$\psi(Z) = \begin{cases} c_1 \sinh\left(\sqrt{K(1-K)}Z\right) + c_2\cosh\left(\sqrt{K(1-K)}Z\right) & \text{if } 0 < K < 1, \\ c_1 + c_2 Z & \text{if } K = 1, \\ c_1\sin\left(\sqrt{K(K-1)}Z\right) + c_2\cos\left(\sqrt{K(K-1)}Z\right) & \text{if } K > 1. \end{cases} \tag{13}$$

This becomes

$$\psi(x) = \begin{cases} c_1\left[\dfrac{\left(Kx+\sqrt{K^2x^2+K(1-K)}\right)^{2(1-K)} - 1}{2\left(Kx+\sqrt{K^2x^2+K(1-K)}\right)^{(1-K)}}\right] + c_2\left[\dfrac{\left(Kx+\sqrt{K^2x^2+K(1-K)}\right)^{2(1-K)} + 1}{2\left(Kx+\sqrt{K^2x^2+K(1-K)}\right)^{(1-K)}}\right] & \text{if } 0 < K < 1, \\ c_1 + c_2\ln\left[2x\right] & \text{if } K = 1, \\ c_1\sin\left(\sqrt{(K-1)}\ln\left[Kx+\sqrt{K^2x^2+K(1-K)}\right]\right) + c_2\cos\left(\sqrt{(K-1)} \times\ln\left[Kx+\sqrt{K^2x^2+K(1-K)}\right]\right) & \text{if } K > 1, \end{cases} \tag{14}$$

in terms of the variable x. Thus, a class of exact solution to the Einstein-Maxwell system is generated for all positive value of K. Solution (14) is given in simple form which is an advantage for physical analysis.

4.2. General Case. Note that (9) can be transformed to a hypergeometric equation. However, it is impossible to express the solutions in terms of elementary function for all K. In general, the solution will be given in terms of special functions. Solutions in a simple form are important for a detailed physical analysis. Hence, first I attempt to obtain a general solution of (9) in a series form using the method of Frobenius and then demonstrate the possibility to extract solutions in terms of polynomials and algebraic functions by imposing restrictions on the parameters.

4.2.1. Series Solution. Since $x = 0$ is a regular point of the differential equation (9), we can apply the method of Frobenius about $x = 0$ to obtain a series solution. Thus, we assume that

$$\psi(x) = \sum_{i=0}^{\infty} a_i x^i \tag{15}$$

is a solution of (9). To express the solution, the coefficients of the series a_i need to determined explicitly. Substituting (15) into (9) we obtain

$$\begin{aligned} &2\left(1-K\right)a_2 + K\left(K-1\right)a_0 \\ &\quad + \left[6\left(1-K\right)a_3 + K\left(K-2-2\beta\right)a_1\right]x \\ &\quad + \sum_{i=2}^{\infty}\{\left(1-K\right)\left(i+1\right)\left(i+2\right)a_{i+2} \\ &\quad + K\left[K-1+i\left(i-2-2\beta\right)\right]a_i\}x^i = 0. \end{aligned} \tag{16}$$

For the validity of (16), we must have

$$2(1-K)a_2 + K(K-1)a_0 = 0, \quad (17a)$$

$$6(1-K)a_3 + K(K-2-2\beta)a_1 = 0, \quad (17b)$$

$$(1-K)(i+1)(i+2)a_{i+2} + K[K-1+i(i-2-2\beta)]a_i = 0, \quad i \geq 2. \quad (17c)$$

It remains to obtain the coefficients a_i from the system (17a)–(17c). Equation (17c) is the linear difference equation governing the structure of the solution. The difference equation (17c) consists of variable, rational coefficients.

First consider the even coefficients $a_0, a_2, a_4, \ldots$. These coefficients generate a pattern

$$a_{2i} = \left(\frac{K}{K-1}\right)^i \frac{1}{(2i)!} \prod_{q=1}^{i} [K-1+(2q-2)(2q-4-2\beta)]\, a_0, \quad (18)$$

where the symbol $\prod$ denotes multiplication. The odd coefficients $a_1, a_3, a_5, \ldots$ can be written in the form

$$a_{2i+1} = \left(\frac{K}{K-1}\right)^i \frac{1}{(2i+1)!} \times \prod_{q=1}^{i} [K-1+(2q-1)(2q-3-2\beta)]\, a_1. \quad (19)$$

Hence, the difference equation (17c) has been solved and all nonzero coefficients are expressible in terms of the leading coefficients a_0 and a_1. Thus, from (15), (18), and (19) the solution of (9) becomes

$$\psi(x) = a_0\left(1+\sum_{i=1}^{\infty}\left(\frac{K}{K-1}\right)^i \frac{1}{(2i)!} \times \prod_{q=1}^{i}[K-1+4(q-1)(q-2-2\beta)]\, x^{2i}\right) + a_1\left(x+\sum_{i=1}^{\infty}\left(\frac{K}{K-1}\right)^i \frac{1}{(2i+1)!} \times \prod_{q=1}^{i}[K-1+(2q-1)(2q-3-2\beta)]\, x^{2i+1}\right), \quad (20)$$

where a_0 and a_1 are arbitrary constants. Therefore, the general solution of (9) for the choice (15) is given by

$$\psi(x) = a_0\psi_1(x) + a_1\psi_2(x), \quad (21)$$

where

$$\psi_1(x) = \left(1+\sum_{i=1}^{\infty}\left(\frac{K}{K-1}\right)^i \frac{1}{(2i)!} \times \prod_{q=1}^{i}[K-1+4(q-1)(q-2-2\beta)]\, x^{2i}\right), \quad (22a)$$

$$\psi_2(x) = \left(x+\sum_{i=1}^{\infty}\left(\frac{K}{K-1}\right)^i \frac{1}{(2i+1)!}\prod_{q=1}^{i} \times [K-1+(2q-1)(2q-3-2\beta)]\, x^{2i+1}\right) \quad (22b)$$

are linearly independent solutions of (9).

It is interesting to observe that when $K = 0$ the series solution (21) reduces to a simple form

$$\psi(x) = a_0 + a_1 x. \quad (23)$$

In this case, the electric field intensity E vanishes and there is no charge.

Note that when $a_0 = 0$ and $a_1 = 1$, the line element (1) takes the form

$$ds^2 = -\left(1-\frac{r^2}{R^2}\right)dt^2 + \left(1-\frac{r^2}{R^2}\right)^{-1} dr^2 + r^2\left(d\theta^2 + \sin^2\theta d\phi^2\right). \quad (24)$$

The above metric corresponds to the familiar isotropic uncharged de Sitter model.

When $a_1 = 0$, the line element (1) takes the particular form

$$ds^2 = -a_0^2 dt^2 + \left(1-\frac{r^2}{R^2}\right)^{-1} dr^2 + r^2\left(d\theta^2 + \sin^2\theta d\phi^2\right). \quad (25)$$

The above metric corresponds to the well-known isotropic uncharged Einstein model.

4.2.2. Terminating Series. The general solution (21) can be expressed in terms of polynomial and algebraic function for restricted values of the parameter K. This will happen because series (22a) and (22b) terminate for restricted values of K. Using this feature, it is possible to generate two sets of solutions in terms of elementary functions by determining specific restrictions on K, as demonstrated below. For simplicity, the difference equation (17c) is used instead of the series (22a) and (22b) to obtain the solutions in terms of elementary functions.

First Solution. Firstly consider the polynomials of even degree. If we set

$$i = 2(j-1), \quad (26a)$$

$$K = 2-(2n-1)^2 + 4n\beta \quad (26b)$$

then for a fixed integer $n > 1$, (17c) become

$$a_{2j} = -\gamma \frac{(n-\beta+j-2)(n-j+1)}{2j(2j-1)} a_{2j-2}, \quad (27)$$

where $\gamma = 4 - 1/[n(n-1-\beta)]$. Observe that (27) implies $a_{2(n+1)} = 0$. It is easy to see that the remaining coefficients $a_{2(n+2)}, a_{2(n+3)}, a_{2(n+4)}, \ldots$ vanish and (27) has the solution

$$a_{2j} = (-\gamma)^j \frac{(n-\beta+j-2)!\,(n-1)!}{(2j)!\,(n-\beta-1)!\,(n-j)!}, \quad 0 \le j \le n, \quad (28)$$

where I have set $a_0 = 1/n(n-\beta-1)$. Therefore, from (15) and, (28), we can express the polynomial in even powers of x as

$$f_1(x) = \sum_{j=0}^{n} (-\gamma)^j \frac{(n-\beta+j-2)!\,(n-1)!}{(2j)!\,(n-\beta-1)!\,(n-j)!} x^{2j} \quad (29)$$

for $K = 2 - (2n-1)^2 + 4n\beta$.

Secondly, consider the polynomials of odd degree. For this case, if we set

$$i = 2(j-1) + 1, \quad (30a)$$

$$K = 2\left[1 - 2n^2 + (2n+1)\beta\right] \quad (30b)$$

then for a fixed integer $n > 0$, (17c) becomes

$$a_{2j+1} = -\mu \frac{(n+j-1-\beta)(n-j+1)}{2j(2j+1)} a_{2j-1}, \quad (31)$$

where $\mu = 4 - 4/[(2n+1)(2n-1-2\beta)]$. Observe that (31) implies $a_{2(n+1)+1} = 0$. Consequently, the remaining coefficients $a_{2(n+2)+1}, a_{2(n+3)+1}, a_{2(n+4)+1}, \ldots$ vanish and (31) has the solution

$$a_{2j+1} = (-\mu)^j \frac{(n-\beta+j-1)!\,(n-1)!}{(2j+1)!\,(n-\beta-1)!\,(n-j)!}, \quad 0 \le j \le n, \quad (32)$$

where I have set $a_1 = 1/n$. Therefore, from (15) and (32), we can express the polynomial in odd powers of x as

$$g_1(x) = \sum_{j=0}^{n} (-\mu)^j \frac{(n-\beta+j-1)!\,(n-1)!}{(2j+1)!\,(n-\beta-1)!\,(n-j)!} x^{2j+1} \quad (33)$$

for $K = 2[1 - 2n^2 + (2n+1)\beta]$.

Polynomial (29) and (33) comprise the first solution of (9) for appropriate values of K.

Second Solution. Assume the second solution of (9) to be of the form

$$\psi(x) = u(x)\left(1 - K + Kx^2\right)^{(3/2)+\beta}, \quad (34)$$

where $u(x)$ is an arbitrary function. Substituting ψ in (9) we obtain

$$\left(1 - K + Kx^2\right)\frac{d^2u}{dx^2} + K(5+2\beta)x\frac{du}{dx} + K(K+2+2\beta)u = 0, \quad (35)$$

which is a linear differential equation in $u(x)$.

Observe that (9) and (35) are of the same type. As in Section 4.2.1, we can first find a general series solution and then two classes of polynomial solution (in even powers of x and in odd powers of x) for (35) using the above technique. Hence, I present the final form of the solution: the polynomial in even powers of x leads to the expression

$$u(x) = \sum_{j=0}^{n-\beta-1} (-\mu)^j \frac{(n+j)!\,(n-\beta-1)!}{(2j)!\,(n-1)!\,(n-\beta-j-1)!} x^{2j} \quad (36)$$

for $K = 2[1 - 2n^2 + (2n+1)\beta]$, where the real constant β is restricted as integer such that $\beta \le n-1$; the polynomial in odd powers of x leads to the result

$$u(x) = \sum_{j=0}^{n-\beta-2} (-\gamma)^j \frac{(n+j)!\,(n-\beta-2)!}{(2j+1)!\,(n-2)!\,(n-\beta-j-2)!} x^{2j+1} \quad (37)$$

for $K = 2 - (2n-1)^2 + 4n\beta$, where the real constant β is restricted as integer such that $\beta \le n-2$ in this case.

Hence, the solutions to (9) becomes:

$$g_2(x) = \left(1 - K + Kx^2\right)^{(3/2)+\beta} \times \sum_{j=0}^{n-\beta-1} (-\mu)^j \frac{(n+j)!\,(n-\beta-1)!}{(2j)!\,(n-1)!\,(n-\beta-j-1)!} x^{2j} \quad (38)$$

for $K = 2[1 - 2n^2 + (2n+1)\beta]$, where β is an integer such that $\beta \le n-1$;

$$f_2(x) = \left(1 - K + Kx^2\right)^{(3/2)+\beta} \times \sum_{j=0}^{n-\beta-2} (-\gamma)^j \frac{(n+j)!\,(n-\beta-2)!}{(2j+1)!\,(n-2)!\,(n-\beta-j-2)!} x^{2j+1} \quad (39)$$

for $K = 2 - (2n-1)^2 + 4n\beta$, where β is an integer such that $\beta \le n-2$.

The algebraic functions (38) and (39) comprise the second solution of (9) for appropriate values of K.

Exact Solution. The solutions generated in Section 4.2.2 can be expressed in terms of two classes of elementary functions. The first category of solution for $\psi(x) = f(x)$ is

$$\begin{aligned} f(x) &= Af_1(x) + Bf_2(x) \\ &= A\sum_{j=0}^{n} (-\gamma)^j \frac{(n-\beta+j-2)!\,(n-1)!}{(2j)!\,(n-\beta-1)!\,(n-j)!} x^{2j} \\ &\quad + B\left(1 - K + Kx^2\right)^{(3/2)+\beta} \\ &\quad \times \sum_{j=0}^{n-\beta-2} (-\gamma)^j \frac{(n+j)!\,(n-\beta-2)!}{(2j+1)!\,(n-2)!\,(n-\beta-j-2)!} x^{2j+1} \end{aligned} \quad (40)$$

for the values

$$\gamma = 4 - \frac{1}{n(n-1-\beta)}, \quad (41a)$$

$$K = 2 - (2n-1)^2 + 4n\beta, \quad \text{where } \beta \le n-2. \quad (41b)$$

The second category of solution for $\psi(x) = g(x)$ is

$$g(x) = Ag_1(x) + Bg_2(x)$$
$$= A\sum_{j=0}^{n}(-\mu)^j \frac{(n-\beta+j-1)!\,(n-1)!}{(2j+1)!\,(n-\beta-1)!\,(n-j)!}x^{2j+1}$$
$$+ B\left(1-K+Kx^2\right)^{(3/2)+\beta} \quad (42)$$
$$\times \sum_{j=0}^{n-\beta-1}(-\mu)^j \frac{(n+j)!\,(n-\beta-1)!}{(2j)!\,(n-1)!\,(n-\beta-j-1)!}x^{2j}$$

for the values

$$\mu = 4 - \frac{4}{(2n+1)(2n-1-2\beta)}, \quad (43a)$$

$$K = 2\left[1 - 2n^2 + (2n+1)\beta\right], \quad \text{where } \beta \le n-1. \quad (43b)$$

In (40) and (42), A and B are arbitrary constants.

The solutions (40) and (42) are given completely in terms of elementary functions: this has the advantage of facilitating the analysis of physical feature of the stellar interior. These solutions are applicable to a charged superdense star with spheroidal geometry. Note that this treatment has combined both charged and neutral cases for a relativistic star: by setting $\beta = 0$ one obtain the solution for neutral case directly.

From these general class of solutions (40) and (42), it is possible to regain particular solutions found in the past. For example, the solutions (40) and (42) reduce to the uncharged models of Maharaj and Leach [14] when $\beta = 0$ which contain the Tikekar [13] superdense neutron star model for $K = -7$. Other explicit functional forms for ψ are obtainable which could be useful to study dense stars. For example, if we set $\beta = 1$ and $K = -4$ $(n = 2)$ then (42) becomes

$$\psi = Ax\left[\frac{1}{2} - \frac{8}{15}x^2 + \frac{64}{375}x^4\right] + 2B\left(5-4x^2\right)^{5/2}. \quad (44)$$

For this case, the line element takes a simple form

$$ds^2 = -\left[\frac{A}{750}\sqrt{1-\frac{r^2}{R^2}}\right.$$
$$\times\left[103 + 144\frac{r^2}{R^2} + 128\frac{r^4}{R^4}\right] \quad (45)$$
$$\left.+2B\left(1+4\frac{r^2}{R^2}\right)\right]^2 dt^2 + \frac{1+4r^2/R^2}{1-r^2/R^2}dr^2$$
$$+ r^2\left(d\theta^2 + \sin^2\theta d\phi^2\right),$$

in terms of the original variable r.

5. Physical Analysis

It is easy to observe from the solutions (40) and (42) that the spheroidal parameter increases with β so that the electric field intensity affect the spheroidal parameter. For vanishing of pressure at the boundary $r = b$ in the solutions (14), (40), and (42), we require $p(b) = 0$, that gives the condition

$$\sqrt{1-\frac{b^2}{R^2}}\left(\frac{(\beta+2)K\left(b^2/R^2\right)}{\left[1-K\left(b^2/R^2\right)\right]}\right)$$
$$\times\left[\frac{\dot{\psi}}{\psi}\right]_{x=\sqrt{1-(b^2/R^2)}} + K - 1 = 0. \quad (46)$$

The generated interior metric matches smoothly with the exterior Reissnar-Nordstrom metric:

$$ds^2 = -\left(1-\frac{2m}{r}+\frac{q^2}{r^2}\right)dt^2 + \left(1-\frac{2m}{r}+\frac{q^2}{r^2}\right)^{-1}dr^2$$
$$+ r^2\left(d\theta^2 + \sin^2\theta d\phi^2\right) \quad (47)$$

across the boundary $r = b$, where m and q are the total mass and charge of the sphere. This generate the relationship between b, β, K, R and the arbitrary constants involved in the solutions ψ as follows:

$$\left(1-\frac{2m}{r}+\frac{q^2}{r^2}\right) = \left[\psi^2\right]_{x=\sqrt{1-(b^2/R^2)}}, \quad (48a)$$

$$\left(1-\frac{2m}{r}+\frac{q^2}{r^2}\right)^{-1} = \frac{1-K\left(b^2/R^2\right)}{1-\left(b^2/R^2\right)}. \quad (48b)$$

The conditions (46) and (48a)-(48b) place the restrictions on the parameters involved in these equations. However, there are sufficient free parameters to satisfy the necessary condition that arise for the model under investigation.

Now, we shall show that the obtained models are physically reasonable by plotting for a particular solution (45) as illustrated in Figures 1–7 by assuming the parameter values $A = -10$, $B = -1$, and $R = 16$ km in the interval $0 \le r \le 1$, where r is in km. The software package Mathematica was utilized to generate the plots for $e^{2\nu}$, $e^{2\lambda}$, ρ, p, E^2, $dp/d\rho$, and p versus ρ. Gravitational potentials are plotted in Figures 1 and 2 which are nonsingular at the center and increasing from center to the boundary. The behaviour of the energy density is plotted in Figure 3, which is positive and monotonically decreasing towards the boundary of the stellar object. The behaviour of matter pressure p is plotted in Figure 4, which is regular, monotonically decreasing throughout the stellar interior. The behaviour of the electric field intensity is described in Figure 5, which is continuous and well behaved. The derivative $dp/d\rho$ is plotted in Figure 6, which is $0 \le dp/d\rho \le 1$ throughout the interior of the stellar object. Therefore, the speed of the sound is less than the speed of the light and causality is maintained. The pressure p verses the density ρ is plotted in Figure 7, and this looks like a straight line so that the equation of state approximate to a linear

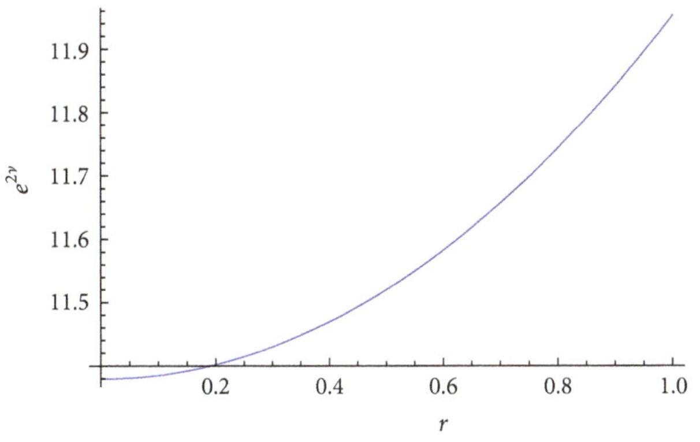

FIGURE 1: Gravitational potential $e^{2\nu}$.

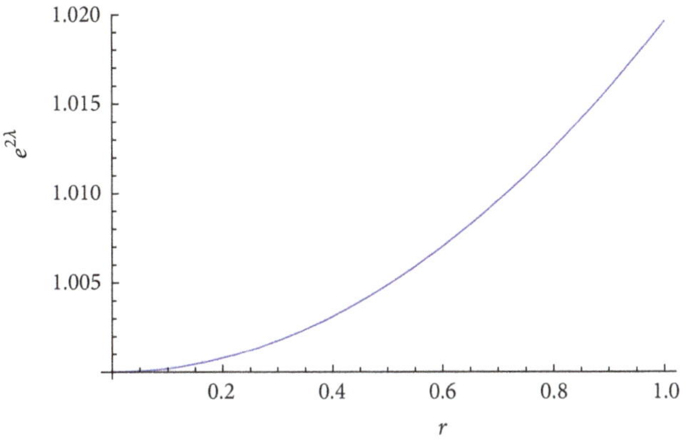

FIGURE 2: Gravitational potential $e^{2\lambda}$.

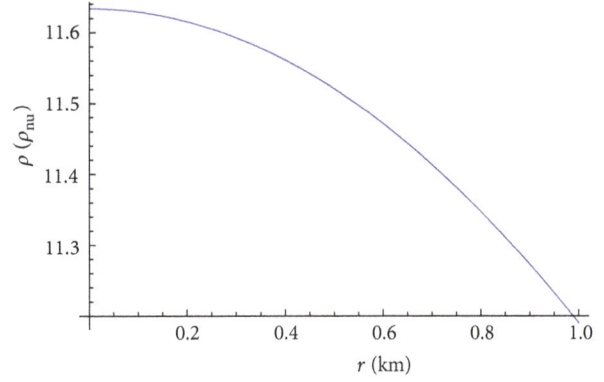

relationship between p and ρ. Thus, we have demonstrated that the particular solution satisfies the requirements for a physically reasonable stellar interior in the context of general relativity.

Numerical values for the physical quantities for the metric (45) for various radius are given in Table 1, where the constant of integration $A = 0$ is assumed for simplicity. The radius of the star b is chosen to be compared with experimental values reported for realistic stars. In all cases, it is noted that at the surface the pressure $p(b) = 0$ (up to 8 decimal places), the density $\rho(b) > 0$ and the charge $q(b) > 0$; at the centre the pressure $p(0) > 0$ and the density $\rho(0) > 0$. The mass of the sphere of radius 7.07 km obtained in Table 1 is comparable with the experimentally determined value 1.44 $M_\odot$ of the strange star SAX J1808.4-3658(SS1) having the same radius [24]. Moreover, similar value of mass was theoretically obtained for SAX J1808.4-3658(SS1) by Dey et al. [25], Sharma and Maharaj [26] and Tikekar and Jotania [27]. A similar mass value for the SAX J1808.4-3658(SS2) was reported by Tikekar and Jotania [27]. Experimental observations suggest a star model for Her. X-1, which is estimated to have a mass between 1.1–1.8 $M_\odot$ and the radius between 6–7.7 km [28]. It is noted that the mass obtained for radius 7.7 km and $R = 9.445343$ km in Table 1 corroborates with this experimentally estimated result for Her. X-1. The mass obtained for 4U 1820–30 is

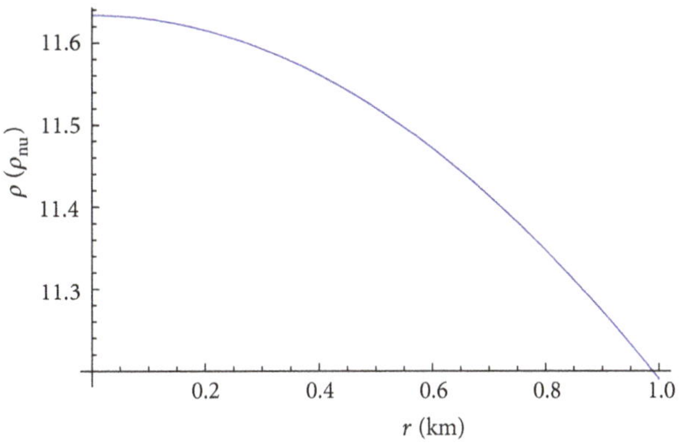

FIGURE 3: Energy density in terms of nuclear density ρ_{nu}.

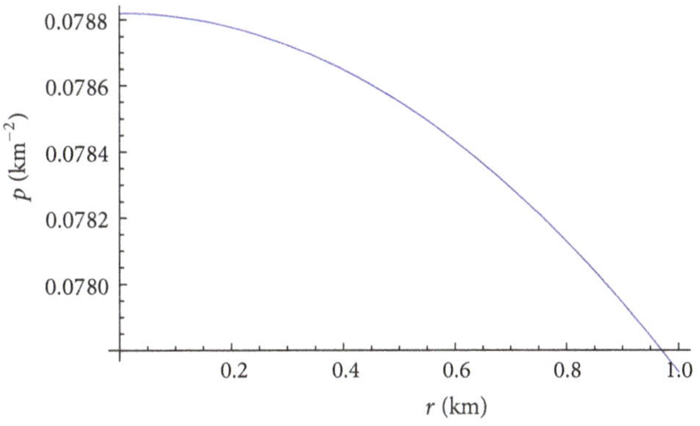

FIGURE 4: Matter pressure.

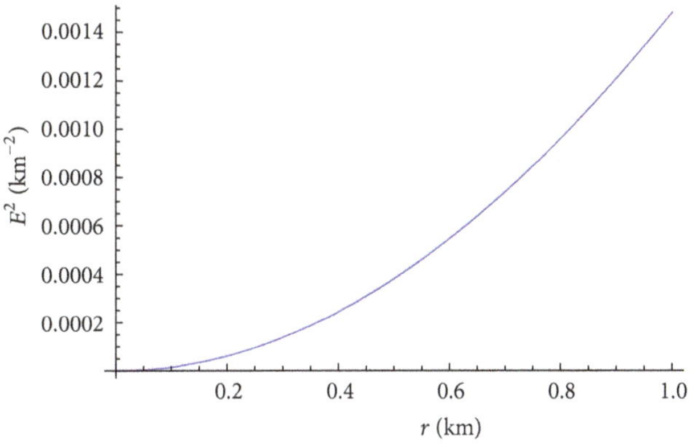

FIGURE 5: Electric field intensity.

similar to the value reported in the past [27]. However, for this particular example of parameter values considered in Table 1, the casuality condition is not satisfied as for the results obtained by Kumar and Gupta[23] and the argument given [23] for this scenario is notable.

6. Conclusion

New classes of solutions are generated to the Einstein-Maxwell system by assuming that the hypersurface $t =$ constant is spheroidal. For a particular choice of electric field intensity

TABLE 1: b, radius of the star; $p(0)$, central pressure; $\rho(b)$, surface density; $\rho(0)$, central density; $q(b)$, surface charge.

Strange star candidate	R (km)	b (km)	$p(0)$ (km^{-2})	$\rho(b)$ (km^{-2})	$\rho(0)$ (km^{-2})	$q(b)$ (km)	Mass ($M_\odot$)
SAX J1808.4-3658 (SS1)	8.672542	7.07	0.465	0.023	0.199	3.478	1.5295
SAX J1808.4-3658 (SS2)	7.789341	6.35	0.576	0.016	0.247	3.883	1.3739
Her. X-1	9.445343	7.7	0.518	0.026	0.222	3.296	1.6658
4U 1820-30	12.266679	10	0.233	0.019	0.099	4.920	2.1633

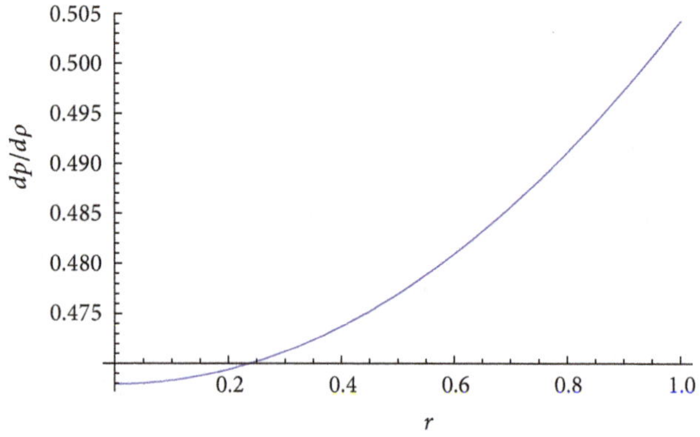

FIGURE 6: The derivative $dp/d\rho$.

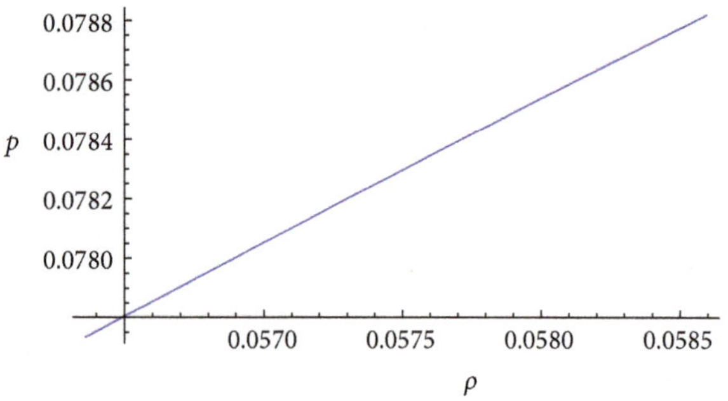

FIGURE 7: Pressure versus density.

($\beta = -1$), a class of solution is found for all positive spheroidal parameter values K. In general, a series solution is generated to the Einstein-Maxwell system and is demonstrated that two classes of solutions (40) and (42) can be extracted in terms of elementary functions. These class of solutions contain de Sitter model, Einstein universe, Tikekar superdense star [13] and Maharaj and Leach neutron star model [14] in the limit of vanishing electric field intensity. For particular parameter values, it is shown geometrically that the model satisfy the necessary physical requirements in the description of a compact object with isotropic matter distribution. Also, the mass values obtained are comparable with experimentally estimated values of realistic stars such as SAX J1808.4-3658(SS1), SAX J1808.4-3658(SS1), Her. X-1, and 4U 1820-30. We believe that the general class of exact solutions found in this paper may assist in more detailed studies of relativistic compact objects.

References

[1] W. B. Bonnor, "The mass of a static charged sphere," *Zeitschrift für Physik*, vol. 160, no. 1, pp. 59–65, 1960.

[2] B. V. Ivanov, "Static charged perfect fluid spheres in general relativity," *Physical Review D*, vol. 65, Article ID 104001, 17 pages, 2002.

[3] R. Sharma, S. Mukherjee, and S. D. Maharaj, "General solution for a class of static charged spheres," *General Relativity and Gravitation*, vol. 33, no. 6, pp. 999–1009, 2001.

[4] S. Thirukkanesh and S. D. Maharaj, "Charged relativistic spheres with generalized potentials," *Mathematical Methods in the Applied Sciences*, vol. 32, no. 6, pp. 684–701, 2009.

[5] Y. K. Gupta and S. K. Maurya, "A class of regular and well behaved charge analogue of Kuchowicz's relativistic super-dense star model," *Astrophysics and Space Science*, vol. 332, no. 2, pp. 415–421, 2011.

[6] R. Sharma, S. Karmakar, and S. Mukherjee, "Maximum mass of a cold compact star," *International Journal of Modern Physics D*, vol. 15, pp. 405–418, 2006.

[7] R. Sharma and S. Mukherjee, "Compact stars: a core-envelope model," *Modern Physics Letters A*, vol. 17, p. 2535, 2002.

[8] R. Sharma and S. Mukherjee, "Her X-1: a quark diquark star?" *Modern Physics Letters A*, vol. 16, p. 1049, 2001.

[9] V. O. Thomas, B. S. Ratanpal, and P. C. Vinodkumar, "Core-envelope models of superdense star with anisotropic envelope," *International Journal of Modern Physics D*, vol. 14, no. 1, pp. 85–96, 2005.

[10] R. Tikekar and V. O. Thomas, "Relativistic fluid sphere on pseudo-spheroidal space-time," *Pramana*, vol. 50, no. 2, pp. 95–103, 1998.

[11] B. C. Paul and R. Tikekar, "A core-envelope model of compact stars," *Gravitation and Cosmology*, vol. 11, pp. 244–248, 2005.

[12] P. C. Vaidya and R. Tikekar, "Exact relativistic model for a superdense star," *Journal of Astrophysics and Astronomy*, vol. 3, no. 3, pp. 325–334, 1982.

[13] R. Tikekar, "Exact model for a relativistic star," *Journal of Mathematical Physics*, vol. 31, no. 10, pp. 2454–2458, 1990.

[14] S. D. Maharaj and P. G. L. Leach, "Exact solutions for the Tikekar superdense star," *Journal of Mathematical Physics*, vol. 37, no. 1, pp. 430–437, 1996.

[15] K. Komathiraj and S. D. Maharaj, "Tikekar superdense stars in electric fields," *Journal of Mathematical Physics*, vol. 48, Article ID 042501, 12 pages, 2007.

[16] L. K. Patel and S. K. Koppar, "A charged analogue of the Vaidya-Tikekar solution," *Australian Journal of Physics*, vol. 40, pp. 441–447, 1987.

[17] R. Tikekar and G. P. Singh, "Interior reissner-nordstrom metric on spheroidal space-times," *Gravitation and Cosmology*, vol. 4, pp. 294–296, 1998.

[18] Y. K. Gupta and M. Kumar, "On the general solution for a class of charged fluid spheres," *General Relativity and Gravitation*, vol. 37, no. 1, pp. 233–236, 2005.

[19] Y. K. Gupta, Pratibha, and S. R. Kumar, "Some nonconformal accelerating perfect fluid plates of embedding class 1 using similarity transformations," *International Journal of Modern Physics A*, vol. 25, p. 1863, 2010.

[20] Y. K. Gupta, Pratibha, and J. Kumar, "A new class of charged analogues of Vaidya–Tikekar type super-dense star," *Astrophysics and Space Science*, vol. 333, no. 1, pp. 143–148, 2011.

[21] N. Bijalwan and Y. K. Gupta, "Closed form Vaidya-Tikekar type charged fluid spheres with pressure," *Astrophysics and Space Science*, vol. 334, pp. 293–299, 2011.

[22] N. Bijalwan, "Exact solutions: neutral and charged static perfect fluids with pressure," *Astrophysics and Space Science*, vol. 337, pp. 161–167, 2012.

[23] J. Kumar and Y. K. Gupta, "A class of new solutions of generalized charged analogues of Buchdahl's type super-dense star," *Astrophysics and Space Science*, vol. 345, pp. 331–337, 2013.

[24] X.-D. Li, I. Bombaci, M. Dey, J. Dey, and E. P. J. van den Heuvel, "Is SAX J1808.4-3658 a strange star?" *Physical Review Letters*, vol. 83, no. 19, pp. 3776–3779, 1999.

[25] M. Dey, I. Bombaci, J. Dey, S. Ray, and B. C. Samanta, "Strange stars with realistic quark vector interaction and phenomenological density-dependent scalar potential," *Physics Letters B*, vol. 438, no. 1-2, pp. 123–128, 1998.

[26] R. Sharma and S. D. Maharaj, "A class of relativistic stars with a linear equation of state," *Monthly Notices of the Royal Astronomical Society*, vol. 375, pp. 1265–1268, 2007.

[27] R. Tikekar and K. Jotania, "On relativistic models of strange stars," *Pramana*, vol. 68, no. 3, pp. 397–406, 2007.

[28] C. Alcock, E. Farhi, and A. Olinto, "Strange stars," *The Astrophysical Journal*, vol. 310, pp. 261–272, 1986.

12

Galactic Cosmic Ray Variability at Two Neutron Monitors: Relation to Kp Index

Kingsley Chukwudi Okpala

University of Nigeria, Nsukka 410002, Nigeria

Correspondence should be addressed to Kingsley Chukwudi Okpala; kingsley.okpala@unn.edu.ng

Academic Editor: José F. Valdés-Galicia

The average characteristics of year-to-year variability of Galactic cosmic ray (GCR) flux measured in one mid-latitude neutron monitor stations (Newark) and high latitude station (Apatity) have been studied under different planetary disturbance (Kp) conditions. The year-to-year variability which oscillates in response to solar cycle was analyzed using Fourier technique and the amplitude of variation was obtained using data for 1980–2005. There is a noticeable trend in the difference between the amplitudes of the year-to-year variation of the two stations. The difference is highest during low Kp conditions and lowest during high Kp condition. There is generally lesser association of GCR with solar wind (SW) flow pressure and density as the Kp index increases. Similar feature is observed with the interplanetary magnetic field IMF (total). These observations have important implications for our present understanding of the effect of solar activity to variability in GCR flux.

1. Introduction

The variability of cosmic ray of Galactic origin near Earth is primarily driven by heliospheric factors (through the interplanetary anisotropy in the heliospheric magnetic field) and geomagnetic factors (by the geomagnetic rigidity cut-off which is latitude dependent). The natures of variability of Galactic cosmic rays (GCR) have been studied in short term scales and long term scales [1–4], and these have revealed significant information about important processes involved in the transportation of GCR from source to the neutron monitor stations. A notable factor is solar activity [5–8]. Neutron monitors (NM) from lower latitudes exhibited higher contribution from the 11-year phase variation which is controlled by the diffusion associated with the change in strength of the interplanetary magnetic field [9]. Kudela and Storini [10] studied the variability of cosmic rays and its association with geomagnetic activity using the disturbance storm time index (Dst) and the southward component of the heliospheric magnetic field (B_z) component of the interplanetary magnetic field for four middle and high cut-off rigidity neutron monitors. They found high negative correlation (−0.86) between GCR intensity and interplanetary magnetic field (IMF) strength for 1 month lag between the two parameters.

Understanding the nature of the variability of cosmic rays especially GCR has continued to attract research interests because of its possible effects on many geophysical processes. Many authors have studied the diurnal and semidiurnal anisotropies [2, 11], annual and semiannual anisotropies [6, 12], and decadal to geologic time scales of GCR flux. Cosmic rays on acceleration through the galaxy become modulated in the heliosphere by the solar wind magnetic field. The differential energy spectrum of GCR in the vicinity of the Earth can be parameterized by the field model which has only one parameter—the modulation potential for a given interstellar spectrum [13]. The geomagnetic field further modulates the GCR flux through the cut-off rigidity which can be adequately described by Stormer's equation:

$$P_c = 1.9M\left(\frac{R_o}{R}\right)^2 \cos^4\lambda_G\,[\mathrm{GV}], \quad (1)$$

where M is the geomagnetic field dipole moment (in 10^{25} Gcm^3), R_0 is the Earth's mean radius, R is the distance from the given location to the dipole centre, and λ_G is the geomagnetic latitude. It is expected that changes in the dipole moment will modulate the flux of Galactic cosmic rays especially in the mid and low latitudes.

McComas et al. [14] used Ulysses and ACE observations to tie the differences in the much less variable solar wind parameters in the polar cap holes (especially density and dynamic pressure) to nearly identical variations in the ecliptic measurement. The differences seen at high latitudes are largely driven by changes in the Sun and its solar wind output and not just the differences in polar carp holes. The weaker solar wind is directly related to the lower average strength of the Sun's open magnetic field which leads to lower supply of mass and energy. The dynamic pressure carries most of the energy of the solar wind in form of bulk flow energy. Leer and Holzer [15] showed that the addition of energy above or below the sonic point is balanced by the increase (or decrease) in the wind speed and mass flux, respectively, which ultimately keeps the solar wind speed roughly the same. So it is expected that the contribution of these parameters (namely, IMF, B_z, and SW speed and density) to GCR modulation will vary in response to the level of irregularities in the heliospheric energy density.

Gleeson and Axford [16] showed that the relationship between the product of solar wind velocity and total magnetic field and the 27-day GCR intensity variation has an evident physical origin which is a hidden effect of the electric field in the cosmic ray transport [4]. Few studies have considered modulations of cosmic rays during solar activity and using certain arrangements of geomagnetic activity level (e.g., [2]), but not much work has been done to understand the association of the solar wind parameters during different levels of interplanetary disturbance. Since a wide range of processes lead to the modulation of GCR flux on the surface of the Earth, search for dominant factors under different interplanetary conditions is needed. In this paper, the GCR flux under different interplanetary condition for two stations is studied with a view to find statistical associations with solar drivers during different interplanetary conditions. The interplanetary disturbance conditions are inferred from the Kp index and subdivided into five groups. The Kp index is obtained from the standardized 3-hour range in the magnetic activity related to a geomagnetically quiet day curve. The Kp is obtained from 13 geomagnetic observatories between 44° and 60° north or south of the dip equator [17].

2. Source(s) of Data

2.1. Cosmic Ray Data. Cosmic ray data consists of pressure corrected hourly data of neutron monitor (NM) count rates from Newark (39.7°N, 75.8°W, and cut-off rigidity 2.4 GV) and Apatity (67.6°N, 33.4°E, and cut-off rigidity 0.65 GV). The Newark data is courtesy of Professor Beiber of the Bartol Research Institute and the Department of Physics and Astronomy, University of Delaware, USA, and the Apatity data is courtesy of Professor Eduard Vashenyuk of the Polar Geophysical Institute of the Russian Academy of Sciences. The data used was for the period 1980–2005.

2.2. Solar Wind Data. Daily mean of solar wind (SW) parameters for this study was obtained from the OMNI database through ftp: from OMNIweb. The solar wind parameters used are total interplanetary magnetic field IMF (B in nT), the z component of the interplanetary magnetic field B_z(nT), the solar wind bulk speed (V in km/sec) in GSE coordinate, solar wind proton density ρ_p (number/cm^3), solar wind flow pressure (P in nPa), and the daily global Kp index. The annual means were set to zero 1990 and were linearly scaled. The global Kp index was obtained from ftp://ftp.ngdc.noaa.gov.

3. Method of Analysis

The data for all the days between 1980 and 2006 were divided according to their daily Kp index. The Kp index was divided into 5 groups, namely, Kp1 (for Kp = 0–10), Kp2 (for Kp = 11–20), Kp3 (for Kp = 21–30), Kp4 (for Kp = 31–40), and Kp5 for (Kp > 40). The Kp scale is a good indicator of global level of geomagnetic activity and has the advantage of providing quantitative measure of the level of quietness (or disturbance) associated with the solar particle radiation within the heliosphere. The greater the Kp index, the more the level of interplanetary magnetic field disturbance; hence Kp1 is equivalent to the quietest conditions in a period and Kp5 is the most disturbed conditions. Days of Forbush decreases have been removed from the analysis to avoid superposed variation that is associated with such days. Forbush decreases are sudden decreases in cosmic rays in the vicinity of the Earth usually associated with transient interplanetary events related to coronal mass ejections from the Sun and/or corotating high speed of solar wind speed. Linear regression analysis was performed on year-to-year basis to understand the effect of the different solar wind parameters on the flux of GCR at each of the NM station. The yearly averages for each group of Kp index were computed as means of all the days having Kp index within the range for that group. Figure 1 shows the yearly means set to zero for 1990 and scaled linearly for the two stations and for all the groups of Kp index discussed earlier.

Since GCR flux variation on an annual scale follows the solar activity oscillation, it is possible to find the harmonic amplitude of its oscillation by performing Fourier transformation to obtain the peak of the fundamental harmonic.

The time dependent harmonic function $F(t)$ with 12 equidistant points (approximating to a solar cycle) in the interval $t = 0$ to $t = \pi$ can be expressed in terms of Fourier series as

$$F(t) = a_o + \sum_{n=1}^{12} a_n \cos(nt) + b_n \sin(nt),$$
$$F(t) = a_o + \sum_{n=1}^{12} (r_n \cos(nt) - \phi_n), \quad (2)$$

where a_o is the mean value of the $F(t)$ for the time interval from $t = 0$ to $t = \pi$, a_n b_n are the coefficients of the nth harmonics, and r_n is the amplitude of the nth harmonic. For

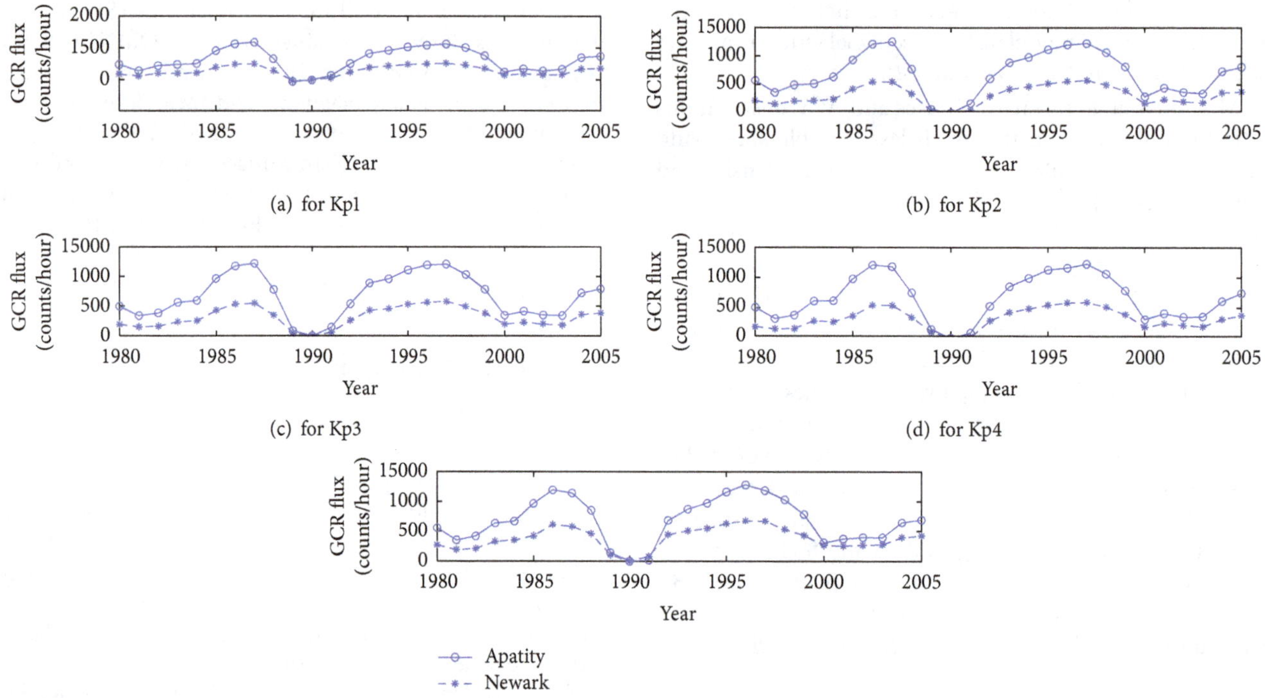

(a) for Kp1

(b) for Kp2

(c) for Kp3

(d) for Kp4

(e) for Kp5

FIGURE 1: Annual means of the GCR flux set to zero for 1990 and scaled linearly.

the present study the fundamental harmonic is of interest. These coefficient are expressed as

$$a_o = \frac{1}{6}\sum_{n=1}^{12} n_i,$$

$$a_n = \frac{1}{6}\sum_{n=1}^{12} n_i \cos(n_i t), \qquad (3)$$

$$b_n = \frac{1}{6}\sum_{n=1}^{12} n_i \sin(n_i t).$$

The amplitude r_n of the nth harmonic is expressed as

$$r_n = \left(a_n^2 + b_n^2\right)^{1/2}. \qquad (4)$$

The amplitudes for the first harmonic for Apatity (r_{Apatity}) and Newark (r_{Newark}) were obtained and the difference (DA) for each Kp group was computed as

$$\text{DA} = r_{\text{Apatity}} - r_{\text{Newark}}. \qquad (5)$$

A plot of the variation of DA for the Kp groups is shown in Figure 2. The annual means of the linearly scaled count rates for Kp1 days for the years considered are well correlated for the two stations and are shown in Figure 3. This trend was observed for all other Kp groupings used in this study. Correlation coefficients for all the five groupings with the solar wind parameters are presented in Tables 1 and 2.

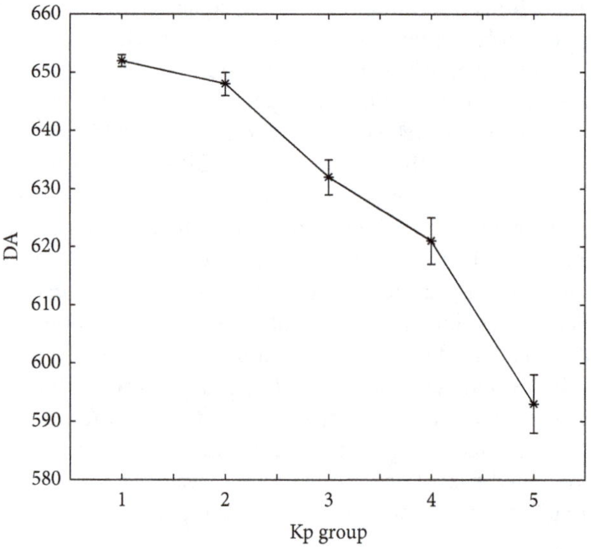

FIGURE 2: DA variation with Kp groups.

4. Discussion of Results

The pressure corrected hourly data of cosmic rays from two neutron monitor stations have been used to obtain the harmonic components of the annual mean of the GCR flux for over two sunspot cycles (1980–2005). Now, low Kp index days imply quiet conditions of the interplanetary medium and conversely, for high Kp index, the implication is disturbed conditions. Therefore the relationship between CR flux and

TABLE 1: Correlation table for Apatity neutron monitor.

	IMF (B)	B_z	SW proton density	SW plasma speed (V)	Flow pressure	VB	Kp group (total number of days)
GCR	−0.72	−0.53	0.60	−0.10	0.49	−0.71	1 (1164)
	−0.82	−0.68	0.43	0.03	0.30	−0.81	2 (2548)
	−0.70	−0.59	0.22	0.26	0.19	−0.65	3 (2182)
	−0.69	−0.06	−0.17	0.28	−0.18	−0.69	4 (1246)
	−0.20	−0.13	−0.12	0.13	−0.17	−0.27	5 (355)

TABLE 2: Correlation table for Newark neutron monitor.

	IMF (B)	B_z	SW proton density	SW plasma speed (V)	Flow pressure	VB	Kp group (total number of days)
GCR	−0.77	−0.55	0.61	−0.18	0.48	−0.74	1 (1164)
	−0.85	−0.70	0.47	0.15	0.37	−0.81	2 (2548)
	−0.71	−0.56	0.28	0.29	0.26	−0.64	3 (2182)
	−0.56	0.12	−0.08	0.30	−0.02	−0.70	4 (1246)
	−0.47	0.35	0.03	0.13	−0.15	−0.50	5 (355)

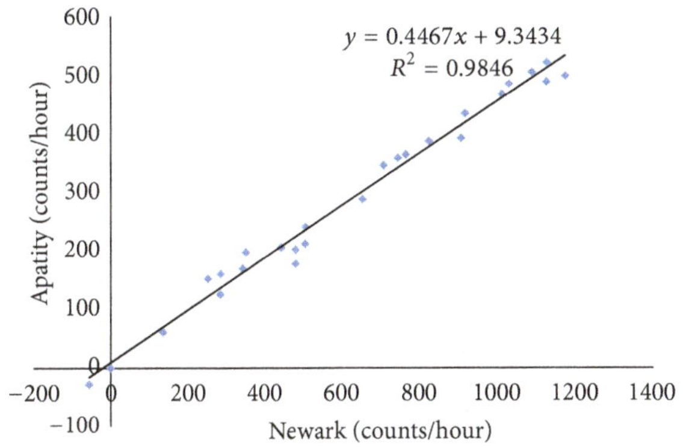

FIGURE 3: Plots of linearly scaled annual mean for Kp1 days.

interplanetary plasma parameters under varying conditions is expected to help us better appreciate the level of modulation that can be associated with key components of the solar wind under different levels of activity on an annual scale. From Figure 1, annual means of the GCR flux for Newark and Apatity showed similar annual variation following the 11-year solar cycle trend for all the levels of interplanetary disturbance with Apatity having higher amplitude because of its low rigidity. The profile of DA shown in Figure 2 as deduced from the amplitude of the Fourier analysis reveals that the difference in amplitude between the stations was lesser with increasing activity index. There appears to be a lesser variability in amplitude with increasing activity for Newark (with lower rigidity cut-off) when compared to Apatity. This trend is significant considering that the correlation between the two stations is very high (Figure 3). This result is particularly significant because the interplanetary disturbance is a measure of the local property of the medium through which the GCR propagates. Since GCR essentially carry signatures of various magnetohydrodynamic waves present in the interplanetary space and geospace, the variation in the two NM stations could therefore be a pointer to magnetospheric effects on the GCR flux possibly via currents such as ring currents and Birkeland currents since these currents are driven by IMF and associated bulk flows of plasma through the magnetosphere. The correlation (Tables 1 and 2) shows a similar trend in the correlation of the GCR with different SW parameters for different interplanetary condition; the difference in the amplitude determined here is likely due to changes in the rigidity cut-off of the stations. The wide variation in the DA for low Kp group of days, conversely, reflects the role of interplanetary conditions which modulate the transfer of energy into the magnetosphere leading to changes in the ever present ring current. However, the ring current (responsible for the main stage of geomagnetic storms) dominates the more disturbed conditions by reducing the rigidity cut-offs of the mid-latitude station (Newark) and consequently a decrease in DA with increasing Kp. The peaks of the annual means of the GCR showed that the highest peaks (Figure 1) are associated with Kp1 while the other groups showed similar peaks though not as high as those for both Newark and Apatity stations. The amplitudes of the year-to-year variation for the two stations expressed as fraction of the mean for each group show that the least amplitude is associated with Kp1 and the highest one with Kp5. However for Apatity station, Kp2 days show a significantly high amplitude, while the same trend is not obvious for Newark.

The Newark GCR flux showed high negative correlation (at 95% confidence) with IMF B (−0.77) which tend to decrease with increasing disturbance activity. The Kp2 group showed higher correlation than the Kp1 days. This trend was also exhibited in Apatity NM (−0.72). The hidden effect of the electric field (VB component) in the transportation of GCR is evident but is mostly tied to the IMF effect rather than the velocity component. This can be inferred from the rather weak correlation between GCR flux and solar wind (SW) speed in the two stations considered. It is necessary to note that VB reflects both the effect of diffusion by the IMF and convection with the SW; thus its effect on the GCR modulation is expected to dominate the observations. Our study shows that this is evident, especially the mid rigidity cut-off station. The strongest association between GCR and SW speed is observed during Kp4 with no particular trend for other disturbance index grouping. It is pertinent to observe that the least correlation of GCR with B_z (and SW proton density) coincides with the maximum correlation between GCR and SW plasma speed (V). The contribution of these parameters (B_z, SW proton density, and plasma speed) to the modulation of GCR within the heliosphere may therefore be strongly dependent on the disturbance field of the solar wind. This corroborates the findings of Leer and Holzer [15] that the addition of energy above or below the sonic point is balanced by the increase in the wind speed and mass flux, respectively, which ultimately keeps the solar wind speed roughly the same. The modulation effects by SW proton density and flow pressure on GCR flux are inversely proportional to the level of disturbance. From Tables 1 and 2, it is evident that these parameters are more associated with the GCR during quiet conditions with the SW proton density showing stronger negative association (−0.61) and the flow pressure (−0.47) under Kp1 condition and ((−0.03) and (−0.07)) under Kp5 conditions, respectively, for Newark. The same trend is observed in Apatity NM station. The B_z component showed fairly high correlation with GCR especially for Kp2 condition. This statistical relationship was weakest for Kp4 in the two stations.

5. Conclusion

(i) The difference in the amplitude of the GCR flux in Newark and Apatity NM stations depends on the level of interplanetary disturbance (Kp). The difference is highest during low Kp conditions and lowest during high Kp condition.

(ii) There is generally lesser association of GCR with SW flow pressure and density as the Kp index increases. Similar trend is observed with the total IMF.

(iii) VB and IMF (total B) correlated well with GCR flux for most of the planetary conditions considered. The correlation was least during Kp 5 condition.

Conflict of Interests

The author declares that there is no conflict of interests regarding the publication of this paper.

Acknowledgment

The author is grateful for the many constructive suggestions from the anonymous reviewers that have greatly impacted on the final form of this paper.

References

[1] A. G. Ananth, S. P. Agrawal, and U. R. Rao, "Study of cosmic ray diurnal variation on a day-to-day basis," *Pramana*, vol. 3, no. 2, pp. 74–88, 1974.

[2] C. M. Tiwari, D. P. Tiwari, A. K. Pandey, and P. K. Shrivastava, "Average anisotropy characteristics of high energy cosmic ray particles and geomagnetic disturbance index Ap," *Journal of Astrophysics and Astronomy*, vol. 26, no. 4, pp. 429–434, 2005.

[3] K. C. Okpala and F. N. Okeke, "Seasonal changes in H-component of quiet day geomagnetic field modulation of galactic cosmic rays," *Nigerian Journal of Space Research*, vol. 7, pp. 86–94, 2010.

[4] R. Modzelewska and M. V. Alania, "The 27-day cosmic ray intensity variations during solar minimum 23/24," *Solar Physics*, vol. 286, no. 2, pp. 593–607, 2013.

[5] H. Mavromichalaki and E. Paouris, "Long-term cosmic ray variability and the CME-index," *Advances in Astronomy*, vol. 2012, Article ID 607172, 8 pages, 2012.

[6] K. C. Okpala and F. N. Okeke, "Investigation of diurnal and seasonal galactic cosmic ray variations on quiet days in two mid latitude stations," *Astroparticle Physics*, vol. 34, no. 12, pp. 878–885, 2011.

[7] A. V. Belov, E. A. Eroshenko, V. A. Oleneva, V. G. Yanke, and H. Mavromichalki, "Long-term behaviour of the cosmic ray anisotropy derived from the worldwide neutron monitor network data," in *Proceedings of the 20th ECRS*, 2006.

[8] I. Usoskin, H. Kanamen, K. Mursula, P. Tanskanen, and G. A. Kovaltsov, "Correlative study of solar activity and cosmic ray intensity," *Journal of Geophysical Research*, vol. 103, pp. 9567–9574, 1998.

[9] S. Y. Oh, Y. Yi, and J. W. Bieber, "Modulation cycles of galactic cosmic ray diurnal anisotropy variation," *Solar Physics*, vol. 262, no. 1, pp. 199–212, 2010.

[10] K. Kudela and M. Storini, "Cosmic ray variability and geomagnetic activity: a statistical study," *Journal of Atmospheric and Solar-Terrestrial Physics*, vol. 67, no. 10, pp. 907–912, 2005.

[11] K. Munakata, T. Kitawada, S. Yasue et al., "Enhanced sidereal diurnal variation of galactic cosmic rays observed by the two-hemisphere network of surface level muon telescopes," *Journal of Geophysical Research A: Space Physics*, vol. 104, no. 2, pp. 2511–2519, 1999.

[12] S. Oh, J. W. Bieber, P. Evenson, J. Clem, Y. Yi, and Y. Kim, "Record neutron monitor counting rates from galactic cosmic rays," *Journal of Geophysical Research: Space Physics*, vol. 118, no. 9, pp. 5431–5436, 2013.

[13] I. G. Usoskin, K. Alanko-Huotari, G. A. Kovaltsov, and K. Mursula, "Heliospheric modulation of cosmic rays: monthly reconstruction for 1951–2004," *Journal of Geophysical Research*, vol. 110, no. A12, p. A12108, 2005.

[14] D. J. McComas, R. W. Ebert, H. A. Elliott et al., "Weaker solar wind from the polar coronal holes and the whole sun," *Geophysical Research Letters*, vol. 35, no. 18, Article ID L18013, 2008.

[15] E. Leer and T. E. Holzer, "Energy addition in the solar wind," *Journal of Geophysical Research*, vol. 85, no. A9, pp. 4681–4688, 1980.

[16] L. J. Gleeson and W. I. Axford, "Cosmic rays in the interplanetary medium," *The Astrophysical Journal*, vol. 149, p. L115, 1967.

[17] NGDC, *Monthly Summary of Geomagnetic Activity*, Issue 2, Geomagnetic Indices Bulletin, 1985.

13

Ion-Acoustic Instabilities in a Multi-Ion Plasma

Noble P. Abraham,[1] Sijo Sebastian,[1] G. Sreekala,[1] R. Jayapal,[1] C. P. Anilkumar,[2] and Venugopal Chandu[1]

[1] *School of Pure & Applied Physics, Mahatma Gandhi University, Priyadarshini Hills, Kottayam, Kerala 686 560, India*
[2] *Equatorial Geophysical Research Laboratory, Indian Institute of Geomagnetism, Krishnapuram, Tirunelveli, Tamil Nadu 627 011, India*

Correspondence should be addressed to Venugopal Chandu; cvgmgphys@yahoo.co.in

Academic Editors: M. S. Dimitrijevic, A. Meli, and S. Naik

We have, in this paper, studied the stability of the ion-acoustic wave in a plasma composed of hydrogen, positively and negatively charged oxygen ions, and electrons, which approximates very well the plasma environment around a comet. Modelling each cometary component (H^+, O^+, and O^-) by a ring distribution, we find that ion-acoustic waves can be generated at frequencies comparable to the hydrogen ion plasma frequency. The dispersion relation has been solved both analytically and numerically. We find that the ratio of the ring speed ($u_{\perp s}$) to the thermal spread (v_{ts}) modifies the dispersion characteristics of the ion-acoustic wave. The contrasting behaviour of the phase velocity of the ion-acoustic wave in the presence of O^- ions for $u_{\perp s} > v_{ts}$ (and vice versa) can be used to detect the presence of negatively charged oxygen ions and also their thermalization.

1. Introduction

Low-frequency electrostatic or longitudinal ion density waves are one of the most fundamental of oscillations in a plasma [1, 2]. In the long-wavelength limit, the ions provide the inertia with the electrons as the source of the restoring force [1]. Ion-acoustic waves also exhibit strong nonlinear properties and are highly Landau damped unless $T_i \ll T_e$, where T_i and T_e are, respectively, the ion and electron temperatures [3–5]. These waves have been observed in both space and laboratory plasmas; they have thus been extensively studied in many types of high-temperature laboratory plasmas [4, 6]. The waves have been invoked to explain wave characteristics observed in Earth's ionosphere [7] and transport in the solar wind, corona, chromosphere [8], and comets [9].

In general a cometary environment contains new born hydrogen and heavier ions, with relative densities depending on the distance from the nucleus. Previous studies have concentrated on positively charged oxygen as the heavier ion species [10]. However, Giotto's observations of the inner coma of comet Halley showed that a new component, namely, negatively charged cometary ions was present, in addition to the usual thermal electrons and ions, fast cometary pickup ions, and so forth, [11]. These negative ions were observed in three broad mass peaks at 7–19, 22–65, and 85–110 amu with O^- being identified unambiguously [11].

A popular model of a cometary environment is the solar wind plasma environment permeated by dilute, drifting ring distribution of electrons and ions with finite thermal spreads [10]. Instabilities driven by an electron velocity ring distributions have been studied by many authors [12–14]. However, ion ring distributions are more important because of the greater amount of free energy available [15].

Instabilities driven by ion ring distributions have also been studied by a number of authors: close to the ion cyclotron frequency, electrostatic ion cyclotron waves propagating either perpendicularly or nearly perpendicularly to the magnetic field can be excited [16, 17]. At still higher frequencies, the magnetic effects on the ions can be neglected and lower-hybrid instabilities driven by the ring ions can occur [17–21]. And at even higher frequencies, the electrons too become unmagnetized and ion-acoustic-like instabilities result.

The frequency of the ion-acoustic wave is comparable with the ion plasma frequency and propagates parallel to the magnetic field. Generally, a combination of warm electrons and cold ions ($T_e > T_i$) with the electrons drifting relative to the ions is the condition required to excite the ion-acoustic instability.

The ion-acoustic wave is one of the more easily observed waves in the plasma environments of comets. For example, ion-acoustic waves in the frequency range of 1.0–1.5 kHz were observed by the ICE spacecraft sent to observe the comet Giacobini-Zinner [9, Figure 3]. Again ion sound waves, with a frequency slightly less than 1 kHz, were detected by the spacecraft Sakigake which observed comet Halley [22, Figure 6].

We have, therefore, studied the stability of the ion-acoustic wave in a five-component plasma: solar wind protons, electrons, cometary hydrogen, and positively and negatively charged oxygen ions. We find that the ratio of the ring speed ($u_{\perp s}$) to the thermal spread (v_{ts}) affects the dispersion characteristics of the ion-acoustic wave. The phase velocity of the ion-acoustic wave depends sensitively on this ratio in the presence of O^- ions. This variation for $u_{\perp s} < v_{ts}$ (and vice versa) is proposed as a tool for detecting the presence of O^- ions and also their thermalization.

2. The Dispersion Relation

As stated above, we intend to study the stability of the ion-acoustic wave in a five-component plasma. The five components are solar wind hydrogen and electrons and the ions of cometary origin picked up by the solar wind. These cometary ions are hydrogen and positively and negatively charged oxygen ions. The solar wind components are modelled by the Maxwellian distributions while the cometary ions are described by ring distributions.

The contributions to the dispersion relation of the plasma components described by the Maxwellian distributions are well known [23]; the same when they are described by ring distributions are given in [15]. The plasma under consideration contains components modelled by both these forms. Hence, combining these contributions, we can write down the dispersion relation for waves of frequency ω and wave vector $\vec{k}$, as

$$D(\omega,\vec{k}) = 1 + \frac{2\omega_{pH^+}^2}{k^2 W_{TH^+}^2}\left[1 + \xi_{H^+} Z(\xi_{H^+})\right] + \frac{2\omega_{pe}^2}{k^2 W_{Te}^2}\left[1 + \xi_e Z(\xi_e)\right] - \sum_{s=H^+,\,O^+,\,O^-} \frac{\left(\omega_{ps}^r\right)^2 \omega}{\left[\omega^2 - k_\perp^2 \left(u_{\perp s} - i v_{ts}\right)^2\right]^{3/2}} = 0. \quad (1)$$

In (1), $\omega_{pj} = [4\pi n_j e_j^2/m_j]^{1/2}$ and $W_{Tj} = [T_j/m_j]^{1/2}$ denote, respectively, the plasma frequency and the thermal velocity for particles of species j. Also n_j, e_j, m_j, and T_j denote, respectively, the density, charge, mass, and temperature of species j. ω_{ps}^r ($s = H^+, O^+, O^-$) denotes the plasma frequency for the ring ions of species s while $u_{\perp s}$ and v_{ts} are, respectively, the ring ion speed and thermal spread for these ions. The plasma dispersion function $Z(\xi)$, arising from the $dv_\parallel$ integration [24], has arguments of $\xi_{H^+} = \omega/kW_{TH^+}$ (for hydrogen ions of solar wind origin) and $\xi_e = (\omega - kv_{ez})/kW_{Te}$ (for electrons). v_{ez} is the drift velocity of the electrons parallel to the magnetic field.

The charge neutrality condition for the plasma under consideration yields

$$\frac{n_{H^+}}{n_e} = \left(1 - \frac{n_{H^+}^r}{n_e} - \frac{n_{O^+}^r}{n_e} + \frac{n_{O^-}^r}{n_e}\right) = (1 - \alpha - \beta + \gamma), \quad (2)$$

with $\alpha = n_{H^+}^r/n_e$, $\beta = n_{O^+}^r/n_e$, and $\gamma = n_{O^-}^r/n_e$, where n, in general, denotes the density of each species.

We derive expressions for the growth rate of the ion-acoustic wave under two limiting conditions, namely, that of a large thermal spreads $((u_{\perp s}/v_{ts}) < 1)$ and small thermal spreads $((u_{\perp s}/v_{ts}) > 1)$. The contributions of the solar wind hydrogen ions and electrons are the same for both cases; the contributions of the ring ions vary depending on their thermal spreads.

We use the asymptotic expansion of the plasma dispersion function $Z(\xi_{H^+})$ for the solar wind hydrogen ions and the power series expansion of $Z(\xi_e)$ for the solar wind electrons. Substituting the appropriate expansions in (1), we finally arrive at the dispersion relation for ion-acoustic waves in an electron-ion plasma with three species of ring ions, as

$$D(\omega,\vec{k}) = \left(1 + k^2\lambda_{De}^2\right) - \frac{\omega_{pH^+}^2}{\omega^2} + i\sqrt{\pi}\frac{1}{k^2\lambda_{De}^2}\frac{1}{kW_{Te}} \times \left\{\frac{n_{H^+}}{n_e}\left(\frac{m_{H^+}}{m_e}\right)^{1/2}\left(\frac{T_e}{T_{H^+}}\right)^{3/2}\omega e^{-\xi_{H^+}^2} + (\omega - kv_{ez})\, e^{-\xi_e^2}\right\} - \sum_{s=H^+,\,O^+,\,O^-}\frac{n_s^r}{n_{H^+}}\frac{m_{H^+}}{m_s} \times \frac{\left(\omega_{pH^+}\right)^2\omega}{\left[\omega^2 - k_\perp^2\left(u_{\perp s} - iv_{ts}\right)^2\right]^{3/2}} = 0. \quad (3)$$

In (3), $\omega = \omega_r + i\gamma$ and $\lambda_{De} = [T_e/(4\pi n_e e^2)]^{1/2}$ is the Debye length.

2.1. Case (i): Large Thermal Spreads. In this subsection we derive expressions for the real frequency and growth/damping rate for ion-acoustic waves when the thermal spreads of the ring ions are large. As a simplifying assumption we also let $\omega_r \approx k_\perp u_{\perp s}$. With the assumption of large thermal spreads

$u_{\perp s} < v_{ts}$, (3) can be simplified to yield an expression for the real frequency ω_r as

$$\omega_r^2 = \frac{(1-\alpha-\beta+\gamma)k^2C_s^2}{(1+k^2\lambda_{De}^2)} \times \left(1+\sum_s \alpha_s\left(\frac{u_\perp}{v_t}\right)^3 \frac{\cos((3/2)\theta)}{\left[1+4(u_\perp/v_t)^2\right]^{3/4}}\right) \quad (4)$$

and the growth/damping rate as

$$\gamma = -\sqrt{\frac{\pi}{8}}\frac{\omega_r^4}{\omega_{pH^+}^2 k^2\lambda_{De}^2}\frac{1}{kC_s}\frac{1}{A} \times \left[\left(\frac{n_{H^+}}{n_e}\right)\left(\frac{T_e}{T_{H^+}}\right)^{3/2} e^{-(\omega_r/kW_{TH^+})^2} - \left(\frac{m_e}{m_{H^+}}\right)^{1/2}\left(\frac{kv_{ez}}{\omega_r}-1\right)e^{-((\omega_r-kv_{ez})/kW_{Te})^2}\right] - \sum_s\left(\frac{\alpha_s}{2}\right)\omega_r\frac{(u_\perp/v_t)^3\sin((3/2)\theta)}{A\left[1+4(u_\perp/v_t)^2\right]^{3/4}}, \quad (5)$$

where $A=[1-\sum_s(\alpha_s/2)(u_\perp/v_t)^3(\cos((3/2)\theta)/[1+4(u_\perp/v_t)^2]^{3/4})]$.

In the above, C_s is the speed of sound defined by $C_s = [T_e/m_{H^+}]^{1/2}$, while

$$\alpha_{H^+} = \frac{n_{H^+}^r}{n_{H^+}}, \qquad \alpha_{O^+} = \frac{m_{H^+}}{m_{O^+}}\frac{n_{O^+}^r}{n_{H^+}}, \qquad \alpha_{O^-} = \frac{m_{H^+}}{m_{O^-}}\frac{n_{O^-}^r}{n_{H^+}}, \qquad \theta = \tan^{-1}\left(\frac{2u_{\perp s}}{v_{ts}}\right). \quad (6)$$

2.2. Case (ii): Small Thermal Spreads. We next consider expressions for the real frequency and the growth/damping rate when the thermal spread is small; that is, $u_{\perp s} > v_{ts}$. These can be extracted from (3) with the additional assumption $\omega_r \approx k_\perp u_{\perp s}$; the final expressions are

$$\omega_r^2 = \frac{(1-\alpha-\beta+\gamma)k^2C_s^2}{(1+k^2\lambda_{De}^2)}\left(1-\sum_s\left(\frac{\alpha_s}{4}\right)\left(\frac{u_\perp}{v_t}\right)^{3/2}\right), \quad (7)$$

$$\gamma = -\sqrt{\frac{\pi}{8}}\frac{\omega_r^2}{kC_s}\frac{1}{(1+k^2\lambda_{De}^2)} \times \frac{\left(1-\sum_s(\alpha_s/4)(u_\perp/v_t)^{3/2}\right)}{\left(1-\sum_s(\alpha_s/16)(u_\perp/v_t)^{3/2}\right)} \times \left[\left(\frac{n_{H^+}}{n_e}\right)\left(\frac{T_e}{T_{H^+}}\right)^{3/2} e^{-(\omega_r/kW_{TH^+})^2} - \left(\frac{m_e}{m_{H^+}}\right)^{1/2}\left(\frac{kv_{ez}}{\omega_r}-1\right)e^{-((\omega_r-kv_{ez})/kW_{Te})^2}\right] - \frac{\sum_s(\alpha_s/8)\omega_r(u_\perp/v_t)^{3/2}}{\left(1-\sum_s(\alpha_s/16)(u_\perp/v_t)^{3/2}\right)}. \quad (8)$$

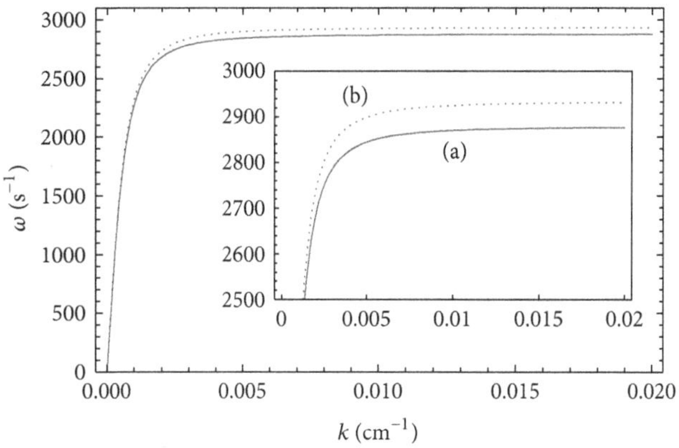

FIGURE 1: Plot of frequency ω versus wave vector $\vec{k}$. Curve (a) is for $u_{\perp s} > v_{ts}$; drift speed $u_{\perp s} = 420 \times 10^5$ cm s^{-1} and thermal spread $v_{ts} = 350\times10^5$ cm s^{-1}; curve (b) is for $u_{\perp s} < v_{ts}$; $u_{\perp s} = 100\times10^5$ cm s^{-1} and $v_{ts} = 320 \times 10^5$ cm s^{-1}.

Thus (4) and (7) are expressions for the frequencies of the ion-acoustic wave under the two limiting cases; (5) and (8) are the corresponding expressions for the growth/damping rate.

3. Results and Discussion

As a check on our results, we note from (4) and (7) the expressions for real frequency that they reduce to the corresponding ones in an electron-ion plasma when the ring ions are absent [25]. Similarly, the expressions for the growth/damping rate (5) and (8) also reduce to that in an electron-ion plasma when the ring ions are absent [25]. However what is interesting from (4) and (7) is that the real frequency is either greater (for $u_{\perp s} < v_{ts}$) or smaller (for $u_{\perp s} > v_{ts}$) than in a single ion plasma (in the absence of ring ions). The electron drift velocity, parallel to the magnetic field, is the source of energy for the instability. Thus, for a given wave vector $\vec{k}$ and ion densities, a greater drift velocity of the electrons is required to drive the wave unstable in the case where the thermal spread of the ring ions is large.

We use parameters similar to that in [10]. The solar wind hydrogen density $n_{H^+} = 4.95$ cm^{-3}, while the hydrogen and electron temperatures are $T_{H^+} = 8\times10^4$°K, $T_e = 2\times10^5$°K. The densities of the ring ions are $n_{H^+}^r = 0.5$ cm^{-3}, $n_{O^+}^r = 0.25$ cm^{-3}, and $n_{O^-}^r = 0.05$ cm^{-3}.

Ion-acoustic waves were observed in the frequency range from 1.0 to 1.5 kHz by the ICE spacecraft at the comet Giocobini-Zinner [9]. Thus to calculate the wave vector $\vec{k}$, a frequency of $\omega = 1.4$ kHz was used in the cold plasma dispersion relation and the wave vector $\vec{k}$ was calculated. $\vec{k}$ was then varied around this value in subsequent calculations.

Figure 1 is a plot of the frequency ω versus wave vector $\vec{k}$, with the frequency ω being obtained by solving (4) and (7). Curve (a) has a drift speed $u_{\perp s} = 420 \times 10^5$ cm s^{-1} and a thermal spread of $v_{ts} = 350 \times 10^5$ cm s^{-1} ($u_{\perp s} > v_{ts}$) while for curve (b) $u_{\perp s} = 100\times10^5$ cm s^{-1}, $v_{ts} = 320\times10^5$ cm s^{-1} ($u_{\perp s} < v_{ts}$). As is evident from the figure the waves are at a higher

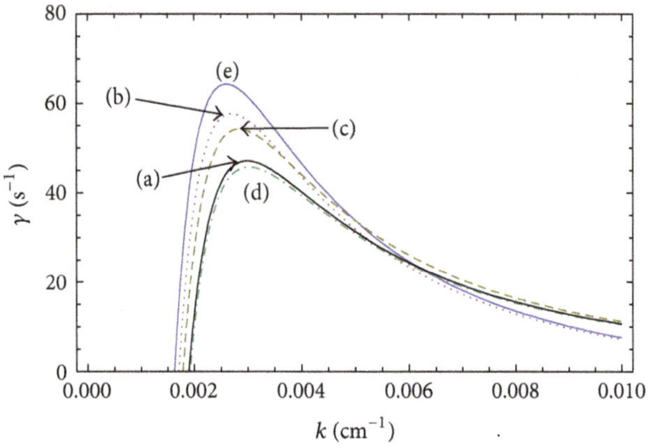

FIGURE 2: Plot of the growth rate γ (5) versus wave vector $\vec{k}$ for $T_e = 2 \times 10^{5\circ}$K, $T_{H^+} = 8 \times 10^{4\circ}$K, $u_{\perp s} = 100 \times 10^5$ cm s^{-1}, $v_{ts} = 320 \times 10^5$ cm s^{-1}, and an electron drift speed $v_{ez} = 2460 \times 10^5$ cm s^{-1}. Curve (a) is for a single ion plasma without any ring ions. Curve (b) has a H^+ ring, curve (c) has an O^+ ring, curve (d) has an O^- ring, and for curve (e) all species of ring ions are present.

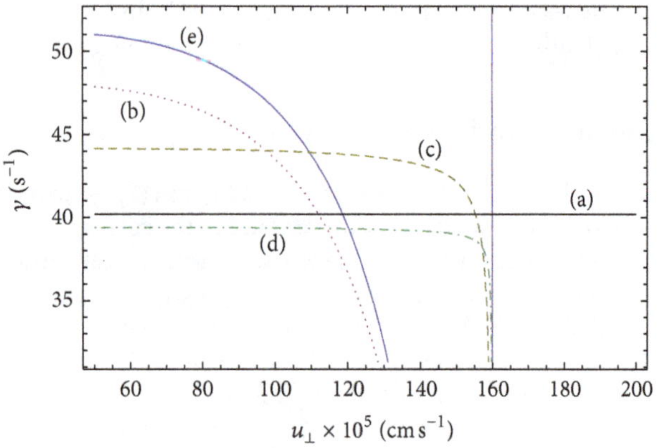

FIGURE 3: Plot of the growth rate γ (5) versus $u_{\perp s}$ ($v_{ts} = 320 \times 10^5$ cm s^{-1}, $\vec{k} = 0.004$ cm^{-1}). Other relevant parameters are the same as in Figure 2. Curve (a) is for a single ion plasma without any ring ions. Curve (b) is for a H^+ ring, curve (c) has an O^+ ring, and curve (d) is for an O^- ring. Curve (e) depicts the case when all species of ring ions are present.

frequency for $u_{\perp s} < v_{ts}$. Since the calculated frequencies are in better agreement with the observed range of frequencies for $u_{\perp s} > v_{ts}$, it is safe to assume that thermalization of the ring ions has not yet occurred. The pickup of new born ions by the solar wind has been described by a three-step process: formation of a ring beam distribution, pitch angle scattering of this initial distribution into a shell, and a slower velocity diffusion that spreads out the shell [26].

We first consider the growth/damping rates when the thermal spread is large.

Figure 2 is a plot of the growth rate γ (5) versus wave vector $\vec{k}$ for $T_e = 2 \times 10^{5\circ}$K, $T_{H^+} = 8 \times 10^{4\circ}$K, $u_{\perp s} = 100 \times 10^5$ cm s^{-1}, $v_{ts} = 320 \times 10^5$ cm s^{-1}, and an electron drift speed $v_{ez} = 2460 \times 10^5$ cm s^{-1}. Curve (a) is for a single ion plasma without ring distributions. Curve (b) has only one type of ring

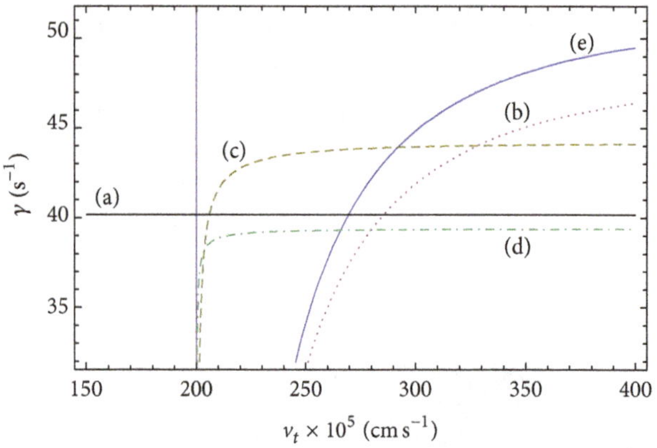

FIGURE 4: Plot of the growth rate γ (5) versus v_{ts} ($u_{\perp s} = 100 \times 10^5$ cm s^{-1}, $\vec{k} = 0.004$ cm^{-1}). Other relevant parameters are the same as in Figure 2. Curve (a) is for a single ion plasma without any ring ions. Curve (b) has a H^+ ring, curve (c) has an O^+ ring, curve (d) has an O^- ring, and for curve (e) all species of ring ions are present.

distribution, namely, the H^+ ring. We find that the growth rate is enhanced. Curve (c) is for the O^+ ring replacing the H^+ ring—there is a slight decrease in the growth rate. When this single ring is the O^- ring, the growth rate is the lowest (curve (d)). The growth rate is also shown when all species of rings are present (curve (e)). The ion-acoustic wave is thus driven unstable by the electrons drifting parallel to the magnetic field (addition of negative ions reduces the density of electrons required to maintain charge neutrality); the ring ions add to the instability, with the lighter ion hydrogen ring being the most effective in enhancing the instability of the wave.

Figures 3 and 4 depict the variation of the growth rate γ (5) versus $u_{\perp s}$ ($v_{ts} = 320 \times 10^5$ cm s^{-1}) (Figure 3) or γ versus v_{ts} ($u_{\perp s} = 100 \times 10^5$ cm s^{-1}) (Figure 4), with the temperatures and electron drift speed being the same as in Figure 2. In both figures curve (a) represents the growth rate in a single ion plasma (ring ions are not present); curve (b) when the H^+ ring alone is present, curve (c) when the O^+ ring alone is present, and curve (d) when an O^- ring alone is present. Curve (e) is for the situation when all species of ring ions are present. While the general nature of the curves is similar to that in Figure 2, the heavier ion rings extend the range of $u_{\perp s}$ and v_{ts} over which the wave is unstable as compared to the lighter ion ring alone being present.

We next consider the case of low thermal spreads ($u_{\perp s} > v_{ts}$).

Figure 5 is a plot of the growth rate (8) versus wave vector $\vec{k}$ for $T_e = 2 \times 10^{5\circ}$K, $T_{H^+} = 8 \times 10^{4\circ}$K, $u_{\perp s} = 420 \times 10^5$ cm s^{-1}, $v_{ts} = 350 \times 10^5$ cm s^{-1}, and an electron drift speed $v_{ez} = 2460 \times 10^5$ cm s^{-1}. Similar to Figure 2, curve (a) is for a single ion plasma without any ring ions. Curve (b) is for the H^+ ring and has the lowest growth rate. Curve (c) is for the O^+ ring replacing the H^+ ring—there is an enhancement in the growth rate, than the single ion plasma. When this single ring is the O^- ring, the growth rate is slightly lowered (curve (d)). Curve (e) shows the growth rate when all the species of rings are present. The ion-acoustic wave is thus

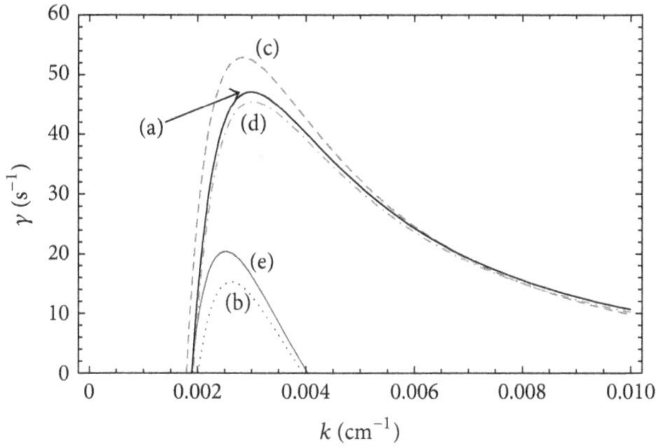

FIGURE 5: Plot of the growth rate γ (8) versus wave vector $\vec{k}$ for $T_e = 2 \times 10^{5\circ}$K, $T_{H^+} = 8 \times 10^{4\circ}$K, $u_{\perp s} = 420 \times 10^5$ cm s^{-1}, $v_{ts} = 350 \times 10^5$ cm s^{-1}, and an electron drift speed $v_{ez} = 2460 \times 10^5$ cm s^{-1}. Curve (a) is for a single ion plasma without any ring ions. Curve (b) has a H^+ ring, curve (c) has an O^+ ring, and curve (d) has an O^- ring. Curve (e) has all species of ring ions.

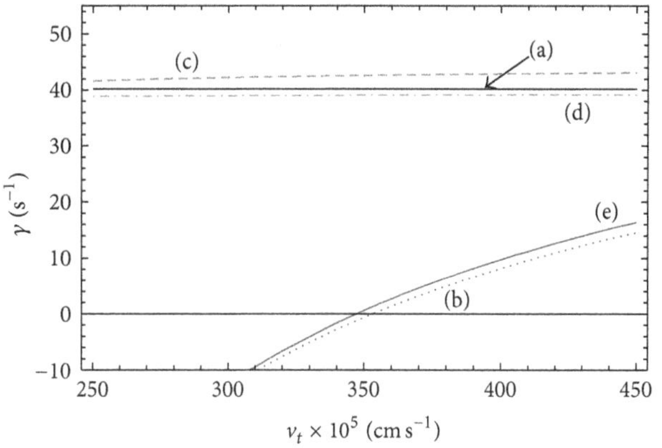

FIGURE 7: Plot of the growth rate γ (8) versus v_{ts} ($u_{\perp s} = 420 \times 10^5$ cm s^{-1}, $\vec{k} = 0.004$ cm^{-1}). Other relevant parameters are the same as in Figure 5. Curve (a) is for a single ion plasma without any ring distributions. Curve (b) has a H^+ ring, curve (c) has an O^+ ring, curve (d) has an O^- ring, and curve (e) has all species of ring ions.

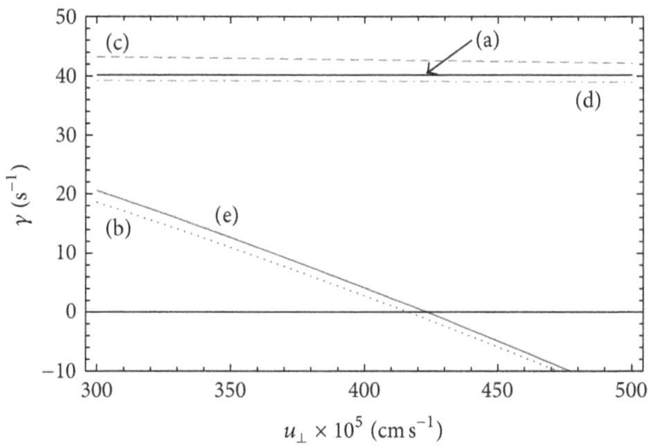

FIGURE 6: Plot of the growth rate γ (8) versus $u_{\perp s}$ ($v_{ts} = 350 \times 10^5$ cm s^{-1}, $\vec{k} = 0.004$ cm^{-1}). Other relevant parameters are the same as in Figure 5. Curve (a) is for a single ion plasma without any ring ions. Curve (b) has a H^+ ring, curve (c) has an O^+ ring, curve (d) has an O^- ring, and curve (e) has all species of ring ions.

again driven unstable by the electrons drifting parallel to the magnetic field (addition of negative ions reduces the density of electrons required to maintain charge neutrality); the ring ions generally reduce the growth with the lighter ion hydrogen ring being the most effective in damping the wave.

Figures 6 and 7 depict the variation of the growth rate γ (8) versus $u_{\perp s}$ ($v_{ts} = 350 \times 10^5$ cm s^{-1}) (Figure 6) or γ versus v_{ts} ($u_{\perp s} = 420 \times 10^5$ cm s^{-1}) (Figure 7), with the temperatures and drift speed being the same as in Figure 5. In both figures, labelling of the curves is similar to that in Figures 3 and 4. While the general nature of the curves is similar to that in Figure 5, the heavier ion rings have much larger ranges of $u_{\perp s}$ and v_{ts} over which the wave is unstable.

In a recent study Misra et al. [27] investigated the propagation characteristics of the dust ion-acoustic (DIA) waves in a collisional, negative ion plasma with immobile, charged dust grains. In the linear regime, they found two modes—the "fast" and "slow" modes. They also found that the Landau damping effect on the fast mode was negligible while the slow mode was stable.

Our dispersion diagram (Figure 1) also reveals the existence of two modes—a "slow" mode (curve (a) of Figure 1 with $u_{\perp s} > v_{ts}$) and a "fast" mode (curve (b), Figure 1 with $u_{\perp s} < v_{ts}$). Also as mentioned in Section 1, ion-acoustic waves were observed in the frequency range of 1.0–1.5 kHz at comet Giacobini-Zinner [9] and at a frequency slightly lesser than 1.0 kHz at comet Halley [22]. The waves are also expected to saturate at frequencies below the local ion plasma frequency [28]. A comparison of the growth rate for the fast (Figures 2, 3, and 4) and the slow (Figures 5, 6, and 7) modes reveals that the growth rate of the fast mode is much larger when compared to that of the slow mode. Thus, while the slow mode in [27] was freely propagating, the slow mode in the plasma under consideration is weakly unstable. It may be noted here that this study uses the full kinetic treatment and hence can also be considered as a generalisation of the fluid treatment in [27].

Finally Figure 8 depicts the variation of the real frequency versus the wave vector $\vec{k}$. The panel at the top is for the fast wave ($u_{\perp s} < v_{ts}$, $u_{\perp s} = 100 \times 10^5$ cm s^{-1}, $v_{ts} = 320 \times 10^5$ cm s^{-1}); curve (a) is for the case when there are no heavier ions ($n_{H^+} = 4.95$ cm^{-3}, $n^r_{H^+} = 0.5$ cm^{-3}) while curve (b) depicts the situation where negatively charged oxygen is also added ($n_{H^+} = 4.95$ cm^{-3}, $n^r_{H^+} = 0.5$ cm^{-3}, and $n^r_{O^-} = 0.2$ cm^{-3}). We find that the addition of O^- increases the frequency and hence the phase velocity of the ion-acoustic wave. The lower panel is for the slow wave ($u_{\perp s} > v_{ts}$, $u_{\perp s} = 420 \times 10^5$ cm s^{-1}, $v_{ts} = 350 \times 10^5$ cm s^{-1})—curve (c) is for the case where only the lighter ions are present ($n_{H^+} = 4.95$ cm^{-3}, $n^r_{H^+} = 0.5$ cm^{-3}) and curve (d) depicts the situation when O^- is also added ($n_{H^+} = 4.95$ cm^{-3}, $n^r_{H^+} = 0.5$ cm^{-3}, and $n^r_{O^-} = 0.2$ cm^{-3}). The decrease in the frequency and thereby the phase velocity is distinctly evident from the figure.

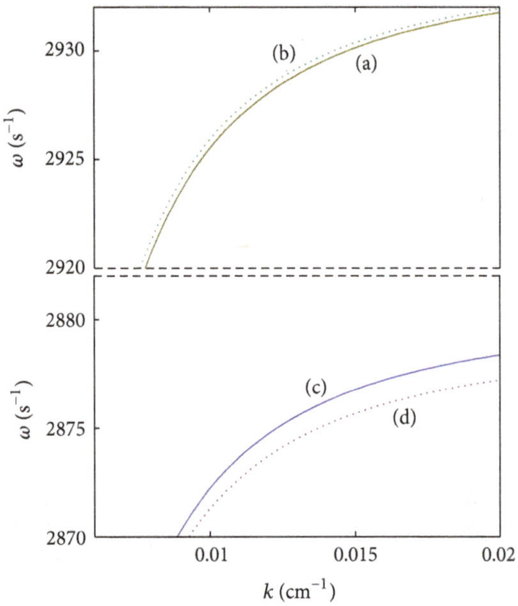

FIGURE 8: Plot of frequency ω versus wave vector $\vec{k}$. The panel at the top is for the fast wave ($u_{\perp s} < v_{ts}$; $u_{\perp s} = 100 \times 10^5\ \text{cm s}^{-1}$, $v_{ts} = 320 \times 10^5\ \text{cm s}^{-1}$) of Figure 1: curve (a) is for the case when there are no heavier ions ($n_{H^+} = 4.95\ \text{cm}^{-3}$, $n^r_{H^+} = 0.5\ \text{cm}^{-3}$) while curve (b) depicts the situation when negatively charged oxygen is also added ($n_{H^+} = 4.95\ \text{cm}^{-3}$, $n^r_{H^+} = 0.5\ \text{cm}^{-3}$, and $n^r_{O^-} = 0.2\ \text{cm}^{-3}$). The lower panel is for the slow wave ($u_{\perp s} > v_{ts}$; $u_{\perp s} = 420 \times 10^5\ \text{cm s}^{-1}$, $v_{ts} = 350 \times 10^5\ \text{cm s}^{-1}$) of Figure 1: curve (c) is for the case where only the lighter ions are present ($n_{H^+} = 4.95\ \text{cm}^{-3}$, $n^r_{H^+} = 0.5\ \text{cm}^{-3}$) and curve (d) depicts the situation when O^- is also added ($n_{H^+} = 4.95\ \text{cm}^{-3}$, $n^r_{H^+} = 0.5\ \text{cm}^{-3}$, and $n^r_{O^-} = 0.2\ \text{cm}^{-3}$). Other relevant parameters are the same as in Figure 1.

Variation of the phase velocity of ion-acoustic waves has been proposed as a diagnostic tool for the detection of charged dust grains in laboratory plasmas [28, 29]. We thus propose to extend this idea—the contrasting behaviour of the phase velocity of the ion-acoustic wave in the presence of O^- ions for $u_{\perp s} > v_{ts}$ (and vice versa) can be used not only to detect the presence of negatively charged oxygen ions but also their thermalization as well.

4. Conclusions

We have studied the stability of the ion-acoustic waves in a five-component plasma, which approximates very well the plasma environment around a comet. The five components are solar wind hydrogen and electrons and the ions of cometary origin. The cometary ions are hydrogen and positively and negatively charged oxygen ions. The solar wind components are modelled by the Maxwellian distributions while the cometary ions are described by ring distributions. We find that the ion-acoustic waves can be generated at frequencies comparable to the observed frequency. The ratio of the ring speed to its thermal spread modifies the dispersion characteristics of the ion-acoustic wave. When the thermal spread is large compared to the ring speed ($v_{ts} > u_{\perp s}$), the ring ions enhance the instability of the ion-acoustic wave. The growth rate is lowered when the thermal spread is smaller in comparison with the ring speed ($u_{\perp s} > v_{ts}$). We also find that heavier ring ions extend the range of $u_{\perp s}$ and v_{ts} over which the wave is unstable when compared to the lighter ring ions. The contrasting behaviour of the phase velocity of the ion-acoustic wave in the presence of O^- ions for $u_{\perp s} > v_{ts}$ (and vice versa) can be used to detect the presence of negatively charged oxygen ions and also their thermalization.

Acknowledgments

The authors thank the referees for their valuable comments. Financial assistance from the University Grants Commission (SAP) and Department of Science and Technology (FIST and PURSE Programs) is gratefully acknowledged.

References

[1] J. Castro, P. McQuillen, and T. C. Killian, "Ion acoustic waves in ultracold neutral plasmas," *Physical Review Letters*, vol. 105, no. 6, Article ID 065004, 2010.

[2] L. Tonks and I. Langmuir, "Oscillations in ionized gases," *Physical Review*, vol. 33, no. 2, pp. 195–210, 1929.

[3] T. H. Stix, *Waves in Plasmas*, American Institute of Physics, New York, NY, USA, 2nd edition, 1992.

[4] Y. Nakamura, H. Bailung, and P. K. Shukla, "Observation of ion-acoustic shocks in a dusty plasma," *Physical Review Letters*, vol. 83, no. 8, pp. 1602–1605, 1999.

[5] Z. Liu, L. Liu, and J. Du, "A nonextensive approach for the instability of current-driven ion-acoustic waves in space plasmas," *Physics of Plasmas*, vol. 16, no. 7, Article ID 072111, 5 pages, 2009.

[6] M. Yamada and M. Raether, "Saturation of the ion-acoustic instability in a weakly ionized plasma," *Physical Review Letters*, vol. 32, no. 3, pp. 99–102, 1974.

[7] M. E. Koepke, "Contributions of Q-machine experiments to understanding auroral particle acceleration processes," *Physics of Plasmas*, vol. 9, no. 5, pp. 2420–2427, 2002.

[8] S. R. Cranmer, A. A. Van Ballegooijen, and R. J. Edgar, "Self-consistent coronal heating and solar wind acceleration from anisotropic magnetohydrodynamic turbulence," *Astrophysical Journal, Supplement Series*, vol. 171, no. 2, pp. 520–551, 2007.

[9] F. L. Scarf, F. V. Coroniti, C. F. Kennel, D. A. Gurnett, W.-H. Ip, and E. J. Smith, "Plasma wave observations at comet Giacobini-Zinner," *Science*, vol. 232, no. 4748, pp. 377–381, 1986.

[10] A. L. Brinca and B. T. Tsurutani, "Unusual characteristics of electromagnetic waves excited by cometary newborn ions with large perpendicular energies," *Astronomy & Astrophysics* , vol. 187, no. 1-2, pp. 311–319, 1987.

[11] P. Chaizy, H. Rème, J. A. Sauvaud et al., "Negative ions in the coma of comet Halley," *Nature*, vol. 349, no. 6308, pp. 393–396, 1991.

[12] T. J. Tataronis and F. Crawford, "Cyclotron harmonic wave propagation and instabilities," *Journal of Plasma Physics*, vol. 4, no. 2, pp. 231–264, 1970.

[13] M. Ashour-Abdalla and C. F. Kennel, "Nonconvective and convective electron cyclotron harmonic instabilities," *Journal of Geophysical Research*, vol. 83, p. 1531, 1978.

[14] P. Sprangle, J. L. Vomvoridis, and W. M. Manheimer, "A classical electron cyclotron quasioptical maser," *Applied Physics Letters*, vol. 38, no. 5, pp. 310–313, 1981.

[15] K. Akimoto, K. Papadopoulos, and D. Winske, "Ion-acoustic instabilities driven by an ion velocity ring," *Journal of Plasma Physics*, vol. 34, no. 3, pp. 467–479, 1985.

[16] J. A. Byers and M. Grewal, "Perpendicularly propagating plasma cyclotron instabilities simulated with a one-dimensional computer model," *Physics of Fluids*, vol. 13, no. 7, pp. 1819–1830, 1970.

[17] J. K. Lee and C. K. Birdsall, "Velocity space ring-plasma instability, magnetized, part I: theory," *Physics of Fluids*, vol. 22, no. 7, pp. 1306–1314, 1979.

[18] S. Seiler, M. Yamada, and H. Ikezi, "Lower hybrid instability driven by a spiraling ion beam," *Physical Review Letters*, vol. 37, no. 11, pp. 700–703, 1976.

[19] H. E. Mynick, M. J. Gerver, and C. K. Birdsall, "Stability regions and growth rates for a two-ion component plasma, unmagnetized," *Physics of Fluids*, vol. 20, no. 4, pp. 606–612, 1977.

[20] C. Cattel and M. Hudson, "Flute mode waves near ω_{LH} excited by ion rings in velocity space," *Geophysical Research Letters*, vol. 9, no. 10, pp. 1167–1170, 1982.

[21] K. Akimoto, K. Papadopoulos, and D. Winske, "Lower-hybrid instabilities driven by an ion velocity ring," *Journal of Plasma Physics*, vol. 34, no. 3, pp. 445–465, 1985.

[22] F. Scarf, "Plasma wave observations at comets Giacobini-Zinner and Halley," in *Plasma Waves and Instabilities at Comets and in Magnetospheres*, B. T. Tsurutani and H. Oya, Eds., American Geophysical Union, Washington, DC, USA, 1989.

[23] R. C. Davidson, "Kinetic waves and instabilities in a uniform plasma," in *Basic Plasma Physics*, A. Galeev and R. Sudan, Eds., North-Holland, New York, NY, USA, 1989.

[24] B. D. Fried and S. D. Conte, *The Plasma Dispersion Function*, Academic Press, New York, NY, USA, 1961.

[25] D. A. Gurnnet and A. Bhattacharjee, *Introduction to Plasma Physics: with Space and Laboratory Applications*, Cambridge University Press, Cambridge, UK, 1st edition, 2005.

[26] J. D. Gaffey Jr, D. Winske, and C. S. Wu, "Time scales for formation and spreading of velocity shells of pickup ions in the solar wind," *Journal of Geophysical Research*, vol. 93, no. A6, pp. 5470–5486, 1988.

[27] A. P. Misra, N. C. Adhikary, and P. K. Shukla, "Ion-acoustic solitary waves and shocks in a collisional dusty negative-ion plasma," *Physical Review E*, vol. 86, no. 5, Article ID 056406, 10 pages, 2012.

[28] S. H. Kim and R. L. Merlino, "Charging of dust grains in a plasma with negative ions," *Physics of Plasmas*, vol. 13, no. 5, Article ID 052118, 7 pages, 2006.

[29] M. Rosenberg and R. L. Merlino, "Ion-acoustic instability in a dusty negative ion plasma," *Planetary and Space Science*, vol. 55, no. 10, pp. 1464–1469, 2007.

14

Numerical Experiments for Nuclear Flashes toward Superbursts in an Accreting Neutron Star

Masa-aki Hashimoto,[1] Reiko Kuromizu,[1] Masaomi Ono,[1] Tsuneo Noda,[2] and Masayuki Y. Fujimoto[3]

[1] *Department of Physics, Kyushu University, Fukuoka 810-8560, Japan*
[2] *Kurume Institute of Technology, Fukuoka 830-0052, Japan*
[3] *Department of Physics, Hokkaido University, Sapporo 060-8810, Japan*

Correspondence should be addressed to Masa-aki Hashimoto; hashimoto@phys.kyushu-u.ac.jp

Academic Editor: Luciano Nicastro

We show that the superburst would be originated from thermonuclear burning ignited by accumulated fuels in the deep layers compared to normal X-ray bursts. Two cases are investigated for models related to superbursts by following thermal evolution of a realistic neutron star: helium flash and carbon flash accompanied with many normal bursts. For a helium flash, the burst shows the long duration when the accretion rate is low compared with the observation. The flash could become a superburst if the burning develops to the deflagration and/or detonation. For a carbon flash accompanied with many normal bursts, after successive 2786 normal bursts during 1.81×10^9 s, the temperature reaches the deflagration temperature. This is due to the produced carbon which amount reaches to ≈ 0.1 in the mass fraction. The flash will develop to dynamical phenomena of the deflagration and/or detonation, which may lead to a superburst.

1. Introduction

Type I X-ray bursts have been identified to the thermonuclear explosions on the surface region of accreting neutron (compact) stars. As a consequence, the phenomenon has been studied from both nuclear reactions and nuclear structure inside the compact stars. However, there still remain many uncertainties concerning the elementary processes associated with the bursts [1–3]. Superbursts have been detected from 13 X-ray bursters by *BeppoSAX* and *RXTE* (see, e.g., Table 2 in [4]). In particular, 4U 1636-536 exhibited four superbursts, where the shortest recurrence time is 1.5 years (http://www.astronomerstelegram.org/?read=2140) [5, 6]. Clearly, the light curve consists of a fast rise and slower power law-like decay [2, 7]. The spectrum hardens during the rise phase to the maximum in luminosity, whereas it softens in the decay phase. This is also reflected in the spectral fits to the time-resolved preburst subtracted from X-ray spectra. Each burst has energy of 10^{42} ergs and duration of a few hours. They are usually best described in terms of a black-body model. The effective temperature increases and decreases during the rise and decay phase, respectively. These superbursts are 1000 times luminous and 1000 times long in the duration compared with the normal bursts though the spectral evolution is similar.

Even now, quantitative explanation and/or numerical simulation of superbursts using the stellar evolution code are limited. For example, Keek and Heger [8] do not self-consistently produce the carbon from hydrogen/helium burning in their calculation but instead accrete the carbon directly onto the neutron star, bypassing the hydrogen/helium burning stages. The superbursts last too long and their energy release is too much to be explained in terms of unstable burning of hydrogen/helium so far considered [9]. Moreover, regular normal X-ray bursts are observed before the occurrence of the superburst [10] that includes the precursor burst [11]. The long rise and decay times of superbursts are consistent with the model of unstable burning in the deep layer below the hydrogen/helium burning region. Therefore, it has been suggested that unstable burning of carbon is the origin of the superbursts [9, 12].

If the accreted material onto the neutron star is pure helium, carbon can be produced when helium is burned stably [9]. Unstable helium burning often involves alpha captures on carbon and carbon does not remain much. This would apply to the helium accretor 4U 1820-30 that shows long periods of high X-ray intensity during which no burst occurs, which is consistent with a period of stable helium burning. Note that unstable carbon burning only reproduces the observed feature in superbursts when we take into account neutrino losses and significant heat flux transported from deeper into the accreting layer of the neutron star [9]. Cumming et al. [2] also show that the observed superburst energy is around 10^{42} ergs and more or less independent of ignition depth because of neutrino emission of the excess energy. While Cumming [13] expected recurrence times of the order of 1-2 years, Strohmayer and Brown [9] obtained a recurrence time of about 10 years.

If the accreted material onto the neutron star is a mixture of hydrogen and helium, either unstable or stable burning of hydrogen/helium can produce carbon. While the amount obtained by numerical calculations has been only limited [14, 15] after bursts, carbon is much more readily produced in stable burning [1]. In the observed sources of superbursts, normal XRBs have been observed with a mean rate of about 3 times per day during the period of the observation [16, 17]. This indicates that at least some amounts of the accreted material have been burned stably before a superburst. Furthermore, the superburst from 4U 1254-69 indicates that much of the accreted fuel burns stably [18]. Detection of superbursts at near Eddington accretion rate would reveal the relation between the recurrence time and remained nuclear fuels, which is studied by using the α parameter [19].

Cumming and Bildsten [12] suggested that a small amount of carbon ($X(^{12}\mathrm{C}) \approx 0.05$–$0.1$) could be enough to trigger a thermonuclear runaway with energy comparable to the superburst if carbon resides in a bath of heavy elements. These heavy elements are the products of unstable burning through the rp-process during the mixed hydrogen/helium burning of XRBs [14, 20]. In this case, the superburst recurrence time would depend on accretion rates, being in the order of a few decades, a year to a decade, or a week to a month, according to the accretion rate which is about 0.1, 0.3, or 1 times the Eddington accretion rate. On the other hand, it is proposed from the analysis of the photospheric radius expansion that an accretion rate may change during the burst by some factors [21]. Changes in accretion rates are also considered from recent study of the outburst of a transient X-ray binary [22]. Therefore, it is reasonable to include accretion rate variations in numerical calculations because the observations indicate that the luminosity of accreting neutron stars is variable. It has been suggested that the high temperature reached during a superburst induced the photodisintegration [23]. As a consequence, they got energy generations comparable to those due to the carbon induced superburst.

Another scenario was proposed by Kuulkers et al. [17]. They suggested that hydrogen left after the burning in the hydrogen/helium layer is reignited by the electron capture that is followed by successive captures of neutrons by heavy nuclei occurred in the deeper layer. The recurrence time of superbursts is estimated to be less than one year [10, 17]. However, large amounts of hydrogen should be remained after bursts to explain the energy release in superbursts. Recent calculations have revealed that hydrogen is completely depleted after the hydrogen/helium burning [14, 15]. On the other hand, an exotic process of the diquark pair formation was proposed to explain superbursts [24].

On the other hand, significant progress has been done so far concerning the construction of model to study the phenomena of neutron star; two dimensional hydrodynamical model calculations of X-ray bursts [25, 26], study of the propagation of deflagration wave of a rapidly rotating neutron star [27], magnetorotational study [28, 29], and rotating hot-spot model examination of rotating neutron star [30] concerning burst ocillations. Unfortunately, it would be insufficient to study X-ray bursts in details beyond spherically symmetric model. Since there is no model in the literature that self-consistently calculates the production of carbon in hydrogen/helium and its subsequent ignition as superburst, we perform evolutionary calculations adopting spherically symmetric models of accreting neutron stars with important physical processes included [31].

In Section 2 physical inputs and our evolutionary code are explained. A helium flush model is presented in Section 3 related to superbursts. In Section 4, we present a model of carbon flash accompanied with many normal bursts of H/He combined burnings and show the possibility of a superburst. Concluding remarks are given in Section 5.

2. Evolution Code of a Neutron Star

The general relativistic evolutionary equations of spherical stars in hydrostatic equilibrium as formulated by Throne [32] are written as

$$\begin{aligned}
\frac{\partial M_{tr}}{\partial r} &= 4\pi r^2 \rho_t, \\
\frac{\partial P}{\partial r} &= -\frac{G M_{tr} \rho_t}{r^2}\left(1+\frac{P}{\rho_t c^2}\right)\left(1+\frac{4\pi r^3 P}{M_{tr} c^2}\right)\mathcal{V}^2, \\
\frac{\partial\left(L_r e^{2\phi/c^2}\right)}{\partial M_r} &= e^{2\phi/c^2}\left(\varepsilon_{\mathrm{n}} - \varepsilon_{\nu} - e^{-\phi/c^2} T \frac{\partial s}{\partial t_\infty}\right), \\
\frac{\partial \ln T}{\partial \ln P} &= \min\left(\nabla_{\mathrm{rad}}, \nabla_{\mathrm{ad}}\right), \\
e^{-\phi/c^2}\frac{\partial Y_i}{\partial t_\infty} &= \alpha_i, \\
\frac{\partial M_{tr}}{\partial M_r} &= \frac{\rho_t}{\rho}\mathcal{V}^{-1}, \\
\frac{\partial \phi}{\partial M_{tr}} &= \frac{G\left(M_{tr} + 4\pi r^3 P/c^2\right)}{4\pi r^4 \rho_t}\mathcal{V}^2,
\end{aligned} \tag{1}$$

where

$$\mathscr{V} \equiv \left(1 - \frac{2GM_{tr}}{c^2 r}\right)^{-1/2}. \quad (2)$$

The basic quantities are defined as follows: ρ: rest mass density, T: temperature, Y_i: abundance of the ith particle, M_r: proper mass inside the radius r, t_∞: Schwarzschild time coordinate (proper time at a distant observer), M_{tr}: total mass inside the radius r, ϕ: gravitational potential, L_r: local luminosity, P: pressure, ρ_t: total nongravitational mass-energy density in mass units, s: specific entropy, $\varepsilon_{\rm n}$: heating rate by nuclear burning, ε_ν: cooling rate by escaping neutrinos, α_i: nuclear reaction rate for the ith particle, and $\nabla_{\rm rad}$ ($\nabla_{\rm ad}$): the radiative (adiabatic) temperature gradient. In the accretion layer, the Eulerian coordinate (the mass fraction coordinate with changing mass $q[= M_r/M(t)]$) is used, which is the most suitable method for computations of stellar structure when stellar mass (M) varies [33]. In our calculation, the mixing length theory of convection with the mixing length equal to the pressure scale height has been used [34]. The radiative zero boundary conditions are imposed at the outer boundary. An outermost mesh point, which is regarded as the photosphere, is given at $q = 1 - 4.1 \times 10^{-20}$ [35].

The above set of general relativistic equations for the evolution of spherically symmetric stars has been solved by using a Henyey-type numerical scheme of implicit method. We adopt the evolution code of a spherically symmetric neutron star [35, 36]. The star is divided into 266 meshes of the Lagrange mass-coordinate. The gravitational mass and radius of the neutron star are initially set to be $1.3\,M_\odot$ and 8.1 km, respectively. Then we have $\log g_s = 14.56$ for the gravitational acceleration at the surface of the neutron star g_s. The depth of the accretion layer is ~10 m and consists of 170 meshes. In the region of the combined hydrogen/helium burning, the interval between meshes is typically 5-6 cm. Physical inputs and an approximation network are almost the same as adopted by Fujimoto et al. [31] except for the changes of reaction rates [37] and alpha-network used in Section 3. It should be noted that our approximation network has been constructed based on the one-zone calculation with use of a large network for the bursts [38, 39]. Our network includes 16 nuclei: ^{1}H, ^{4}He, ^{12}C, ^{14}O, ^{15}O, ^{16}O, ^{17}F, ^{22}Mg, ^{30}S, ^{56}Ni, ^{60}Ni, ^{60}Zn, ^{64}Zn, ^{64}Ge, ^{68}Ge, and ^{68}Se. This network can be used till the hydrogen decreases until around 10% in mass fractions [39].

Although input physics has been described in detail [35], we briefly write it again. This is because input physics is crucial to determine the neutron star properties and it is closely related to the study of elementary processes of high density matter. As for the equation of state of outer part of the neutron star ($\rho < 5 \times 10^7\,{\rm g\,cm^{-3}}$), an ideal gas plus radiation is assumed with the electron degeneracy and the Coulomb liquid correction included [40]. For the inner part, the equation of state has been constructed by Richardson et al. [41] based on Canuto [42, 43]. Neutrino emissivities include bremsstrahlung of nucleons [44] and electron-ion [45], and electron-positron pair, photo, and plasmon processes [46].

Opacities for iron- and neutron-rich material include the radiative opacities for ^{56}Fe by Malone [47] and the thermal conductivities [48–50]. For the opacities of lighter elements the analytical approximations by Iben [51] are adopted to the radiative ones [52, 53], to the thermal conductivity for a nonrelativistic electron gas of [54], and to that for a relativistic one [55], respectively. Screening effects of nuclear reactions are taken from Dewitt et al. [56]. In the present paper, we include the crustal heating [57],

$$Q_i = 6.03 \times \dot{M}_{-10} q_i 10^{33}\ {\rm ergs\,g^{-1}}, \quad (3)$$

where i is the number of the reaction, $\dot{M}_{-10}$ is the mass accretion rate in units of $10^{-10}\,M_\odot\,{\rm yr}^{-1}$, and q_i is the number of the effective heat energy per nucleon in MeV for the ith reaction which is tabulated in their paper.

We assume that material is accreted with the same entropy at the stellar surface, neglecting the surface effects caused by the accretion flow. Within the framework of the spherical symmetry, the kinetic energy of the falling material has little influence on the structure in the layer as deep as the burning shell, since the radial motion will be dissipated in the surface layers [35].

Initial models have been constructed through the continuous accretion ($\dot{M}$ = constant) without nuclear burning until the steady state is achieved, where the nonhomologous part of the gravitational energy release vanishes [35].

3. Simple Features of Models Concerning Superbursts—Helium Flash Model

Superbursts observed in 4U 1820-30 are considered to be accreted by pure helium, while 4U 1636-536, 4U 1735-44, and 4U 1254-690 are accreted by hydrogen/helium [58]. The difference of the accretion matter may affect the mechanism of superbursts. Therefore, we first examine rather simplified models of superbursts, which are triggered by a helium flash. We note that deep helium ignition towards superbursts was discussed by Kuulkers et al. [59] on the X-ray binary 4U 0614+091. For the carbon flash, we also investigate realistic situation accompanied with many normal bursts in Section 4.

The superburst of 4U 1820-30 was observed in the duration of 2.5 hr, where the burst energy is 1.4×10^{42} ergs [9], and the peak luminosity is $L_{\rm peak} \simeq 3.4 \times 10^{38}\,{\rm ergs^{-1}}$ [17]. To follow a pure helium flash, we use an alpha-network consisted of alpha-nuclei up to ^{56}Ni, where the nuclear reaction rates are taken from Angulo et al. [37]. This network includes not only (α,γ) reactions but also (α,p) reactions; the latter in competition with the former is assumed to follow (p,γ) reactions instantaneously. From Figure 1, we can see that although $L_{\rm peak}$ is consistent with the observation for a high accretion rate, $\dot{M} = 3 \times 10^{-8}$, and intermediate rate, $3 \times 10^{-9}\,M_\odot\,{\rm yr}^{-1}$, the duration of the burst is too short to explain the burst energy. It should be noted that $L_{\rm peak}$ attains the Eddington limit as described in Hanawa and Fujimoto [36]. On the other hand, for a low accretion rate, $\dot{M} = 3 \times 10^{-10}\,M_\odot\,{\rm yr}^{-1}$, we have obtained the burst energy around 10^{42} ergs which lasts more than 3 hr (see Figure 1). It is noted

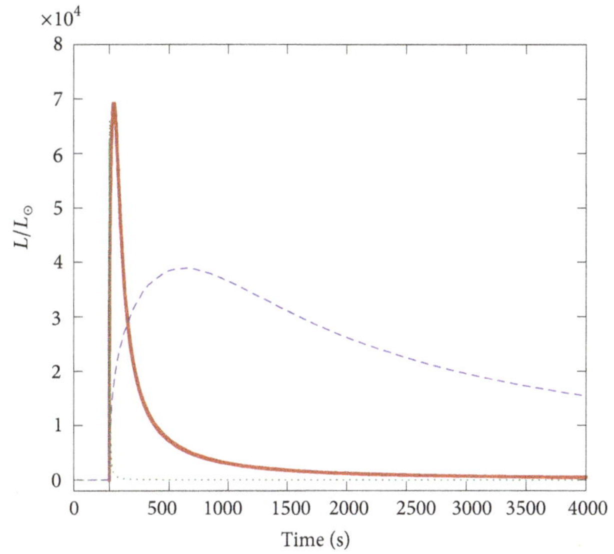

FIGURE 1: Light curves for three representative accretion rates, $\dot{M} = 3 \times 10^{-8}$ (dotted line), 3×10^{-9} (solid line), and $3 \times 10^{-10}\, M_\odot\,\mathrm{yr}^{-1}$ (dashed line). Time is set to be zero at the beginning of each burst.

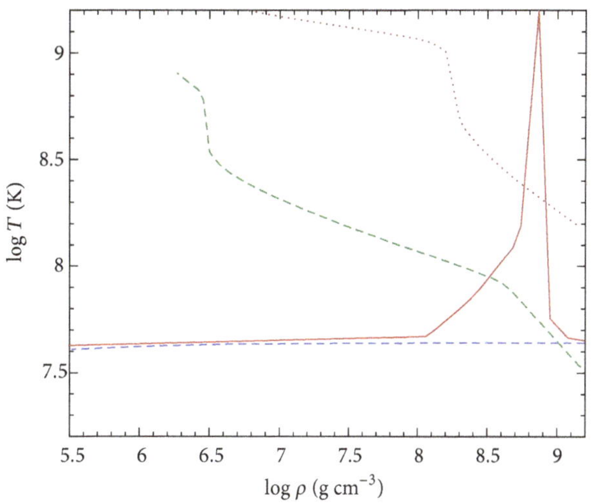

FIGURE 2: Temperature distributions for $\dot{M} = 3 \times 10^{-10}\, M_\odot\,\mathrm{yr}^{-1}$ in the initial state (lower dotted line) and to the stage of the maximum L_n (solid line). The dashed line is the ignition curve ($Y = 0.1$) and the upper dotted line indicates that $\tau_\mathrm{dyn} = \tau_{3\alpha}$.

that the ignition pressures in units of dyn cm^{-2} for the above accretion rates are $10^{22.7}$, $10^{24.1}$, and $10^{26.5}$, respectively. Since $\dot{M}$ of the superburst in 4U 1820-30 is estimated to be several times $10^{-9}\, M_\odot\,\mathrm{yr}^{-1}$ [13], we recognize that simple models based on the single burst with an accretion rate assumed for the accreting neutron star are inconsistent with the observation (see Section 4). However, since the ultracompact source 4U 0614+91 likely accretes helium at $10^{-10}\, M_\odot\,\mathrm{yr}^{-1}$ with a superburst [59], we could carefully study the helium flash.

Figure 2 shows the temperature distribution for $\dot{M} = 3 \times 10^{-10}\, M_\odot\,\mathrm{yr}^{-1}$ against the density of the initial state (dotted line) and that of the stage at the maximum nuclear luminosity $L_\mathrm{n,max}$ (solid line), where $L_\mathrm{n} = \int \varepsilon_\mathrm{n} dM_r$ denotes the nuclear luminosity. The dashed line indicates the ignition curve of the 3α reaction $\varepsilon_{3\alpha} = \varepsilon_\mathrm{rad}$ for the helium mass fraction $Y = 0.1$, with the nuclear energy generation rate of 3α reaction $\varepsilon_{3\alpha}$ and the radiative energy loss rate ε_rad [14]. Although this criterion should be carefully reconsidered [60], we adopt the present one for simplicity. The dotted line is the deflagration temperature defined by equating the dynamical time scale and the nuclear heating time scale ($\tau_\mathrm{dyn} = \tau_\mathrm{n}$) with

$$\tau_\mathrm{dyn} = \frac{H_p}{c_s}, \qquad \tau_\mathrm{n} = \frac{C_p T}{\varepsilon_\mathrm{n}}, \tag{4}$$

where H_p ($\equiv -dr/d\ln P$) is the pressure scale height, c_s is the sonic velocity, and C_p is the specific heat under the constant pressure. It is remarkable that the temperature in the layers of $\log\rho = 8.8$ exceeds $\log T = 8.5$ for $L_\mathrm{n,max}$: the flash may become the deflagration. It needs to perform a dynamical calculation to elucidate how the deflagration develops inside the accretion layers. Although we cannot represent the proper L_peak consistent with the observation of $\dot{M}$, the helium flash in low accretion rate could become a possible site of the superburst.

4. Model Accompanied with Normal Bursts of Combined Hydrogen and Helium Burnings

In the previous section, we have shown that simple helium burst models cannot explain the observed superbursts. New model is needed to produce enough carbon for a superburst to occur. Therefore, we present a sequence of calculations until the amount of carbon increases enough. We first adopt an accretion rate of $5 \times 10^{-9}\, M_\odot\,\mathrm{yr}^{-1}$. And to acquire more carbon, we only change it to $1 \times 10^{-9}\, M_\odot\,\mathrm{yr}^{-1}$, which save the computational time. To increase the temperature in the distribution, we change the accretion rate to the first one. Furthermore, we also raise the crustal heating by a factor of ten to save the computational time.

4.1. Gross Features toward the Carbon Flash. Let us make an initial model to simulate a superburst with the accretion rate of $5 \times 10^{-9}\, M_\odot\,\mathrm{yr}^{-1}$. This accretion rate is considered to be in the reasonable range of $0.1 \le \dot{M}/\dot{M}_\mathrm{Edd} \le 0.3$ with $\dot{M}_\mathrm{Edd} = 1.7 \times 10^{-8}\, M_\odot\,\mathrm{yr}^{-1}$ for the observed sources [61]. Note that superbursts at near-Eddington rate are suggested for GX 17+2 [19]. With the nuclear burning suppressed, we construct an initial temperature distribution by continuous accretion with this specified $\dot{M}$. Mass fractions in the accretion matter are assumed to be H (73.0%), ^{4}He (25.0%), ^{14}O (0.7%), and ^{15}O (1.3%). The bottom of the accretion layer is set to be pure ^{56}Ni, which is equivalent to ^{56}Fe in the present purpose. The steady state concerning the accretion rate is assumed; that is, $\dot{M} = \dot{M}_{^{56}\mathrm{Ni}}$; this means that accreted matter increases the mass of the layer of pure ^{56}Ni; that is, accreting matter becomes promptly ^{56}Ni and omits nuclear burning processes.

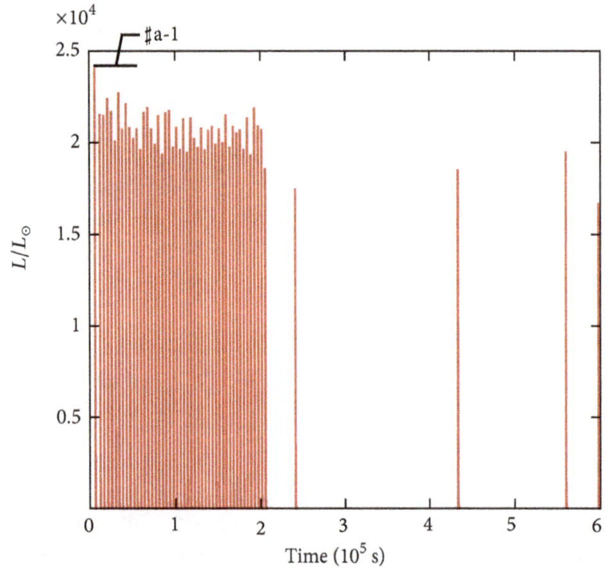

FIGURE 3: Light curves from the beginning of accretion to $t = 6 \times 10^5$ s. The left upper mark of ♯a-1 shows the light curve in Figure 4.

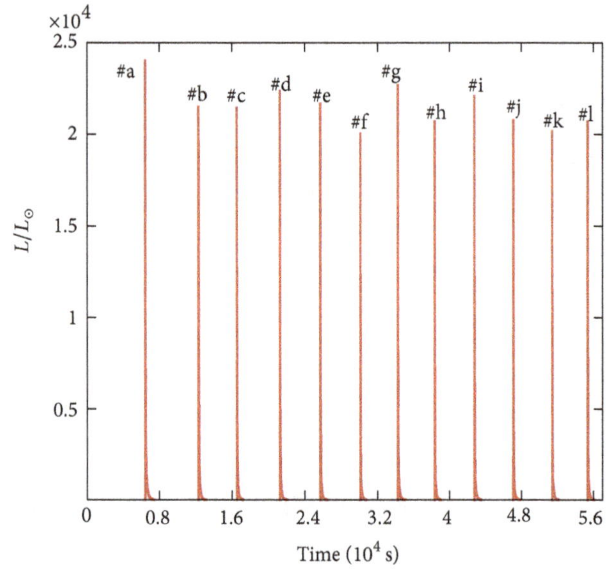

FIGURE 4: Light curves from the beginning to $t = 5.7 \times 10^4$ s. Marks of ♯a-l indicate each burst from the beginning of accretion (t = 0).

At the end, we obtain the isothermal temperature distribution of $\log T \sim 8.42$.

We repeatedly calculate normal bursts (combined hydrogen/helium burning) with use of the evolutionary code which includes the approximation reaction network. Though hydrogen consumption after a flash with use of this network might have been underestimated by ~10% for $\log P \sim 22.8$ [14], remaining hydrogen is less than ~1% in the bottom layer of normal bursts due to the convection. Therefore, in the deep region related to a superburst ($\log P > 23$), where rp-process does not work anymore, our network can also be used except for detailed abundance distribution. Figure 3 shows the bursts from the beginning of accretion to 6×10^5 s with the accretion rate $5 \times 10^{-9}\ M_{\odot}\,\mathrm{yr}^{-1}$. In Figure 4, we show the twelve bursts till 5.7×10^4 s which should be compared with those of Woosley et al. [15] having the solar initial composition and $1.75 \times 10^{-9}\ M_{\odot}\,\mathrm{yr}^{-1}$. In view of the fact that accretion rates are different from each other, both cases produce the regular bursts. In our case, these regular bursts continue 2.1×10^5 s and then the recurrence time becomes longer. Since there exists significant amount of produced materials inside the deep region, the heat transported to the inner part of the neutron star results in the lengthened recurrence intervals of bursts.

The totally calculated evolutionary time is 1.81×10^9 s and the total number of normal bursts amounts to 2786. The time sequence of the bursts is illustrated in Figure 5. The time interval Δt and the number of bursts are given in Table 1 for the individual period specified in terms of $\dot{M}$. Figures 6–8 show the light curve during the intervals (3–11), (11–15), and (17.6–18.1) $\times 10^8$ s. Critical bursts are marked by ♯1–7, whose profiles are discussed in Section 4.2. Important epochs leading to a superburst are specified as follows. The epoch $*\alpha$ is 930 s after the burst at 2×10^8 s during "period 1," and

TABLE 1: Time catalogue corresponding to the accretion rates for three periods. Δt is the time interval between the periods.

Period	1	2	3
$\dot{M}$ ($M_{\odot}\,\mathrm{yr}^{-1}$)	5×10^{-9}	1×10^{-9}	5×10^{-9}
Time (10^8 s)	0–12	12–16.8	16.8–18.1
Δt (10^8 s)	12	4.8	1.3
Number of bursts	1237	1416	133
Specific stages of bursts	$*\alpha$, ♯1–6	$*\beta$	♯7, $*\gamma$, $*\delta$

the epoch $*\beta$ is 854 s after the burst at 1.5×10^9 s during "period 2." The epoch $*\gamma$ corresponds to the end of the last burst ♯7 and $*\delta$ is just before the ^{12}C+^{12}C ignition which is 5500 s after the epoch $*\gamma$.

Figure 9 shows temperature profiles against density and pressure, respectively, at the epochs $*\alpha - *\delta$. The ignition curve and deflagration line of the ^{12}C+^{12}C reaction are also shown in Figure 9. We note that in the context of detonations in superbursts, comparing $\tau_{^{12}\mathrm{C}+^{12}\mathrm{C}}$ to the dynamical time is discussed by Weinberg and Bildsten [62]. The two convexities near $\log \rho$ = 6-7 are due to the effect of the unstable hydrogen/helium burning. The energy generation rate is shown in Figure 10. Hot-CNO cycle, $^{14}\mathrm{O}(e^+, \nu)\ ^{14}\mathrm{N}(p, \gamma)\ ^{15}\mathrm{O}(e^+, \nu)\ ^{15}\mathrm{N}(p, \alpha)\ ^{12}\mathrm{C}(p, \gamma)\ ^{13}\mathrm{N}(p, \gamma)^{14}\mathrm{O}$, has produced the energy of 10^{14} ergs^{-1} for $\log P \leq 22$ [38]. Around $\log P$ = 27-28, the energy generation rates of $*\alpha$, $*\beta$, and $*\gamma$ before the ^{12}C+^{12}C ignition $*\delta$ become small because there remains small amount of fuel as the result of both the rp-process and steady burning of the ^{12}C+^{12}C reaction. It should be noted that the curve of ε_{n} for $*\delta$ falls down steeply around $\log P$ = 26–26.5 because of the significant decrease in abundances due to the sudden development of the convection as illustrated in Figure 14.

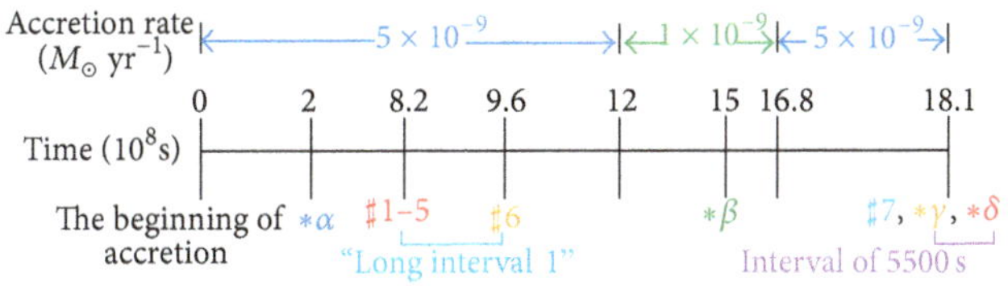

FIGURE 5: Time sequences of the models accompanied with normal bursts. When we start the accretion on the neutron star, time is set to be zero.

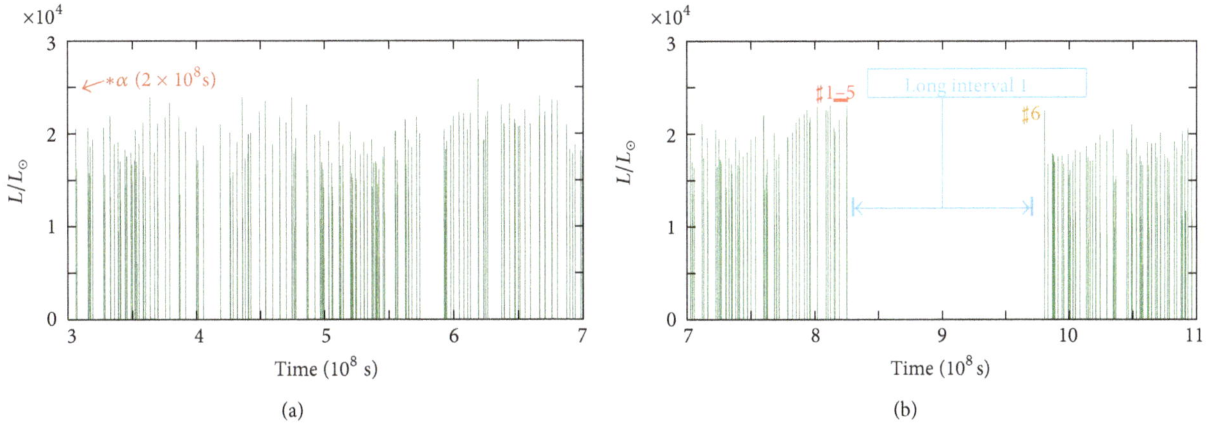

FIGURE 6: Light curves at (3–7) × 10^8 s (a) and that from (7–11) × 10^8 s (b). *α corresponds to the same epoch as shown by the mark in Figure 9. The interval between (8.25–9.75) × 10^8 s is named the "log interval 1." Marks of ♯1–5 and ♯6 are successive numbers and a sign of each burst before and after "long interval 1," respectively.

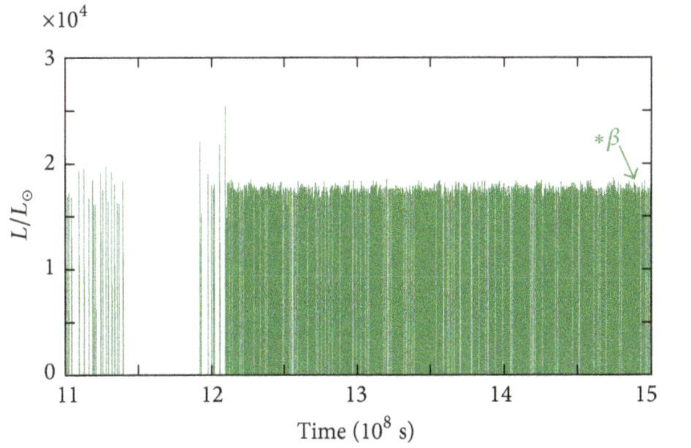

FIGURE 7: Light curves at (11–15) × 10^8 s. *β indicates the same epoch as the mark in Figure 9.

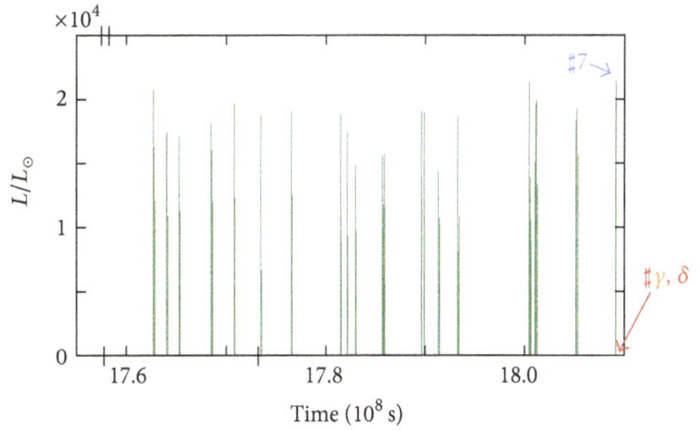

FIGURE 8: Light curves at (17.6–18.1) × 10^8 s. ♯7 indicates the last burst before the ignition of the carbon flash (*γ and *δ).

For "period 1," the accretion rate is set to be 5 × 10^{-9} $M_\odot$ yr^{-1} from the beginning of the accretion to 1.2 × 10^9 s. The number of bursts occurred in this interval is 1237. The temperature distribution inside the deep accreted layers ($\log\rho \geq 8$ and $\log P \geq 25$) remains isothermal of $\log T \sim 8.44$ (see the temperature distribution at 2 × 10^8 s in Figure 9). We can recognize that unstable combined hydrogen/helium burning has been generated for $\log\rho$ = 6–6.5 and $\log P$ = 22.5–23.5. Figure 11 shows the composition distribution at 2 × 10^8 s (*α). The range of the pressure is equal to that in Figure 9. Around $\log P$ = 22-23, the rp-process produces both ^{68}Ge and ^{64}Zn. Though there is no remained fuel of hydrogen [14], remained helium produces ^{12}C for $\log P > 24$ (see the solid line in Figure 11) due to the steady burning. The increase in ^{56}Ni is ascribed to the numerical diffusion of the initial distribution and convective mixing at the beginning of the accretion for $\log P > 26$; thus, we can consider that the mass fraction of ^{56}Ni in $26 \leq \log P \leq 27.5$ should be added to ^{68}Ge. For "period 1," the carbon burning is stable ($\log P \sim 28$, $\log T \sim 8.4$ and $X(^{12}C) \leq 0.01$) in the sense that the increase in T does not reach the ignition curve. In the bottom of the accretion layer ($\log P \sim 28$), carbon

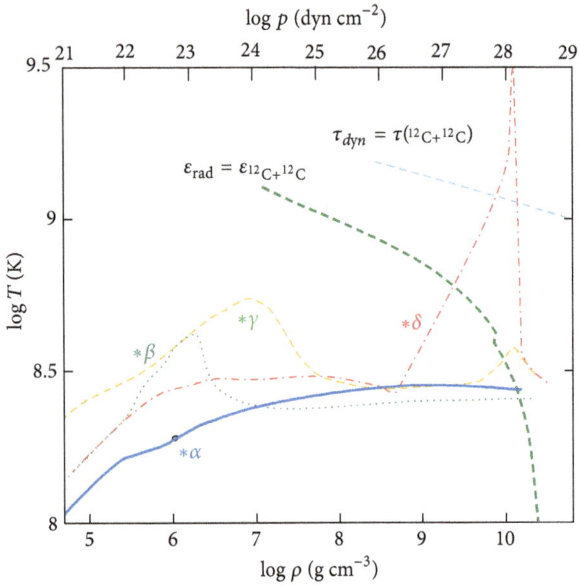

FIGURE 9: Temperature distribution against the density (bottom) and the pressure (top) at 2×10^8 s (solid line; $*\alpha$), 1.5×10^9 s (dotted line; $*\beta$), 1.81×10^9 s (dashed line; $*\gamma$), and 5500 s from the $*\gamma$ (dot-dashed line; δ). At the stage $*\delta$, the $^{12}C+^{12}C$ reaction is ignited. Two lines on the right are ignition curves of $^{12}C+^{12}C$ reaction (dashed line, $\varepsilon_{rad} = \varepsilon_{^{12}C+^{12}C}$), and deflagration line (dashed line, $\tau_{dyn} = \tau_{^{12}C+^{12}C}$), with $X(^{12}C) = 0.1$, $X(^{56}Ni) = 0.9$.

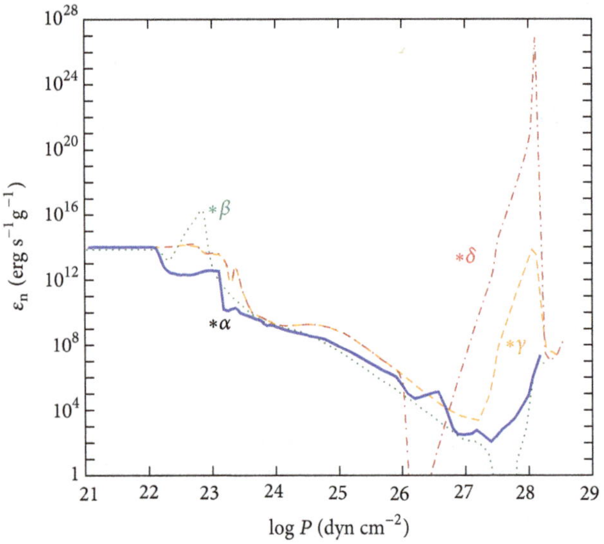

FIGURE 10: Nuclear energy generation rate versus pressure for four cases as shown in Figure 9.

sometimes burns intensively and the temperature of the layer has increased to $\log T \geq 8.6$. The amount of carbon is too small ($X(^{12}C) < 0.01$) to trigger the nuclear flash. The nuclear energy generation for $*\alpha$ is supplied by helium burning for $\log P = 23$–25, while it is supplied by reactions such as $^{12}C(\alpha,\gamma)^{16}O$ and $^{12}C+^{12}C$ for $\log P = 25$–27.5 (Figure 10). The crustal heating help ε_n to increase again for $\log P > 27.5$.

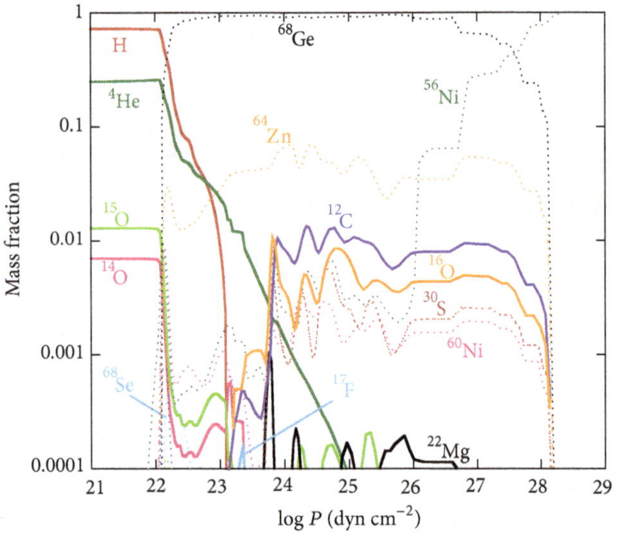

FIGURE 11: Composition distribution at 2×10^8 s ($*\alpha$). For $\log P = 22$-23, the rp-process occurred and ^{68}Ge and ^{64}Zn are produced. For the relation between lines and composition, see Table 2.

TABLE 2: Correspondence between lines and compositions.

Solid line	H	^4He	^{12}C	^{14}O	^{15}O	^{16}O	^{17}F	^{22}Mg
Dotted line	^{30}S	^{56}Ni	^{60}Zn	^{60}Ni	^{64}Ge	^{64}Zn	^{68}Se	^{68}Ge

After $t = 1.2 \times 10^9$ s, the accretion rate is artificially changed to $1 \times 10^{-9} M_\odot \, yr^{-1}$. This results in the increase in ^{12}C abundance. We keep this accretion rate in the interval of 4.8×10^8 s, and the 1416 bursts occur during this interval (period 2). The temperature distribution at 1.5×10^9 s for this accretion rate is shown by the dotted line in Figure 9 (denoted by $*\beta$). Figure 12 shows the composition distribution at this stage ($*\beta$). For "period 2," hydrogen is consumed completely in the region of $\log P > 23$ and the remained helium increases ^{12}C appreciably. The rp-process first produces ^{68}GE and ^{64}Zn and afterwards also ^{60}Ni and ^{56}Ni. The appreciable decrease in ε_n corresponds to the dip of the composition distribution around $\log P \sim 27.4$ in Figure 12.

We have changed again the accretion rate to $\dot{M} = 5 \times 10^{-9} M_\odot \, yr^{-1}$ and keep it for 1.3×10^8 s (period 3). The number of bursts is 133 in this interval. Though the temperature distribution has become lower compared with "period 1," if we continue the calculations of normal bursts, the temperature distribution should be recovered to that in "period 1." Therefore, we increase the crustal heating by a factor of 10 to save the computational time. Although this artificial change appears to be unreal, from the point of nuclear physics, physical process concerning the crustal heating has been rather uncertain; the rate is often multiplied some factors (see, e.g., [8, 63]). We note that the heating rates (ergs/g) change up to 20 times for the pressures shown in tables of Haensel and Zdunik [57], and we have implemented the crustal heating using the tables as the heating source.

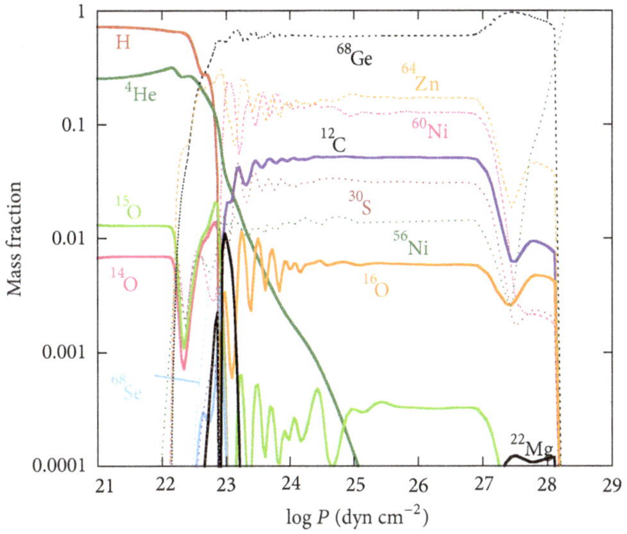

FIGURE 12: Composition distribution at 1.5×10^{9} s ($*\beta$). The rp-process forms ^{68}Ge, ^{64}Zn and ^{60}Ni.

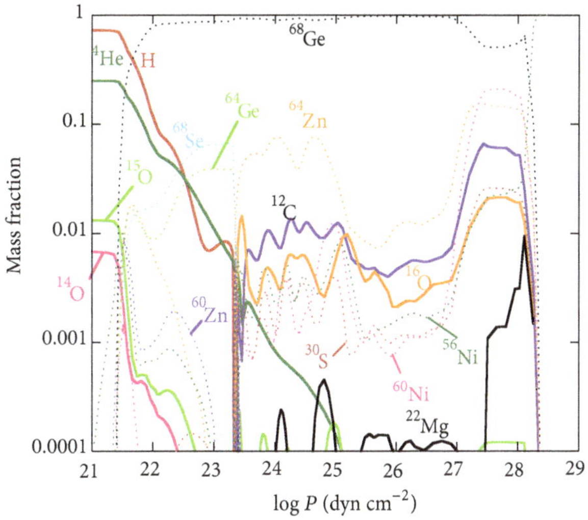

FIGURE 13: Composition distribution at 1.81×10^{9} s ($*\gamma$). For log P = 21.6-23.3, the amount of ^{68}Se and ^{64}Ge becomes large temporarily due to the hydrogen/helium mixed burning.

Figure 13 shows the composition distribution at the stage ($*\gamma$). As in the case of "period 1," hydrogen is consumed in the bottom region of the hydrogen/helium burning and the remained helium forms ^{12}C continuously below the region. After 5500 s ($*\delta$) from the end ($*\gamma$) of the last burst (♯7), we obtain the carbon flash. Finally, the ^{12}C+^{12}C reaction is successfully ignited in log P = 28 at the last epoch (1.81×10^{9} s). The composition distributes uniformly for log P = 26–28 due to the convection as shown in Figure 14. We note that the ^{12}C+^{12}C flash does not yet develop to the peak of the total nuclear energy generation rate (Figure 19).

4.2. Characteristic Features of Key Bursts Associated with Many Normal Bursts. Let us discuss the profiles and/or sequences of the bursts for each period. In Figure 6(b), there is a significant interval at 8.25–9.75×10^{8} s. We call it the "long interval 1." We pay attention to five bursts (♯1–5) before the "long interval 1" and the first burst (♯6) after the interval.

Five bursts ♯1–5 extracted from Figure 6 are shown in Figure 15. "Long interval 1" begins at 8.252×10^{8} s. The light curve of the burst ♯1 is shown in Figures 16 and 17, where we set the time to be zero when the burst ♯1 starts. The roman numerals specify the epochs: I: just before the convection, II: peak of the total nuclear energy generation rate $L_{n,max}$, and III: 1/10 of the maximum luminosity L_{max}. In Figure 16, the epoch "IV" for the burst ♯5 corresponds to 220 s after the onset of the burst.

The shapes of the light curves and the products for the bursts ♯1–4 and ♯6 are almost same. The burst ♯5 is the longest burst compared with other bursts of ♯1–4 (Figure 16) and ♯6. At "IV" of the burst ♯5, the hydrogen burns in log $P \leq 22.5$, and ^{60}Zn is produced where the convection occurs for log P = 22.2-23.2 that leads to the exhaustion of hydrogen. The heat from this burning region of the convection is the reason for the long burst as seen in Figure 16, and the consumption

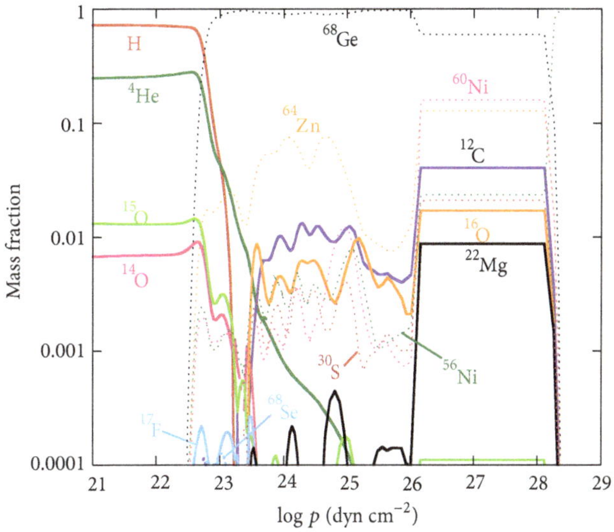

FIGURE 14: Composition distribution just after the ^{12}C+^{12}C ignition ($*\delta$).

of fuels leads to the long interval after ♯5 (Figures 6(b) and 15). This feature can be seen in other sequences of bursts. At the epoch "IV" of the burst ♯3 (2240 s after the onset of the burst), the convection also occurs between log P = 22.5–23 where hydrogen is completely consumed. Therefore, as seen in Figure 15 the interval to the next burst ♯4 becomes long compared to other intervals except for the "long interval 1," because the accumulated hydrogen has been consumed. The same situation also occurs for the rather long interval 1.14–1.19×10^{9} s in Figure 7.

The luminosity (Figure 17) in the burst $*\beta$ of $\dot{M} = 1 \times 10^{-9} M_{\odot}\,\mathrm{yr}^{-1}$ is a little small compared with the bursts ♯1–5 and ♯6 of $\dot{M} = 5 \times 10^{-9} M_{\odot}\,\mathrm{yr}^{-1}$. Though the products of ^{68}Ge, ^{68}Se, ^{64}Zn, and ^{22}Mg are the same as those in

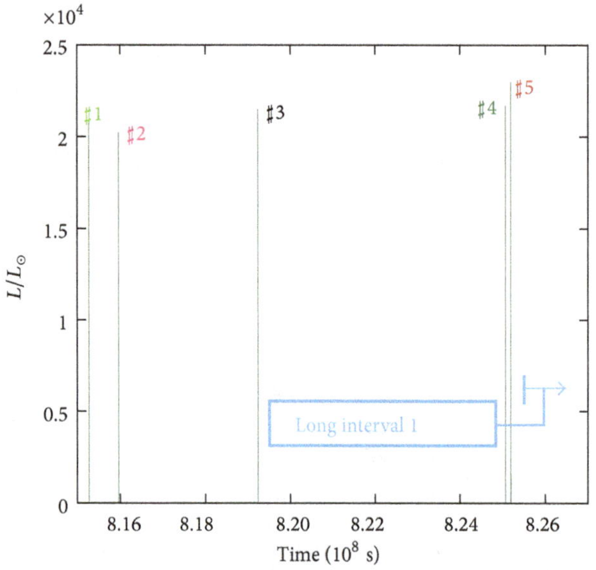

FIGURE 15: Light curves at 8.15–8.27×10^8 s. Each number from ♯1–5 corresponds to each burst. “Long interval 1” begins after the elapsed time 8.252×10^8 s (♯5). Detailed profiles of the burst ♯1 and 5 are shown in Figure 16.

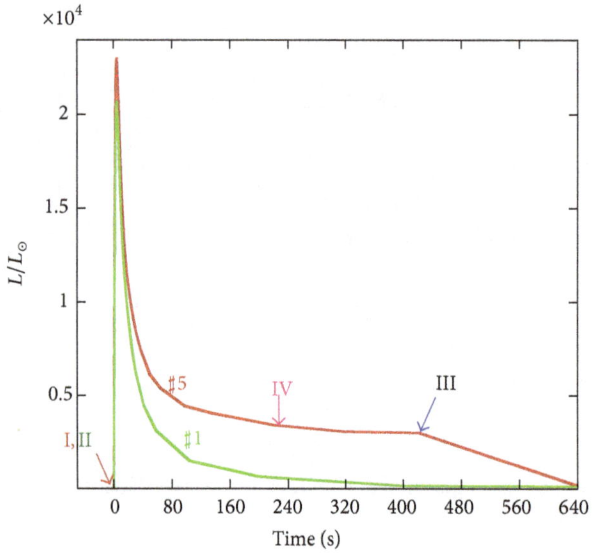

FIGURE 16: Light curves of the burst ♯1 and burst ♯5 that correspond to Figure 15. IV is the time of 220 s after the onset of the burst. The burst ♯5 is the longest burst compared with other bursts ♯1–4. For comparison, the burst ♯1 is also shown.

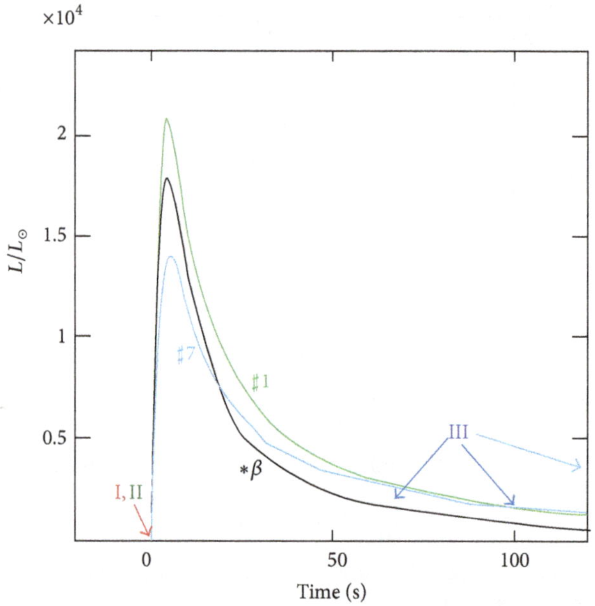

FIGURE 17: Light curves of the bursts ♯1, $*\beta$, and ♯7. Roman numerals are corresponded with the time of I: just before the start of convection, II: $L_{n,max}$, and III: 1/10 of L_{max}. The epoch III in ♯7 is 154 s after the onset of the burst.

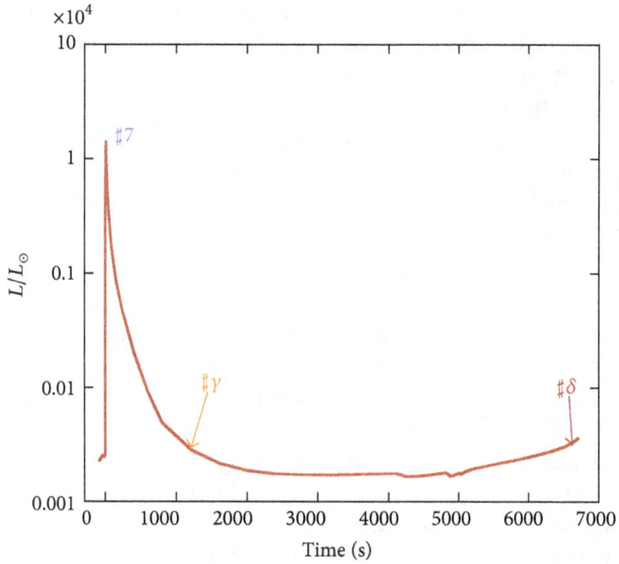

FIGURE 18: Light curve of the burst ♯7 and the onset of the carbon flash. ♯γ and ♯δ correspond to the same epoch as shown by the mark in Figure 9.

the other bursts, the amount of heavy elements like ^{60}Ni and ^{64}Zn are produced more than those of the other bursts. The temperature distribution of the inner accretion layer ($\log P = 24$–28) is lower than that of the other bursts with $\dot{M} = 5 \times 10^{-9} M_{\odot}\,\mathrm{yr}^{-1}$ (see Figure 9, where T in the stage of $*\beta$ for $\log P > 24$ is low due to the conduction). We can consider that due to the increased conduction the luminosity in the bursts decreases because the heat from the combined hydrogen/helium burning has flowed into the core. This phenomena can be understood from the idea of *watershed* introduced by Fujimoto et al. [35]. As illustrated in Figure 17, the last burst ♯7 before the carbon ignition is weaker in strength than that of the burst $*\beta$, because the hydrogen/helium burning is ignited at the deeper region ($\log P = 23.3$), which can be seen from the ash of the rp-process ^{68}Se in Figure 13, compared to the other bursts ($\log P = 22.9$) as seen in Figures 11 and 12.

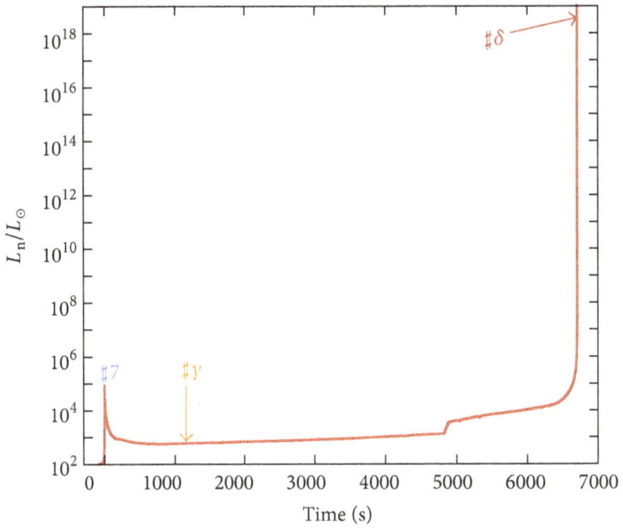

FIGURE 19: Total nuclear energy generation rate for the burst ♯7 and the onset of the carbon flash.

The light curve and the total nuclear energy generation rate, L_n, from the burst ♯7 to the onset of the carbon flash are shown in Figures 18 and 19. Just after the burst, $*\gamma$, the convection occurs at the narrow region of log P = 27.9-28 (Figure 13). At ≃ 5300 s in Figure 19, the convection begins to spread to lower pressure region, and it extends from log P = 28 to 26 as shown in Figure 14. We note that L_n at $*\delta$ is not attained to the $L_{n,max}$. However, since the temperature in log $P \geq 27.8$ has increased beyond the deflagration temperature defined by $\tau_{dyn} = \tau_{^{12}C+^{12}C}$ as seen from Figure 9, we can infer that the flash should develop to a dynamical phenomena of deflagration. Finally, we stop the calculation of the flash due to the numerical difficulty in the assumption of the hydrostatical equilibrium; nevertheless, we insist that this flash could become a superburst.

5. Concluding Remarks

We have presented the two cases for the models of superbursts: single helium flash and carbon flash accompanied with many normal bursts. For the helium flash, the burst has a long duration time although the accretion rate is different from the observation. We suggest that the helium flash could originate the superburst if the burning develops to the deflagration and/or detonation. For a single carbon flash, the temperature does not reach the deflagration temperature due to the limit of heat conduction.

For the carbon flash accompanied with many normal bursts, we carried out the successive 2786 normal bursts up to the time of 1.81×10^9 s using the observed accretion rates. We showed the profiles of the several normal bursts and the onset of the carbon flash after the normal bursts. Since there exists significant amount of produced elements in the deeper than the region where hydrogen/helium burning occurred, the heat transported to the core by them lengthens the recurrence intervals of bursts. We have also shown that the recurrence time becomes longer than other bursts when the little burning occurred for the bottom of the hydrogen/helium burning layer after the main burning had ceased. This leads to the consumption of nuclear fuels of H and He and increases the products of rp-process. We have recognized that "long interval 1" is a remarkable case for the delay to the next burst due to the above reasons. Before the carbon flash, we find various profiles of the light curve and different intervals between bursts; Quantitative comparison with the observations should shed a new insight into the X-ray burst research. Since normal bursts have been observed before the superbursts, our scenario is consistent with the observations. We conclude that a carbon flash accompanied with many normal bursts should trigger a superburst. It is interesting to apply the above idea to a superburst of a helium accretor like 4U1820-30.

For the helium flash shown in Section 3, accretion rate to produce a superburst is rather low compared with observations. If we repeat normal helium flashes with the observed $\dot{M}$ as demonstrated in Section 4.2, a superburst could be triggered in a deep region of the accreting layers. Finally, we remark that to get a detailed history toward the superburst, it is needed to calculate many bursts using the code of the neutron star evolution with a large nuclear reaction network coupled.

Conflict of Interests

The authors declare that there is no conflict of interests regarding the publication of this paper.

References

[1] H. Schatz, L. Bildsten, A. Cumming, and M. Ouellette, "Nuclear physics in normal X-ray bursts and superblasts," *Nuclear Physics A*, vol. 718, pp. 247–254, 2003.

[2] A. Cumming, J. Macbeth, J. J. M. in't Zand, and D. Page, "Long type I x-ray bursts and neutron star interior physics," *The Astrophysical Journal*, vol. 646, pp. 429–451, 2006.

[3] A. Parikh, J. José, G. Sala, and C. Iliadis, "Nucleosynthesis in type I X-ray bursts," *Progress in Particle and Nuclear Physics*, vol. 69, no. 1, pp. 225–253, 2013.

[4] L. Keek, A. Heger, and J. J. M. in't Zand, "Superburst models for neutron stars with hydrogen- and helium-rich atmospheres," *The Astrophysical Journal*, vol. 752, no. 2, 150, 2012.

[5] R. Wijnands, "Recurrent very long type I X-ray bursts in the low-mass X-ray binary 4U 1636-53," *The Astrophysical Journal Letters*, vol. 554, no. 1, article L59, 2001.

[6] T. E. Strohmayer and C. B. Markwardt, "Evidence for a millisecond pulsar in 4U 1636-53 during a superburst," *Astrophysical Journal Letters*, vol. 577, no. 1, pp. 337–345, 2002.

[7] A. Cumming and J. Macbeth, "The thermal evolution following a superburst on an accreting neutron star," *Astrophysical Journal Letters*, vol. 603, no. 1, pp. L37–L40, 2004.

[8] L. Keek and A. Heger, "Multi-zone models of superbursts from accreting neutron stars," *Astrophysical Journal*, vol. 743, no. 2, article 189, 2011.

[9] T. E. Strohmayer and E. F. Brown, "A remarkable 3 hour thermonuclear burst from 4U 1820-30," *The Astrophysical Journal*, vol. 566, pp. 1045–1059, 2002.

[10] E. Kuulkers, "A superburst from GX 3+1," *Astronomy and Astrophysics*, vol. 383, no. 1, pp. L5–L8, 2002.

[11] L. Keek, "Photospheric radius expansion in superburst precursors from neutron stars," *The Astrophysical Journal*, vol. 756, article 130, 2012.

[12] A. Cumming and L. Bildsten, "Carbon flashes in the heavy-element ocean on accreting neutron stars," *Astrophysical Journal Letters*, vol. 559, no. 2, pp. L127–L130, 2001.

[13] A. Cumming, "Models of type I X-RAY bursts from 4U 1820-30," *The Astrophysical Journal*, vol. 595, no. 2, pp. 1077–1085, 2003.

[14] O. Koike, M. Hashimoto, R. Kuromizu, and S. Fujimoto, "Final products of the rp-process on accreting neutron stars," *The Astrophysical Journal*, vol. 603, no. 1, pp. 242–251, 2004.

[15] S. E. Woosley, A. Heger, and A. Cumming, "Models for type I X-ray bursts with improved nuclear physics," *The Astrophysical Journal Supplement Series*, vol. 151, no. 1, p. 75, 2004.

[16] R. Cornelisse, E. Kuulkers, J. J. M. in't Zand, F. Verbunt, and J. Heise, "A four-hours long burst from Serpens X-1," *Astronomy & Astrophysics*, vol. 382, pp. 174–177, 2002.

[17] E. Kuulkers, J. J. M. In't Zand, and M. H. van Kerkwijk, "A half-a-day long thermonuclear X-ray burst from KS 1731-260," *Astronomy and Astrophysics*, vol. 382, no. 2, pp. 503–512, 2002.

[18] J. J. M. In't Zand, E. Kuulkers, F. Verbunt, J. Heise, and R. Cornelisse, "A superburst from 4U 1254-69," *Astronomy and Astrophysics*, vol. 411, no. 3, pp. L487–L491, 2003.

[19] J. J. M. in't Zand, R. Cornelisse, and A. Cumming, "Superbursts at near-Eddington mass accretion rates," *Astronomy & Astrophysics*, vol. 426, no. 1, pp. 257–265, 2004.

[20] H. Schatz, L. Bildsten, A. Cumming, and M. Wiescher, "The rapid proton process ashes from stable nuclear burning on an accreting neutron star," *Astrophysical Journal Letters*, vol. 524, no. 2, pp. 1014–1029, 1999.

[21] H. Worpel, D. K. Galloway, and D. J. Price, "Evidence for accretion rate change during type i X-ray bursts," *Astrophysical Journal*, vol. 772, no. 2, article 94, 2013.

[22] A. Bahramian, C. O. Heinke, and G. R. Sivakoff, "Discovery of the third transient X-ray binary in the galactic globular cluster Terzan 5," *The Astrophysical Journal*, vol. 780, no. 2, article 127, 2014.

[23] H. Schatz, L. Bildsten, and A. Cumming, "Photodisintegration-triggered nuclear energy release in superbursts," *Astrophysical Journal Letters*, vol. 583, no. 2, pp. L87–L90, 2003.

[24] M. Sinha, M. Dey, S. Ray, and J. Dey, "Super bursts and long bursts as surface phenomena of compact objects," *Monthly Notices of the Royal Astronomical Society*, vol. 337, no. 4, pp. 1368–1372, 2002.

[25] C. M. Malone, A. Nonaka, A. S. Almgren, J. B. Bell, and M. Zingale, "Multidimensional modeling of type I X-ray bursts. I. Two-dimensional convection prior to the outburst of a pure ^{4}He accretor," *The Astrophysical Journal*, vol. 728, no. 2, article 118, 2011.

[26] C. M. Malone, M. Zingale, A. Nonaka, A. S. Almgren, and J. B. Bell, "Multidimensional modeling of type I X-ray bursts. II. Two-dimensional convection in a mixed H/He accretor," *The Astrophysical Journal*, vol. 788, article 115, 2014.

[27] Y. Cavecchi, A. L. Watts, J. Braithwaite, and Y. Levin, "Flame propagation on the surfaces of rapidly rotating neutron stars during type I X-ray bursts," submitted, http://arxiv.org/abs/1212.2872.

[28] R. V. E. Lovelace, A. K. Kulkarni, and M. M. Romanova, "Torsional magnetic oscillations in type I X-ray bursts," *The Astrophysical Journal*, vol. 656, no. 1, pp. 393–398, 2007.

[29] D. J. B. Payne and A. Melatos, "Magnetic burial and the harmonic content of millisecond oscillations in thermonuclear X-ray bursts," *The Astrophysical Journal*, vol. 652, no. 1, p. 597, 2014.

[30] R. Artigue, D. Barret, F. K. Lamb, K. H. Lo, and M. C. Miller, "Testing the rotating hotspot model using X-ray burst oscillations from 4U 1636–536," *Monthly Notices of the Royal Astronomical Society*, vol. 433, no. 1, pp. L64–L68, 2013.

[31] M. Fujimoto, T. Hanawa, I. Iben Jr., and M. B. Richardson, "Thermal evolution of accreting neutron stars. II—long X-ray bursts as a probe into the interior," *The Astrophysical Journal*, vol. 315, pp. 198–208, 1987.

[32] K. S. Thorne, "The relativistic equations of stellar structure and evolution," *Astrophysical Journal*, vol. 212, pp. 825–831, 1977.

[33] D. Sugimoto, K. Nomoto, and Y. Eriguchi, "Stable numerical method in computation of stellar evolution," *Progress of Theoretical Physics Supplement*, vol. 70, pp. 115–131, 1981.

[34] M. Y. Fujimoto and D. Sugimoto, "Helium shell flashes and evolution of accreting white dwarfs," *The Astrophysical Journal*, vol. 257, pp. 291–302, 1982.

[35] M. Fujimoto, T. Hanawa, I. Iben Jr., and M. B. Richardson, "Thermal evolution of accreting neutron stars," *The Astrophysical Journal*, vol. 278, pp. 813–824, 1984.

[36] T. Hanawa and M. Y. Fujimoto, "Thermal response of neutron stars to shell flashes," *Publication of the Astronomical Society of Japan*, vol. 36, pp. 199–214, 1984.

[37] C. Angulo, M. Arnould, and M. Rayet, "A compilation of charged-particle induced thermonuclear reaction rates," *Nuclear Physics A*, vol. 656, no. 1, pp. 3–183, 1999.

[38] R. K. Wallace and S. E. Woosley, "Explosive hydrogen burning," *Astrophysical Journal*, vol. 45, p. 389, 1981.

[39] T. Hanawa, D. Sugimoto, and M. A. Hashimoto, "Nucleosynthesis in explosive hydrogen burning and its implications in ten-minute interval of X-ray bursts," *Publications of the Astronomical Society of Japan*, vol. 35, pp. 491–506, 1983.

[40] W. L. Slattery, G. D. Doolen, and H. E. Dewitt, "Improved equation of state for the classical one-component plasma," *Physical Review A*, vol. 21, no. 6, pp. 2087–2095, 1980.

[41] M. B. Richardson, H. M. Van Horn, K. F. Ratcliff, and R. C. Malone, "Neutron star evolutionary sequences," *The Astrophysical Journal*, vol. 255, pp. 624–653, 1982.

[42] V. Canuto, "Equation of state at ultrahigh sensities," *Annual Review of Astronomy and Astrophysics*, vol. 12, pp. 167–214, 1974.

[43] V. Canuto, "Equation of state at ultrahigh densities," *Annual Review of Astronomy and Astrophysics*, vol. 13, pp. 335–380, 1975.

[44] B. L. Friman and O. V. Maxwell, "Neutrino emissivities of neutron stars," *The Astrophysical Journal*, vol. 232, pp. 541–557, 1979.

[45] G. G. Festa and M. A. Ruderman, "Neutrino-pair bremsstrahlung from a degenerate electron gas," *Physical Review*, vol. 180, no. 5, pp. 1227–1231, 1969.

[46] G. Beaudet, V. Petrosian, and E. E. Salpeter, "Energy losses due to neutrino processes," *The Astrophysical Journal*, vol. 150, p. 979, 1967.

[47] R. C. Malone, *Cooling of superfluid neutron stars [Ph.D thesis]*, Cornell University, Ithaca, NY, USA, 1974.

[48] G. Baym, C. Pethick, and D. Pines, "Electrical conductivity of neutron star matter," *Nature*, vol. 224, pp. 674–675, 1969.

[49] E. Flowers and N. Itho, "Transport properties of dense matter," *The Astrophysical Journal*, vol. 206, pp. 218–242, 1976.

[50] N. Itoh, N. Matsumoto, M. Seki, and Y. Kohyama, "Neutrino-pair bremsstrahlung in dense stars. II—crystalline lattice case," *The Astrophysical Journal*, vol. 279, pp. 413–418, 1984.

[51] I. Iben Jr., "Thermal pulses; p-capture, alpha-capture, s-process nucleosynthesis; and convective mixing in a star of intermediate mass," *The Astrophysical Journal*, vol. 196, part 1, pp. 525–547, 1975.

[52] A. N. Cox and J. N. Stewart, "Rosseland opacity tables for population I compositions," *The Astrophysical Journal Supplement*, vol. 19, p. 243, 1970.

[53] A. N. Cox and J. N. Stewart, "Rosseland opacity tables for population II compositions," *The Astrophysical Journal*, vol. 19, Supplement, p. 261, 1970.

[54] W. B. Hubbard and M. Lampe, "Thermal conduction by electrons in stellar matter," *The Astrophysical Journal Supplement Series*, vol. 18, p. 297, 1969.

[55] V. Canuto, "Electrical conductivity and conductive opacity of a relativistic electron gas," *The Astrophysical Journal*, vol. 159, pp. 641–652, 1970.

[56] H. E. Dewitt, H. C. Gravoske, and M. S. Cooper, "Screening factors for nuclear reactions. I. General theory," *The Astrophysical Journal*, vol. 181, pp. 439–456, 1973.

[57] P. Haensel and J. L. Zdunik, "Non-equilibrium processes in the crust of an accreting neutron star," *Astronomy and Astrophysics*, vol. 227, no. 2, pp. 431–436, 1990.

[58] D. K. Galloway, M. P. Muno, J. M. Hartman, D. Psaltis, and D. Chakrabarty, "Thermonuclear (type i) X-ray bursts observed by the rossi X-ray timing explorer," *Astrophysical Journal*, vol. 179, no. 2, pp. 360–422, 2008.

[59] E. Kuulkers, J. J. M. in't Zand, and J.-L. Atteia, "What ignites on the neutron star of 4U 0614+091?" *Astronomy & Astrophysics*, vol. 514, article A65, 2010.

[60] L. Bildsten, *Thermonuclear Burning on Rapidly Accreting Neutron Stars*, Kluwer Academic Publishers, Dordrecht, The Netherlands, 1998.

[61] E. P. J. van den Heuvel, in *Proceedings of the 2nd BeppoSAX Meeting*, E. P. J. van den Heuvel, J. J. M. In't Zand, and R. A. M. J. Wijers, Eds., 2003.

[62] N. N. Weinberg and L. Bildsten, "Carbon detonation and shock-triggered helium burning in neutron star superbursts," *Astrophysical Journal Letters*, vol. 670, no. 2, pp. 1291–1300, 2007.

[63] Y. Matsuo, H. Tsujimoto, and T. Noda, "Effects of a new triple-α reaction on x-ray bursts of a helium-accreting neutron star," *Progress of Theoretical Physics*, vol. 126, article 1177, 2011.

15

On Robe's Circular Restricted Problem of Three Variable Mass Bodies

Jagadish Singh[1] and Oni Leke[2]

[1] *Department of Mathematics, Faculty of Science, Ahmadu Bello University Zaria, PMB 2222, Samaru-Zaria, Kaduna, Nigeria*
[2] *Department of Mathematics, College of Science, University of Agriculture, PMB 2373, North-Bank, Makurdi, Nigeria*

Correspondence should be addressed to Oni Leke; lekkyonix4ree@yahoo.com

Academic Editors: M. Biesiada, K. Bolejko, M. S. Dimitrijevic, M. Jamil, and E. Saridakis

This paper investigates the motion of a test particle around the equilibrium points under the setup of the Robe's circular restricted three-body problem in which the masses of the three bodies vary arbitrarily with time at the same rate. The first primary is assumed to be a fluid in the shape of a sphere whose density also varies with time. The nonautonomous equations are derived and transformed to the autonomized form. Two collinear equilibrium points exist, with one positioned at the center of the fluid while the other exists for the mass ratio and density parameter provided the density parameter assumes value greater than one. Further, circular equilibrium points exist and pairs of out-of-plane equilibrium points forming triangles with the centers of the primaries are found. The out-of-plane points depend on the arbitrary constant κ, of the motion of the primaries, density ratio, and mass parameter. The linear stability of the equilibrium points is studied and it is seen that the circular and out-of-plane equilibrium points are unstable while the collinear equilibrium points are stable under some conditions. A numerical example regarding out-of-plane points is given in the case of the Earth, Moon, and submarine system. This study may be useful in the investigations of dynamic problem of the "ocean planets" Kepler-62e and Kepler-62f orbiting the star Kepler-62.

1. Introduction

The classical restricted three-body problem (RTBP) constitutes one of the most important problems in dynamical astronomy. The study of this problem is of great theoretical, practical, historical, and educational relevance. The investigation of this problem in its several versions has been the focus of continuous and intense research activity for more than two hundred years. The study of this problem in its many variants has had important implications in several scientific fields including, among others, celestial mechanics, galactic dynamics, chaos theory, and molecular physics. The RTBP is still a stimulating and active research field that has been receiving considerable attention of scientists and astronomers because of its applications in dynamics of the solar and stellar systems, lunar theory, and artificial satellites.

A different kind of restricted three-body problem was formulated by Robe [1], a set up in which the first primary is a rigid spherical shell filled with homogenous, incompressible fluid of density ρ_1, and the second primary is a mass point outside the shell and moving around the first primary in a Keplerian orbit, while the infinitesimal mass is a small sphere of density ρ_3 moving inside the shell and is subject to the attraction of the second primary and the buoyancy force due to the fluid.

In estimating buoyancy force, Robe [1] assumed that the pressure field of the fluid ρ_1 has spherical symmetry around the center of the shell, and he considered only one out of the three components of the pressure field, which is due to the own gravitational field of the fluid ρ_1.

A. R. Plastino and A. Plastino [2] took into account all these components of pressure field. But in their study, they assumed the hydrostatic equilibrium figure of the first primary as Roche's ellipsoid. They found that when the density parameter D is zero, every point inside the fluid is an equilibrium point; otherwise, the ellipsoid's center is the only equilibrium point. They also examined the linear stability of equilibrium points. Hallan and Rana [3] investigated the existence of all equilibrium point and their stability in the Robe's [1] problem. It was seen that the Robe's elliptic restricted three-body problem has only one equilibrium point for all values of the density parameter and the mass parameter, while the Robe's circular restricted three-body problem can have two, three, or infinite numbers of equilibrium points. As regards to the stability of these equilibria, they found that the collinear equilibrium points are stable while triangular and circular points are always unstable. Recently, Kaur and Aggarwal [4] investigated the Robe's problem of 2 + 2 bodies and applied it to the study of the motion of two submarines in the Earth-Moon system. Singh and Hafsah [5] examined the Robe's circular restricted three-body problem when the first primary is a fluid in the shape of an oblate spheroid and the second primary is a triaxial rigid body.

The classical restricted three-body problem assumes that the masses of celestial bodies are constant. However, the phenomenon of isotropic radiation or absorption in stars led scientists to formulate the restricted problem of three bodies with variable mass. As an example, we could mention the motion of rockets, black holes formation, motion of a satellite around a radiating star surrounded by a cloud and varying its mass due to particles of the cloud, and comets loosing part or all of their mass as a result of roaming around the Sun (or other stars) due to their interaction with the solar wind which blows off particles from their surfaces. The problem of the motion of astronomical objects with variable mass has many interesting applications in stellar, galactic, and planetary dynamics.

The study of two bodies with variable masses seems to have been first investigated by Dufour [6] where he examined the astronomical phenomena of variable mass relating the secular variation of lunar acceleration with the increase of the Earth's mass due to the impact of meteorites. Later, Gylden [7] wrote the differential equations of motion for the problem when the masses are subject to variation. The integrable case to this differential equation was then given by Meshcherskii [8, 9]. The problem was later known as the Gylden-Meshcherskii problem. A characterization of this problem was studied by Singh and Leke [10]. The effect of the isotropic variation of the mass of the star in a planetary system and the possible ejection of a planet from the system were studied by Veras et al. [11]. Recently, Singh and Leke [12] investigated the existence and stability of equilibrium points in the Robe's restricted three-body problem with variable masses.

Besides the Gylden-Meshcherskii problem, there are other different cases of two bodies with variable masses, which are classified according to the presence or absence of reactive forces, to whether the bodies move in an inertial frame or not, and so on (see [13]). For instance, when the particles are at rest in an inertial coordinate system, this case may be used to study the orbits of a celestial body moving through a static atmosphere, whose particles attach to it or detach from it as it moves. The restricted three-body problem with nonisotropic variation of the masses has been studied by Bekov [14], Bekov et al. [15], and Letelier and Da Silva [16]. A simple example of this kind of problem is the system of two variable primaries and a rocket. In this case, it is the thrust from the rocket that defines the force that acts on the test particle, in addition to the gravitational attraction from the two primaries, while the rocket does not affect the orbits of the primaries.

In this paper, the existence and stability of equilibrium points under the frame of the Robe problem [1], when the participating bodies vary their masses at the same rate, is studied. Here, we assume that the primaries move in a stationary medium, from which they absorb or lose mass; the first primary being a fluid in the shape of a sphere and the test particle which is a small sphere located inside the fluid also gain or lose mass to the fluid. Hence, there is no need to assume a rigid spherical shell. This study may be useful in the investigations of dynamic the problem of water-planetary system discovered by Kepler spacecraft. These "ocean planets" are orbiting the star Kepler-62 and are designated Kepler-62e and Kepler-62f. The existence of these Earth-size planets covered completely by a water envelope (water planets) has long fascinated scientists and the general public. The model of this problem can also be used to study the small oscillation of the Earth's inner core taking into account the Moon's attraction during the course of evolution.

This paper is orginzed as follows: Section 2 contains the equations of motion; the equilibrium points are investigated in Section 3; Section 4 investigates the linear stability of the equilibrium points; Section 5 discusses the obtained results and the conclusions.

2. Equations of Motion

Let m_1 be the mass of the first primary which is a fluid in the shape of a sphere of radius $\mathfrak{R}$ with center at M_1 having density δ_1 and volume V_1. Also, let m_2 be the mass of the second primary with center at M_2 which describes a circular orbit around the first one. Both masses are assumed to vary with time as they travel in a static medium which acts as a sink or source of mass. Now, let m_3 be the mass of the test particle whose mass is very small compared with the masses of the primaries, with center at M_3, having density δ_3 and volume V_3. We suppose that its mass varies with time, also as it moves about in the fluid, it gains or loses mass to the medium. Let the positions vector between the center of the fluid and the centers of the second and the test particle be $\vec{r}_{12}$ and $\vec{r}_{13}$, respectively, and let that between the test particle and the second primary be $\vec{r}_{23}$. Following Robe [1] and knowing that the masses, distances, and densities vary with time, the forces acting on the third body are the force of attraction of m_2; the gravitational force $\vec{F}_A$ exerted by the fluid, that is $\vec{F}_A = -(4\mathfrak{R}^3/3r_{13}^3)\pi G\delta_1 m_3\overrightarrow{M_1M_3}$; and the buoyancy force exerted by the fluid which is $\vec{F}_B = (4\pi\mathfrak{R}^3/3r_{13})G\delta_1^2(m_3/\delta_3)\overrightarrow{M_3M_1}$.

We adopt a rotating coordinate system $Oxyz$ with origin at the center of mass, O, of the primaries, Ox pointing towards the second primary, and Oxy being the orbital plane of m_2. The equations of motion of the test particle, taking into account the forces acting on it, have the following form [1, 14]:

$$\ddot{x} - 2\omega\dot{y} = \omega^2 x + \dot{\omega} y - \frac{4\pi\mathfrak{R}^3 (x - x_1)}{3r_{13}^3} \times G\delta_1 \left(1 - \frac{\delta_1}{\delta_3}\right) - \frac{\mu_2 (x - x_2)}{r_{23}^3} - \frac{\dot{m}_3}{m_3} (\dot{x} - \omega y),$$

$$\ddot{y} + 2\omega\dot{x} = \omega^2 y - \dot{\omega}x - \frac{4\pi\mathfrak{R}^3 y}{3r_{13}^3} \times G\delta_1\left(1 - \frac{\delta_1}{\delta_3}\right) - \frac{\mu_2 y}{r_{23}^3} - \frac{\dot{m}_3}{m_3}(\dot{y} - \omega x),$$

$$\ddot{z} = \frac{4\pi\mathfrak{R}^3 z}{3r_{13}^3}G\delta_1\left(1 - \frac{\delta_1}{\delta_3}\right) - \frac{\mu_2 z}{r_{23}^3} - \frac{\dot{m}_3\dot{z}}{m_3}, \quad (1)$$

where $r_{13}^2 = (x - x_1)^2 + y^2 + z^2$, $r_{23}^2 = (x - x_2)^2 + y^2 + z^2$.

The barycentric coordinates x_1 and x_2 are connected with the distance between the primaries by the following equations:

$$\mu_1(t) = \mu\frac{x_2}{r_{12}}, \qquad \mu_2(t) = -\mu\frac{x_1}{r_{12}}, \quad (2)$$

where $\mu_1(t) = Gm_1(t)$, $\mu_2(t) = Gm_2(t)$, $m_1(t) = \delta_1 V_1$; $V_1 = 4\pi\mathfrak{R}^3/3$, while G is the gravitational constant and the over dot denotes differentiation with respect to time t.

Now, in order to obtain useful dynamical predictions, we transform (x, y, z, t) to the autonomous form (ξ, η, ζ, τ). Following [15], the time dependence of the masses is described [15] by the function $\mu(t)$:

$$x = \left(\frac{\mu_0}{\mu}\right)^3\xi, \qquad y = \left(\frac{\mu_0}{\mu}\right)^3\eta,$$

$$z = \left(\frac{\mu_0}{\mu}\right)^3\zeta, \qquad \frac{dt}{d\tau} = \left(\frac{\mu_0}{\mu}\right)^5,$$

$$\omega(t) = \omega_0\left(\frac{\mu}{\mu_0}\right)^5, \qquad \mu_i = \mu_{0i}\frac{\mu}{\mu_0}, \quad (3)$$

$$m_3 = m_{03}\frac{\mu}{\mu_0}, \qquad r_{i3} = \rho_{i3}\left(\frac{\mu_0}{\mu}\right)^3,$$

$$r_{12} = \rho_{12}\left(\frac{\mu_0}{\mu}\right)^3, \qquad (i = 1, 2).$$

As in [14], the particular solutions for the case with variable parameters μ_1, μ_2, and m_3 in the form of the Eddington-Jeans laws with indices $n = 3$ and $n = 6$ is expressed with the help of the function μ:

$$\dot{\mu} = \alpha_1\mu_1^n, \qquad \dot{\mu} = \alpha_2\mu_2^n, \qquad \dot{m}_3 = \alpha m_3^n, \quad (4)$$

where α, α_1, and α_2 are constants. The exponent $n = 1$ falls in the stellar range while $n = 2$ and $n = 3$ result, respectively, in the first and second law of Meshcherskii [8, 9] mass variations. Equations of (4) indicate that the laws of variation of the three masses are the same.

Also, the dynamical system has the particular solution of the following type [15, 17]:

$$r_{12}\mu m_3^2 = \kappa C^2, \quad (5)$$

where $\kappa > 0$ is a constant, and $C \neq 0$ is a constant of the area integral.

Finally, in addition, we assume that the densities of the fluid and the test particle vary such that

$$\delta_1(t) = \mu\frac{\delta_{01}}{\mu_0}, \qquad \delta_3(t) = \mu\frac{\delta_{03}}{\mu_0}, \quad (6)$$

where δ_{01} and δ_{03} are the densities of the medium and the test particle, respectively, at initial time t_0.

Substituting (3) to (6) in (1) and reducing it throughout by μ^7/μ_0^7, we get

$$\xi'' - 2\omega_0\eta' = \Omega_\xi, \qquad \eta'' + 2\omega_0\xi' = \Omega_\eta, \qquad \zeta'' = \Omega_\zeta, \quad (7)$$

where

$$\Omega = \frac{\kappa\omega_0^2(\xi^2 + \eta^2 + \zeta^2)}{2} - \frac{\omega_0^2\zeta^2}{2} + \frac{\mu_{02}}{\rho_{23}} - G\frac{D}{2}\left[(\xi - \xi_1)^2 + \eta^2 + \zeta^2\right],$$

$$\rho_{13}^2 = (\xi - \xi_1)^2 + \eta^2 + \zeta^2,$$

$$\rho_{23}^2 = (\xi - \xi_2)^2 + \eta^2 + \zeta^2, \quad (8)$$

$$\xi_1 = \frac{-\mu_{20}}{\mu_0}\rho_{12}, \qquad 2\xi_2 = \frac{\mu_{10}}{\mu_0}\rho_{12},$$

$$D = \frac{4\pi}{3}\delta_{01}\left(1 - \frac{\delta_{01}}{\delta_{03}}\right).$$

Here, ρ_{12} is constant and connects the parameter κ by the relation

$$\rho_{12}\mu_0 = \kappa C^2. \quad (9)$$

Equations (5) and (9) indicate that the ratio of the product of the distances between the center of the primaries, mass of the test particle and the sum of the masses with the gravitational constant to the constant of the area integral, always remains a constant in both the autonomous and the non autonomous systems.

Now, we choose units for the distance and time, such that at initial time t_0, $\rho_{12} = 1$, $\omega_0 = C = 1$, respectively. Putting these in (9), for the unit of sum of the masses, we get $\mu_0 = G = \kappa$.

Next, without loss of generality, we introduce the mass parameter defined as $\upsilon = \mu_{20}/\mu_0$ $(0 < \upsilon < 1)$ and also assume that the pressure field of the fluid of density δ_{01} maintains a spherical symmetry around the center of the fluid such that $\rho_{13} = \mathfrak{R}$. Also, we have $\xi_1 = -\upsilon$ and $\xi_2 = 1 - \upsilon$. With the help of these units, the system of (7) takes the following form:

$$\xi'' - 2\eta' = \Omega_\xi, \qquad \eta'' + 2\xi' = \Omega_\eta, \qquad \zeta'' = \Omega_\zeta, \quad (10)$$

where

$$\Omega(\xi, \eta, \zeta) = \frac{\kappa(\xi^2 + \eta^2)}{2} + \frac{(\kappa - 1)\zeta^2}{2} + \frac{\kappa\upsilon}{\rho_{23}} - \frac{\kappa D}{2}\rho_{13}^2,$$

$$\rho_{13}^2 = (\xi + \upsilon)^2 + \eta^2 + \zeta^2, \quad (11)$$

$$\rho_{23}^2 = (\xi + \upsilon - 1)^2 + \eta^2 + \zeta^2,$$

and the dash signifies differentiation with respect to the new time τ.

Equations (10) are the autonomized equations of motion of the test particle of our problem. These equations are different from that in [18] and analogous to the equations in Hallan and Rana [3] only differing due to the second term that appears in the force function and the parameter κ.

3. The Equilibrium Points

The equilibrium points represent stationary solutions of the RTBP. These solutions are the singularities of the manifold of the components of the velocity and the coordinates and are found by setting $\xi' = \eta' = \zeta' = \xi'' = \eta'' = \zeta'' = 0$ in the equations of motion (10). That is, they are the solutions of the equations $\Omega_\xi = \Omega_\eta = \Omega_\zeta = 0$, which are

$$\xi - \frac{\upsilon(\xi+\upsilon-1)}{\rho_{23}^3} - D(\xi+\upsilon) = 0,$$
$$\left(1 - \frac{\upsilon}{\rho_{23}^3} - D\right) = 0, \quad \eta \neq 0, \tag{12}$$
$$\left(\frac{\kappa-1}{\kappa} - \frac{\upsilon}{\rho_{23}^3} - D\right) = 0, \quad \zeta \neq 0.$$

The solutions are categorized as follows.

(1) The solutions of first equation of (12) with $\eta = \zeta = 0$ yield the collinear equilibrium points. These points lie on the line joining the center of the first and second primary.

(2) When the first and the second equation of (12) are solved with $\zeta = 0$, we get the circular points. These points lie in the spherical fluid and form circles.

(3) The solutions of first and the third equations of (12) with $\eta = 0$ results in the out-of-plane equilibrium points. These solutions are valid provided they lie inside the fluid. We shall consider them in Sections 3.1, 3.2, and 3.3, respectively.

3.1. Collinear Points. From first equation of (12), with $\eta = \zeta = 0$, we get

$$(\xi+\upsilon)\left[(1-D)\xi^2 + (\upsilon - 2 + 2D - 2D\upsilon)\xi + 1 + 2D\upsilon - D\upsilon^2 - D\right] = 0. \tag{13}$$

Hence, (13) has three roots, the first being $\xi = -\upsilon$ and is always a solution whether $D = 0$ or is not, for $0 < \kappa < \infty$. The two remaining roots are found by considering the second equality in (13), which gives

$$\xi_{2,3} = \frac{-2+\upsilon+2D(1-\upsilon) \pm \sqrt{\Delta}}{2(D-1)}, \tag{14}$$

where $\Delta = \upsilon(4D+\upsilon-4)$.

Now, the solutions exist only for $D > 1$, however, the second solution is greater than $1-\upsilon$ and consequently will lie outside the shell, so we ignore it. The third solution is

$$\xi_3 = \frac{-2+\upsilon+2D(1-\upsilon) - \sqrt{\Delta}}{2(D-1)} \tag{15}$$

and is less than $1-\upsilon$, the ξ-coordinate of the second primary. Hence, there are two collinear equilibrium points which lie on the line joining the centers of the primaries. Therefore, the coordinate $(-\upsilon, 0, 0)$ is always an equilibrium point. For $D > 1$, there exist an equilibrium point $(\xi_3, 0, 0)$ which lies to the left or right of the first primary depending upon whether $D < 1 + 2\upsilon$ or $D > 1 + 2\upsilon$. When $D = 1 + 2\upsilon$, the only equilibrium point is the center of the shell.

3.2. Circular Points. These solutions are found by solving the first and second equations of (12) with $\zeta = 0$. Solving the second equation of (12) gives

$$D = 1 - \frac{\upsilon}{\rho_{23}^3}. \tag{16}$$

Substituting (16) in the first equation of (12) yields $D = 1-\upsilon$ and consequently $\rho_{23}^3 = 1$.

Hence, we have the solution

$$(\xi+\upsilon-1)^2 + \eta^2 = 1, \qquad D = 1-\upsilon, \tag{17}$$

which gives the coordinate of any point on the circle (17) with center $(1-\upsilon, 0, 0)$ which is the center of the second primary and radius one which is the distance between the centers of the fluid and the second primary. Thus, the solution gives us an infinite number of equilibrium points, provided they lie inside the fluid.

3.3. Out-of-Plane Points. The out-of-plane equilibrium points are found by solving first and third equations of (12) with $\eta = 0$. Solving first for ρ_{23} in the third equation of (12), we get

$$\rho_{23} = \left(\frac{\kappa\upsilon}{\kappa - 1 - \kappa D}\right)^{1/3}. \tag{18}$$

Substituting (18) in the first equation of (12) and simplifying results in

$$\xi = (\kappa-1)(\upsilon-1) + \kappa D. \tag{19}$$

Knowing that $\rho_{23}^3 = (\xi+\upsilon-1)^2 + \zeta^2$, substituting (18) and (19) in it, and solving for ζ, we get

$$\zeta = \pm\sqrt{\left(\frac{\kappa\upsilon}{\kappa-1-\kappa D}\right)^{2/3} - \kappa^2(1-\upsilon-D)^2}. \tag{20}$$

Equations (19) and (20) give the position $(\xi, 0, \zeta)$ of real out-of-plane equilibrium points provided $\kappa \geq 1/(1-D)$. Should $\kappa < 1/(1-D)$, then no real out-of-plane points exist as (20) turns out to be imaginary or complex quantity. When $\kappa = 1$, these points fully coincide with that of Hallan and Rana [3].

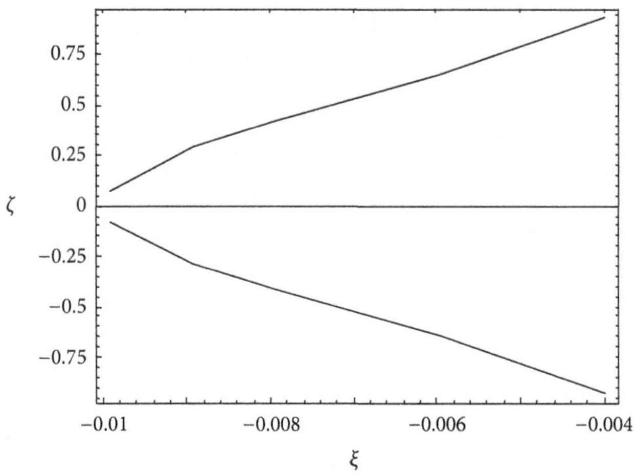

FIGURE 1: Out-of-plane points for, $\upsilon = 0.01$ and $D = -0.001$.

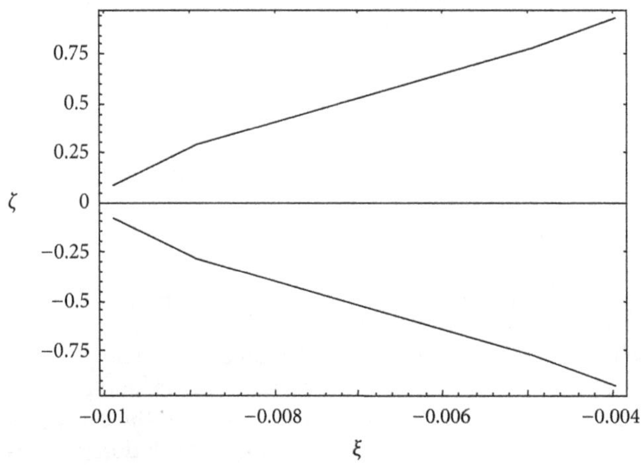

FIGURE 2: Out-of-plane points for, $\upsilon = 0.01$ and $D = 0$.

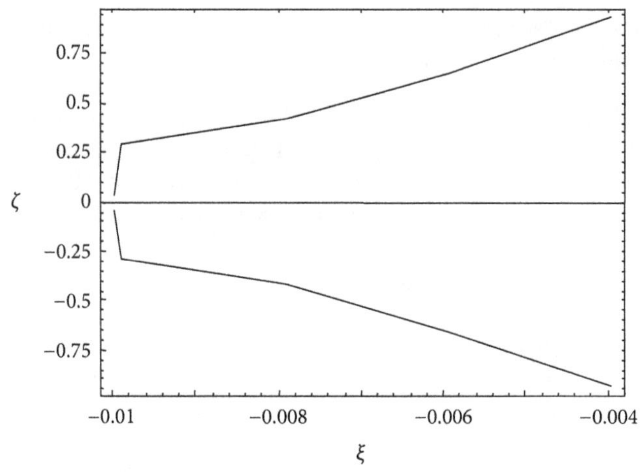

FIGURE 3: Out-of-plane points for $\upsilon = 0.01$ and $D = 0.001$.

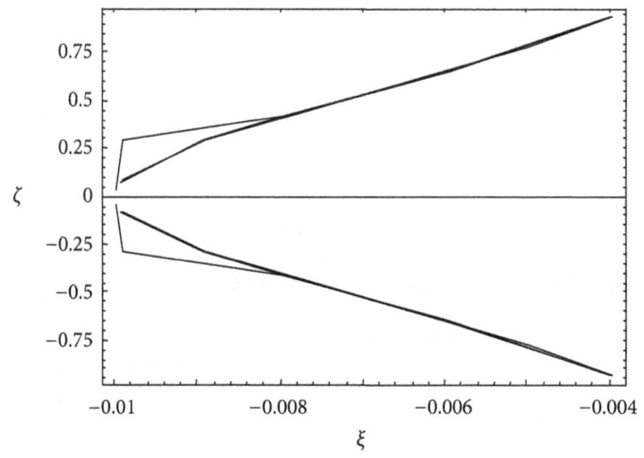

FIGURE 4: Out-of-plane points for, $\upsilon = 0.01$ and $\kappa \in [0.999001, 1.0111]$.

When $D = 0$, (i.e., $\delta_{01} = \delta_{03}$), the coordinates of the out-of-plane points become the same with that in Singh and Leke [12]. Hence, it is seen that the equilibrium points are fully analogous to those found by Hallan and Rana [3] except for the out-of-plane equilibrium points which are affected by the parameter κ, the density parameter, and the mass ratio. The positions of the out-of-plane points are given in Tables 1, 2 and 3 and their graphical representations in Figures 1, 2, and 3 for a test particle in the Earth-Moon system, when the density parameter D is negative, zero, and positive, respectively.

We summarize our numerical effort as follows. In Table 1, when $D = -0.001$ this implies that $\delta_{03} < \delta_{01}$. In this case for $0 < \kappa \leq 0.999$ and $1.009 < \kappa < \infty$, out-of-plane points do not exist, but however exist in the interval $0.999001 \leq \kappa \leq 1.999$. In Table 2, $D = 0$ and so $\delta_{03} = \delta_{01}$. In this case the out-of-plane points exist only when $1.001 \leq \kappa \leq 1.01$ and do not exist in the remaining entire range of κ, while for Table 3, $D = 0.001$ and so $\delta_{03} > \delta_{01}$. Here, real out-of-plane solutions exist, for the values of the parameter κ, in the interval $1.002 \leq \kappa \leq 1.0111$ and are nonexistent for any value of κ outside this range. Hence, though κ has a large range of values, however, the physically meaningful range is $\kappa \in [0.999001, 1.0111]$, at which the out-of-plane points exist (see Figure 4). However, these ranges may differ for different density parameters D, which are determined by the densities of the fluid and the test particle.

4. Stability of Equilibrium Points

To examine the stability of an equilibrium configuration, that is, its ability to restrain the body motion in its vicinity, we apply small displacement u, v, w to the coordinates (ξ_0, η_0, ζ_0) of the third body, to the positions, $\xi = \xi_0 + u$, $\eta = \eta_0 + v$ and $\zeta = \zeta_0 + w$. If its motion rapidly departs from the vicinity of the point, we call such a position of equilibrium an unstable one. If however the body merely oscillates about the point, it is said to be a stable position.

Now, we linearize (10) to obtain the variational equations:

$$\begin{aligned} u'' - 2v' &= \left(\Omega^0_{\xi\xi}\right) u + \left(\Omega^0_{\xi\eta}\right) v + \left(\Omega^0_{\xi\zeta}\right) w, \\ v'' + 2u' &= \left(\Omega^0_{\xi\eta}\right) u + \left(\Omega^0_{\eta\eta}\right) v + \left(\Omega^0_{\eta\zeta}\right) w, \\ w'' &= \left(\Omega^0_{\xi\zeta}\right) u + \left(\Omega^0_{\eta\zeta}\right) v + \left(\Omega^0_{\zeta\zeta}\right) w, \end{aligned} \tag{21}$$

TABLE 1: The out-of-plane points for $\upsilon = 0.01$ and $D = -0.001$.

κ	ξ	$\pm\zeta$	Comments
0.01	0.98009	0.0232859 + 0.0403325i	Real out of plane point do not exist
0.5	0.4945	Complex	—
0.999	Real and negative	Complex	—
0.999001	—	215.369	Real out-of-plane points exist
1	−0.001	1.91758	—
1.001	−0.011	1.3988	—
1.002	−0.002982	1.12303	—
1.003	−0.003973	0.932023	Real out-of-plane point exist
1.004	−0.004964	0.781521	—
1.005	−0.005955	0.652458	—
1.006	−0.006946	0.533794	—
1.007	−0.007937	0.416357	—
1.008	−0.008928	0.28657	—
1.009	−0.009919	0.0755768	—
$1.009 < \kappa < \infty$	Real and negative	Imaginary	Real out-of-plane point do not exist

TABLE 2: The out-of-plane points for $\upsilon = 0.01$ and $D = 0$.

κ	ξ	$\pm\zeta$	Comments
0.01	0.9801	Complex	Real out of plane point do not exist
0.5	0.9801	—	—
0.999	0.00099	—	—
0.999001	0.00098901	—	—
1	0	Infinity	Infinite remote solution
1.001	−0.00099	1.9138	Real out-of-plane points exist
1.002	−0.00198	1.39986	—
1.003	−0.00297	1.12416	—
1.004	−0.00396	0.933202	—
1.005	−0.00495	0.782774	—
1.006	−0.00594	0.653819	—
1.007	−0.00693	0.535323	—
1.008	−0.00792	0.418177	—
1.009	−0.00891	0.289043	—
1.0091	−0.009009	0.274411	—
1.01	−0.0099	0.0839973	—
$1.01 < \kappa < \infty$	Real and negative	Imaginary	Real out-of-plane point do not exist

where the superscript 0 indicates that the partial derivatives are to be evaluated at the equilibrium points.

4.1. Collinear Points. Robe [1] discussed the stability of the equilibrium point at the center of the shell, and Hallan and Rana [3] have also discussed that in the case of the noncollinear points when $D < 0$. Hence, we shall discuss here the stability of the equilibrium point $(\xi_3, 0, 0)$ near the center of the fluid. To do this, we let solutions of the first two equations of (21) be $u = A\exp(\lambda\tau)$, $v = B\exp(\lambda\tau)$, where A, B, and λ are constants.

Finding first and second derivatives of the solutions, substituting them in the first two equations of (21), and simplifying, we obtain the matrix which has a nonzero solution when

$$\begin{vmatrix} \left(\lambda^2 - \Omega^0_{\xi\xi}\right) & \left(2\lambda + \Omega^0_{\xi\eta}\right) \\ \left(2\lambda - \Omega^0_{\xi\eta}\right) & \left(\lambda^2 - \Omega^0_{\eta\eta}\right) \end{vmatrix} = 0. \tag{22}$$

Expanding the determinant, we get

$$\lambda^4 - \left(\Omega^0_{\xi\xi} + \Omega^0_{\eta\eta} - 4\right)\lambda^2 + \Omega^0_{\xi\xi}\Omega^0_{\eta\eta} - \left(\Omega^0_{\xi\eta}\right)^2 = 0. \tag{23}$$

This is the characteristic equation corresponding to the variational equations (21) when motion is considered in the $\xi\eta$-plane.

TABLE 3: The out-of-plane points for $\upsilon = 0.01$ and $D = 0.001$.

κ	ξ	$\pm\zeta$	Comments
0.01	0.98011	Complex	Real out-of-plane point do not exist
0.5	0.4955	—	—
0.999	0.001989	—	—
0.999001	0.00198801	—	—
1	0.001	—	—
1.001	0.000011	—	—
1.002	−0.000978	1.92083	Real out-of-plane point exist
1.003	−0.001967	1.40162	—
1.004	−0.002956	1.12573	—
1.005	−0.003945	0.934716	—
1.006	−0.004934	0.784303	—
1.007	−0.005923	0.655425	—
1.008	−0.006912	0.537083	—
1.009	−0.007901	0.420232	—
1.0091	−0.0079999	0.408219	—
1.01	−0.0099	0.291786	—
1.0111	−0.0099779	0.0393474	—
$1.0112 < \kappa < \infty$	Real and negative	Complex	Real out-of-plane point do not exist

Now, the values of the second order partial derivatives computed at the point $(\xi_3, 0, 0)$, with the substitution $\eta = \zeta = 0$ are as follow:

$$\begin{aligned}\Omega^0_{\xi\xi} &= \kappa\left(1 - D + 4A_1\right),\\ \Omega^0_{\eta\eta} &= \kappa\left(1 - D - 2A_1\right),\\ \Omega^0_{\xi\zeta} &= \Omega^0_{\eta\zeta} = \Omega^0_{\xi\eta} = 0,\\ \Omega^0_{\zeta\zeta} &= \kappa\left(\frac{\kappa - 1}{\kappa} - 2A_1 - D\right),\end{aligned} \tag{24}$$

where $A_1 = (4\upsilon(D-1)^3)/[\upsilon + \sqrt{\upsilon(4D+\upsilon-4)}]^3$.

When $\kappa = 1$, the equations in system (24) fully coincide with those of Hallan and Rana [3].

Substituting (24) in the variational equations (21), at once gives

$$\begin{aligned}u'' - 2v' &= \kappa\left(1 - D + 4A_1\right)u,\\ v'' + 2u' &= \kappa\left(1 - D - 2A_1\right)v,\end{aligned} \tag{25}$$

$$w'' = \kappa\left(\frac{\kappa - 1}{\kappa} - 2A_1 - D\right)w. \tag{26}$$

Now, (26) is independent of (25) and depicts that the motion parallel to the ζ-axis is stable provided $1 \le \kappa < 1/(1 - 2A_1 - D)$.

Now, the characteristic equation (23) using (24) becomes

$$\begin{aligned}&\lambda^4 + \lambda^2\left\{2\kappa\left(\frac{2}{\kappa} - 1 + D - A_1\right)\right\}\\ &\quad + \kappa^2\left(D - 1 - 4A_1\right)\left(D - 1 + 2A_1\right) = 0,\end{aligned} \tag{27}$$

where $D > 1$, $A_1 > 0$.

The roots of (28) are

$$\Lambda_n = \frac{-P \pm \sqrt{\Delta}}{2}, \quad (n = 1, 2), \tag{28}$$

where

$$\begin{aligned}P &= 2\kappa\left(\frac{2}{\kappa} - 1 + D - A_1\right),\\ \Delta &= 4\left\{4\kappa\left(D - A_1\right) - 4\left(\kappa - 1\right) + 9\kappa^2 A_1^2\right\},\end{aligned} \tag{29}$$

Δ is the discriminant of (27) and is always positive for any κ. When $\kappa = 1$, the value of Δ fully coincides with that of Hallan and Rana [3] which ought to be $\Delta = 4\{4(D - A_1) + 9A_1^2\}$.

Now, since $\Delta > 0$, if in (28) the quantity in parenthesis is positive; that is, $P > 0$, which occurs when

$$0 < \kappa < \frac{2}{1 + A_1 - D}, \tag{30}$$

we see that $P > \sqrt{\Delta}$. Hence from (28), we have $-P \pm \sqrt{\Delta} < 0$, and so both values of Λ_n are negative and consequently, the roots (28) are distinct pure imaginary. Therefore, the equilibrium point $(\xi_3, 0, 0)$ is stable provided equation (30) holds; otherwise, it is unstable.

4.2. Circular Points. These equilibrium points exist only for $D = 1 - \upsilon$. The coordinates of any point on the circle $(\xi + \upsilon - 1)^2 + \eta^2 = 1$, $\zeta = 0$ are of the form $(1 - \upsilon - \cos\phi, \sin\phi, 0)$.

The partial derivatives at these points are

$$\begin{aligned}
\Omega^0_{\xi\xi} &= 3\kappa\upsilon\cos^2\phi,\\
\Omega^0_{\eta\eta} &= 3\kappa\upsilon\sin^2\phi,\\
\Omega^0_{\zeta\zeta} &= -1,\\
\Omega^0_{\xi\eta} &= -3\kappa\upsilon\sin\phi\cos\phi,\\
\Omega^0_{\eta\zeta} &= \Omega^0_{\xi\zeta} = 0.
\end{aligned} \tag{31}$$

When these are substituted in the variational equations (21) with $\zeta = 0$, we have

$$\begin{aligned}
u'' - 2v' &= 3\kappa\upsilon\cos\phi\,(u\cos\phi - v\sin\phi),\\
v'' + 2u' &= -3\kappa\upsilon\sin\phi\,(u\cos\phi - v\sin\phi),\\
w'' &= -w.
\end{aligned} \tag{32}$$

The last equation of (32) shows that motion is stable along the ζ-axis. The characteristic equation of the first two equations of (32) is

$$\lambda^2\left(\lambda^2 - 3\kappa\upsilon + 4\right) = 0. \tag{33}$$

Its roots are $\lambda_{1,2} = 0$, $\lambda_{3,4} = \pm\sqrt{-4 + 3\kappa\upsilon}$, and so the equilibrium points are unstable due to multiple zero roots.

4.3. Out-of-Plane Points. For the stability of the out-of-plane equilibrium points, we consider the following partial derivatives:

$$\begin{aligned}
\Omega^0_{\xi\xi} &= 1 + \frac{3\kappa^2(1-\upsilon-D)^2(\kappa-1-\kappa D)^{5/3}}{(\kappa\upsilon)^{2/3}},\\
\Omega^0_{\eta\eta} &= 1,\\
\Omega^0_{\xi\eta} &= \Omega^0_{\eta\zeta} = 0,\\
\Omega^0_{\xi\zeta} &= \frac{-3\zeta\kappa(1-\upsilon-D)(\kappa-1-\kappa D)^{5/3}}{(\kappa\upsilon)^{2/3}},\\
\Omega^0_{\zeta\zeta} &= \frac{3\zeta^2(\kappa-1-\kappa D)^{5/3}}{(\kappa\upsilon)^{2/3}}.
\end{aligned} \tag{34}$$

The characteristic equation in this case is gotten by substituting the trial solutions $u = A\exp(\lambda\tau)$, $v = B\exp(\lambda\tau)$, $w = C\exp(\lambda\tau)$ in the variational equations (21) to get

$$\lambda^6 + p\lambda^4 + q\lambda^2 + r = 0, \tag{35}$$

where

$$\begin{aligned}
p &= 4 - \Omega^0_{\xi\xi} + \Omega^0_{\eta\eta} + \Omega^0_{\zeta\zeta},\\
q &= \Omega^0_{\eta\eta}\left(\Omega^0_{\xi\xi} + \Omega^0_{\zeta\zeta}\right)\\
&\quad + \Omega^0_{\zeta\zeta}\left[\Omega^0_{\xi\xi} - 4\left(\Omega^0_{\xi\eta}\right)^2\right] - \left(\Omega^0_{\xi\zeta}\right)^2 - \left(\Omega^0_{\eta\zeta}\right)^2,\\
r &= \Omega^0_{\eta\eta}\left(\Omega^0_{\xi\zeta}\right)^2 + \left(\Omega^0_{\eta\zeta}\right)\left[\Omega^0_{\eta\zeta}\Omega^0_{\xi\xi} - 2\Omega^0_{\xi\eta}\Omega^0_{\xi\zeta}\right]\\
&\quad + \Omega^0_{\zeta\zeta}\left[\left(\Omega^0_{\xi\eta}\right)^2 - \Omega^0_{\xi\xi}\Omega^0_{\eta\eta}\right].
\end{aligned} \tag{36}$$

Substituting (34) in (35), we at once have

$$\begin{aligned}
&\lambda^6 + [5 - 3\kappa(1-D)]\lambda^4\\
&\quad + \left[1 - \frac{3(\kappa-1-\kappa D)}{\upsilon^{2/3}}\right.\\
&\qquad \left.\times\left\{2\upsilon^{2/3} - 3\kappa^{4/3}(1-\upsilon-D)^2(\kappa-1-\kappa D)^{2/3}\right\}\right]\lambda^2\\
&\quad + 3(1+\kappa D-\kappa)\\
&\qquad \times\left[1 - \frac{\kappa^{4/3}(\kappa-1-\kappa D)(1-\upsilon-D)^2}{\upsilon^{2/3}}\right] = 0.
\end{aligned} \tag{37}$$

Its roots are

$$\begin{aligned}
\lambda_{1,2} &\longrightarrow \pm\left[\frac{p}{3} + \frac{(p^2-q)}{3}\left(\frac{2}{N}\right)^{1/3} + \frac{N^{1/3}}{32^{1/3}}\right]^{1/2},\\
\lambda_{3,4} &\longrightarrow \pm\left[\frac{p}{3} - \frac{p^2}{32^{2/3}N^{1/3}} + \frac{ip^2}{2^{1/3}\sqrt{3}N^{1/3}} + \frac{q}{2^{2/3}N^{1/3}}\right.\\
&\qquad \left. - \frac{i\sqrt{3q}}{2^{2/3}N^{1/3}} + \frac{N^{1/3}}{22^{1/3}\sqrt{3}} + \frac{N^{1/3}}{62^{1/3}}\right]^{1/2},\\
\lambda_{5,6} &\longrightarrow \pm\left[\frac{p}{3} - \frac{p^2}{32^{2/3}N^{1/3}} - \frac{ip^2}{2^{2/3}\sqrt{3}N^{1/3}} + \frac{q}{2^{2/3}N^{1/3}}\right.\\
&\qquad \left. + \frac{i\sqrt{3q}}{2^{2/3}N^{1/3}} + \frac{N^{1/3}}{22^{1/3}\sqrt{3}} - \frac{N^{1/3}}{62^{1/3}}\right]^{1/2},
\end{aligned} \tag{38}$$

where $N = 2p^3 - 9pq + 27r + \sqrt{4(3q-p^2)^3 + (2p^3 - 9pq + 27r)^2}$, $p = 5 - 3\kappa(1-D)$, $q = 1 - (3(\kappa-1-\kappa D)/\upsilon^{2/3})\{2\upsilon^{2/3} - 3\kappa^{4/3}(1-\upsilon-D)^2(\kappa-1-\kappa D)^{2/3}\}$, $r = 3(1+\kappa D-\kappa)[1 - \kappa^{4/3}(\kappa-1-\kappa D)(1-\upsilon-D)^2/\upsilon^{2/3}]$.

The roots (37) are computed numerically for motion of a test particle (a submarine) under the gravitational attraction of the Earth-Moon system when $D > 0$ and $D < 0$ ($\delta_{01} < \delta_{03}$ and $\delta_{01} > \delta_{03}$, resp.) for $0 < \kappa < \infty$. Therefore, we take $\upsilon = 0.01$ and $\kappa \in [0.999001, 1.0111]$ with the following values in each case following Kaur and Aggarwal [4].

TABLE 4: The characteristic roots $\lambda_{1,2} = \pm\omega_1$, $\lambda_{3,4} = \pm\omega_2$, $\lambda_{5,6} = \pm\omega_3$ for $\upsilon = 0.01$, $D = -2572.699$.

κ	ω_1	ω_2	ω_3
0.999001	$3526.63 + 3526.09i$	$2.13681i$	$3526.63 - 3526.09i$
1	$3529.28 + 3528.73i$	$2.13716i$	$3529.28 - 3528.73i$
1.001	$3531.93 + 3531.38i$	$2.13752i$	$3531.93 - 3531.38i$
1.002	$3534.57 + 3534.03i$	$2.13787i$	$3534.57 - 3534.03i$
1.003	$3537.22 + 3536.67i$	$2.13823i$	$3537.22 - 3536.67i$
1.004	$3539.86 + 3539.32i$	$2.13859i$	$3539.86 - 3539.32i$
1.009	$3553.08 + 3552.53i$	$2.14036i$	$3553.08 - 3552.53i$
1.01	$3555.72 + 3555.17i$	$2.14071i$	$3555.72 - 3555.17i$
1.0111	$3558.62 + 3558.08i$	$2.1411i$	$3558.62 - 3558.08i$

In the case when $\delta_{01} < \delta_{03}$, we take

mass of test particle $m_3 = 1848632\,\text{kg}$,

density of salt water $\delta_{01} = 1027\,\text{kg/m}^3$,

density of test particle $\delta_{03} = 1204.39\,\text{kg/m}^3$.

In new units, we have

$$\delta_{01} = 9317.7, \delta_{03} = 10927.11, \text{and so } D = 1372.4.$$

Similarly, when $\delta_{01} > \delta_{03}$, we take

mass of test particle $m_3 = 1449832\,\text{kg}$,

density of salt water $\delta_{01} = 1027\,\text{kg/m}^3$,

density of submarine $\delta_{03} = 804.79\,\text{kg/m}^3$.

In new units, we have

$$\delta_{01} = 9317.7, \delta_{03} = 7301.65\ , \text{and so } D = -2572.699.$$

Aside from these examples, we also consider the case when $D = \pm 0.001$ (see Tables 1 and 3), so that a wider generalization can be reached regarding the characteristic roots (38) which consequently determines whether the equilibrium point is a stable one or not.

Using the software package *Mathematica*; the six characteristic roots are presented in Tables 4, 5, 6, and 7 numerically for $\upsilon = 0.01$, for different density parameters and a wide range of the parameter κ. We seek to find the case where all the six roots are pure imaginary quantities or complex figures with negative real parts. If this happens, then the solutions will be bounded and motion will be stable; otherwise, they will be unstable.

From these tables, we see that that for a specific set of values of these parameters at least one of the roots among all has a positive real part or a complex root with the existence of a positive real part. Therefore, this causes the solutions to be unbounded and consequently producing unstable equilibrium points. Hence, we conclude that the out-of-plane equilibrium points are unstable equilibrium points due to a positive root and positive real part in complex roots. This agrees with the result of Singh [5].

The equilibrium solutions of the nonautonomous system with variable coefficients are in general unstable points according to the Lyapunov's theorem of stable solutions [19].

5. Discussion and Conclusion

We have derived the equations of motion and established the possible equilibrium points of the third body of infinitesimal mass in a setup of Robe's [1] restricted three-body problem when the three participating bodies all vary their masses arbitrarily at the same rate and the density of the fluid and third body also vary as the masses. We find that the nonautonomous equations of motion are different from that of the restricted problem of three variable mass bodies derived by Bekov et al. [15], while the autonomized equations of our study are also different from those of Robe [1] and Hallan and Rana [3] due to the assumptions we have introduced here. The autonomized system of (6) is different from that in Hallan and Rana [3] only due to an additional term which appears due to the parameter κ. When $\kappa = 1$, the equations become fully analogous to theirs.

The equilibrium points are sought, and it is seen that the point at the center of the fluid is always an equilibrium point of the Robe problem. An equilibrium point near the center of the fluid, points on the circle (circular points), and two out-of-plane points on the $\xi\zeta$-plane also exist, with all similar to that found in Hallan and Rana [3]; though the later have several points in our case while there exists only a pair in that of Hallan and Rana [3].

The linear stability of the equilibrium points of the autonomized have been studied and the outcomes are analogous with the stability results in Hallan and Rana [3], in the sense that the equilibrium points collinear with the centers of the fluid and the second primary are stable under some given conditions which depend on the mass ratio, density parameter, and the parameter κ. The circular are unstable due to the presence of multiple zero roots, while the out-of-plane equilibrium points are also unstable due to a positive root and a positive real part of the complex roots.

In our recent paper, Singh and Leke [12], the motion of a test particle around the equilibrium points was generalized to include the effect of mass variations of the primaries which vary isotropically in accordance with the unified Meshcherskii law, when the motion of the primaries is determined by the Gylden-Meshcherskii problem. Here, ejection or attachment form or to the surfaces of the primaries do not create reactive forces. Also, we have taken the first primary as

TABLE 5: The characteristic roots $\lambda_{1,2} = \pm\omega_1, \lambda_{3,4} = \pm\omega_2, \lambda_{5,6} = \pm\omega_3$ for $v = 0.01$ and $D = -0.001$.

κ	ω_1	ω_2	ω_3
0.999001	0.0000547723	$2.1781 \times 10^{-7} + i$	$2.1781 \times 10^{-7} - i$
1	0.0541419	$0.0214162 + 1.00021i$	$0.0214162 - 1.00021i$
1.001	0.0756118	$0.0379596 + 1.00065i$	$0.0379596 - 1.00065i$
1.002	0.0913455	$0.0529917 + 1.00124i$	$0.0529917 - 1.00124i$
1.003	0.103949	$0.0670958 + 1.00195i$	$0.0670958 - 1.00195i$
1.004	0.114444	$0.080533 + 1.00276i$	$0.080533 - 1.00276i$
1.009	0.148257	$0.141477 + 1.00796i$	$0.141477 - 1.00796i$
1.01	0.152474	$0.152799 + 1.00919i$	$0.152799 - 1.00919i$
1.0111	0.156379	$0.165005 + 1.01059i$	$0.165005 - 1.01059i$

TABLE 6: The characteristic roots $\lambda_{1,2} = \pm\omega_1, \lambda_{3,4} = \pm\omega_2, \lambda_{5,6} = \pm\omega_3$ for $v = 0.01$ and $D = 0.001$.

κ	ω_1	ω_2	ω_3
0.999001	$0.0336136 + 0.980468i$	$0.000207226 - 0.0789312i$	$0.0322915 - 1.02011i$
1	$0.0187794 + 0.989012i$	$0.000045675 - 0.0553234i$	$0.0183652 - 1.01118i$
1.001	$0.0000595752 + 0.999965i$	$1.42435 \times 10^{-11} - 0.00173207i$	$0.0000595711 - 1.00003i$
1.002	0.0540899	$0.0213661 + 1.00021i$	$0.0213661 - 1.00021i$
1.003	0.0755428	$0.0378711 + 1.00065i$	$0.0378711 - 1.00065i$
1.004	0.0912669	$0.0528684 + 1.00123i$	$0.0528684 - 1.00123i$
1.009	0.137552	$0.117813 + 1.00566i$	$0.117813 - 1.00566i$
1.01	0.14327	$0.129617 + 1.00677i$	$0.129617 - 1.00677i$
1.0111	0.148654	$0.142293 + 1.00805i$	$0.142293 - 1.00805i$

TABLE 7: The characteristic roots $\lambda_{1,2} = \pm\omega_1, \lambda_{3,4} = \pm\omega_2, \lambda_{5,6} = \pm\omega_3$ for $v = 0.01$ and $D = 1372.4$.

κ	ω_1	ω_2	ω_3
0.999001	$2426.1 + 1401.08i$	$1400.71 - 2426.74i$	$0.962167 - 1.66652i$
1	$2427.92 + 1402.13i$	$1401.76 - 2428.56i$	$0.962327 - 1.6668i$
1.001	$2429.74 + 1403.18i$	$1402.81 - 2430.38i$	$0.962488 - 1.66708i$
1.002	$2431.56 + 1404.23i$	$1403.86 - 2432.2i$	$0.962648 - 1.66735i$
1.003	$2433.38 + 1405.28i$	$1404.91 - 2434.02i$	$0.962808 - 1.66763i$
1.004	$2435.2 + 1406.33i$	$1405.96 - 2435.84i$	$0.962967 - 1.66791i$
1.009	$2444.29 + 1411.58i$	$1411.21 - 2444.93i$	$0.963764 - 1.66929i$
1.01	$2446.1 + 1412.63i$	$1412.26 - 2446.74i$	$0.963923 - 1.66956i$
1.0111	$2448.1 + 1413.78i$	$1413.41 - 2448.74i$	$0.964098 - 1.66987i$

a rigid spherical shell filled with a fluid of constant density and volume and containing the test particle, while in the present study, we have assumed that the first primary is a fluid in the shape of a sphere with nonisotropic mass and density variation, given that, the test particle is contained in the fluid. The second primary and the test particle both have their masses varying arbitrarily with time at the same rate as the first primary.

In the previous study, the autonomized dynamical system with constant coefficients is gotten, only when the shell is empty or when the densities of the medium and the test particle are equal, while in the present study, such limitation do not arise. In the present study, we found two collinear equilibrium points on the line joining the centers of the fluid and the second primary with one at the center of the fluid and the other away from it. Further, circular equilibrium points exist on the $\xi\eta$-plane and pairs of out-of-plane points which depend on the arbitrary constant $\kappa \in (0, \infty)$, and density and mass parameters are found on $\xi\zeta$-plane. However, in the previous study, there is only one collinear equilibrium point located at the center of the rigid shell and a pair of the out-of plane equilibrium points which exist only for $\kappa > 1$.

The linear stability analysis however turns out to be same as the equilibrium points on the line collinear with the centers of the primary of the autonomized system which are conditionally stable; while the equilibrium points on the $\xi\zeta$-plane and the circular points are unstable.

The result of our research work can be summarized as follows. The restricted problem under the framework of the Robe's [1] problem with three variable mass bodies, which vary arbitrarily with time at the same rate, has the equilibrium points which; are, the points near the center of the fluid, points on the circle (circular points), and pairs of out-of-plane points. These equilibrium points are analogous to the problem studied by Hallan and Rana [3] and their stability results are similar, though we have assumed that the first primary is a spherical fluid and that the masses of the test particle and primaries vary with time. The idea of the submarine's mass changing with time may perhaps seem impossible because it is a rigid body. However, from Archimedes' principle, a submarine floating or submerging depends on buoyancy which is controlled by the ballast tanks found between the submarine's inner and outer hulls. Submarine resting on the water surface has positive buoyancy which means it is less dense than water. When this happens, the ballast tanks are empty. For it to submerge, vents on top of the tanks are opened and water floods in thereby making the submarine denser than the sea water. In this case, the submarine has negative buoyancy. For it to float again, the water in the tanks will be forced out and the submarine becomes less dense than the water and eventually floats. In this study, the buoyancy force depends on the mass of the submarine m_3, the densities of the submarine, and the medium, as well as the volume of the submarine. And so by the simple relation, connecting mass, density, and volume, we understand that, when water is allowed into the ballast, as the mass of the submarine is changing so is the density until it becomes denser than the medium and sinks. When water is let out of the tanks, the mass and density of the submarine reduce and as a matter of fact, it will float when the medium is denser. Neutral buoyancy is attained when the weight of the submarine equals the amount of water it displaces. The submarine will neither rise nor sink in this state. Hence, with changing mass comes a changing density while the volume remains a constant.

This study may be useful in the investigations of the dynamic problem of Earth-size planets covered completely by a water envelope (water planets), and also the study of the small oscillation of the Earth's inner core taking into account the Moon's attraction during the course of evolution. The problem discussed in this paper is highly idealized and therefore calls for more research.

References

[1] H. A. G. Robe, "A new kind of 3-body problem," *Celestial Mechanics and Dynamical Astronomy*, vol. 16, no. 3, pp. 343–351, 1977.

[2] A. R. Plastino and A. Plastino, "Robe's restricted three-body problem revisited," *Celestial Mechanics and Dynamical Astronomy*, vol. 61, pp. 197–206, 1995.

[3] P. P. Hallan and N. Rana, "The existence and stability of equilibrium points in the Robe's restricted three-body problem," *Celestial Mechanics and Dynamical Astronomy*, vol. 79, no. 2, pp. 145–155, 2001.

[4] B. Kaur and R. Aggarwal, "Robe's problem: its extension to 2 + 2 bodies," *Astrophysics and Space Science*, vol. 339, no. 2, pp. 283–294, 2012.

[5] J. Singh and L. M. Hafsah, "Robe's circular restricted three-body problem under oblate and triaxial primaries," *Earth, Moon, and Planets*, vol. 109, pp. 1–11, 2012.

[6] M. Dufour, *Comptes Rendus Hebdomadairesde l'Accademie de Sciences*, pp. 840–842, 1886.

[7] H. Gylden, "Die Bahnbewegungen in einem Systeme von zwei Körpern in dem Falle, dass die Massen Veränderungen unterworfen sind," *Astronomische Nachrichten*, vol. 109, no. 1-2, pp. 1–6, 1884.

[8] I. V. Meshcherskii, "Ein Specialfall des Gyldén'schen Problems (A. N. 2593)," *Astronomische Nachrichten*, vol. 132, no. 9, pp. 129–130, 1893.

[9] I. V. Meshcherskii, "Ueber die Integration der Bewegungsgleichungen im Probleme zweier Körper von veränderlicher Masse," *Astronomische Nachrichten*, vol. 159, pp. 229–242, 1902.

[10] J. Singh and O. Leke, "Stability of the photogravitational restricted three-body problem with variable masses," *Astrophysics and Space Science*, vol. 326, pp. 305–314, 2010.

[11] D. Veras, M. C. Wyatt, A. J. Mustill, A. Bonsor, and J. J. Eldridge, "The great escape: how exoplanets and smaller bodies desert dying stars," *Monthly Notices of the Royal Astronomical Society*, vol. 417, no. 3, pp. 2104–2123, 2011.

[12] J. Singh and O. Leke, "Existence and stability of equilibrium points in the Robe's restricted three-body problem with variable masses," *International Journal of Astronomy and Astrophysics*, vol. 3, pp. 113–122, 2013.

[13] E. P. Razbitnaya, "The problem of two bodies with variable masses—classification of different cases," *Soviet Astronomy*, vol. 29, pp. 684–687, 1985.

[14] A. A. Bekov, "Integrable cases of Hamilton-Jacobi equation and the restricted, rectilinear three-body problem with variable masses," *Akademiya Nauk Kazakhskoi SSR, Trudy Astrofizicheskogo Instituta*, vol. 12, p. 47, 1987.

[15] A. A. Bekov, A. N. Beysekov, and L. T. Aldibaeva, "On the dynamics of non-stationary binary stellar systems with non-isotropic mass flow," *Astronomical & Astrophysical Transactions*, vol. 24, pp. 311–316, 2005.

[16] P. S. Letelier and T. A. da Silva, "Solutions to the restricted three-body problem with variable mass," *Astrophysics and Space Science*, vol. 332, no. 2, pp. 325–329, 2011.

[17] B. E. Gelf'gat, *Modern Problems of Celestial Mechanics and Astrodynamics*, Nauka, Moscow, Russia, 1973.

[18] A. A. Bekov, "Libration points of the restricted problem of three bodies of variable mass," *Soviet Astronomy*, vol. 32, no. 1, pp. 106–107, 1988.

[19] M. L. Krasnov, A. I. Kiselyov, and G. I. Makarenko, *A Book of Problems in Ordinary Differential Equations*, MIR, Moscow, Russia, 1983.

16

Invariant Imbedding and the Radiation Transfer in a Plane-Parallel Inhomogeneous Atmosphere

Arthur G. Nikoghossian

Ambartsumian Byurakan Astrophysical Observatory, Byurakan, 378433 Aragatsotn, Armenia

Correspondence should be addressed to Arthur G. Nikoghossian; nikoghoss@yahoo.com

Academic Editors: M. S. Dimitrijevic, M. Kueppers, and S. Wedemeyer-Bohm

The invariant imbedding technique is applied to the problems of radiation transfer in a plane-parallel inhomogeneous atmosphere. All the parameters which describe the elementary event of scattering and the distribution of the energy sources are allowed to vary with depth. Mathematically, the considered standard problems of the theory are reduced to initial-value problems which are better adapted to capabilities of the modern high speed computers. The reflectance of an atmosphere is shown to play a prominent role in describing the diffusion process since all the other characteristics of the radiation field are expressed through it. Three transfer problems frequently encountered in astrophysical applications are discussed: the radiation diffusion in the source-free medium, in a medium with arbitrarily distributed energy sources, as well as the problem of finding the statistical mean quantities, characteristics of the multiple scattering in the atmosphere.

1. Introduction

The fast progress in observational capabilities of astrophysical instruments enables to obtain a fairly detailed picture of investigated phenomena in cosmic objects. The high-resolution spectra available nowadays afford an opportunity to study different types of inhomogeneities, the theoretical interpretation of which encounters, in general, much difficulty. Additional difficulties appear in the line-formation problems when one has to take proper account of the multiple scattering effects. This is due to coupling set in between various volumes of the radiating medium and the redistribution of radiation over frequency and directions. The classical treatment of such problems usually leads to integrodifferential equations with the conditions specified at the boundaries of the medium.

The mathematical complexity of these problems formulated for *homogeneous* media, stimulated to develop a variety of analytical techniques applicable to one or another specific class of the radiation transfer problems. There exists a vast literature on the field, particularly in astrophysical context. Of different methods concerning our discussion the most important is Ambartsumian's invariance principle [1, 2], which overcomes the above difficulties by finding the requisite intensity of emerging radiation without prior knowledge of the radiation field in the entire atmosphere. An alternative approach developed by Bellman [3] and Sobolev [4, 5] is based on extensive use of the so-called "surface" resolvent function. The idea of this approach, in its turn, goes back to Kreĭn [6].

From the pioneering works treating the transfer problems is in inhomogeneous absorbing and scattering atmospheres we note here the papers by Preisendorfer [7, 8] and Busbridge [9]. Later the theory was developed by Sobolev [4, 10] and Yanovitskij [11]. The further progress of the theory is based on Ambartsumian's method of addition of layers [1, 12] generalised by the present author [13–16] over the case of *inhomogeneous* media. In these papers we proposed also a new approach for solving the linear radiation transfer problems which assumes a preliminary determination of the global optical properties of an atmosphere for a family of atmospheres with different optical thicknesses. This appreciably facilitates finding of the internal field of radiation. For instance, in the simplest scalar one-dimensional case, knowledge of the reflection coefficient alone is sufficient to determine the radiation field inside the medium without solving any new equations [14].

In this paper we use the invariant imbedding technique [17–19] to inhomogeneous atmospheres with the plane-parallel geometry and reformulate the classical boundary-value problems to reduce them to initial-value problems. The numerical solution of resulting problems for integrodifferential equations is easy to obtain on the modern high speed electronic computers and, what is important, they are usually numerically stable. We show that, again, as in the scalar case, one needs to solve only one such problem for the reflectance of the medium since the other quantities of interest are found from explicit formulas.

It is important to note that under inhomogeneous atmosphere in this paper we mean an atmosphere in which any parameter defining the elementary processes of absorption and diffusion (the profile of the absorption coefficient, the probability of reradiation in scattering, and the law of redistribution of radiation over directions and frequencies as well as the role of absorption in the continuum) can vary with depth. In theoretical treatments one usually takes the averaged in some sense values of these parameters and restricts oneself to taking account of the depth-dependent distribution of the internal energy sources due to changes in characteristic thermodynamic parameters within the medium [20, 21].

The outline of the paper is as follows. We begin in Section 2 by treating the source-free problem for a plane-parallel inhomogeneous atmosphere of finite optical thickness. The case of isotropic scattering with complete redistribution over frequency is discussed for expository reason. In Section 3, we consider the problem of the radiation transfer in an atmosphere with arbitrarily distributed energy sources. It is shown that the radiation field in this case can be found without solving any new equation. Next section is devoted to statistical description of the radiation diffusion process. The mean number of scattering events and the average time of the photons diffusion in the atmosphere are found. The obtained results are discussed in the final section.

2. The Problem of Diffuse Reflection and Transmission

Consider the radiation transfer through a plane-parallel inhomogeneous atmosphere of finite optical thickness τ_0 in the centre of the spectral line. For simplicity, the scattering process is assumed isotropic with complete redistribution of radiation over frequencies. We limit ourselves by assuming the depthdependence for only the scattering coefficient. The interested reader can with only small effort write down the proper equations for a more general situation. The broadening of the spectral line is generally described by the Voigt profile of the absorption coefficient $\alpha(x) = H(x, a)$, where a is the Voigt parameter and x is the so-called dimensionless frequency measured by the displacement from the centre of the line in the units of the Doppler widths. The normalisation factor of the Voigt function is $A = 1/\sqrt{\pi}$. The role of absorption in the continuum is specified by the parameter β

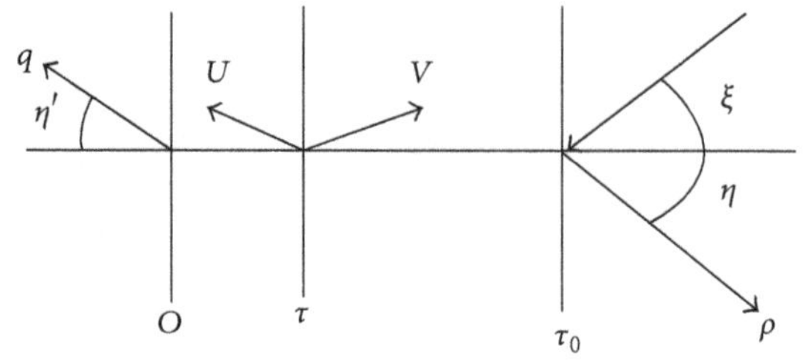

FIGURE 1: Schematic illustration of the radiation transfer in a finite plane-parallel medium illuminated from the side of boundary $\tau = \tau_0$.

which is the ratio of the absorption coefficient in continuum to that in the centre of the spectral line. We introduce the notation $\lambda(\tau)$ for the depth-dependent probability of the photon reradiation during elementary event of scattering.

We begin by determining the global optical characteristics of the medium when it is illuminated from the side of the boundary $\tau = \tau_0$ (Figure 1). We denote the azimuth-averaged reflection coefficient by $\rho(x', \eta; x, \xi, \tau_0)$, where x and ξ are the frequency and cosine of the angle of the incident photon and x', η are the similar quantities for the reflected photon. It is introduced in such a way that ρ/ξ has a probabilistic meaning.

The function ρ satisfies the equation [22, 23]

$$\frac{d\rho}{d\tau_0} = -\left[\frac{\gamma(x')}{\eta} + \frac{\gamma(x)}{\xi}\right]\rho(x', \eta; x, \xi, \tau_0) + \frac{\tilde{\lambda}(\tau_0)}{2}\varphi(x', \eta, \tau_0)\varphi(x, \xi, \tau_0), \tag{1}$$

where $\tilde{\lambda} = A\lambda$, $\gamma(x) = \alpha(x) + \beta$ is assumed to be independent of the optical depth,

$$\varphi(x, \xi, \tau_0) = \alpha(x) + \int_{-\infty}^{\infty} \alpha(x')\,dx' \times \int_0^1 \rho(x', \eta'; x, \xi, \tau_0)\frac{d\eta'}{\eta'}, \tag{2}$$

and $\rho(x', \eta; x, \xi, 0) = 0$.

Proceeding to the transmission coefficient, $q(x', \eta; x, \xi, \tau_0)$, we introduce the notation $\sigma(x', \eta; x, \xi, \tau_0)$ for its diffuse part, so that

$$\alpha(x)\, q(x', \eta; x, \xi, \tau_0) = \eta\delta(x' - x)\,\delta(\eta - \xi)\exp\left[-\frac{\gamma(x)}{\xi}\tau_0\right] + \alpha(x)\,\sigma(x', \eta; x, \xi, \tau_0). \tag{3}$$

The function σ is determined from equation

$$\frac{d\sigma}{d\tau_0} = -\frac{\gamma(x)}{\xi}\sigma(x', \eta; x, \xi, \tau_0) + \frac{\tilde{\lambda}(\tau_0)}{2}\varphi(x, \xi, \tau_0)\,\psi(x', \eta, \tau_0) \tag{4}$$

with

$$\psi\left(x',\eta,\tau_0\right)=\int_{-\infty}^{\infty}\alpha\left(x''\right)dx''\int_0^1 q\left(x',\eta;x'',\eta',\tau_0\right)\frac{d\eta'}{\eta'}, \tag{5}$$

or

$$\psi\left(x',\eta,\tau_0\right)=\exp\left[-\frac{\gamma(x)}{\eta}\tau_0\right]+\int_{-\infty}^{\infty}\alpha\left(x''\right)dx''\int_0^1 \sigma\left(x',\eta;x'',\eta',\tau_0\right)\frac{d\eta'}{\eta'}, \tag{6}$$

and $\sigma(x',\eta;x,\xi,0)=0$ as an initial condition.

Equations (1) and (4) are obtained by using the invariant imbedding standard procedure, that is, by adding an infinitely thin layer to the boundary $\tau=\tau_0$ and then letting its thickness tend to zero in the limit (see for details [14]). We use the same procedure below in obtaining the internal field of radiation.

Let us begin with the quantity $U(x',\eta,\tau;x,\xi,\tau_0)$ which specifies the probability that the incident photon will be found after multiple scattering at depth τ as a photon within the frequency and direction intervals $(x',x'+dx';\eta,\eta+d\eta)$. Here, again, it is expedient to separate out its diffuse part u by analogy to that for transmission coefficient:

$$\alpha(x)U\left(x',\eta,\tau;x,\xi,\tau_0\right)=\eta\delta(\eta-\xi)\delta\left(x-x'\right)\times\exp\left[-\frac{\gamma(x)}{\xi}(\tau_0-\tau)\right]+\alpha(x)u\left(x',\eta,\tau;x,\xi,\tau_0\right). \tag{7}$$

Performing the invariant imbedding procedure we arrive at

$$\frac{du}{d\tau_0}=-\frac{\gamma(x)}{\xi}u\left(x',\eta,\tau;x,\xi,\tau_0\right)+\frac{\tilde{\lambda}(\tau_0)}{2}\varphi(x,\xi,\tau_0)\Psi\left(x',\eta,\tau;\tau_0\right), \tag{8}$$

where

$$\Psi\left(x',\eta,\tau;\tau_0\right)=\exp\left[-\frac{\gamma(x)}{\eta}(\tau_0-\tau)\right]+\int_{-\infty}^{\infty}\alpha\left(x''\right)dx''\times\int_0^1 u\left(x',\eta,\tau;x'',\eta',\tau_0\right)\frac{d\eta'}{\eta'} \tag{9}$$

and $u(x',\eta,\tau;x,\xi,\tau)=0$.

It is obvious that $u(x',\eta,0;x,\xi,\tau_0)=\sigma(x',\eta;x,\xi,\tau_0)$ and $\Psi(x',\eta,0;\tau_0)=\psi(x',\eta,\tau_0)$, so that on solving (8) one also finds the transmission coefficient for a family of media. When one needs to determine the function u for different depths but for an atmosphere with a fixed optical thickness, that is, in order to solve the ordinary transfer equations, knowledge of the transmission coefficient makes it possible to deal again with the initial-value problem. Note also that once the reflectance of the atmosphere is determined (i.e., the function φ is known), one can use (8) to derive an explicit formula for the function u.

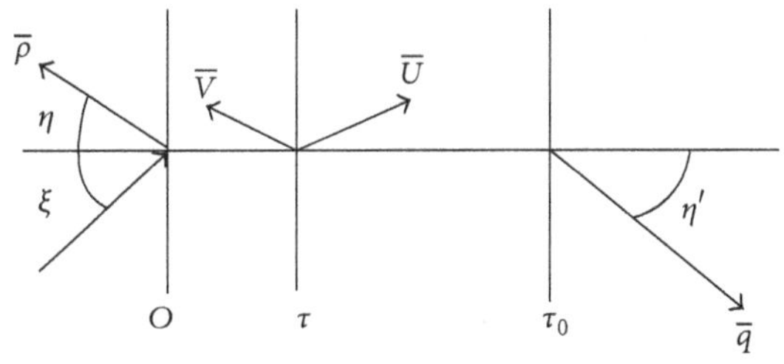

FIGURE 2: Schematic illustration of the radiation transfer in a finite plane-parallel medium illuminated from the side of boundary $\tau=0$.

Now we will show that knowledge of the functions $\rho(x',\eta;x,\xi,\tau_0)$ and $u(x',\eta,\tau;x,\xi,\tau_0)$ is essentially enough to find the other quantities of interest. For instance, on the base of simple physical arguments, we find

$$V\left(x',\eta,\tau;x,\xi,\tau_0\right)=\int_{-\infty}^{\infty}\alpha\left(x''\right)dx''\times\int_0^1\rho\left(x',\eta;x'',\eta',\tau_0\right)U\left(x'',\eta',\tau;x,\xi,\tau_0\right)\frac{d\eta'}{\eta'}. \tag{10}$$

Up to now we have considered the case of a medium illuminated from the side of the boundary $\tau=\tau_0$. Let the medium be illuminated now from the opposite side (see Figure 2). We shall see that the quantities ρ and u found above fully determine the optical characteristics and the internal field of radiation also in this problem (the quantities applying to this case will be supplied by an overhead bar). For example, the invariant imbedding method leads to the following equation for the reflection coefficient $\overline{\rho}(x',\eta;x,\xi,\tau_0)$:

$$\frac{d\overline{\rho}}{d\tau_0}=\psi(x,\xi,\tau_0)\psi\left(x',\eta,\tau_0\right) \tag{11}$$

with the initial condition $\overline{\rho}(x',\eta;x,\xi,0)=0$. It is easily seen that solution of (11) is reduced to computing the ordinary integral. As for the transmission coefficient, the reversibility principle of optical phenomena implies $\overline{q}(x',\eta;x,\xi,\tau_0)=q(x,\xi;x',\eta,\tau_0)$.

The radiation field inside the medium found with the same method is reduced to computing simple integrals. For instance, the function $\overline{V}(x',\eta,\tau;x,\xi,\tau_0)$, which specifies the intensity of radiation directed to the boundary $\tau=0$, satisfies equation

$$\frac{d\overline{V}}{d\tau_0}=\frac{\lambda(\tau_0)}{2}\psi(x,\xi,\tau_0)\Psi\left(x',\eta,\tau,\tau_0\right) \tag{12}$$

with the initial condition $\overline{V}(x', \eta, \tau; x, \xi, \tau) = 0$. Here again, since the right-hand side of (12) is known, its solution is equivalent to computing the following integral:

$$\overline{V}\left(x', \eta, \tau; x, \xi, \tau_0\right) = \frac{1}{2}\int_{\tau}^{\tau_0} \lambda\left(\tau_0'\right)\psi\left(x, \xi, \tau_0'\right)\psi\left(x', \eta, \tau_0'\right)d\tau_0'. \quad (13)$$

Knowledge of $\overline{V}$ allows, in its turn, finding the last of requisite quantities yielding the intensity of radiation in the opposite direction:

$$\overline{U}\left(x', \eta, \tau; x, \xi, \tau_0\right) = \psi\left(x, \xi, \tau\right) + \int_{\tau}^{\tau_0} \alpha\left(x''\right)dx'' \times \int_0^1 \rho\left(x', \eta; x'', \eta', \tau_0\right)\overline{V}\left(x'', \eta', \tau; x, \xi, \tau_0\right)\frac{d\eta'}{\eta'}. \quad (14)$$

This completes the solution of the source-free problem. We saw that this requires solving only two initial-value problems (1) and (8) for integrodifferential equations. Let us turn next to the solution of another problem frequently encountered in astrophysical applications.

3. Internal Energy Sources

Once the problem of diffuse reflection and transmission is solved, one may obtain much easier the solution of another standard problem concerning the radiation field in an atmosphere containing energy sources. One of the various parameters whose distribution influenced the observed spectra is the power of the internal energy sources, which are specified by the values of thermodynamic parameters and are generally distributed nonuniformly in the atmosphere. Thus, the problem of determining the effects of an inhomogeneous distribution of the internal energy sources naturally arises in any realistic astrophysical problem of the spectra interpretation.

Consider a three-dimensional plane-parallel and inhomogeneous atmosphere of finite optical thickness containing energy sources of the power $B(\tau, x, \eta)$. We denote the intensities of radiation emerging from the medium through the boundaries $\tau = \tau_0$ and $\tau = 0$ by $I_1(x, \xi, \tau_0)$ and $I_2(x, \xi, \tau_0)$, respectively. These quantities are easy to find taking into account the reversibility principle of the optical phenomena applied to the functions U and V, which now can be interpreted as the probabilities that the photons moving at the depth τ to the right and to the left in Figure 1 will escape the atmosphere from the boundary $\tau = \tau_0$. Then we may write

$$I_1\left(x, \xi, \tau_0\right) = \int_{-\infty}^{\infty} dx' \int_0^1 d\eta \times \int_0^{\tau_0} B\left(\tau, x', \eta\right)W\left(x', \eta, \tau; x, \xi, \tau_0\right)d\tau, \quad (15)$$

where $W = U + V$; that is,

$$W\left(x', \eta, \tau; x, \xi, \tau_0\right) = \int_{-\infty}^{\infty} dx'' \int_0^1 \chi\left(x', \eta; x'', \eta', \tau\right)U\left(x, \xi, \tau; x'', \eta', \tau_0\right)\frac{d\eta'}{\eta'},$$

$$\chi\left(x', \eta; x, \xi, \tau\right) = \xi\delta\left(x - x'\right)\delta\left(\eta - \xi\right) + \alpha\left(x\right)\rho\left(x', \eta; x, \xi, \tau\right). \quad (16)$$

Similarly, having in mind the alternative probability meaning of the functions $\overline{U}$ and $\overline{V}$, we find

$$I_2\left(x, \xi, \tau_0\right) = \int_{-\infty}^{\infty} dx' \int_0^1 d\eta \times \int_0^{\tau_0} B\left(\tau, x', \eta\right)\overline{W}\left(x', \eta, \tau; x, \xi, \tau_0\right)d\tau, \quad (17)$$

where

$$\overline{W}\left(x', \eta, \tau; x, \xi, \tau_0\right) = \int_{-\infty}^{\infty} dx'' \int_0^1 \chi\left(x', \eta; x'', \eta', \tau\right)\overline{V}\left(x, \xi, \tau; x'', \eta', \tau_0\right)\frac{d\eta'}{\eta'}. \quad (18)$$

We now proceed to the problem of determining the radiation field inside the treated atmosphere. To this end, we introduce the notation $I^-(x, \eta, \tau; \tau_0)$ for the intensity of radiation of frequency x at the optical depth τ directed to the boundary $\tau = 0$ at the angle $\cos^{-1}\eta$. Similarly, the intensity directed to the boundary $\tau = \tau_0$ is denoted by $I^+(x, \eta, \tau; \tau_0)$. The angles in both cases are referenced from the normal to the corresponding boundary. The customary invariant imbedding procedure yields

$$\frac{dI^-\left(x, \eta, \tau; \tau_0\right)}{d\tau_0} = B\left(\tau_0, x, \eta\right)\Psi\left(x, \eta, \tau; \tau_0\right) + \frac{\lambda\left(\tau_0\right)}{2}\alpha\left(x\right)\int_{-\infty}^{\infty}\alpha\left(x'\right)dx' \times \int_0^1 U\left(x, \eta, \tau : x', \eta', \tau_0\right)I_1\left(x', \eta', \tau_0\right)\frac{d\eta'}{\eta'} \quad (19)$$

with the condition $I^-(x, \eta, \tau;\ 0) = 0$. It is discernable that all the quantities appearing in the right-hand side of (19) are known so that the problem is simply reduced to calculating of the ordinary integral.

Once $I^{-}(x,\eta,\tau;\tau_0)$ is known, it is easy to find the last of the requisite quantities, $I^{+}(x,\eta,\tau;\tau_0)$, since they are related by a simple formula with an obvious physical significance:

$$I^{+}(x,\eta,\tau;\tau_0) = I_1(x,\eta,\tau) + \int_{-\infty}^{\infty} \alpha(x')\,dx' \times \int_0^1 \rho(x,\eta;x',\eta',\tau)\, I^{-}(x',\eta',\tau;\tau_0)\,\frac{d\eta'}{\eta'}. \tag{20}$$

Thus, solving the diffuse reflection and transmission problem of the preceding section is an important prelude to solving the radiation transfer problem in atmospheres containing energy sources. Finally, we remind that the approach used here provides a solution to the problem for a family of atmospheres of different optical thicknesses.

4. The Statistical Description of the Radiation Diffusion

In treating different astrophysical problems, one often needs to estimate various statistical averages characterizing the radiation diffusion in an atmosphere. This facilitates better understanding of the physical essence of a number of effects predicted by the mathematical solution of the problem. The statistical investigation of multiple scattering makes it possible to determine a number of important physical characteristics of an atmosphere such as the mean radiation density and the mean degree of excitation of the atoms. One of the important statistical averages is the mean number of scattering events (MNS) underwent by the photon during its travel in the atmosphere. Of the extensive literature on this topic we mention here Ambartsumian's pioneering work [1, 24], in which for determining this quantity for homogeneous atmosphere the following formula was proposed:

$$N = \lambda \frac{\partial \ln I}{\partial \lambda}, \tag{21}$$

where I is the radiation intensity. As it was shown in [25], this formula is valid for any flux of "moving" photons (i.e., not for those subsequently destroyed in the medium). The problem of estimating the MNS in the case of inhomogeneous media was examined in [14], and it was shown that the procedure of the formal differentiation over λ remains in force also in this general case despite the fact that now the scattering coefficient varies with depth. This may be concluded by observing the way this function enters in the proper equations.

Consider, for instance, the statistics of multiple scattering of the photon of frequency x incident on the boundary $\tau = \tau_0$ of the medium at the angle $\cos^{-1}\xi$. We are interested in the MNS for three types of photons: reflected, transmitted, and destroyed in the medium independent of their final frequency and direction. Denoting the probabilities for each of these processes by $\widehat{\rho}(x,\xi,\tau_0)$, $\widehat{q}(x,\xi,\tau_0)$, and $\widehat{s}(x,\xi,\tau_0)$, it easy to derive from (1)–(4)

$$\frac{d\widehat{\rho}}{d\tau_0} = -\frac{\gamma(x)}{\xi}\widehat{\rho}(x,\xi,\tau_0) - \left[1 - \frac{\widetilde{\lambda}(\tau_0)}{2}\varphi_0(\tau_0)\right]\varphi(x,\xi,\tau_0) + \alpha(x) - \beta\,\widehat{\rho}(x,\xi,\tau_0), \tag{22}$$

$$\frac{d\widehat{q}}{d\tau_0} = -\frac{\gamma(x)}{\xi}\widehat{q}(x,\xi,\tau_0) + \frac{\widetilde{\lambda}(\tau_0)}{2}\psi_0(\tau_0)\,\varphi(x,\xi,\tau_0), \tag{23}$$

$$\frac{d\widehat{s}}{d\tau_0} = -\frac{\gamma(x)}{\xi}\widehat{s}(x,\xi,\tau_0) - \left[1 - \lambda(\tau_0) + \frac{\widetilde{\lambda}(\tau_0)}{2}\phi_0(\tau_0)\right]\varphi(x,\xi,\tau_0) + \beta\left[1 + \widehat{\rho}(x,\xi,\tau_0)\right], \tag{24}$$

where we have introduced the notations

$$\varphi_0(\tau_0) = \int_{-\infty}^{\infty}\alpha(x)\,dx\int_0^1 \varphi(x,\eta,\tau_0)\frac{d\eta}{\eta},$$
$$\psi_0(\tau_0) = \int_{-\infty}^{\infty}\alpha(x)\,dx\int_0^1 \psi(x,\eta,\tau_0)\frac{d\eta}{\eta}, \tag{25}$$
$$\phi_0(\tau_0) = \int_{-\infty}^{\infty}\alpha(x)\,dx\int_0^1 \widehat{s}(x,\eta,\tau_0)\frac{d\eta}{\eta}.$$

Once (1), (4) (or (8)) are solved, the zeroth moments of the functions φ and ψ can be regarded as known. To solve (24), we need to find also $\phi_0(\tau_0)$. One can easily derive Volterra-type integral equation for this function from the same (24):

$$\phi_0(\tau_0) = \frac{1}{2}\int_0^{\tau_0}\lambda(\tau)\,L(\tau_0-\tau)\,\phi_0(\tau)\,d\tau + G(\tau_0), \tag{26}$$

where the kernel-function L given by

$$L(\tau) = \int_{-\infty}^{\infty}\alpha(x)\,dx\int_0^1 \varphi(x,\xi,\tau)\exp\left(-\frac{\gamma(x)}{\xi}\tau\right)\frac{d\xi}{\xi} \tag{27}$$

is well known in the radiative transfer theory and

$$G(\tau) = \int_0^{\tau_0}[1-\lambda(\tau)]\,L(\tau)\,d\tau + \beta F(\tau), \tag{28}$$

$$F(\tau) = \int_{-\infty}^{\infty}\alpha(x)\,dx\int_0^1\left[1+\widehat{\rho}(x,\xi,\tau)\right]\exp\left(-\frac{\gamma(x)}{\xi}\tau\right)\frac{d\xi}{\xi}. \tag{29}$$

Further, the formal differentiation of (22) and (23) over λ allows obtaining separate equations for the expected number of scattering events, $N_*(x,\xi,\tau_0)$, $N_0(x,\xi,\tau_0)$ correspondingly for the reflected and transmitted photons:

$$\frac{dN_*}{d\tau_0} = -\frac{\gamma(x)}{\xi}N_*(x,\xi,\tau_0) + \frac{\tilde{\lambda}(\tau_0)}{2}\left[f_0(\tau_0)+\varphi_0(\tau_0)\right]\varphi(x,\xi,\tau_0) - \left[1-\frac{\tilde{\lambda}(\tau_0)}{2}\varphi_0(\tau_0)\right]f(x,\xi,\tau_0) - \beta N_*(x,\xi,\tau_0),$$

$$\frac{dN_0}{d\tau_0} = -\frac{\gamma(x)}{\xi}N_0(x,\xi,\tau_0) + \frac{\tilde{\lambda}(\tau_0)}{2}\left[g_0(\tau_0)+\psi_0(\tau_0)\right]\varphi(x,\xi,\tau_0) + \frac{\tilde{\lambda}(\tau_0)}{2}\psi_0(\tau_0)f(x,\xi,\tau_0), \tag{30}$$

where $f(x,\xi,\tau_0) = \lambda\partial\varphi(x,\xi,\tau_0)/\partial\lambda$, $f_0(\tau_0) = \lambda\partial\varphi_0(\tau_0)/\partial\lambda$, and $g_0(\tau_0) = \lambda\partial\psi_0(\tau_0)/\partial\lambda$. As the initial conditions, we have $N_*(x,\xi,0) = N_0(x,\xi,0) = 0$. The ratios $N_*/\widehat{\rho}$ and $N_0/\widehat{q}$ obviously give the requisite MNS for the reflected and transmitted photons.

Let us turn further to the photons which are destroyed in the course of multiple scattering in the atmosphere. The invariant imbedding technique must be now applied by counting the number of scattering events for each elementary process appearing in adding a complementary layer to the initial medium. The generating function approach developed in [25] allows finding the expected number of scattering events, N_a, for this type of photons from

$$\frac{dN_a}{d\tau_0} = -\frac{\gamma(x)}{\xi}N_a(x,\xi,\tau_0) + \frac{\tilde{\lambda}(\tau_0)}{2}\left[h_0(\tau_0)+\phi_0(\tau_0)+1-\lambda(\tau_0)\right]\times\varphi(x,\xi,\tau_0) + \left[\frac{\tilde{\lambda}(\tau_0)}{2}\phi_0(\tau_0)+1-\lambda(\tau_0)\right]f(x,\xi,\tau_0) + \beta\left[1+\widehat{\rho}(x,\xi,\tau_0)+N_*(x,\xi,\tau_0)\right], \tag{31}$$

where $h_0(\tau_0) = \lambda\partial\phi_0(\tau_0)/\partial\lambda$ and the initial condition is $N_a(x,\xi,0) = 0$. It is readily seen that $\langle N\rangle = N_* + N_0 + N_a$ represents the MNS for the photons of frequency x incident on the medium at the angle $\cos^{-1}\xi$ irrespective of whether or not they are subsequently destroyed in the medium or leave it. Taking into account that $\widehat{\rho}+\widehat{q}+\widehat{s} = 1$ and $\varphi_0+\psi_0+\phi_0 = 2\sqrt{\pi}$ and introducing the notation

$$\Phi(\tau_0) = f_0(\tau_0)+g_0(\tau_0)+h_0(\tau_0) = \int_{-\infty}^{\infty}\alpha(x)\,dx\int_0^1\langle N(x,\eta,\tau_0)\rangle\frac{d\eta}{\eta}, \tag{32}$$

we arrive at

$$\frac{d\langle N\rangle}{d\tau_0} = -\frac{\gamma(x)}{\xi}\langle N(x,\xi,\tau_0)\rangle + \left[\frac{\tilde{\lambda}(\tau_0)}{2}\Phi(\tau_0)+1\right]\varphi(x,\xi,\tau_0) + \beta\left[1+\widehat{\rho}(x,\xi,\tau_0)\right] \tag{33}$$

with the initial condition $\langle N(x,\xi,0)\rangle = 0$.

As above in the case of $\widehat{s}(x,\xi,\tau_0)$ (24), we deal with the similar initial-value problem which can be solved if only the zeroth moment $\Phi(\tau_0)$ is determined. It is easy to show that this function satisfies integral equation (26) with the only difference that now the free term is

$$G(\tau_0) = \int_0^{\tau_0}L(\tau)\,d\tau + \beta F(\tau_0), \tag{34}$$

where the functions L and F are given by (27), (29).

We see that knowledge of only reflection function is sufficient to find the MNS for all the incident photons independent of their future "fate." Equations (24), (25) easily yield the numerical solution. Some analytical results can be obtained in the specific case of homogeneous medium [25, 26], for which all the above-introduced functions dependent on frequency and direction are, in fact, the functions of the combined variable $\gamma(x)/\xi$. Note also that taking derivatives in these equations to zero we are led to the results previously derived in the mentioned paper for semi-infinite media. Comparing (33) and (24) for $\beta = 0$, that is, neglecting the role of absorption in continuum, we find $\langle N(x,\xi,\tau_0)\rangle = \widehat{s}(x,\xi,\tau_0)/(1-\lambda)$.

Let us pursue our considerations further and apply the same approach to determine the continuously distributed random quantities describing the radiation diffusion. As an illustration, we consider here the problem of finding the average time spent by the photon on multiple scattering in the medium. Because of the important role which this statistical mean quantity plays in astrophysical applications, it was a subject of investigations by a number of researchers [27–30]. In the general case, when the photons are destroyed not only during scattering but also in flight, this average makes it possible to gauge the relative importance of the energy dissipation in the medium and its flow through a boundary. Another important application of this average is associated with the problem usually arising in the presence of the nonstationary sources of energy in atmosphere, when one needs to reveal whether the radiative equilibrium is established or not [31].

Turning directly to our problem, we note that mathematically the only case of interest is that for which the photon spends time only on travelling the path between two successive scattering events. With regard to the mean time spent by the diffusing photon while the atoms are in the excited state, it can be taken into account when necessary by simple multiplication of the MNS and the average time required by each of the atoms for the reemission process. This is admissible because these two random variables are statistically independent.

For convenience, we will measure the time intervals in the units of $t = 1/nck_{\nu_0}$, where n is the number of scattering particles in $1\,\text{cm}^3$ and k_{ν_0} is the absorption coefficient in the centre of the line, calculated for one atom. It is easy to see that t represents the time required to travel the mean free path between two successive scattering events for a photon in the line centre if there is no absorption in continuum. We denote the dimensionless time by ω. In determining the time averages, we, again, as above, consider three types of photons: reflected, transmitted, and destroyed in the course of multiple scattering. Here we confine ourselves to relatively detailed treatment of the problem for the first of these categories of photons. In the case of two other types, we present only the final results.

Starting with the process of reflection, let us introduce the generalised reflection coefficient $\tilde{\rho}(x', \eta; x, \xi; \tau_0, \omega)$ which is the time-dependent analogue of that defined in Section 2 and concerns the photons reflected in the time interval $(\omega, \omega + d\omega)$. Invariant imbedding approach allows to write

$$\begin{aligned} \frac{d\tilde{\rho}}{d\tau_0} &+ \left(\frac{1}{\eta} + \frac{1}{\xi}\right)\frac{d\tilde{\rho}}{d\omega} \\ &= -\left[\frac{\gamma(x')}{\eta} + \frac{\gamma(x)}{\xi}\right]\tilde{\rho}(x', \eta; x, \xi; \tau_0, \omega) \\ &+ \frac{\tilde{\lambda}(\tau_0)}{2}\left\{\alpha(x)\alpha(x')\delta(\omega)\right. \\ &\quad + \alpha(x)\int_{-\infty}^{\infty}\alpha(x'')\,dx'' \\ &\quad \times \int_0^1 \tilde{\rho}(x', \eta; x'', \eta'; \tau_0, \omega)\frac{d\eta'}{\eta'} \\ &\quad + \alpha(x')\int_{-\infty}^{\infty}\alpha(x'')\,dx'' \\ &\quad \times \int_0^1 \tilde{\rho}(x'', \eta'; x, \xi; \tau_0, \omega)\frac{d\eta'}{\eta'} \\ &\quad + \int_0^{\omega} d\omega' \int_{-\infty}^{\infty}\alpha(x'')\,dx'' \\ &\quad \times \int_0^1 \tilde{\rho}(x', \eta; x'', \eta'; \tau_0, \omega')\frac{d\eta'}{\eta'} \\ &\quad \times \int_{-\infty}^{\infty}\alpha(x''')\,dx''' \\ &\quad \left.\times \int_0^1 \tilde{\rho}(x''', \eta''; x, \xi; \tau_0, \omega - \omega')\frac{d\eta''}{\eta''}\right\}, \end{aligned} \tag{35}$$

where δ is the Dirac δ-function.

The method of characteristic functions applied to this equation is equivalent to performing the Laplace transformation [27]. For the Laplace transform of the time-dependent reflection coefficient

$$T(x', \eta; x, \xi; \tau_0, s) = \int_0^{\infty} \tilde{\rho}(x', \eta; x, \xi; \tau_0, \omega)\,e^{-s\omega}\,d\omega, \tag{36}$$

one can write

$$\begin{aligned} \frac{dT}{d\tau_0} = &-\left\{\left[\frac{\gamma(x')}{\eta} + \frac{\gamma(x)}{\xi}\right] - s\left(\frac{1}{\xi} + \frac{1}{\eta}\right)\right\} \\ &\times T(x', \eta; x, \xi; \tau_0, s) \\ &+ \frac{\lambda(\tau_0)}{2}\varpi(x', \eta, \tau_0, s)\,\varpi(x, \xi, \tau_0, s), \end{aligned} \tag{37}$$

where

$$\begin{aligned} \varpi(x, \xi, \tau_0, s) = \alpha(x) &+ \int_{-\infty}^{\infty}\alpha(x')\,dx' \\ &\times \int_0^1 T(x', \eta'; x, \xi, \tau_0, s)\frac{d\eta'}{\eta'}. \end{aligned} \tag{38}$$

Taking $s = 0$, we go back to (1). For finding the required average time for reflected photons, we need the derivative $\Omega_*(x, \eta; x, \xi; \tau_0) = dT/ds|_{s=0}$. It follows from (37) that

$$\begin{aligned} \frac{d\Omega_*}{d\tau_0} = &-\left[\frac{\gamma(x')}{\eta} + \frac{\gamma(x)}{\xi}\right]\Omega_*(x', \eta; x, \xi; \tau_0) \\ &-\left(\frac{1}{\xi} + \frac{1}{\eta}\right)\tilde{\rho}(x', \eta; x, \xi; \tau_0) \\ &+ \frac{\lambda(\tau_0)}{2}\left[\varphi(x', \eta, \tau_0)\tilde{f}(x, \xi, \tau_0)\right. \\ &\qquad \left. + \tilde{f}(x', \eta, \tau_0)\varphi(x, \xi, \tau_0)\right], \end{aligned} \tag{39}$$

where

$$\tilde{f}(x, \xi, \tau_0) = \int_{-\infty}^{\infty}\alpha(x')\,dx'\int_0^1 \Omega_*(x', \eta; x, \xi, \tau_0)\frac{d\eta'}{\eta'}. \tag{40}$$

It is not difficult to show that this equation can be obtained by formal differentiating (1) with respect to β and exchanging the sign. Similar result for homogeneous atmosphere was obtained for the first time in [27]. Thus, now we arrive at an important generalization of this result by showing that it remains valid also for inhomogeneous media. As above in the case of MNS, the differentiation procedure holds true only for reflected and transmitted photons. It is obvious that the ratio Ω_*/ρ gives the detailed information on the average time spent by reflected photons dependent on their initial and final frequency and directional characteristics.

We are interested here by the average time, $\widehat{\Omega}_*(x,\xi,\tau_0)$, spent by all the reflected photons irrespective of their final frequency and direction. Integration of (39) over x' and η yields

$$\begin{aligned}\frac{d\widehat{\Omega}_*}{d\tau_0} &= -\frac{\gamma(x)}{\xi}\widehat{\Omega}_*(x,\xi,\tau_0) + \frac{1}{\xi}\widehat{\rho}(x,\xi,\tau_0) \\ &+ \frac{\widetilde{\lambda}(\tau_0)}{2}\widetilde{f}_0(\tau_0)\varphi(x,\xi,\tau_0) \\ &- \left[1 - \frac{\widetilde{\lambda}(\tau_0)}{2}\varphi_0(\tau_0)\right]\widetilde{f}(x,\xi,\tau_0) \\ &+ 1 + \widehat{\rho}(x,\xi,\tau_0) - \beta\widehat{\Omega}_*(x,\xi,\tau_0),\end{aligned} \tag{41}$$

where $\widetilde{f}_0(\tau_0) = -\partial\varphi_0(\tau_0)/\partial\beta$ and $\widehat{\Omega}_*(x,\xi,0) = 0$.

By analogous manner we obtain the mean expected time $\widehat{\Omega}_0(x,\xi,\tau_0)$ for all the transmitted photons if the boundary $\tau = \tau_0$ of the atmosphere is illuminated by photons with (x,ξ) characteristics:

$$\begin{aligned}\frac{d\widehat{\Omega}_0}{d\tau_0} &= -\frac{\gamma(x)}{\xi}\widehat{\Omega}_0(x,\xi,\tau_0) + \frac{1}{\xi}\widehat{q}(x,\xi,\tau_0) \\ &+ \frac{\widetilde{\lambda}(\tau_0)}{2}\left[\widetilde{g}_0(\tau_0)\varphi(x,\xi,\tau_0) + \psi_0(\tau_0)\widetilde{f}(x,\xi,\tau_0)\right],\end{aligned} \tag{42}$$

where $\widetilde{g}_0(\tau_0) = -\partial\psi_0(\tau_0)/\partial\beta$ and the initial condition is $\widehat{\Omega}_0(x,\xi,0) = 0$.

As for the photons destroyed in the atmosphere, derivation of appropriate equation for the expected time Ω_a cannot be found by direct differentiation and must be obtained by employing the imbedding technique. As a result, we obtain

$$\begin{aligned}\frac{d\widehat{\Omega}_a}{d\tau_0} &= -\frac{\gamma(x)}{\xi}\widehat{\Omega}_a(x,\xi,\tau_0) + \frac{1}{\xi}\widehat{s}(x,\xi,\tau_0) \\ &+ \frac{\widetilde{\lambda}(\tau_0)}{2}\left[\widetilde{h}_0(\tau_0)\varphi(x,\xi,\tau_0) + \phi_0(\tau_0)\widetilde{f}(x,\xi,\tau_0)\right] \\ &+ \left[1 - \lambda(\tau_0)\right]\widetilde{f}(x,\xi,\tau_0) + \beta\widehat{\Omega}_*(x,\xi,\tau_0),\end{aligned} \tag{43}$$

where we introduced notations $\widetilde{h}_0(\tau_0) = -\partial\phi_0(\tau_0)/\partial\beta$ and the initial condition $\widehat{\Omega}_a(x,\xi,0) = 0$. Once (41)–(43) are solved, the ratios $\widehat{\Omega}_*/\widehat{\rho}$, $\widehat{\Omega}_0/\widehat{q}$, and $\widehat{\Omega}_a/\widehat{s}$ give the requisite values of the average time for three categories of photons. Further, $\langle\Omega\rangle = \widehat{\Omega}_* + \widehat{\Omega}_0 + \widehat{\Omega}_a$ represents the average time for incident photons diffusion in the atmosphere irrespective of whether they escape from the medium or are destroyed in it. From (41)–(43) we obtain

$$\begin{aligned}\frac{d\langle\Omega\rangle}{d\tau_0} &= -\frac{\gamma(x)}{\xi}\langle\Omega(x,\xi,\tau_0)\rangle \\ &+ \frac{\widetilde{\lambda}(\tau_0)}{2}\widetilde{\Phi}(\tau_0)\varphi(x,\xi,\tau_0) + 1 + \widehat{\rho}(x,\xi,\tau_0),\end{aligned} \tag{44}$$

where

$$\begin{aligned}\widetilde{\Phi}(\tau_0) &= \widetilde{f}_0(\tau_0) + \widetilde{g}_0(\tau_0) + \widetilde{h}_0(\tau_0) \\ &= \int_{-\infty}^{\infty}\alpha(x)\,dx\int_0^1\langle\Omega(x,\eta,\tau_0)\rangle\frac{d\eta}{\eta}\end{aligned} \tag{45}$$

and initial condition $\langle\Omega(x,\xi,0)\rangle = 0$. As for the function $\widetilde{\Phi}(\tau_0)$, it satisfies integral equation (26) with the free term $G(\tau_0) = \beta F(\tau_0)$.

Equation (44) implies that again, as in the case of MNS, knowledge of the reflection coefficient alone ensures the temporal description of the diffusion process in the atmosphere. In fact, solution of similar initial-value problems for integrodifferential equations gives a detailed statistical description on the multiple scattering in inhomogeneous media. As it was pointed out, in the treated special case of completely incoherent scattering, the form of the proper functions and equations can be simplified by introducing the combined variables of the $\gamma(x)/\xi$ type. However, we give preference to separation of these arguments, which is better suited for numerical calculations. Note in conclusion that in the particular case of homogeneous media comparison of (24), (33), and (44) yields the well-known relation between different statistical mean quantities $(1-\lambda)\langle N\rangle + \lambda\beta\langle\Omega\rangle = \widehat{s}$.

5. Concluding Remarks

We discussed two frequently encountered model problems of the radiative transfer in a plane-parallel inhomogeneous medium and showed that they can be mathematically reduced to the solution of only one initial-value problem for integrodifferential equations for reflectance (1) with the subsequent evaluation of several ordinary integrals. This allows overcoming the well-known difficulties specific to the boundary-value problems, to which the classical formulation of the physical problems usually leads. For simplicity, the inhomogeneity of the medium was explicitly indicated only in the scattering coefficient, while all other parameters controlling the elementary scattering and absorption processes can be also allowed to be dependent on optical depth. All the requisite quantities are found for a family of atmospheres of different optical thicknesses.

The same approach was applied in finding the MNS and the average time of the photons travel in the medium for different types of photons. Solution of the resulting integrodifferential equations gives the detailed statistical description of the radiation diffusion process depending on the angular and frequency parameters of the incident, reflected, transmitted, and destructed photons. The statistical mean quantities are of special interest for applications concerning all the diffusing photons independent of that whether they are destroyed in the medium or escape from it. These quantities are completely determined by the reflection properties of the medium. It is apparent that by analogous manner one can find the statistical averages for any discretely or continuously distributed random quantities describing the radiation diffusion process.

Summarizing the results obtained in the paper we observe that the starting point in all of the considered problems is the determination of the reflectance of an atmosphere. In other words, knowledge of only the reflection coefficient makes it possible to gain complete insight into the field of radiation inside the medium and the statistical properties of the diffusion process. This is of great importance in view of that this coefficient is defined from a separate equation and that it is an observable and measurable quantity in contrast to the source function which plays an important role in the classical theory of the radiative transfer. In fact, the obtained results can be regarded as a generalization, in some sense, of Ambartsumian's invariance idea to the finite inhomogeneous atmosphere.

From pure mathematical point of view, the proposed approach facilitates solving the traditional and frequently used model problems to a large extent. It is based on obvious physical arguments, so it is intuitively clear, universal, and easy to use.

Conflict of Interests

The author declares that there is no conflict of interests regarding the publication of this paper.

References

[1] V. A. Ambartsumian, *Scientific Works*, vol. 1, Izdatel'stvo Akademii Nauk Armianskoi SSR, Yerevan, Armenia, 1960 (Russian).

[2] V. A. Ambartsumian, "Diffuse reflection of light by a foggy medium," *Doklady Akademii Nauk SSSR*, vol. 38, pp. 229–232, 1943.

[3] R. Bellman, "Functional equations in the theory of dynamic programming. VII: a partial differential equation for the Fredholm resolvent," *Proceedings of the American Mathematical Society*, vol. 8, pp. 435–440, 1957.

[4] V. V. Sobolev, "Radiation diffusion in a semi-infinite medium," *Doklady Akademii Nauk SSSR*, vol. 116, p. 45, 1957.

[5] V. V. Sobolev, "On the theory of the radiation diffusion in stellar atmospheres," *Astronomicheskii Zhournal*, vol. 31, p. 573, 1959.

[6] M. G. Kreĭn, "On a new method of solution of linear integral equations of first and second kinds," *Doklady Akademii Nauk SSSR*, vol. 100, pp. 413–416, 1955.

[7] R. W. Preisendorfer, "Functional relations for the r and t operators on plane-parallel media," *Proceedings of the National Academy of Sciences of the United States of America*, vol. 44, no. 4, pp. 323–327, 1958.

[8] R. W. Preisendorfer, *Hydrologic Optics*, vol. 4, U.S. Department of Commerce, Honolulu, Hawaii, USA, 1976.

[9] I. W. Busbridge, "On inhomogeneous stellar atmospheres," *The Astrophysical Journal*, vol. 133, pp. 198–209, 1961.

[10] V. V. Sobolev, *Scattering of the Light in Atmospheres of Planets*, Nauka, Moscow, Russia, 1972 (Russian).

[11] E. G. Yanovitskij, *Light Scattering in Inhomogeneous Atmospheres*, Springer, 1997.

[12] V. A. Ambartsumian, "On the one-dimensional case of the problem of scattering and absorbing medium of finite optical thickness," *Izvestiya Akadermii Nauk Armyanskoy SSR*, vol. 1-2, 1944.

[13] A. G. Nikoghossian, "Radiative transfer in one-dimensional inhomogeneous atmospheres," *Astronomy and Astrophysics*, vol. 422, no. 3, pp. 1059–1066, 2004.

[14] A. G. Nikoghossian, "Radiative transfer in inhomogeneous atmospheres. I," *Astrophysics*, vol. 47, no. 1, pp. 104–116, 2004.

[15] A. G. Nikoghossian, "Radiative transfer in inhomogeneous atmospheres. II," *Astrophysics*, vol. 47, no. 2, pp. 248–259, 2004.

[16] A. G. Nikoghossian, "Radiative transfer in inhomogeneous atmospheres. III," *Astrophysics*, vol. 47, no. 3, pp. 412–421, 2004.

[17] R. Bellman, R. Kalaba, and G. M. Wing, "Invariant imbedding and mathematical physics. I: particle processes," *Journal of Mathematical Physics*, vol. 1, no. 4, pp. 280–308, 1960.

[18] R. Bellman, R. Kalaba, and M. Prestrud, *Invariant Imbedding and Radiative Transfer in Slabs of Finite Thickness*, Elsevier, New York, NY, USA, 1963.

[19] J. Casti and R. Kalaba, *Imbedding Methods in Applied Mathematics*, Addison-Wesley, 1973.

[20] R. G. Athay, *Transport in Spectral Lines*, D. Reidel, Dordrecht, The Netherlands, 1972.

[21] D. Mihalas, *Stellar Atmospheres*, Freeman, San Francisco, Calif, USA, 1978.

[22] A. G. Nikoghossian, "On some trends in the progress of astrophysical radiative transfer," *Light Scattering Reviews*, vol. 8, pp. 377–426, 2013.

[23] A. G. Nikoghossian, "Solution of linear radiative transfer problems in plane-parallel atmospheres. III," *Astrophysics*, vol. 56, no. 1, pp. 130–141, 2013.

[24] V. A. Ambartsumian, "On the number of scatterings during diffusion of photons in foggy medium," *Doklady Akademii Nauk Armyanskoj SSR*, vol. 8, p. 101, 1948.

[25] A. G. Nikoghossian, "The statistical description of radiation field on the basis of invariance principle," *Astrophysics*, vol. 21, no. 2, p. 323, 1984.

[26] V. V. Sobolev, *Radiative Transfer and the Spectra of Celestial Bodies*, Nauka, Moscow, Russia, 1969 (Russian).

[27] A. G. Nikoghossian, "Statistical description of radiation field on the basis of the invariance principle. III: average time of photon travel in the scattering medium," *Astrophysics*, vol. 24, no. 1, pp. 89–99, 1986.

[28] W. M. Irvine, "The formation of absorption bands and the distribution of photon optical paths in a scattering atmosphere," *Bulletin of the Astronomical Institutes of the Netherlands*, vol. 17, p. 266, 1964.

[29] W. M. Irvine, "The distribution of photon optical paths in a scattering atmosphere," *The Astrophysical Journal*, vol. 144, pp. 1140–1147, 1966.

[30] V. V. Ivanov, "The mean free path of a photon in a scattering medium," *Astrophysics*, vol. 6, 1970.

[31] V. V. Sobolev, *A Treatise on Radiative Transfer*, Van Nostrand, Princeton, UK, 1963.

17

Delta T: Polynomial Approximation of Time Period 1620–2013

M. Khalid, Mariam Sultana, and Faheem Zaidi

Department of Mathematical Sciences, Federal Urdu University of Arts, Sciences & Technology, Karachi 75300, Pakistan

Correspondence should be addressed to M. Khalid; khalidsiddiqui@fuuast.edu.pk

Academic Editor: Josep M. Trigo-Rodríguez

The difference between the Uniform Dynamical Time and Universal Time is referred to as ΔT (delta T). Delta T is used in numerous astronomical calculations, that is, eclipses,and length of day. It is additionally required to reduce quantified positions of minor planets to a uniform timescale for the purpose of orbital determination. Since Universal Time is established on the basis of the variable rotation of planet Earth, the quantity ΔT mirrors the unevenness of that rotation, and so it changes slowly, but rather irregularly, as time passes. We have worked on empirical formulae for estimating ΔT and have discovered a set of polynomials of the 4th order with nine intervals which is accurate within the range of ±0.6 seconds for the duration of years 1620–2013.

1. Introduction

The expression "timescale" is quite frequently used in astronomical contexts. To define it in astronomical terms, it may be put as a way of measuring time based on a particular periodic natural phenomenon. Two main distinct groups of timescales are used in astronomy. The first group of timescales is based on second which are known as International Atomic Time (IAT). It is the standard for the SI (System International) second. The SI-based timescales are comparatively new in the history of timekeeping, since they depend on atomic clocks that were first put to regular use in the 1950s era. Prior to that, all timescales were associated somewhat with the rotation of the Earth.

Timescales that rely on the rotation of the Earth are used for astronomical purposes as well. A relevant example would be a telescope pointing that relies on the geographic orientation of the observer. Universal Time UT mostly refers to the specific timescale $UT1$. Historically, Universal Time (earlier known as the Greenwich Mean Time) has been achieved from Greenwich sidereal time using a general expression. However, UT is not fit for the computation of positions of the Moon, Sun, and planets using gravitational theories of their respective movements. Such theories prohibit variations in the rate of the rotation of Earth on its axis. Modern astrodynamical theories of the motions of the Sun, the Moon, and the planets are based on an evenly increasing and uniform timescale referred to as Terrestrial Time TT. TT runs a little ahead of $UT1$(a refined measure of mean solar time at Greenwich) by an amount known as delta $T = TT - UT1$. As Earth's rotation does not decelerate at a uniform rate, nontidal effects make it inconceivable to predict the precise values of ΔT in the distant past or remote future. Unfortunately, estimating the standard error in ΔT before 1600 AD is a tough task. It depends on several factors including the accuracy of determining ΔT from previous eclipse records and designing the models of physical processes creating changes in Earth's rotation.

In the recent past, various polynomial representations for the ΔT values for the last few centuries have been suggested by Meeus [1], Islam et al. [2], and Meeus and Simons [3]. In this paper, we aim to present set of the 4th degree polynomials for delta T with the least possible absolute error within ±0.6 seconds for the duration of 1620–2013.

2. Literature Review

Proving that the Earth rotates was no easy task for the scientists of former times. This problem goes back to the 17th century, when Halley [4] found that quadratic terms had to be added to the Moon's mean longitude to match the times recorded for ancient eclipses. Laplace [5] announced that the acceleration term was due to perturbations from the Earth's orbital eccentricity. Adams [6] then determined that Laplace had not included many higher order terms, which reduced Laplace's final result to about half of Halley's empirical value. To explain the source for the remaining observed effect, Ferrell [7] and Delaunay [8] independently assigned this discrepancy to tidal interactions between the Earth and the Moon. Newcomb [9] analyzed variations of the Earth's rotation to explain some of the lunar residuals, but he could not obtain verification of these variations from inner planet data (Newcomb [10]).

In the latter part of the 19th century, Chandler discovered minute variations in Earth's axis of rotation or polar wobble (Chandler [11, 12]). Newcomb then announced that these, and possibly other irregular variations of the Earth's rotation, might be an explanation of the residuals in the lunar mean longitude (Newcomb [10]).

In all these cases, calculated positions were ahead of the observed positions by amounts proportional to the frequencies. Astronomers came to believe that discrepancies in ephemerides were not due to errors in the expressions for the mean longitude but were due to unmodeled irregularities and a deceleration of the Earth's rotation, on which UT depends. Spencer Jones [13] examined residuals in the mean longitudes of the Sun, Moon, and two other planets and then concluded that the error was due to a slow deceleration in Earth's rotation (Spencer Jones [14]). Since Newton's theory of gravitation requires a nonaccelerating uniform timescale for the computation of orbital motions and because the Earth's rotation was assumed to be decelerating, astronomers thought a timescale determined by the orbits themselves would be the uniform scale they needed. Thus, their proposition was a timescale called Ephemeris Time (ET), based on orbital motions, to be used for all dynamical calculations. In 1984, ET was replaced by Terrestrial Dynamical Time (TDT) as an independent argument for apparent geocentric ephemerides. The unit of TDT is a day of 86 400 SI seconds at mean sea level. For practical purposes, TDT is TAI + 32.184 seconds. In 1991, TDT was renamed Terrestrial Time (TT). TT is considered to be a uniform timescale and used as the time argument for the predictions of the astronomical events in dynamical theories (Seidelmann [15]; Islam et al. [2]).

Eclipse predictions are computed in TDT. To convert TDT predictions to UT, the difference between TDT and $UT1$ must be in our knowledge. This parameter is known as $\Delta T = TDT - UT1$ or delta$T$.

The mathematical modeling of the Earth's delta T has been ongoing for some time and there are multiple models to choose from. Before the advancement of AD 1600, values of ΔT had to rely on historical records of naked-eye observations of eclipses and occultations. Stephenson and a few other researchers have identified hundreds of these types of observations in early European, Middle Eastern, and Chinese annals, manuscripts, canons, and records. Moreover, some of these scientists have fit a lot of records with simple polynomials to achieve best fits for describing the value of ΔT for the centuries occurring before AD 1600 (Morrison and Ward [16]; Stephenson and Fatoohi [17]; Stephenson and Fatoohi [18]; Stephenson and Morrison [19]; Stephenson and Morrison [20]; Stephenson [21]; Morrison and Stephenson [22]; Steele and Stephenson [23]). Despite their relatively low precision and inaccuracy, these data represent our only record of the value of ΔT during the past several millennia.

Close-to-accurate values of ΔT only exist sometime after the invention of the telescope (1610). A careful analysis of telescopic timings of stellar occultation by the Moon permits the direct measurement of ΔT during this time period. In fact, all values of ΔT before 1955 depend on observations of the Moon, either via solar eclipses or via lunar occultations (Seidelmann [15]). Nowadays, UT is representative of the observed rotation angle of the Earth relative to an inertial reference frame formed by extragalactic radio sources. Its measurement, with the help of several observatories, is coordinated by the International Earth Rotation and Reference Systems Service (IERS).

Numerous polynomial representations for the ΔT values of the last few centuries have been suggested in the recent past to evade the need of incorporating lengthy tables in a computer programs. Meeus [1] presents a 12th order polynomial liable for the time span of 1800–1997 with the maximum absolute error of 2.3 seconds and two 9th or 10th order polynomials covering a similar time span with a maximum absolute error of 1.069 SI seconds (Meeus [1]). Montenbruck and Pfleger [24] provide seven 3rd order polynomials to accommodate the period between 1825 and 2000 with maximum absolute error of 2.13 SI seconds. Meeus and Simons [3] give fourth-order polynomials with eight segments to cover the period between 1620 and 2000 with maximum absolute error of 3.2 SI seconds (Meeus and Simons [3]; Islam et al. [2]). Islam et al. [2] presented eight 4th degree polynomials to the entire curve of ΔT with precisely the same intervals as were used by Meeus and Simons [3] compromising maximum absolute error of ±0.7 seconds on time span from 1620 to 2000 AD.

3. New Proposed Polynomial Approximation

The data used in this research was excerpted from the pages $K8$ and $K9$ of the *Astronomical Almanac* published by the Nautical Almanac Offices of the US Naval Observatory and the Bulletins issued by the International Earth Rotation Service, IERS, Paris. For a new set of polynomials, we apply method of the least square on the curve of delta T to get appropriate polynomials for epoch 1620–2013 (Figure 1). The quantity u, as defined by Meeus and Simons [3], is

$$u = k + \frac{(\text{year} - 2000)}{100}. \tag{1}$$

In this equation, the purpose of the k quantities is basically to make the independent variable u as small as possible during a certain interval of time, so we simply call it the "scaling factor."

The empirical formulae that we found here can calculate the value of ΔT at any instant of the years 1620–2013 AD. The 4th degree polynomial approximation of ΔT is

$$\Delta T = a_o + a_1 u + a_2 u^2 + a_3 u^3 + a_4 u^4, \tag{2}$$

where a_o, a_1, a_2, a_3, and a_4 are coefficients of respective polynomials taken in seconds, given in Table 1.

Table 1: Values of scaling factor k with coefficients of the 4th degree polynomial.

Duration	k	$a0$	$a1$	$a2$	$a3$	$a4$	Max. err.
1620–1672	3.670	76.541	−253.532	695.901	−1256.982	627.152	0.5709
1673–1729	3.120	10.872	−40.744	236.890	−351.537	36.612	0.5989
1730–1797	2.495	13.480	13.075	8.635	−3.307	−128.294	0.5953
1798–1843	1.925	12.584	1.929	60.896	−1432.216	3129.071	0.4643
1844–1877	1.525	6.364	11.004	407.776	−4168.394	7561.686	0.5894
1878–1904	1.220	−5.058	−1.701	−46.403	−866.171	5917.585	0.5410
1905–1945	0.880	13.392	128.592	−279.165	−1282.050	4039.490	0.5495
1946–1989	0.455	30.782	34.348	46.452	1295.550	−3210.913	0.4279
1990–2013	0.115	55.281	91.248	87.202	−3092.565	8255.422	0.2477

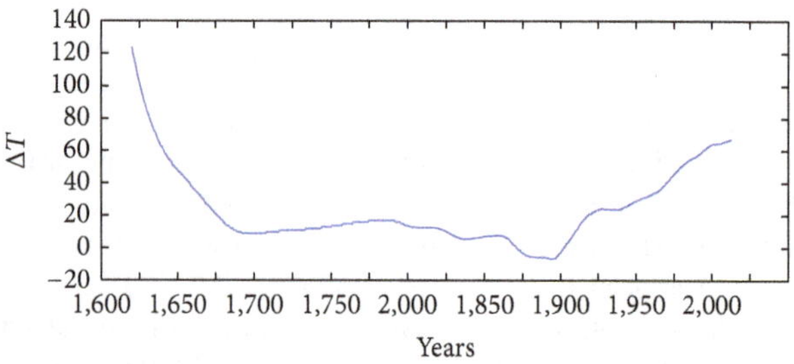

Figure 1: Curve shows variation in delta T with epoch 1620–2013.

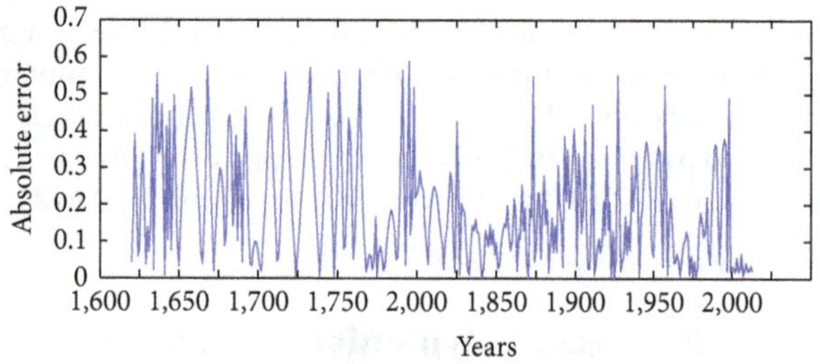

Figure 2: Graph shows absolute variation in ΔT (the 4th degree polynomial) with epoch 1620–2013.

4. Discussion of Results and Conclusion

The value of scaling parameter k and coefficients of the 4th degree polynomial with nine intervals that we have calculated are displayed in Table 1. It is clear from the table that the values of k and coefficients are only valid for adjacent intervals. We choose the 4th degrees polynomials in order to compare our results with previous approximations (Meeus and Simons [3] and Islam et al. [2]).

Figure 2 shows error analysis of the 4th degree polynomial with respect to its nine segments. Error analysis revealed that, in the 4th degree polynomial, minimum error of 0.000168 occurs in the year 1712, while maximum error of 0.598961 appears in the year 1692.

The maximum error found in this study is so minute that it turns out to be less than all sets of polynomials presented in Montenbruck and Pfleger [24], Meeus and Simons [3], and Islam et al. [2]. Therefore, it is concluded that a set of nine fourth degree polynomials has proven to be the most appropriate set covering a time span of 1620–2013 AD with ±0.6 seconds of accuracy. The time accuracy of astronomical events can predict the timescale in a much better fashion with this polynomial approximation. The uses include the solving of predicaments ranging from astronomical software to eliminating the steps in the search of the long table of the observed values of ΔT during the aforementioned period.

Conflict of Interests

The authors declare that there is no conflict of interests regarding the publication of this paper.

Acknowledgments

The authors would like to thank Ms. Wishaal Khalid and other, anonymous, reviewers for their valuable comments and suggestions for the improvement of the quality of the paper.

References

[1] J. Meeus, "The effect of delta T on astronomical calculation," *British Astronomical Association*, vol. 108, no. 3, pp. 154–156, 1998.

[2] S. Islam, M. Sadiq, and Q. Shahid, "Assessing polynomial approximation for delta T," *Journal of Basic & Applied Science*, vol. 4, no. 1, pp. 1–4, 2008.

[3] J. Meeus and L. Simons, "Polynomial approximation of Delta T," *British Astronomical Association*, vol. 110, pp. 323–324, 2000.

[4] E. Halley, "Some account of the ancient state of the city of Palmyra, with short remarks upon the inscriptions found there," *Philosophical Transactions*, vol. 19, no. 215–235, pp. 160–175, 1695.

[5] P. S. Laplace, *Mécanique céleste*, Memoires de l'Academie des Sciences de Paris, 1786.

[6] J. C. Adams, "On the secular variation of the moon's mean motion," *Philosophical Transactions of the Royal Society A*, vol. 143, pp. 397–406, 1853.

[7] W. Ferrell, "Note on the influence of the tides in causing an apparent secular acceleration of the moon's mean motion," *Proceedings of the American Academy of Arts and Sciences*, vol. 6, pp. 379–392, 1864.

[8] C. E. Delaunay, "Memoire sur une nouvelle méthode pour la determination du movement de la lune," *Comptes Rendus de l'Académie des Sciences*, vol. 61, 1865.

[9] S. Newcomb, *Researches on the Motion of the Moon, Part I*, Washington Observations for 1875 (Washington: US Gov. Printing Off.), 1878.

[10] S. Newcomb, "Researches on the motion of Moon: part II," *Astronomical Papers of the American Ephemeris and Nautical Almanac*, vol. 11, 1912.

[11] S. C. Chandler, "On the variation of latitude ‚I," *The Astronomical Journal*, vol. 11, no. 248, pp. 59–61, 1891.

[12] S. C. Chandler, "On the variation of latitude VI," *Astronomical Journal*, vol. 12, pp. 65–72, 1892.

[13] H. Spencer Jones, "The rotation of the earth, and the secular accelerations of the sun, moon and planets," *Monthly Notices of the Royal Astronomical Society*, vol. 99, pp. 541–558, 1939.

[14] H. Spencer Jones, "The rotation of the earth," *Monthly Notices of the Royal Astronomical Society*, vol. 87, pp. 4–31, 1926.

[15] P. K. Seidelmann, *Explanatory Supplement to the Astronomical Almanac*, pp. 39–93, University Science Books, Mill Valley, Calif, USA, 1992.

[16] L. V. Morrison and C. G. Ward, "Analysis of transit of Mercury," *Monthly Notices of the Royal Astronomical Society*, vol. 173, pp. 183–206, 1975.

[17] F. R. Stephenson and L. J. Fatoohi, "Solar and lunar eclipse measurements by medieval muslim astronomers I. Background," *Journal for the History of Astronomy*, vol. 25, p. 99, 1994.

[18] F. R. Stephenson and L. J. Fatoohi, "Solar and lunar eclipse measurements by medieval muslim astronomers I I. Observations," *Journal for the History of Astronomy*, vol. 26, pp. 227–236, 1995.

[19] F. R. Stephenson and L. V. Morrison, "Long-term changes in the rotation of the Earth: 700 B.C. to A.D. 1980," *Philosophical Transactions of the Royal Society of London A*, vol. 313, pp. 47–70, 1984.

[20] F. R. Stephenson and L. V. Morrison, "Long-term fluctuations in the Earth's rotation: 700 BC to AD 1990," *Philosophical Transactions of the Royal Society of London A*, vol. 351, pp. 165–202, 1990.

[21] F. R. Stephenson, *Historical Eclipses and Earth's Rotation*, Cambridge University Press, Cambridge, UK, 1997.

[22] L. V. Morrison and F. R. Stephenson, "Historical values of the Earth's clock error Delta T and the calculation of eclipses," *Journal for the History of Astronomy*, vol. 35, no. 2, pp. 327–336, 2004.

[23] J. M. Steele and F. R. Stephenson, "Astronomical evidence for the accuracy of clocks in Pre-Jesuit China," *Journal for the History of Astronomy*, vol. 29, pp. 35–48, 1998.

[24] P. Montenbruck and T. Pfleger, *Astronomy on the Personal Computer*, Springer, Berlin, Germany, 2000.

18

R-Process Nucleosynthesis in MHD Jet Explosions of Core-Collapse Supernovae

Motoaki Saruwatari,[1] Masa-aki Hashimoto,[1] Ryohei Fukuda,[1] and Shin-ichiro Fujimoto[2]

[1] *Department of Physics, Kyushu University, Hakozaki, Fukuoka 812-8581, Japan*
[2] *Department of Electronic Control, Kumamoto National College of Technology, Kumamoto 861-1102, Japan*

Correspondence should be addressed to Masa-aki Hashimoto; hashimoto@phys.kyushu-u.ac.jp

Academic Editors: W. Cui, Y.-Z. Fan, and J. F. Valdés-Galicia

We investigate the *r*-process nucleosynthesis during the magnetohydrodynamical (MHD) explosion of a supernova in a helium star of 3.3 $M_\odot$, where effects of neutrinos are taken into account using the leakage scheme in the two-dimensional (2D) hydrodynamic code. Jet-like explosion due to the combined effects of differential rotation and magnetic field is able to erode the lower electron fraction matter from the inner layers. We find that the ejected material of low electron fraction responsible for the *r*-process comes out from just outside the neutrino sphere deep inside the Fe-core. It is found that heavy element nucleosynthesis depends on the initial conditions of rotational and magnetic fields. In particular, the third peak of the distribution is significantly overproduced relative to the solar system abundances, which would indicate a possible *r*-process site owing to MHD jets in supernovae.

1. Introduction

Study of the *r*-process has been developed considerably keeping pace with the terrestrial experiments of nuclear physics far from the stability line of nuclides [1]. In particular, among the three peaks, which correspond to the elements of ^{80}Se, ^{130}Te, and ^{195}Pt, in the abundance pattern for the solar system *r*-elements, the transition from the second to third peak elements has been stressed by nuclear physicists [2]. Although supernovae could be one of the astrophysical sites of the *r*-process [2, 3], explosion mechanism is not still completely resolved, where supernova explosions are originated from the gravitational collapse of massive stars of $M \geq 8\,M_\odot$ [4, 5]. However it is unclear whether neutron-rich elements could be ejected or not during the shock wave propagation. As far as the one-dimensional calculations, almost all realistic numerical simulations concerning the collapse-driven supernovae of $M \geq 13\,M_\odot$ have failed to explode the outer layer above the Fe-core due to drooping of the energetic shock wave propagation [6, 7]. Although there exist calculations for 8 and 11 $M_\odot$ stars to explode, the explosion energies are very weak [8–15]. Therefore, a plausible site/mechanism of the *r*-process has not yet been clarified.

On the other hand, models of magnetorotational explosion (MRE) for core-collapse supernovae have been presented as a supernova mechanism [16–19] since both rapid rotations and/or strong magnetic fields could be resulted for neutron stars after the explosions. Furthermore, MRE with a realistic magnetic field configuration has been investigated [20–22]. In their series of papers, it has been shown that magnetorotational instability plays a critical role concerning the explosion energies which would explain the explosion energies of Type II and Ib supernovae. However, changes in the electron fractions and/or the heavy element nucleosynthesis have not been discussed well. Therefore, it should be studied whether MHD explosions affect the *r*-process even within a qualitative method.

Two-dimensional (2D) magnetohydrodynamical (MHD) calculations have been performed with the various initial parameters concerning rotation and magnetic field [13, 23–27]. The ZEUS-2D code [28] has been modified to include an equation of state [29], electron captures with a simple scheme of neutrino (ν-) transport [23]. Adopting these achievements, Nishimura et al. [30] have performed 2D/MHD calculations to study possibilities of the *r*-process during the supernova explosion of massive stars under the assumption of adiabatic explosion. They have shown the pattern of distributions of the r-elements of the solar system abundances. However, they have also found that the electron fractions (Y_e) increase significantly enough to destroy the r-process elemental distributions if the neutrino capture processes are included, where the processes were obtained from the results of spherical explosion calculations with the realistic neutrino transport included [31]. The problem is remained whether the adopted method for neutrino captures can be legitimate or not; effects of neutrino transport have not been included at all, and instead Nishimura et al. [30] have used the profiles of densities and temperatures obtained from the adiabatic calculations. It should be done to check their results by including the neutrino transport and to study the explosion energies even under extreme parameters of rotation/magnetic fields. In the meanwhile, Winteler et al. [32] have shown the r-process nucleosynthesis with the use of results of a magnetorotationally driven supernova simulation, where they have performed 3D calculations with a 3D spectral scheme for neutrinos. Although they have obtained enough r-elements, their MHD simulation has finished at around 31 ms after the bounce. It would be useful to show results of two-dimensional MHD calculations with longer simulation time and/or higher resolution for calculations of the r-process. It is noted that Winteler et al. [32] have emphasized the possible importance of the early appearance of r-process matter in low metallicities which could be originated by MHD explosion of supernovae.

In the present paper, we give the calculational results of the MHD explosion for the He-star of $3.3\,M_\odot$ untill the final simulation time $t_f \simeq 600$ ms. Although these results have already been briefly reported [33], we show the details of simulation procedure and their results to compensate the contents. For the MHD calculations, five models are adopted for the initial configuration of rotation and magnetic fields. In the initial magnetic field, we assume very strong magnetic fields which validity is not known. In the observations of magnetar with the magnetic field 1.6×10^{14} G [34], they have suggested a rather massive progenitor mass from the age of all the early type stars. In the complete online magnetar catalog cited by Mori et al. [34], magnetic field of 26 magnetars ranges 10^{14}–10^{16} G which have been obtained from the analysis of the decrease in the rotational period under the assumption of magnetic dipole braking in a vacuum. These observations encourage us to set the initial condition of a strong magnetic field. Contrary to the previous investigation of the r-process under adiabatic MHD explosion [30], we include the effects of neutrinos using a leakage scheme [35–40] with some modifications explained in Section 2. Finally, we investigate the possibility of the r-process in the MHD jets with use of our large nuclear reaction network. We find the region that produces the r-process elements having the particular distribution of low Y_e.

In Section 2, our supernova models that include the neutrino effects are given, and we also explain the initial models and r-process networks. The results of the r-process nucleosynthesis calculations are presented in Section 3. We summarize our results in Section 4, discuss remained problems, and propose future works in Section 5.

2. Supernova Models

2.1. MHD Equations. Ideal MHD equations are enumerated as follows [13, 39]:

$$\begin{aligned}
&\frac{d\rho}{dt} + \rho\nabla\cdot\mathbf{v} = 0,\\
&\rho\frac{d\mathbf{v}}{dt} = -\nabla P - \rho\nabla\Phi + \frac{1}{4\pi}(\nabla\times\mathbf{B})\times\mathbf{B},\\
&\frac{\partial\mathbf{B}}{\partial t} = \nabla\times(\mathbf{v}\times\mathbf{B}),\\
&\nabla^2\Phi = 4\pi G\rho,\\
&\rho\frac{d(e/\rho)}{dt} = -P\nabla\cdot\mathbf{v} + Q^+ - Q^-,
\end{aligned} \tag{1}$$

where ρ is the density, $\mathbf{v}$ is the velocity, P is the pressure, $\mathbf{B}$ is the magnetic field, e is the internal energy density, and Q^+ and Q^- are the neutrino (ν) heating and cooling rates, respectively. The gravitational potential Φ is solved from the Poisson solver [28].

2.2. Neutrino Leakage Scheme. Neutrino luminosity (L_ν) at a neutrino sphere can be estimated from the average ν-energy $\bar{\epsilon}_{\nu,\mathrm{esc}}$ (see (15) later) that escapes freely:

$$L_\nu = \int_V \bar{\epsilon}_{\nu,\mathrm{esc}}\frac{n_\nu}{\tau_{\mathrm{esc}}}dV, \tag{2}$$

where n_ν is the ν-number density and τ_{esc} is the escape time for a neutrino to reach the ν-sphere R_ν, that is, obtained from the leakage scheme [38, 41] in terms of the ν-mean free path (λ_{tot}) defined by

$$\int_{R_\nu}^{\infty}\frac{1}{\lambda_{\mathrm{tot}}}dr = \frac{2}{3}. \tag{3}$$

The mean free path of neutrinos is given as

$$\lambda_{\mathrm{tot}}^{-1} = \frac{\rho Y_N}{m_u}\sigma_{\mathrm{coh}} + \frac{\rho Y_p}{m_u}\sigma_{\mathrm{sc},p} + \frac{\rho Y_n}{m_u}\sigma_{\mathrm{sc},n} + \frac{\rho Y_n}{m_u}\sigma_{ab}, \tag{4}$$

where m_u is the atomic mass unit and Y_n, Y_p, and Y_N are the number fractions relative to baryons for neutrons, protons, and nuclei, respectively. The values of $\sigma_{\mathrm{sc},p}$, $\sigma_{\mathrm{sc},n}$, and σ_{coh} are the cross sections for scattering on protons, neutrons, and nuclei, respectively.

Contrary to the original leakage scheme, we do not adopt free stream approximation for neutrinos outside R_ν. Since the minimum size of the mesh interval in our hydrodynamic calculations is 10^5 cm, we prescribe the time step so that for each time step of Δt, neutrinos run by $c\Delta t$, which is typically 10^4 cm in our simulations.

The terms of Q^+ and Q^- are calculated in the following [42, 43]. Outside the neutrino sphere, the two terms are, respectively,

$$\begin{aligned}
Q^+ &= \sigma_{ab}n_nF_\nu,\\
Q^- &= \Gamma_pn_p + \Gamma_nn_n + R_{ee} + R_\gamma,
\end{aligned} \tag{5}$$

where $n_n(n_p)$, F_ν, and σ_{ab} are the number density of free neutrons (protons), energy flux of neutrinos at each point calculated by the equation of continuity, and the ν-absorption cross section by free neutrons ($n + \nu_e \rightarrow p + e^-$), that is, the most important heating source and a function of average energy and density, respectively [44]. Neutrino production (emission) rates, Γ_p, Γ_n, and R_{ee}, R_γ are explained below (see (8) and the equations below (18)). Inside the neutrino sphere, we calculate only the term Q^- as follows:

$$Q^- = \bar{\epsilon}_{\nu,\mathrm{esc}} \frac{n_\nu}{\tau_{\mathrm{esc}}}. \tag{6}$$

For the $\bar{\nu}$-absorption by free protons, we can get both Q^+ and Q^- by replacing physical quantities between ν and $\bar{\nu}$.

2.3. Neutrino Processes and Physical Inputs. The change in electron fraction Y_e is given by (e.g., Kotake et al. [39])

$$\frac{dY_e}{dt} = -\Gamma_p Y_p - \Gamma_N Y_N - \frac{c}{\lambda_\nu} Y_n + \frac{c}{\lambda_{\bar{\nu}}} Y_p, \tag{7}$$

where λ_ν is the mean free path of electron neutrino and $\lambda_{\bar{\nu}}$ is that of antineutrino [44]. The last two terms in the right hand side, which manifest the effects of ν-radiation, play an essential role to change Y_e after around 200 ms measured from the bounce.

The electron capture rate by a proton ($p + e^- \rightarrow n + \nu_e$) with $Q_p = 1.3$ MeV is obtained from Epstein and Pethick [45]:

$$\Gamma_p = \frac{1}{2\pi^3 \hbar} \frac{G_F^2 C_V^2 \left(1 + 3a^2\right)}{\left(h^3 c^3\right)^2} I_p, \tag{8}$$

$$I_p = \int_Q^{\mu_e} dE_e E_e^2 E_\nu^m f_e \left(1 - f_\nu\right). \tag{9}$$

Here $m = 2$, $Q = Q_p + \mu_\nu$, and f_α is the Fermi-Dirac distribution of particles α $(= e, \nu)$:

$$f_\alpha = \frac{1}{1 + e^{(\epsilon_\alpha - \mu_\alpha)/T_\alpha}}, \tag{10}$$

where ϵ_α and μ_α in units of the Boltzmann constant are the single particle energy and the chemical potential, respectively. We note that outside the equilibrium region between neutrinos and baryons; significant thermal deviation comes out between temperatures of neutrinos and baryons, that is, $T_\nu \neq T_m$. The fundamental constant of G_F is the Fermi coupling, $C_V = 0.97$ the pseudovector coupling, and C_A the axial vector coupling ones. We set the ratio $|C_A/C_V| = 1.27$. The electron capture rate by a nucleus of the atomic number Z with the Q-value Q_N is

$$\Gamma_N = \frac{12}{7} \frac{1}{2\pi^3 h} \frac{G_F^2 C_A^2 (Z - 20)}{\left(h^3 c^3\right)^2} I_N, \tag{11}$$

where I_N is obtained from (9) with the values of $m = 2$ and $Q = Q_N + \mu_\nu$. Note that this capture process is assumed to be inhibited above the neutron number $N = 40$ due to the effects of shell blocking [45–47]. We should note that the temperature effects and correlations are responsible for removing the inhibition [47].

Energies of emitted neutrinos by the individual electron captures are given by

$$E_{\nu,p} = \frac{J_p}{I_p}, \qquad E_{\nu,N} = \frac{J_N}{I_N}, \tag{12}$$

where J_p and J_N are obtained from I_p and I_N with $m = 3$, respectively [45, 48]. The average energy of neutrinos emitted by electron captures is written as follows:

$$\bar{\epsilon}_\nu = \frac{E_{\nu,p} \dot{Y}_p + E_{\nu,N} \dot{Y}_N}{\dot{Y}_p + \dot{Y}_N}. \tag{13}$$

This average energy is added to obtain the ν-energy density at the next time step of simulations.

Inside the ν-sphere, T_ν and μ_ν are evaluated in terms of n_ν and the ν-energy density e_ν:

$$\begin{aligned} n_\nu &= \frac{T_\nu^3}{(\hbar c)^3} \int 4\pi \epsilon_\nu^2 f_\nu d\epsilon_\nu, \\ e_\nu &= \frac{T_\nu^4}{(\hbar c)^3} \int 4\pi \epsilon_\nu^3 f_\nu d\epsilon_\nu, \end{aligned} \tag{14}$$

where the Fermi-Dirac distribution is assumed for neutrinos. Outside the ν-sphere, neutrino radiation can be approximated to be the black body one with $\mu_\nu = 0$. The average ν-energy is written as follows;

$$\bar{\epsilon}_{\nu,\mathrm{esc}} = \frac{F_3}{F_2} T_{\nu,\mathrm{sp}}, \tag{15}$$

where $T_{\nu,\mathrm{sp}}$ is the ν-temperature at the ν-sphere and the Fermi integrals, F_2 and F_3, are, respectively:

$$F_2 = \int_0^\infty \frac{x^2}{1 + e^x} dx, \qquad F_3 = \int_0^\infty \frac{x^3}{1 + e^x} dx. \tag{16}$$

We can approximate these terms by the analytic formula given by Epstein and Pethick [45]. Since $T_\nu \geq T_m$ inside the ν-sphere, the baryon energy density e_m increases due to the energy flow from neutrinos:

$$\frac{de_m}{dt} = \frac{c}{\lambda_\nu} e_\nu, \tag{17}$$

where λ_ν is the mean free path of relevant neutrinos [44]. In this region, we replace $T_{\nu,\mathrm{sp}}$ by T_m to obtain $\bar{\epsilon}_{\nu,\mathrm{esc}}$ in (15), because neutrino temperature decreases due to the diffusive effect around the ν sphere, which would underestimate the ν-energy. This is our first modification of the original leakage scheme.

The similar procedure for the positron capture rate Γ_n can be applied to anti-neutrinos of the reaction, $n + e^+ \rightarrow p + \bar{\nu}_e$, with the substitutions

$$\mu_e \longrightarrow -\mu_e, \qquad \mu_\nu \longrightarrow \mu_{\bar{\nu}}, \qquad T_\nu \longrightarrow T_{\bar{\nu}}. \tag{18}$$

Other neutrino processes are as follows:

$$e^+ + e^- \longrightarrow \nu_x + \overline{\nu_x},$$
$$\gamma \longrightarrow \nu_x + \overline{\nu_x}, \quad (19)$$
$$x = e, \tau, \mu.$$

Here, x means electron, τ, and μ-neutrinos. These processes are important for a late stage of the explosion, because the most neutrino luminosity for the late stage of the explosion comes from these processes. The rates of these processes have been obtained by Ruffert et al. [36, 37]. While electron neutrinos emitted by these processes contribute to neutrino cooling and heating, μ, τ-neutrinos do only neutrino cooling. The emission rate of ν_e or $\overline{\nu_e}$ by electron-positron pair annihilation is given by

$$R_{ee}(\nu_e, \overline{\nu_e}) = \frac{(C_1 + C_2)_{\nu_e,\overline{\nu_e}}}{36} \frac{\sigma_0 c}{(m_e c^2)^2} \epsilon_{e^-} \cdot \epsilon_{e^+} \times \left(1 - f_{\nu_e}(\epsilon)\right)_{ee} \left(1 - f_{\overline{\nu_e}}(\epsilon)\right)_{ee}, \quad (20)$$

where $\sigma_0 = 1.76 \times 10^{-44}$ cm^2, and ϵ_{e^-} or ϵ_{e^+} indicates electron/positron energy density. The weak interaction constants are $(C_1 + C_2) = (C_V - C_A)^2 + (C_V + C_A)^2$, and $(1 - f_{\nu_e}(\epsilon))_{ee}$ is the blocking factor in the neutrino phase space and approximately expressed by

$$\left(1 - f_{\nu_e}(\epsilon)\right)_{ee} = \left(1 + \exp\left(-\left(\frac{1}{2}\frac{F_4(\eta_e)}{F_3(\eta_e)} + \frac{1}{2}\frac{F_4(-\eta_e)}{F_3(-\eta_e)} - \eta_{\nu_e}\right)\right)\right)^{-1}, \quad (21)$$

where $F_n(\eta)$ means fermi integral:

$$F_n(\eta) = \frac{T^{n+1}}{(\hbar c)^{n+1}} \int_\eta^\infty \frac{x^n}{1 - e^{x-\eta}} dx, \quad (22)$$

with $\eta = \mu_e/T$.

For the production of ν_μ, $\overline{\nu_\mu}$ and ν_τ, $\overline{\nu_\tau}$, the corresponding rate is

$$R_{ee}(\nu_x) = \frac{(C_1 + C_2)_{\nu_x,\overline{\nu_x}}}{9} \frac{\sigma_0 c}{(m_e c^2)^2} \times \epsilon_{e^-} \cdot \epsilon_{e^+} \left(1 - f_{\nu_x}(\epsilon)\right)_{ee}^2, \quad (23)$$

where $(C_1 + C_2)_{\nu_x,\overline{\nu_x}} = (C_V - C_A)^2 + (C_V + C_A - 2)^2$.

The rate of creation of ν_e or $\overline{\nu_e}$ by the decay of transversal plasmons can be written with sufficient accuracy as

$$R_\gamma(\nu_e, \overline{\nu_e}) = \frac{\pi^3}{3\alpha^*} C_V^2 \frac{\sigma_0 c^2}{(m_e c^2)^2} \frac{T^8}{(hc)^6} \gamma^6 \exp(-\gamma) \times (1 + \gamma) \left(1 - f_{\nu_e}(\epsilon)\right)_\gamma \left(1 - f_{\overline{\nu_e}}(\epsilon)\right)_\gamma, \quad (24)$$

and the corresponding rate for producing ν_x becomes

$$R_\gamma(\nu_x) = \frac{4\pi^3}{3\alpha^*} (C_V - 1)^2 \frac{\sigma_0 c^2}{(m_e c^2)^2} \frac{T^8}{(hc)^6} \gamma^6 \times \exp(-\gamma)(1 + \gamma)\left(1 - f_{\nu_x}\right)_\gamma. \quad (25)$$

The fine structure constant, $\alpha^* = 1/137.036$, and $\gamma = 5.5565 \times 10^{-2} \sqrt{(1/3)(\pi^2 + 3\eta_e^2)}$, and $(1 - f_{\nu_x})_\gamma$ is the blocking factor,

$$\left(1 - f_{\nu_x}\right)_\gamma = \left[1 + \exp\left(-\left(1 + \frac{1}{2}\frac{\gamma^2}{1 + \gamma} - \eta_{\nu_x}\right)\right)\right]^{-1}. \quad (26)$$

2.4. Neutrinos outside the Neutrino Sphere. Neutrino number density n_ν on each mesh can be calculated as follows:

$$\frac{dn_\nu}{dt} = \Gamma_p n_p + \Gamma_n n_n + R_{ee} + R_\gamma - n_{\nu,\mathrm{esc}}, \quad (27)$$

where $n_{\nu,\mathrm{esc}}$ indicates number density of protons, neutrons, and escaping neutrino density estimated by escaping time scale (see (28)), respectively. For the second modification of the leakage scheme, the last term in the right hand side in (27) is calculated in the neutrino sphere as follows:

$$n_{\nu,\mathrm{esc}} = \frac{n_\nu}{\beta \tau_{\mathrm{esc}}}, \quad (28)$$

where $\beta\tau_{\mathrm{esc}}$ is the neutrino escape time scale and it is estimated as follows,

$$\tau_{\mathrm{esc}} = \max\left(\frac{\Delta R}{c}, \frac{3\Delta R^2}{\pi^2 c \lambda_{\mathrm{tot}}}\right), \quad (29)$$

where ΔR is the distance from each point to the neutrino sphere R_ν. The factor β is introduced and it could take the value in the range 1–5 [49, 50]. In our case, we set the value to be $\beta = 2\sqrt{3}$, considering the isotropic diffusion of neutrinos from around the neutrino sphere. The escaping neutrino density is added to the neutrino density at the neutrino sphere. Outside the neutrino sphere, we consider streaming neutrinos (see the below equation of continuity) except for slightly absorbed neutrinos. Therefore, we calculate equation of continuity:

$$\frac{\partial n_\nu}{\partial t} + \frac{1}{r}\frac{\partial (n_\nu r c)}{\partial r} = \Gamma_p n_p + \Gamma_n n_n + R_{ee} + R_\gamma - n_{ab}, \quad (30)$$

where n_{ab} is the absorbed neutrino density calculated from the heating rate $(Q^+ n_\nu / e_\nu)$. We should note that Kotake et al. [23, 39] set the neutrino fraction to be zero outside the ν-sphere. We solved the continuity equation outside the ν-sphere, which affects the location of the sphere. Furthermore, they utilized a postprocessing approach for the heating term. The heating term may change the dynamics. Due to the heating, jet formation may become preferable.

In Figure 1, we show the neutrino luminosities and electron fractions calculated by the method explained in

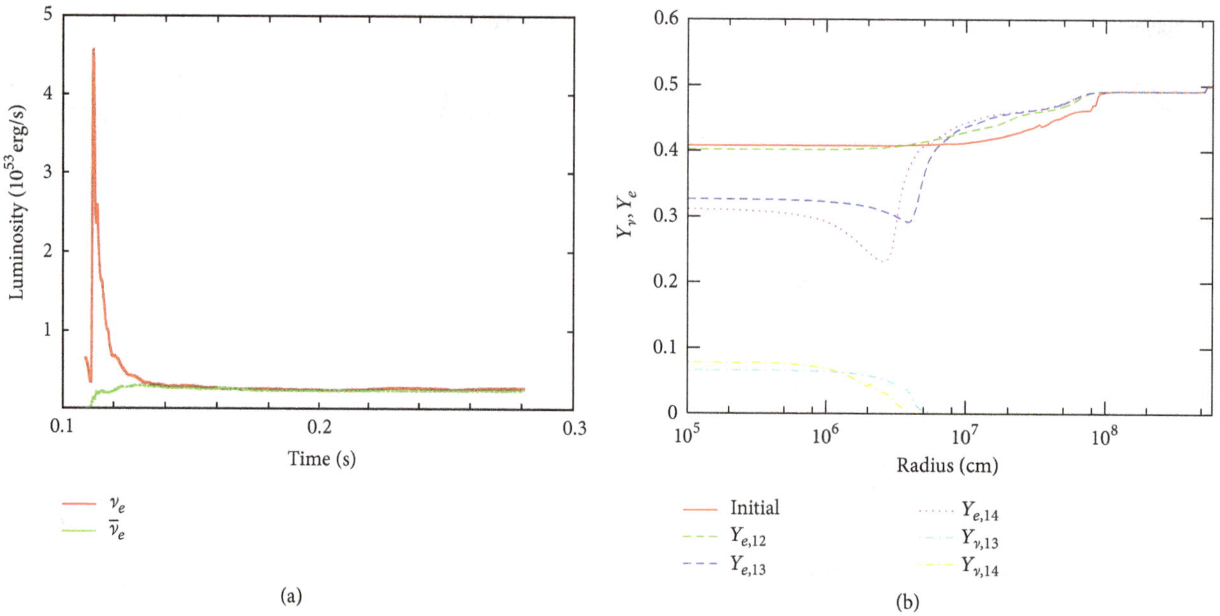

FIGURE 1: (a) It shows the evolution of the neutrino luminosity after the collapse for a model 1. (b) It shows changes in electron fraction (Y_e) when the central density becomes 10^{12}, 10^{13}, and 10^{14} g cm^{-3}, respectively, and neutrino fraction (Y_ν) when the central density becomes 10^{13} and 10^{14} g cm^{-3}, respectively. Our results could approximate those of Liebendörfer et al. [31].

this section which corresponds to a model of spherically symmetry (model 1 in Table 1). They are compared with the figures (Figures 8 and 1(d)) in Liebendörfer et al. [31]. Considering the simple scheme of the neutrino transport, our results could approximate more accurate ones which adopt the detailed neutrino transport scheme.

2.5. Initial Models and r-Process Networks. In all computations, spherical coordinates (r, θ) are adopted. The computational region is set to be $0 \le r \le 4000$ km and $0 \le \theta \le \pi/2$, where the included mass in the precollapse models amounts to $1.42\,M_\odot$. The first quadrant of the meridian section is covered with $300(r) \times 30(\theta)$ mesh points (Fe-core plus some amounts of Silicon-rich layer).

To acquire information of mass elements, nine thousand tracer particles are distributed on each mesh point within the region of $0.449 \le Y_e \le 0.49$ from the center to the place of $1.3\,M_\odot$ ($r = 2200$ km) at the beginning of the collapse. Five initial models are prepared as shown in Table 1. We adopt cylindrical properties of the angular velocity Ω and the toroidal component of the magnetic field B_ϕ as follows:

$$\Omega(X, Z) = \Omega_0 \frac{X_0^2}{X^2 + X_0^2} \frac{Z_0^4}{Z^4 + Z_0^4},$$
$$B_\phi(X, Z) = B_0 \frac{X_0^2}{X^2 + X_0^2} \frac{Z_0^4}{Z^4 + Z_0^4}, \qquad (31)$$

where X and Z are the distances from the rotational axis and the equatorial plane. X_0 and Z_0 are model parameters. Both Ω_0 and B_0 are the initial values at $X = 0$ and $Z = 0$. Here, we add a model 5 in addition to the initial models adopted by Nishimura et al. [30]. This model has the largest value of Ω_0 among five models. As shown in the next section, among the five models, model 5 can eject material with very low electron fraction Y_e.

Here, we note that Winteler et al. [32] have used a shellular rotation law as their initial model of $15\,M_\odot$ [51]. The value of $T/|W| = 7.63 \times 10^{-3}$ may not be so different from those of our models. On the other hand, their initial distribution of the magnetic field is assumed to be purely poloidal field having $E_m/|W| = 2.63 \times 10^{-8}$. The difference of the distribution of the magnetic field could affect the formation of jet along the polar direction.

Although significant improvements could have been performed for the nuclear data of nucleosynthesis calculations, we utilize the same nuclear reaction networks that have been constructed for the r-process calculations by Nishimura et al. [30]. This is because our purpose is to investigate some qualitative effects by including the neutrino transport on the r-process. Let us briefly explain the nuclear data included in the networks. The networks have been extended toward the neutron-rich side till the neutron-drip line. Each network consists of about 4000 nuclear species up to $Z = 100$. Included reactions are two-body ones, (n, γ), (p, γ), (α, γ), (p, n), (α, p), (α, n), plus three-body one, $(3\alpha, \gamma)$, and their inverses. Two kinds of the network, called FRDM and ETFSI, have been constructed. The mass formula of FRDM is constructed by the Nilsson-Struntinsky model considering effects of shell and microscopic part. ETFSI approach is a semiclassical approximation to the Hartree-Fock method in which the shell corrections are calculated with the integral version of the Strutinsky theorem. Reaction rates are constructed based on experimental data if available which are

TABLE 1: Initial parameters of precollapse models.

Model	$T/\|W\|$ (%)	$E_m/\|W\|$ (%)	X_0^*	Z_0^*	Ω_0 (s^{-1})	B_0 (G)
Model 1	0	0	0	0	0	0
Model 2	0.5	0.1	1	1	5.2	5.4×10^{12}
Model 3	0.5	0.1	0.5	1	7.9	1.0×10^{13}
Model 4	0.5	0.1	0.1	1	42.9	5.2×10^{13}
Model 5	1.5	0.1	0.1	1	72.9	5.2×10^{13}

Note: $X_0^* = X_0/10^8$ cm and $Z_0^* = Z_0/10^8$ cm. Models 1 to 4 have the same initial parameters as those adopted by Nishimura et al. [30]. We add a model 5 that has the largest value of a parameter of Ω_0 among five models, where other parameters are the same as those of the model 4.

TABLE 2: Calculated quantities that are crucial in the r-process.

Model	t_b	t_f	$T/\|W\|_f$	$E_m/\|W\|_f$	E^*_{exp}	$M_{ej}/M_\odot$	$M_{rej}/M_\odot$
Model 1	111	283	0	0	0.023	—	—
Model 2	125	311	6.91	0.053	0.127	—	—
Model 3	129	329	8.74	0.116	0.164	—	—
Model 4	133	433	8.80	0.142	1.13	0.111	—
Model 5	180	624	15.3	0.339	0.484	0.022	5.90×10^{-3}

Note: t_b indicates the time (ms) at the bounce. The calculations are stopped at the time t_f (ms). The ratios $T/|W|_f$ and $E_m/|W|_f$ are expressed in %. $E^*_{exp} = E_{exp}/10^{51}$ ergs. M_{ej} is the sum of the ejected tracer particles. M_{rej} is the ejected mass of the r-element for $A \geq 63$.

supplemented by theoretical data with inverse reaction rates and partition functions with use of FRDM or ETFSI.

3. Explosion Models, Distribution of Electron Fraction, and r-Process Calculations

We investigate hydrodynamical stages of the collapse, bounce, and propagation of the shock wave with use of ZEUS-2D code using a simple neutrino transport scheme as shown in Section 2. Our results of MHD calculations are summarized in Table 2, where E_{exp} is the explosion energy when the shock reaches the edge of the Fe-core and M_{ej} is the mass summed over the ejected tracer particles. We note that the explosion does not occur for model 1 to model 3. While the jet-like explosion occurs along the equator (up to 40° from the equator) in model 4, a collimated jet emerged from the rotational axis in model 5 (Figure 2). A protoneutron star remains after the jet-like explosion. During the explosion, temperature exceeds 10^{10} K around the original layers of the Si + Fe core, where the nuclear statistical equilibrium is realized.

In model 4, the equatorial region is ejected as shown in Figure 2 having rather high value of $Y_e \simeq 0.50$ (Figure 3(a)). In model 5, materials are ejected with the jets along the polar regions, whose total angle is subtended over 20° from the axis (see Figure 2). The corresponding evolutions of Y_e relevant to the r-process are shown in Figure 4. The lowest value of $Y_e < 0.20$ is found around the polar region as seen in Figure 3(b).

Figure 5 shows the ejected mass against Y_e in the range $0.05 \leq Y_e \leq 0.50$. In model 4, the ejecta with $Y_e > 0.40$ comes from the Si-rich layers along the equatorial region, which is attributed to the enhanced centrifugal force relative to the magnetic one. We recognize that as against the spherical explosion, Y_e decreases significantly for model 5, due to the collimated jet along the rotational axis. This is because neutrino luminosity is low by a factor of ten compared to that of spherical explosion shown in Figure 1(a). This can be seen in Figure 6, where the density along polar axis is low compared to the case of model 4. Therefore, the reaction responsible for the increase in Y_e, $n + \nu_e \rightarrow p + e^-$ becomes ineffective.

We note that we distribute 9000 tracer particles on each mesh point. To check the change in distribution of Y_e owing to that of tracer particles for model 5, (1) we scatter 15000 tracer particles inside the computational region and (2) nine thousand particles between 500 km and 2200 km. As a result, we have confirmed that deficiency of Y_e between 0.2–0.3 is the same for all cases. For case (1), tiny amounts of Y_e appears below $Y_e = 0.2$. Therefore, nucleosynthesis results qualitatively do not depend on the method how to distribute tracer particles.

We calculate the r-process nucleosynthesis for the explosion model 5. Before the nucleosynthesis calculation, we have assumed abundances to be in nuclear statistical equilibrium state (NSE) as has been done [44, 52]. The NSE code is used just after the temperature drops 10^{10} K to around 9×10^9 K. Then, the nuclear reaction network of the r-process has been operated till the temperature decreases to $2 - 3 \times 10^9$ K ($t \sim 600$ ms) using the results of the MHD calculations. After that, network calculations are performed until $T \sim 10^7$ K ($t \sim 10$ s) with the method in Flower and Hoyle [53]. We include neutrino captures in our nucleosynthesis network that has been applied for tracer particles. The capture rates ($n + \nu_e \rightarrow p + e^-$, $p + \bar{\nu}_e \rightarrow n + e^+$) that were not incorporated by Nishimura et al. [30] are obtained from neutrino luminosity, neutrino flux, and other hydrodynamical information as we have described in the previous sections. Concerning the nucleosynthesis, we can get actually this information after the end of NSE stage ($t \sim t_f$) under the assumption of constant neutrino luminosity, since L_ν does not change appreciably

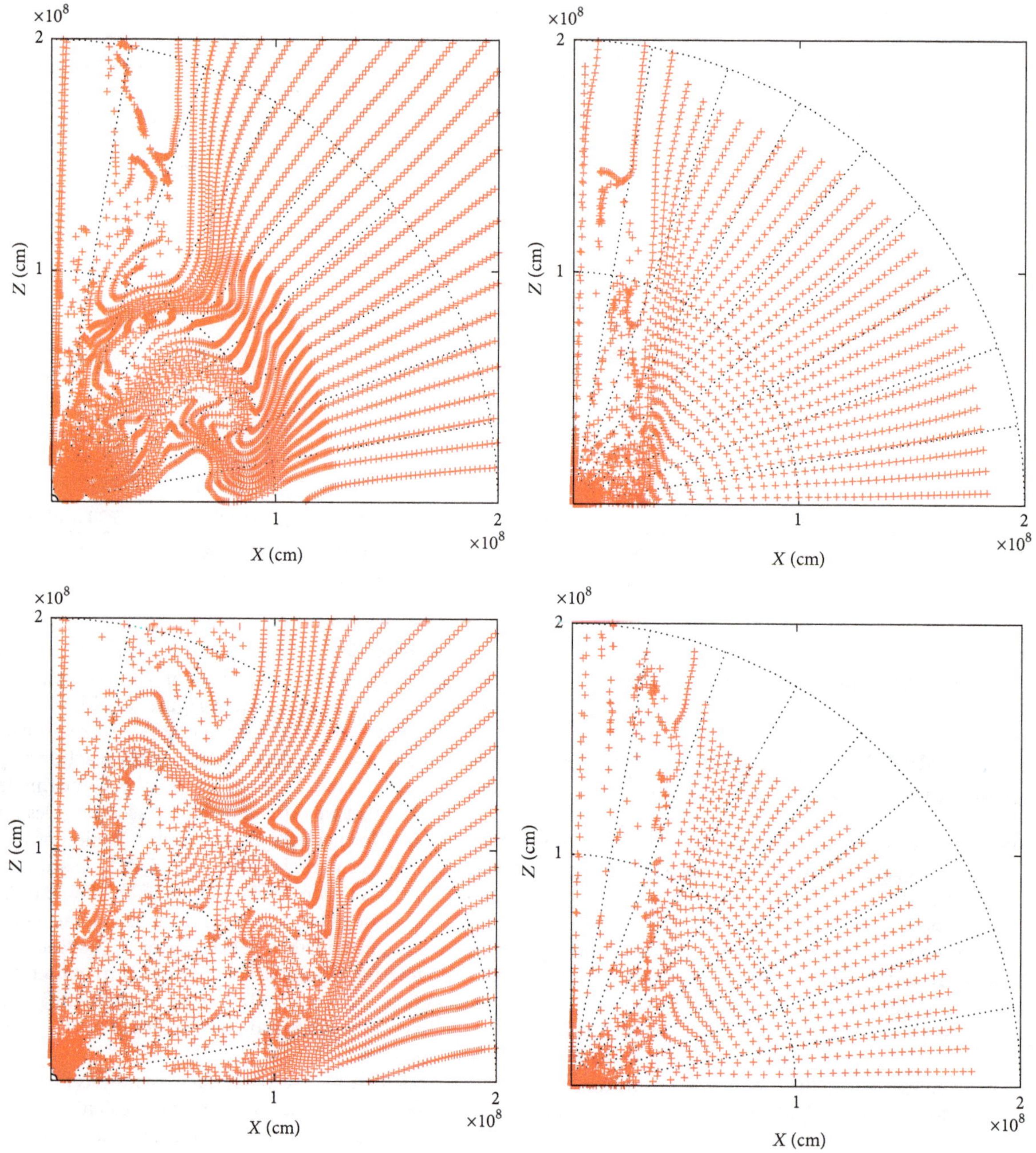

FIGURE 2: Snapshots of tracer particles at t = 100 ms (top) and t = 200 ms (bottom) after the bounce in models 4 (left) and 5 (right).

after t~200 ms. We note that a difference between the neutron and proton single-particle energies in a dense medium may change neutrino capture cross sections significantly [54]. As a consequence, it is shown that the luminosities of all neutrino flavors are reduced while the spectral differences between electron neutrinos and antineutrinos are increased [55]. These changes in weak interaction processes should be examined by including them for the hydrodynamical simulations.

In case of the rapid rotation and strong magnetic field (model 4), barely jet-like explosion is obtained in the direction of the equatorial region (Figure 6(a)). It is, however, impossible to reproduce the r-elements even up to the second peak of the solar r-process abundance pattern, because Y_e of the ejected materials distributes in the range of high values of $Y_e \geq 0.4$. In the model 5, where we have adopted a special initial configuration of concentrated magnetic field with strong differential rotation, jet-like explosion emerges in the direction of the rotational axis (Figure 6(b)). The difference is that model 5 has larger value of the angular velocity compared to model 4 by a factor of 1.7. We compared

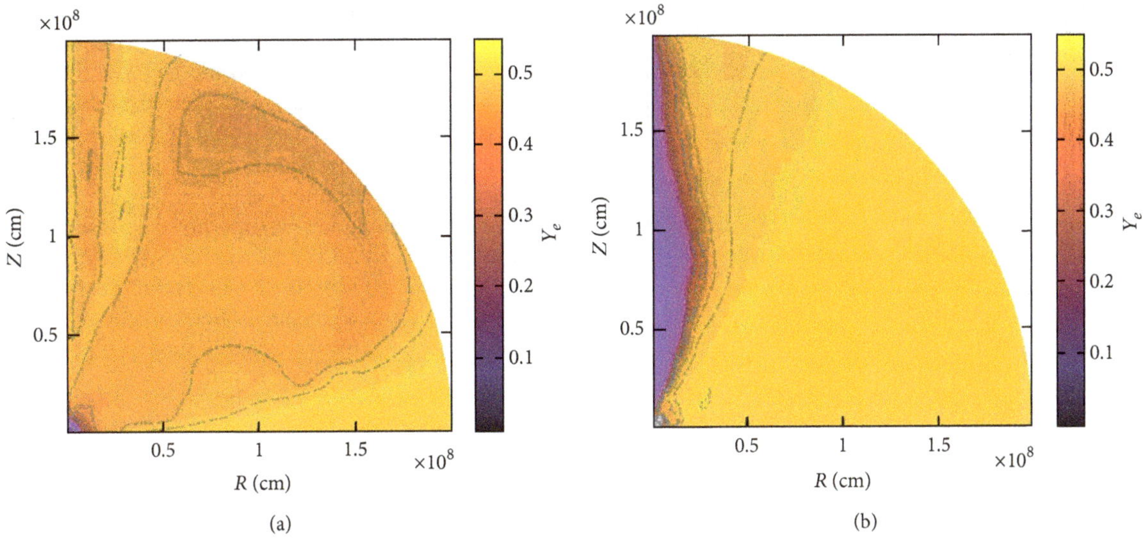

FIGURE 3: Contour of Y_e over the range 0.1–0.5 at the final stage of calculation in model 4 (a) and model 5 (b).

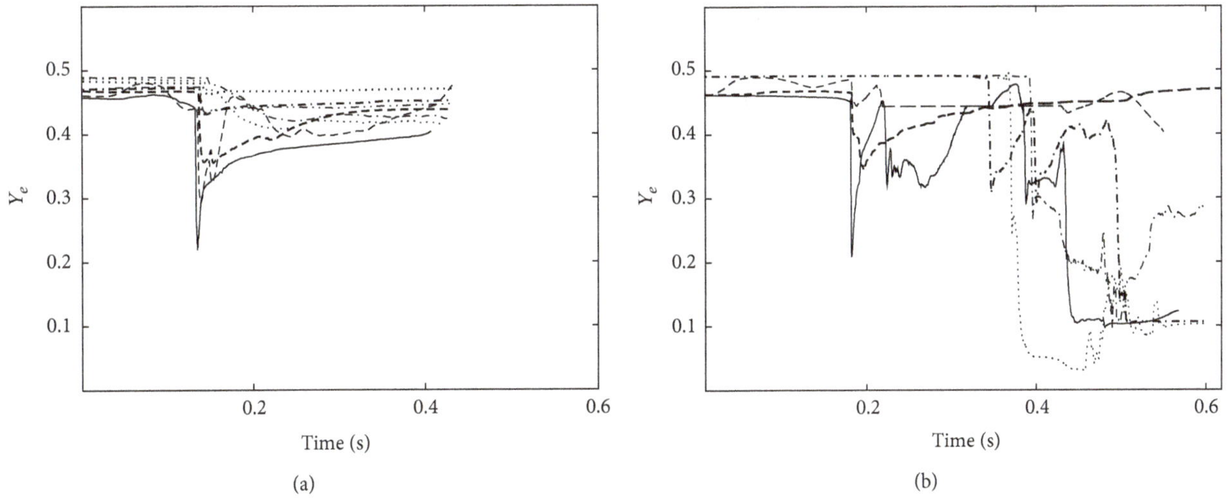

FIGURE 4: Time evolution of Y_e in model 4 (a) and model 5 (b). For model 4, the explosion resembles to spherical one and all Y_e have increased above 0.4 due to the neutrino capture process.

the produced heavy elements with the solar r-elements in Figure 7, where results for two different mass formulas are shown. Model 5 may present a site for reproducing only the third peak with the first and second peaks underproduced.

4. Summary and Discussion

We have barely shown a qualitative possibility of the r-process nucleosynthesis during MHD explosion in a massive star of $13\,M_\odot$. Initial models have been constructed changing the distributions of rotation and magnetic fields parametrically [30].

We include neutrino effects by using the leakage scheme. This scheme treats neutrino effects approximately, where we assume Fermi-Dirac distribution for neutrinos. Furthermore, we add modification about the physical process just outside the neutrino sphere (15) and timescale of neutrino drift (28) and (29) in addition to the original leakage scheme. Validity for 2D calculations cannot be assured for this scheme, because flow to the θ direction is not included. If jets are so strong that the corresponding density becomes low, increase in Y_e would be suppressed compared to the 1D case. Therefore, we can say that the discussion of Nishimura et al. [30] with use of an analytical formula of 1D results is inaccurate.

Other schemes have been developed for solving neutrino transport [56]. For example, IDSA (Isotropic Diffusion Source Approximation) solves Boltzmann equation approximately, and the result obtained by using this scheme is consistent with one dimensional simulations, where

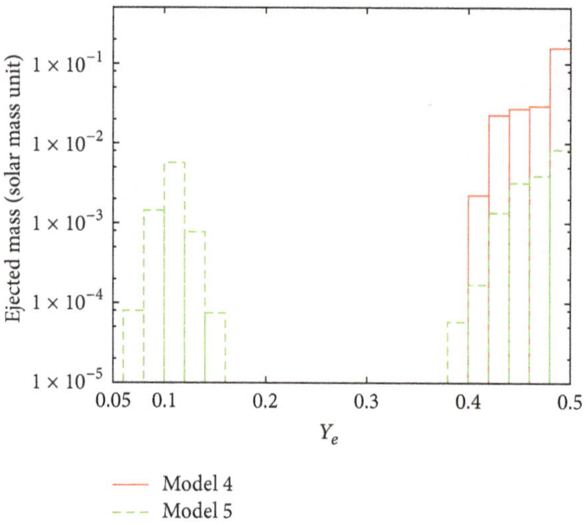

FIGURE 5: Ejected mass as a function of Y_e at around $t_f = 600$ ms (model 5). The thick lines indicate the results of model 4 ($t_f = 433$ ms). Two separate distributions in Y_e are due to different neutrino irradiation just outside the neutrino sphere.

the Fermi-Dirac distribution has been found to be a good approximation [56].

We have calculated MHD simulations by varying two parameters in initial models (the distribution of rotation and magnetic field). The previous adiabatic simulations without neutrino effects have succeeded in explosion except for the model 3 [30]. However, in the present case, only two models with neutrino effects whose distribution of magnetic field and rotation concentrated in the central region result in explosion. Generally, for strong and concentrated rotation case with some magnetic field, a collimated jet-like explosion occurs. For relatively weak and concentrated rotation model, large region around from rotational axis to equatorial axis is blasted, which is assisted with the magnetic field effect. In the present study, ejected region is different from that of Nishimura et al. [30]. Model 5 shows that deep region is ejected in comparison with Nishimura's model 4. It means that the produced composition depends crucially on rotation parameter. Model 5 may present an appropriate site for reproducing only the third peak relative to the first and second peaks insufficiently built up. Contrary to the negative conclusion against the possible r-process in the previous study [30], we can show at least the elemental distributions of the r-elements as far as the third peak of the solar pattern is concerned. This is due to the lower neutrino luminosities from 200 to 500 ms after the bounce. At the same time, we have shown the possibility for the lower Y_e materials to be ejected significantly if the neutrino transport mechanism works appropriately.

Observations of γ-ray bursts associated with supernovae are rare in the present observations [57]. Considering the relations between γ-ray bursts and the MHD explosions, the new site of the r-process to produce significantly only the third peak of the solar abundance pattern should be also rare. The nuclear process to produce the abundances after the third peak would have some relations to our MHD jet model. Since γ-ray bursts should have continued after the formation of the first star, a new model beyond our jet model (e.g., Winteler et al. [32]) and their motivation of the r-process for low metallicities would give a clue about nuclear cosmochronology represented by ^{232}Th which half life is as long as the age of the universe [52].

We could conclude that supernova explosions of massive stars associated with the r-process cannot be excluded under some assumed conditions; a progenitor has special distributions of rotational/magnetic fields inside the stellar core; simple neutrino transport scheme such as a leakage scheme can be applied.

5. Future Work

We propose that both the supernova mechanism and r-process nucleosynthesis still remain to be some crucial problems. This might be one of the reason why whole consistency for the origin of elements has not been understood. We should include the following effects for the r-process calculations.

5.1. Detailed Neutrino Transport Scheme. In the present study, we use a simple leakage scheme for description of neutrino transport. Leakage scheme can describe neutrino transport in very easy way compared to the Boltzmann equation solver. Neutrino effect, however, is very sensitive to dynamics [58], and our method may not be inadequate for the 2D calculations. We need to adopt a more detailed neutrino transport scheme such as that of IDSA which may be capable of applying to multi-dimensional simulations. Due to the difficulty of multi-dimensional simulations with Boltzmann equation solver, IDSA simulation will play a more important role [13, 56].

5.2. Neutrino Oscillation. We do not consider neutrino oscillation because transport with neutrino oscillation is difficult to handle. Since mean energies of τ and μ neutrinos are higher than those of electron neutrino, the effect may cause significant effects for neutrino heating [59]. Furthermore, it has been shown that the charged current weak interaction processes affect the luminosities of neutrino flavors which are related to neutrino oscillations [55]. These fundamental processes should be studied with hydrodynamical simulations.

5.3. SASI. Standing accretion shock instability (SASI) has been focused on hydrodynamic simulations [11–13]. It is considered to help the shock strength increase in heating regions. While SASI is in a good sense for explosion mechanism, it may play minus role for heavy element nucleosynthesis. Although a strong convection may occur behind the shock front by SASI, the convection effect tends to average the Y_e distribution [60].

5.4. Distribution of Magnetic Field. We consider only toroidal magnetic field. If we input polar magnetic field as an initial parameter [61], even a weak rotation model may lead to

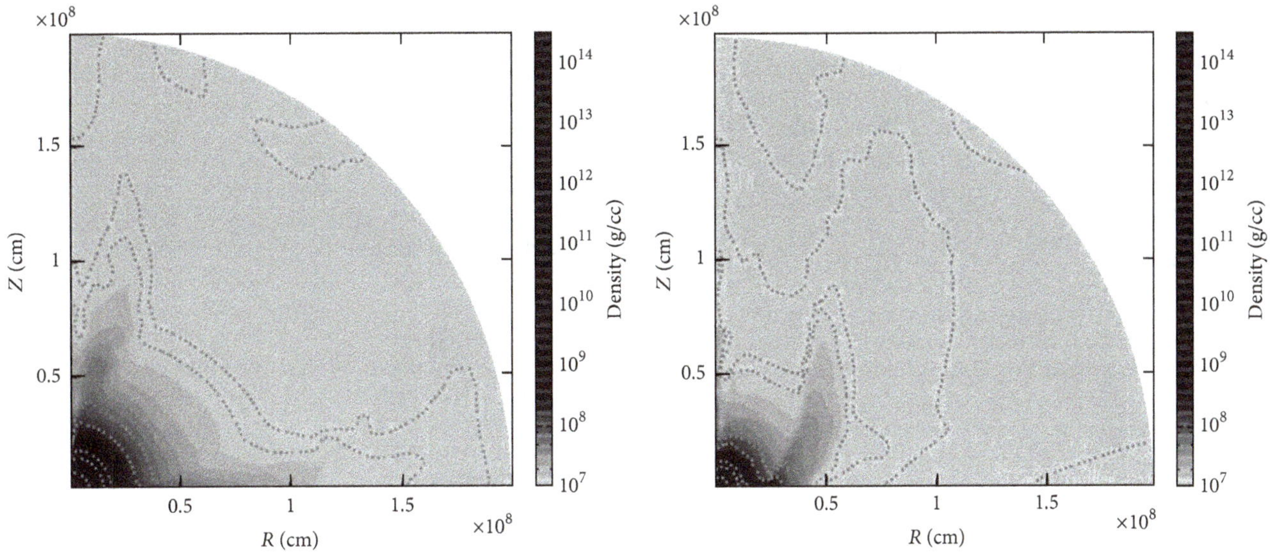

(a) (b)

FIGURE 6: Density contour over 10^7–10^{14} g cm^{-3} at $t = 200$ ms after the bounce in model 4 (a) and model 5 (b). Model 5 indicates a jet in which density is rather low along the polar axis.

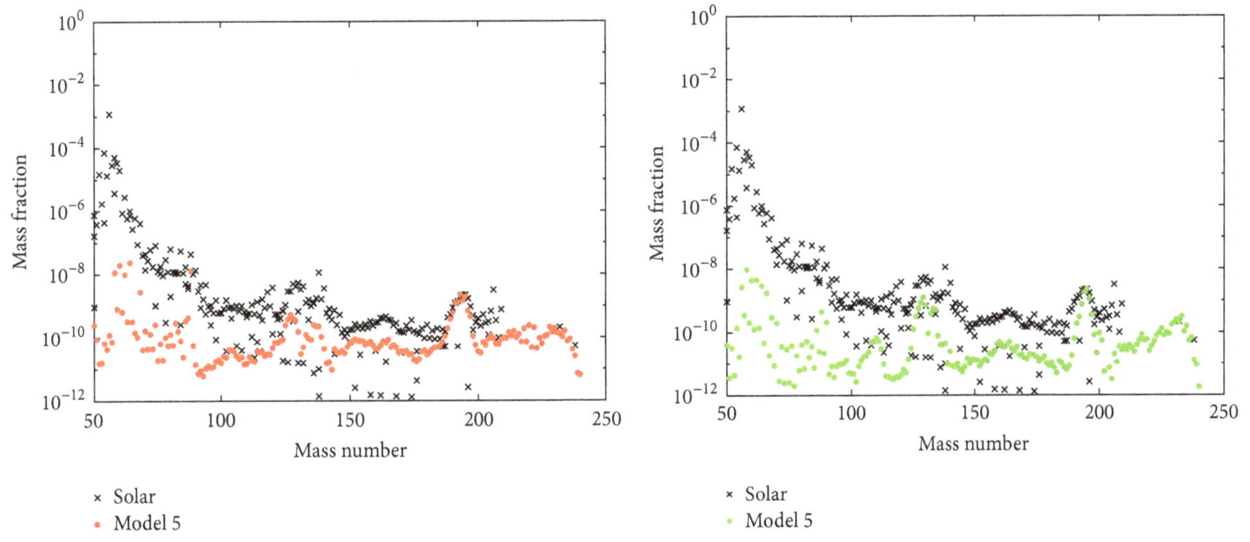

(a) (b)

FIGURE 7: Nucleosynthesis calculation results using ETFSI (a) and FRDM (b). The third peak is reproduced for both mass formula. However, the slope of the abundance pattern of FRDM is too steep around $A \simeq 200$ to compare with the solar pattern. The last peak of $A > 200$ cannot be reproduced in both cases.

succeed in explosion and draw up low Y_e matter from deep inside the region. In this point, MRE explosion with a realistic magnetic field configuration should be investigated [20–22] to see the effects on the r-process.

5.5. Simulations of Other Massive Star Models. We have adopted a 3.3 $M_{\odot}$ He-core model, because it has been convenient to set up an initial model. We need to simulate for higher stellar masses (e.g., Ono et al. [62]). Neutrino luminosity may rise for more massive models. Increasing neutrino luminosity could unfortunately cause neutrino capture process furthermore and heavier nuclei may not be synthesized.

5.6. Magnetar. We find that the produced protoneutron star has very strong magnetic field (10^{14}–10^{16} G). This might suggest the formation of magnetar [63–72]. However, in the present calculation with a numerical scheme adopted, the MRI cannot be resolved [21]; it is hopeful to study whether there exists any relation between the formation of magnetar and heavy elemental synthesis.

Acknowledgment

This work has been supported in part by a Grant-in-Aid for Scientific Research (19104006, 21540272, 24540278, and 25400281) of the Ministry of Education, Culture, Sports, Science, and Technology of Japan. The authors would like to thank the reviewers for improving the draft.

References

[1] M. Arnould, S. Goriely, and K. Takahashi, "The r-process of stellar nucleosynthesis: astrophysics and nuclear physics achievements and mysteries," *Physics Reports*, vol. 450, no. 4-6, pp. 97–213, 2007.

[2] F. K. Thielemann, A. Arcones, R. Käppeli et al., "What are the astrophysical sites for the *r*-process and the production of heavy elements? " *Progress in Particle and Nuclear Physics*, vol. 66, no. 2, pp. 346–353, 2011.

[3] Y. Z. Qian, "The r-process: current understanding and future tests," in *Proceedings of the 1st Argonne/MSU/JINA/INT RIA Wokshop*, 2005.

[4] M. Hashimoto, "Supernova nucleosynthesis in massive stars," *Progress of Theoretical Physics*, vol. 94, no. 5, pp. 663–736, 1995.

[5] A. Heger, C. L. Fryer, S. E. Woosley, N. Langer, and D. D. H. Hartmann, "How massive single stars end their life," *Astrophysical Journal Letters*, vol. 591, no. 1, pp. 288–300, 2003.

[6] H. T. Janka, R. Buras, K. Kifonidis, A. Marek, and M. Rampp, "Core-collapse supernovae at the threshold," *Springer Proceeding in Physics*, vol. 99, pp. 253–262, 2005.

[7] K. Sumiyoshi, S. Yamada, H. Suzuki, H. Shen, S. Chiba, and H. Toki, "Postbounce evolution of core-collapse supernovae: long-term effects of the equation of state," *Astrophysical Journal Letters*, vol. 629, no. 2, pp. 922–932, 2005.

[8] A. Burrows, E. Livne, L. Dessart, C. D. Ott, and J. Murphy, "A new mechanism for core-collapse supernova explosions," *Astrophysical Journal Letters*, vol. 640, no. 2, pp. 878–890, 2006.

[9] F. S. Kitaura, H. T. Janka, and W. Hillebrandt, "Explosions of O-Ne-Mg cores, the Crab supernova and subluminous type II-P supernovae," *Astronomy and Astrophysics*, vol. 450, no. 1, pp. 345–350, 2006.

[10] A. Burrows, L. Dessart, E. Livne, C. D. Ott, and J. Murphy, "Simulations of magnetically driven supernova and hypernova explosions in the context of rapid rotation," *The Astrophysical Journal*, vol. 664, no. 1, p. 416, 2007.

[11] S. W. Bruenn, A. Mezzacappa, W. R. Hix et al., "The explosion of a rotating star as a supernova mechanism," in *American Institute of Physics Conference Series*, G. Giobbi, A. Tornambe, G. Raimondo et al., Eds., vol. 1111, pp. 593–601, 2009.

[12] A. Marek and H. T. Janka, "Delayed neutrino-driven supernova explosions aided by the standing accretion-shock instability," *Astrophysical Journal Letters*, vol. 694, no. 1, pp. 664–696, 2009.

[13] Y. Suwa, K. Kotake, T. Takiwaki, S. C. Whitehouse, M. Liebendörfer, and K. Sato, "Explosion geometry of a rotating 13m⊙ star driven by the sasi-aided neutrino-heating supernova mechanism," *Publications of the Astronomical Society of Japan*, vol. 62, no. 6, pp. L49–L53, 2010.

[14] T. Fischer, S. C. Whitehouse, A. Mezzacappa, F. K. Thielemann, and M. Liebendörfer, "Protoneutron star evolution and the neutrino-driven wind in general relativistic neutrino radiation hydrodynamics simulations," *Astronomy & Astrophysics*, vol. 517, article A80, 25 pages, 2010.

[15] T. Takiwaki, K. Kotake, and Y. Suwa, "Three-dimensional hydrodynamic core-collapse supernova simulations for an 11.2 $M_\odot$ star with spectral neutrino transport," *Astronomical Journal*, vol. 738, no. 2, article 165, 13 pages.

[16] G. S. Bisnovaty-Kogan, "Mechanism of core-collapse supernovae & simulation results from the CHIMERA code," *Astronomicheskii Zhurnal*, vol. 47, p. 813, 1970.

[17] N. V. Ardeljan, G. S. Bisnovatyi-Kogan, and S. G. Moiseenko, "Explosion mechanisms of supernovae: the magnetorotational model," *Physics-Uspekhi*, vol. 40, pp. 1076–1079, 1997.

[18] N. V. Ardeljan, G. S. Bisnovatyi-Kogan, and S. G. Moiseenko, "Magnetorotational mechanism: 2D simulation," in *The Local Bubble and Beyond Lyman-Spitzer-Colloquium: Proceedings of the IAU Colloquium No. 166*, vol. 506 of *Lecture Notes in Physics*, pp. 145–148, 1998.

[19] G. S. Bisnovatyi-Kogan, N. V. Ardeljan, and S. G. Moiseenko, "Magnetorotational explosions: supernovae and jet formation," *Memorie della Societa Astronomica Italiana*, vol. 73, pp. 1134–1143, 2002.

[20] N. V. Ardeljan, G. S. Bisnovatyi-Kogan, and S. G. Moiseenko, "Magnetorotational supernovae," *Monthly Notices of the Royal Astronomical Society*, vol. 359, no. 1, pp. 333–344, 2005.

[21] S. G. Moiseenko, G. S. Bisnovatyi-Kogan, and N. V. Ardeljan, "A magnetorotational core-collapse model with jets," *Monthly Notices of the Royal Astronomical Society*, vol. 370, no. 1, pp. 501–512, 2006.

[22] G. S. Bisnovatyi-Kogan, S. G. Moiseenko, and N. V. Ardeljan, "Different magneto-rotational supernovae," *Astronomy Reports*, vol. 12, no. 12, pp. 997–1008, 2008.

[23] K. Kotake, H. Sawai, S. Yamada, and K. Sato, "Magnetorotational effects on anisotropic neutrino emission and convection in core-collapse supernovae," *Astrophysical Journal Letters*, vol. 608, no. 1, pp. 391–404, 2004.

[24] K. Kotake, S. Yamada, K. Sato, K. Sumiyoshi, H. Ono, and H. Suzuki, "Gravitational radiation from rotational core collapse: effects of magnetic fields and realistic equations of state," *Physical Review D*, vol. 69, no. 12, Article ID 124004, 2004.

[25] S. Yamada and H. Sawai, "Numerical study on the rotational collapse of strongly magnetized cores of massive stars," *Astrophysical Journal Letters*, vol. 608, no. 2, pp. 907–924, 2004.

[26] S. Yamada, K. Kotake, and T. Yamasaki, "The role of neutrinos, rotations and magnetic .elds in collapse-driven supernovae," *New Journal of Physics*, vol. 6, article 79, pp. 1–24, 2004.

[27] T. Takiwaki, K. Kotake, S. Nagataki, and K. Sato, "Magneto-driven shock waves in core-collapse supernovae," *Astrophysical Journal Letters*, vol. 616, no. 2, pp. 1086–1094, 2004.

[28] J. M. Stone and M. L. Norman, "ZEUS-2D: a radiation magnetohydrodynamics code for astrophysical flows in two space dimensions. II: the magnetohydrodynamic algorithms and tests," *Astrophysical Journal Supplement*, vol. 80, p. 791, 1992.

[29] H. Shen, H. Toki, K. Oyamatsu, and K. Sumiyoshi, "Relativistic equation of state of nuclear matter for supernova and neutron star," *Nuclear Physics A*, vol. 637, no. 3, pp. 435–450, 1998.

[30] S. Nishimura, K. Kotake, M. A. Hashimoto et al., "r-process nucleosynthesis in magnetohydrodynamic jet explosions of core-collapse supernovae," *Astrophysical Journal Letters*, vol. 642, no. 1, pp. 410–419, 2006.

[31] M. Liebendörfer, A. Mezzacappa, F. K. Thielemann, B. Messer, W. R. Hix, and S. Bruenn, "Probing the gravitational well: no supernova explosion in spherical symmetry with general relativistic Boltzmann neutrino transport," *Physical Review D*, vol. 63, no. 10, Article ID 103004, 13 pages, 2001.

[32] C. Winteler, R. Käppeli, A. Perego et al., "Magnetorotationally driven supernovae as the origin of early galaxy r-process elements?" *Astrophysical Journal Letters*, vol. 750, no. 1, article L22, 2012.

[33] M. Saruwatari, M. Hashimoto, K. Kotake, and S. Yamada, "R-process nucleosynthesis during the magnetohydrodynamics explosions of a massive star," in *Proceedings of the 10th Internal Symposium on Origin of Matter and Evolution of Galaxies (OMEG '10)*, vol. 1269 of *AIP Conference Proceedings*, pp. 409–411, Osaka, Japan, March, 2010.

[34] K. Mori, E. Gotthelf, S. Zhang et al., "NuSTAR discovery of A 3.76 s transient magnetar near sagittarius A," *The Astrophysical Journal*, vol. 770, no. 2, article L23.

[35] R. Epstein, "The generation of gravitational radiation by escaping supernova neutrinos," *The Astrophysical Journal*, vol. 223, pp. 1037–1042, 1978.

[36] M. Ruffert, H. T. Janka, and G. Schafer, "Coalescing neutron stars—a steptowards physical models I. Hydrodynamical evolution and gravitational-wave emission," *Astronomy & Astrophysics*, vol. 311, pp. 532–566, 1996.

[37] A. Staudt and H. V. Klapdor-Kleingrothaus, "Calculation of beta-delayed fission rates of neutron rich nuslei far off stability," *Nuclear Physics A*, vol. 549, no. 2, pp. 254–264, 1992.

[38] K. A. van Riper and J. M. Lattimer, "Stellar core collapse. I: infall epoch," *The Astrophysical Journal*, vol. 249, pp. 270–289, 1981.

[39] K. Kotake, S. Yamada, and K. Sato, "Anisotropic neutrino radiation in rotational core collapse," *Astrophysical Journal Letters*, vol. 595, no. 1, pp. 304–316, 2003.

[40] S. Rosswog and M. Liebendörfer, "High-resolution calculations of merging neutron stars. II: neutrino emission," *Monthly Notices of the Royal Astronomical Society*, vol. 342, no. 3, pp. 673–689.

[41] K. A. van Riper, "Stellar core collapse. II: inner core bounce and shock propagation," *The Astrophysical Journal*, vol. 257, no. 15, pp. 793–820, 1982.

[42] H. A. Bethe, "Supernova mechanisms," *Reviews of Modern Physics*, vol. 62, no. 4, pp. 801–866, 1990.

[43] H. T. Janka, "Conditions for shock revival by neutrino heating in core-collapse supernovae," *Astronomy and Astrophysics*, vol. 368, no. 2, pp. 527–560, 2001.

[44] D. L. Tubbs and D. N. Schramm, "Neutrino opacities at high temperatures and densities," *The Astrophysical Journal*, vol. 201, pp. 467–488, 1975.

[45] R. I. Epstein and C. J. Pethick, " Lepton loss and entropy generation in stellar collapse," *The Astrophysical Journal*, vol. 243, pp. 1003–1012, 1981.

[46] G. M. Fuller, W. A. Fowler, and M. J. Newmann, "Stellar weak interaction rates for intermediate-mass nuclei. II: A = 21 to A = 60," *The Astrophysical Journal*, vol. 252, pp. 715–740, 1982.

[47] K. Langanke, G. Martínez-Pinedo, J. M. Sampaio et al., "Electron capture rates on nuclei and implications for stellar core collapse," *Physical Review Letters*, vol. 90, no. 24, Article ID 241102, pp. 1–4, 2003.

[48] S. L. Shapiro and S. A. Teukolsky, *Black Holes, White Dwarfs, and Neutron Stars*, Wiley, 1983.

[49] S. A. Bludman, I. Lichtenstadt, and G. Hayden, "Homologous collapse and deleptonization of an evolved stellar core," *The Astrophysical Journal*, vol. 261, pp. 661–676, 1982.

[50] J. Cooperstein, "Neutrinos in supernovae," *Physics Reports*, vol. 163, no. 1-3, pp. 95–126, 1988.

[51] A. Heger, S. E. Woosley, and H. C. Spruit, "Presupernova evolution of differentially rotating massive stars including magnetic fields," *Astrophysical Journal*, vol. 626, no. 1, pp. 350–363, 2005.

[52] D. D. Clayton, *Principles of Stellar Evolution and Nucleosynthesis*, McGraw-Hill, New York, NY, USA, 1968.

[53] W. A. Fowler and F. Hoyle, "Neutrino processes and pair formation in massive stars and supernovae," *Astrophysical Journal Supplement*, vol. 9, p. 201, 1964.

[54] L. F. Roberts, S. Reddy, and G. Shen, "Medium modification of the charged-current neutrino opacity and its implications," *Physical Review C*, vol. 86, no. 6, Article ID 065803, 10 pages, 2012.

[55] G. Martinez-Pinedo, T. Fischer, A. Lohs, and L. Huther, "Charged-current weak interaction processes in hot and dense matter and its impact on the spectra of neutrinos emitted from proto-neutron star cooling," *Physical Review Letters*, vol. 109, no. 25, Article ID 251104, 2012.

[56] M. Liebendörfer, S. C. Whitehouse, and T. Fischer, "The isotropic diffusion source approximation for supernova neutrino transport," *Astrophysical Journal Letters*, vol. 698, no. 2, pp. 1174–1190, 2009.

[57] M. Ono, M. A. Hashimoto, S. I. Fujimoto, K. Kotake, and S. Yamada, "Explosive nucleosynthesis in magnetohydrodynamical jets from collapsars," *Progress of Theoretical Physics*, vol. 122, no. 3, pp. 755–777, 2009.

[58] H. T. Janka and E. Müller, "Neutrino heating, convection, and the mechanism of Type-II supernova explosions," *Astronomy & Astrophysics*, vol. 306, p. 167, 1996.

[59] Y. Suwa, K. Kotake, T. Takiwaki, M. Liebendorfer, and K. Sato, "Impacts of collective neutrino oscillations on supernova explosions," *Astronomical Journal*, vol. 749, no. 2, article 98, 17 pages.

[60] S. I. Fujimoto, K. Kotake, M. A. Hashimoto, M. Ono, and N. Ohnishi, "Explosive nucleosynthesis in the neutrino-driven aspherical supernova explosion of a non-rotating 15 M∞ star with solar metallicity," *Astrophysical Journal*, vol. 738, no. 1, article 61, 2011.

[61] S. I. Fujimoto, K. Kotake, S. Yamada, M. A. Hashimoto, and K. Sato, "Magnetohydrodynamic simulations of a rotating massive star collapsing to a black hole," *Astrophysical Journal Letters*, vol. 644, no. 2, pp. 1040–1055, 2006.

[62] M. Ono, S. Nagataki, H. Ito et al., "Matter mixing in aspherical core-collapse supernovae: a search for possible conditions for conveying 56Ni into high velocity regions," *The Astrophysical Journal*, vol. 773, no. 2, article 161, 2013.

[63] V. V. Usov, "Millisecond pulsars with extremely strong magnetic fields as a cosmological source of γ-ray bursts," *Nature*, vol. 357, pp. 472–474, 1992.

[64] C. Thompson, "A model of gamma-ray bursts," *Monthly Notices of the Royal Astronomical Society*, vol. 270, no. 3, p. 480, 1994.

[65] E. G. Blackman and I. Yi, "On fueling gamma-ray bursts and their afterglows with pulsars," *Astrophysical Journal Letters*, vol. 498, no. 1, pp. L31–L35, 1998.

[66] J. C. Wheeler, I. Yi, P. Höflich, and L. Wang, "Asymmetric supernovae, pulsars, magnetars, and gamma-ray bursts," *The Astrophysica l Journal*, vol. 537, no. 2, pp. 810–823, 2000.

[67] B. Zhang and P. Meszaros, "Gamma-ray burst afterglow with continuous energy injection: signature of a highly magnetized millisecond pulsar," *The Astrophysical Journal Letters*, vol. 552, no. 1, article L35, 2001.

[68] T. A. Thompson, P. Chang, and E. Quataert, "Magnetar spin-down, hyperenergetic supernovae, and gamma-ray bursts," *Astrophysical Journal Letters*, vol. 611, no. 1, pp. 380–393, 2004.

[69] N. Bucciantini, E. Quataert, J. Arons, B. D. Metzger, and T. A. Thompson, "Magnetar-driven bubbles and the origin of collimated outflows in gamma-ray bursts," *Monthly Notices of the Royal Astronomical Society*, vol. 380, no. 4, pp. 1541–1553, 2007.

[70] N. Bucciantini, E. Quataert, J. Arons, B. D. Metzger, and T. A. Thompson, "Relativistic jets and long-duration gamma-ray bursts from the birth of magnetars," *Monthly Notices of the Royal Astronomical Society*, vol. 383, no. 1, pp. L25–L29, 2008.

[71] N. Bucciantini, E. Quataert, B. D. Metzger, T. A. Thompson, J. Arons, and L. Del Zanna, "Magnetized relativistic jets and long-duration GRBs from magnetar spin-down during core-collapse supernovae," *Monthly Notices of the Royal Astronomical Society*, vol. 396, no. 4, pp. 2038–2050, 2009.

[72] B. D. Metzger, T. A. Thompson, and E. Quataert, "Proto-neutron star winds with magnetic fields and rotation," *Astrophysical Journal Letters*, vol. 659, no. 1, pp. 561–579, 2007.

19

An Extensive Photometric Investigation of the W UMa System DK Cyg

M. M. Elkhateeb,[1,2] M. I. Nouh,[1,2] E. Elkholy,[1,2] and B. Korany[1,3]

[1]Astronomy Department, National Research Institute of Astronomy and Geophysics, Helwan, Cairo 11421, Egypt
[2]Physics Department, College of Science, Northern Border University, Arar 1321, Saudi Arabia
[3]Physics Department, Faculty of Applied Science, Umm Al-Qura University, Makkah 715, Saudi Arabia

Correspondence should be addressed to M. I. Nouh; abdo_nouh@hotmail.com

Academic Editor: Theodor Pribulla

DK Cyg ($P = 0.4707$) is a contact binary system that undergoes complete eclipses. All the published photoelectric data have been collected and utilized to reexamine and update the period behavior of the system. A significant period increase with rate of 12.590×10^{-11} days/cycle was calculated. New period and ephemeris have been calculated for the system. A long term photometric solution study was performed and a light curve elements were calculated. We investigated the evolutionary status of the system using theoretical evolutionary models.

1. Introduction

The eclipsing binary DK Cyg (BD +330 4304, 10.37–10.93 m_v) is a well known contact binary system with a period of about 0.4707 days. It was discovered as variable star earlier by Guthnick and Prager [1], so their epoch of intensive observations is very long. The earliest photographic light curve classified the system as a W Ursae Majoris type. Rucinski and Lu [2] carried out the first spectroscopic observations and estimated the mass ratio as $q = 0.325$ and classified the system as an A-subtype contact system with spectral type of A8V. Visual light curves were published by Piotrowski [3] and Tsesevitch [4] from Klepikova's observations.

First photoelectric observations for the system were carried out by Hinderer [5], while Binnendijk [6] observed the system photoelectrically in B- and V-bands and derived least squares orbital solution. The system DK Cyg was classified in the General Catalogue of Variable Stars as A7V [7], while Binnendijk [6] adopted it as A2. Mochnacki and Doughty [8] showed that the color index of the system judged its spectral type and found that the spectral type of the system is more likely to be about F0 to F2. Because the system DK Cyg is a summer object in the Northern hemisphere with 11.5-hour period and short durations of night, it is bound to remain ill-observed [9]. Only four complete light curves by Binnendijk [6], Paparo et al. [10], Awadalla [9], and Baran et al. [11] are published. Photoelectric observations and new times of minima have been carried out by many authors: Borkovits et al. [12], Sarounová and Wolf [13], Drozdz and Ogloza [14], Hübscher et al. [15], Dogru et al. [16], Hubscher et al. [17], Hubscher et al. [18], Erkan et al. [19], Diethelm [20], Dogru et al. [21], Simmons [22], Diethelm [23], and Diethelm [24].

In the present paper we are going to perform comprehensive photometric study for the system DK Cyg. The structures of the paper are as follows: Section 2 deals with the period change, Section 3 is devoted to the light curve modeling, and Section 4 presents the discussion and conclusion reached.

2. Period Change

Although the period variation of contact binary systems of the W UMa-type is a controversial issue of binary star astrophysics, the cause of the variations (long as well as short term) is still a mystery for a discussion of possible physical mechanisms [25]. Magnetic activity cycle is one of the main mechanisms that caused a period variation together with the mass exchange between the components of each system. Kaszas et al. [26] stated that the long term period variation may be interpreted by a perturbation of the third companion or surface activity of the system components.

Observations by Binnendijk [6] showed a change of the secondary minimum depth and a new linear light element was derived. Period study by Paparo et al. [10] showed that the orbital period of the system DK Cyg increases and the first parabolic light elements were calculated, which confirmed the light curve variability. Kiss et al. [25] updated the linear ephemeris of DK Cyg, while Awadalla [9] recalculated a new quadratic element for the system and confirmed the light curve variability suggested by Paparo et al. [10]. Wolf et al. [27] used a set of 101 published times of minimum covering the interval between 1926 and 2000 in order to update the quadratic element calculated by Awadalla [9]. They showed that the period increases by the rate 11.5×10^{-11} days/cycle. Borkovits et al. [28] follow the period behavior of the system using set of published minima from HJD 2424760 to HJD 2453302.

In this paper we studied the orbital period behavior of the system DK Cyg using the $(O - C)$ diagram based on more complete data set collected from the literatures and databases of BAV, AAVSO, and BBSAG observers. Part of our collected data set was given by Kreiner et al. [29]; unpublished Hipparcos observations and main part were downloaded from website (http://astro.sci.muni.cz/variables/ocgate/); Table 1 listed only those minima not listed on the mentioned websites. A total of 195 minima times were incorporated in our analysis covering about 86 years (66689 orbital revolutions) from 1927 to 2013. It is clear that our set of data added about 94 of new minima and increases the interval limit of the orbital period study about 13 years more than the data of Wolf et al. [27], which may give more accurate insight on the period behavior of the system. The different types of the collected minima (i.e., photographic, visual, photoelectric, and CCD) were weighted according to their type. The residual $(O - C)$'s were computed using Binnendijk [6] ephemeris (1) and represented in Figure 1. No distinction has been made between primary and secondary minima:

$$\text{Min}\, I = 2437999.5838 + 0.47069055 * E. \tag{1}$$

It can be seen from the figure that the behavior of the orbital period of the system DK Cyg shows a parabolic distribution which is generally interpreted by the transfer of mass from one component to the other of binary. Reasonable linear least squares fit of the data available improved the light elements given in (1) to

$$\text{Min}\, I = \underset{\pm 0.0192}{2437999.5961} + \underset{\pm 0.0243}{0^d.47069206} * E. \tag{2}$$

The linear element yields a new period of $P = 0^d.47069206$ days which is longer by 0.13 seconds with respect to the value given by Binnendijk [6]. Quadratic least squares fit gives

$$\begin{aligned}\text{Min}\, I &= \underset{\pm 0.0192}{2437999.5803} + \underset{\pm 0.0237}{0^d.47069064} * E \\ &\quad + \underset{\pm 2.5654\times 10^{-12}}{6.284 \times 10^{-11}} \times E^2.\end{aligned} \tag{3}$$

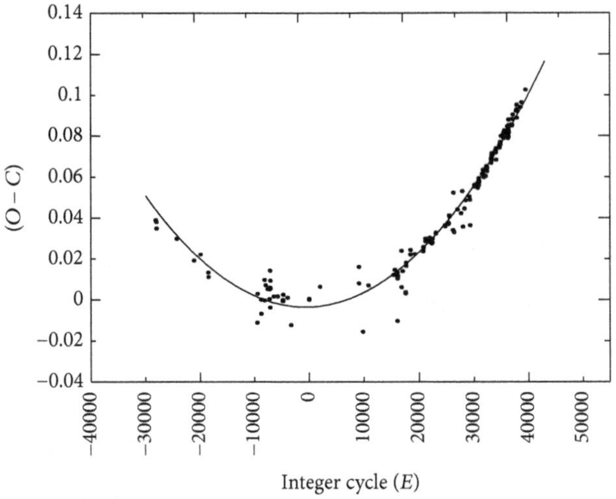

FIGURE 1: Period behavior of DK Cyg.

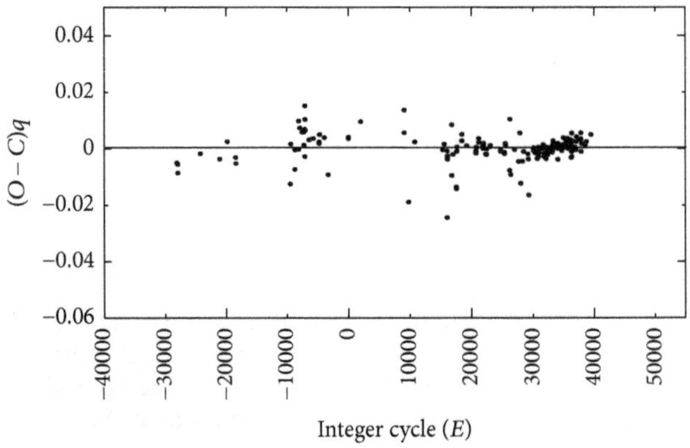

FIGURE 2: Calculated residuals from the quadratic ephemeris.

The rate of period increasing resulting from the quadratic elements (3) is $dP/dE = 12.568 \times 10^{-11}$ days/cycle or 9.746×10^{-8} days/year or 0.84 seconds/century. More future systematic and continuous photometric observations are needed to follow a continuous change in the orbital period of the system DK Cyg which may show a periodic behavior. The fourth column of Table 1 represents the quadratic residuals $(O - C)q$ calculated using the new element of (2) and represented in Figure 2. All published linear and quadratic elements together with that resulting from our calculations are listed in Table 2. It is noted from the table that the quadratic term resulting from our calculations has slightly higher value than that calculated by Awadalla [9], Wolf et al. [27], and Borkovits et al. [28]. This can be interpreted by the increasing of the set of minima in our study compared to the one they used (nearly double); also we covered an interval larger than the one they used.

3. Light Curve Modeling

Light curve modeling for the system DK Cyg by Mochnacki and Doughty [8] using Binnendijk [6] observations in V-band showed nonmatching between the theoretical curve and

TABLE 1: Times of minimum light for DK Cyg.

HJD	Method	E	$(O-C)$	$(O-C)q$	References	HJD	Method	E	$(O-C)$	$(O-C)q$	References
2434179.4690	Vis	−8116	0.00970	0.00971	1	2451749.4450	pe	29212	0.04885	−0.00370	3
2447758.4352	Pe	20733	0.02423	−0.00099	2	2451777.6740	ccd	29272	0.03642	−0.01636	5
2447790.4437	Pe	20801	0.02577	0.00037	2	2452163.6590	ccd	30092	0.05517	−0.00074	5
2447963.6620	Pe	21169	0.02995	0.00354	3	2452245.5592	ccd	30266	0.05521	−0.00137	5
2447963.8960	Pe	21169.5	0.02860	0.00220	3	2452253.5613	ccd	30283	0.05557	−0.00108	5
2448265.1380	Pe	21809.5	0.02865	0.00046	4	2452441.8384	ccd	30683	0.05645	−0.00176	5
2448265.1382	Pe	21809.5	0.02885	0.00066	4	2452512.4415	pe	30833	0.05597	−0.00284	7
2448272.1987	Pe	21824.5	0.02899	0.00076	4	2452525.6231	ccd	30861	0.05824	−0.00069	5
2448297.6160	Pe	21878.5	0.02900	0.00062	3	2452526.5644	ccd	30863	0.05816	−0.00078	5
2448302.7930	Pe	21889.5	0.02841	−0.00001	3	2452811.8062	ccd	31469	0.06148	0.00013	5
2448308.2078	Pe	21901	0.03027	0.00182	4	2453223.4286	ccd	32343.5	0.06500	−0.00006	8
2448308.2079	Pe	21901	0.03037	0.00192	4	2453228.3681	ccd	32354	0.06225	−0.00274	8
2448336.4491	Pe	21961	0.03013	0.00152	4	2453246.4950	ccd	32392.5	0.06756	0.00242	8
2449988.5840	ccd	25471	0.04120	0.00182	5	2453247.4346	ccd	32394.5	0.06578	0.00063	8
2450003.6456	ccd	25503	0.04070	0.00122	5	2453285.3260	ccd	32475	0.06659	0.00110	8
2450313.8240	ccd	26162	0.03403	−0.00765	5	2453286.2657	ccd	32477	0.06491	−0.00059	8
2450341.6130	ccd	26221	0.05229	0.01041	5	2453302.2672	ccd	32511	0.06293	−0.00271	8
2450397.6060	ccd	26340	0.03311	−0.00917	5	2454799.5505	ccd	35692	0.07959	0.00004	9
2450692.7400	ccd	26967	0.04414	−0.00030	5	2455043.8381	ccd	36211	0.07882	−0.00310	10
2451000.0990	pe	27620	0.04221	−0.00452	5	2455062.6680	ccd	36251	0.08107	−0.00105	11
2451095.6600	ccd	27823	0.05303	0.00557	5	2455088.5544	ccd	36306	0.07953	−0.00285	10
2451160.5980	ccd	27961	0.03573	−0.01222	5	2455810.6029	ccd	37840	0.08870	−0.00096	12
2451379.4820	pe	28426	0.04863	−0.00101	3						

(1) Szafraniec [30]; (2) Hubscher et al. [31]; (3) Wolf et al. [27]; (4) Hipparcos observations (unpublished); (5) Baldwin and Samolyk [32]; (6) Kiss et al. [25]; (7) Borkovits et al. [33]; (8) Borkovits et al. [28]; (9) Gerner [34]; (10) Menzies [35]; (11) Samolyk [36]; (12) Simmons [22].

TABLE 2: The light elements of DK Cyg.

JD	Period	Quadratic term	References
2437999.5838	0.470690550		Binnendijk [6]
2437999.5828	0.470690660	5.390×10^{-10}	Paparo et al. [10]
2437999.5825	0.470690730	5.760×10^{-11}	Awadalla [9]
2451000.0999	0.470692900		Kiss et al. [25]
2437999.5825	0.470690640	5.750×10^{-11}	Wolf et al. [27]
2451000.1031	0.470693909	5.862×10^{-11}	Borkovits et al. [28]
2437999.5961	0.470692060		Present work
2437999.5803	0.470690640	6.284×10^{-11}	Present work

the observations. The photometric mass ratio calculated from their accepted model was $q_{ph} = 0.33 \pm 0.02$, while the spectroscopic value estimated using radial velocity study by Rucinski and Lu [2] is $q_{sp} = 0.325 \pm 0.04$. Baran et al. [11] estimated an alternating model of spectroscopic and photometric data based on iterative solutions. Their model shows a better fit by introducing a cool spot on the surface of the more luminous component and adopted the third light as free parameter in the computations. On the other hand Rucinski and Lu [2] stated that they did not find any evidence for the existence of a third component in the system during their spectroscopic study. They refer the probability for the presence of a third star in the system to the $(O-C)$ diagram [29], which shows sinusoidal variation. This evidence is weak because only one cycle is covered up to date.

In the present work we used complete light curves published by Binnendijk [6], Paparo et al. [10], Awadalla [9], and Baran et al. [11] in V-band through a long term photometric solution study in order to estimate the physical parameters of the system and to follow its evolutionary status. The collected light curves showed a flat-bottom minima and O'Connell effect. Observations by Paparo et al. [10] and Awadalla [9] displayed some scattering specially at the two maxima of their light curves. Also the observed light curve

Table 3: Photometric solutions for DK Cyg.

Parameter	Binnendijk [6]	Paparo et al. [10]	Awadalla [9]	Baran et al. [11]
A	5500	5500	5500	5500
i (°)	80.59 ± 0.12	80.22 ± 0.21	80.83 ± 0.27	79.97 ± 0.06
$g_1 = g_2$	0.32	0.32	0.32	0.32
$A_1 = A_2$	0.5	0.5	0.5	0.5
$q\ (M_2/M_1)$	0.306*	0.306*	0.306*	0.306*
$\Omega_1 = \Omega_2$	2.4064 ± 0.002	2.4077 ± 0.005	2.3325 ± 0.004	2.3886 ± 0.001
Ω_{in}	2.4794	2.4794	2.4794	2.4794
Ω_{out}	2.2888	2.2888	2.2888	2.2888
T_1 (°K)	7500*	7500*	7500*	7500*
T_2 (°K)	6767 ± 4	6726 ± 7	6726 ± 9	6759 ± 2
r_1 pole	0.4696 ± 0.0007	0.4694 ± 0.0013	0.4861 ± 0.0014	0.4735 ± 0.0003
r_1 side	0.5091 ± 0.0010	0.5087 ± 0.0018	0.5328 ± 0.0021	0.5145 ± 0.0004
r_1 back	0.5400 ± 0.0013	0.5395 ± 0.0024	0.5716 ± 0.0029	0.5471 ± 0.0005
r_2 pole	0.2792 ± 0.0008	0.2789 ± 0.0014	0.2980 ± 0.0017	0.2835 ± 0.0003
r_2 side	0.2932 ± 0.0010	0.2929 ± 0.0017	0.3166 ± 0.0022	0.2985 ± 0.0004
r_2 back	0.3414 ± 0.0019	0.3407 ± 0.0034	0.3982 ± 0.0068	0.3522 ± 0.0008
Spot **A** of star **1**				
Colatitude	130*	130*	130*	130*
Longitude	180*	180*	180*	180*
Spot radius	33.61 ± 0.230	30.74 ± 0.437	27.23 ± 1.14	35.014 ± 0.09
Temp. factor	0.796 ± 0.003	0.840 ± 0.007	0.924 ± 0.01	0.819 ± 0.001
Spot **A** of star **2**				
Colatitude	120*	120*	120*	120*
Longitude	290*	290*	290*	290*
Spot radius	32.99 ± 3.60	29.42 ± 1.34	33.08 ± 1.24	29.44 ± 1.20
Temp. factor	1.01 ± 0.01	1.01 ± 0.01	1.17 ± 0.01	1.02 ± 0.002
$\sum(O-C)^2$	0.0229	0.02909	0.02458	0.0453

*Not adjusted.

by Awadalla [9] shows sudden increase in the light level at secondary minimum with respect to the other collected curves.

Photometric analysis for the studied light curves of the system DK Cyg was carried out using Mode 3 (overcontact) of WDint56a Package [43] based on the 2009 version of Wilson and Devinney (W-D) code with Kurucz model atmospheres [44–46]. The observed light curves were analyzed using all individual observations. Appropriate gravity darkening and bolometric albedo exponents were assumed for the convective envelope. We adopted $\mathbf{g_1} = \mathbf{g_2} = 0.32$ [47] and $\mathbf{A_1} = \mathbf{A_2} = 0.5$ [48]. Bolometric limb darkening values are adopted using the table of van Hamme [49]. Temperature of the primary star was adopted according to Baran et al. [11] model ($\mathbf{T_1} = \mathbf{7500}$°K).

The adjustable parameters are the mean temperature of the secondary component $\mathbf{T_2}$, orbital inclination **i**, and the potential of the two components $\mathbf{\Omega} = \mathbf{\Omega_1} = \mathbf{\Omega_2}$, while the spectroscopic mass ratio ($\mathbf{q_{sp}} = 0.306$) by Baran et al. [11] was fixed for all calculated models together with the primary star's temperature ($\mathbf{T_1}$).

We started modeling using as initial values the parameters of Baran et al. [11] solution based on cool spot on the luminous components and a third light as a free parameter. The used parameters show disagreement between the theoretical and observed light curves, except for Baran et al. observations. Regarding the conclusion of Rucinski and Lu [2], which stated a weak evidence of presence of a third component, we tried to construct a spotted model without the third light. We constructed a model including two spots; the first one is a cool spot located on surface of the more massive component, while the other is a hot spot located on the surface of the other component. The accepted model reveals good agreement between theoretical and observed light curves for all collected data. Table 3 lists the calculated parameters for the four light curves, while Figure 3 represented the theoretical light curves according to the accepted solution together with the reflected points in V-band. The $\sum(O-C)^2$ values in Table 3 are indicative of comparisons in future studies, since the number of observations and the accuracy are not the same in the four light curves. Absolute physical parameters for each component of the system DK Cyg were calculated based on the results of the radial velocity data of Baran et al. [11] and our new photometric solution for each light curve. The calculated parameters are listed in Table 3. The results show that the primary component is more massive and hotter than

TABLE 4: Absolute physical parameters for DK Cyg.

Parameter	Binnendijk [6]	Paparo et al. [10]	Awadalla [9]	Baran et al. [11]
$M_{1\odot}$	1.7363 ± 0.0709	1.7358 ± 0.0709	1.7679 ± 0.0722	1.7438 ± 0.0712
$M_{2\odot}$	0.5313 ± 0.0217	0.5312 ± 0.0217	0.5410 ± 0.0221	0.5336 ± 0.0218
$R_{1\odot}$	1.7037 ± 0.0696	1.7029 ± 0.0695	1.7635 ± 0.0720	1.7178 ± 0.0701
$R_{2\odot}$	1.0129 ± 0.0414	1.0118 ± 0.0413	1.0811 ± 0.0441	1.0285 ± 0.0420
$T_{1\odot}$	1.2980 ± 0.0530	1.2980 ± 0.0530	1.2980 ± 0.0530	1.2980 ± 0.053
$T_{2\odot}$	1.1712 ± 0.0478	1.1641 ± 0.0475	1.1641 ± 0.0475	1.1698 ± 0.0478
M_{1_bol}	2.4617 ± 0.1005	2.4627 ± 0.1005	2.3868 ± 0.0974	2.4438 ± 0.1000
M_{2_bol}	4.0375 ± 0.1648	4.0662 ± 0.1660	3.9224 ± 0.1601	4.0094 ± 0.1637
$L_{1\odot}$	8.2285 ± 0.3359	8.2208 ± 0.3356	8.8163 ± 0.3600	8.3653 ± 0.3415
$L_{2\odot}$	1.9276 ± 0.0787	1.8772 ± 0.0766	2.1431 ± 0.0875	1.9780 ± 0.0808

Note: subscripts 1 and 2 mean primary and secondary component, respectively.

TABLE 5: Physical parameters of the five A-type contact binaries.

Star name	Parameters						References
	$M_1(M_\odot)$	$M_2(M_\odot)$	$R_1(R_\odot)$	$R_2(R_\odot)$	$L_1(L_\odot)$	$L_2(L_\odot)$	
YY CrB	1.404	0.339	1.427	0.757	2.58	6.68	1
AW UMa	1.6	0.121	1.786	0.739	7.47	0.804	2
EQ Tau	1.214	0.541	1.136	0.787	1.31	0.6	3
RR Cen	1.82	0.38	2.1	1.05	8.89	2.2	4
V566 Oph	1.41	0.34	1.45	0.77	4.46	1.23	5

(1) Essam et al. [38], (2) Elkhateeb and Nouh [39], (3) Elkhateeb and Nouh [40], (4) Yang et al. [41], and (5) Degirmenci [42].

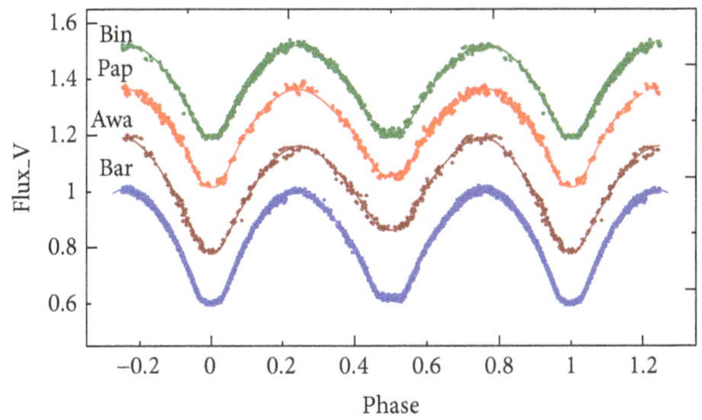

FIGURE 3: Observed and synthetic light curves of Binndijik [6] (Bin), Paparo et al. [10] (Pap), Awadalla [9] (Awa), and Baran et al. [11] (Bar), for the system DK Cyg.

the secondary component. A three-dimensional geometrical structure for the system DK Cyg is displayed in Figure 4 using the software Package Binary Maker 3.03 [50] based on the calculated parameters resulting from our models.

4. Discussion and Conclusion

Studying of the period behavior of the system DK Cyg based on all available published times of minima, covering 86 yr of observations including 195 times of light minima, shows a continuous period increase with the rate dP/dE = 12.590×10^{-11} days/cycle or 9.763×10^{-8} days/year or 0.84 seconds/century. New linear and quadratic elements were calculated using all available published data and yield a new period of P = 0.47069203 days. A long term photometric study was performed using published observations by Binnendijk [6], Paparo et al. [10], Awadalla [9], and Baran et al. [11]. More systematic and continuous photometric observations for the system DK Cyg are needed to confirm a continuous change in the period and follow its light curve variation.

One of the difficulties for W UMa binaries is to use stellar models of single stars to investigate the evolutionary status of these systems. However, using these theoretical models may give approximate view about the evolutionary status of the system.

We used the physical parameters listed in Table 4 to investigate the current evolutionary status of DK Cyg. In Figures 5 and 6, we plotted the components of DK Cygon on the mass-luminosity (M-L) and mass-radius (M-R) relations along with the evolutionary tracks computed by Girardi et al. [51] for both zero age main sequence stars (ZAMS) and terminal age main sequence stars (TAMS) with metallicity z = 0.019. As it is clear from the figures, the primary component of the system is located nearly on the ZAMS for both the M-L and M-R relations. The secondary component is above the TAMS track for M-L and the M-R relations. For the sake of comparison, we plotted sample of A-type contact binaries listed in Table 5. The components of DK Cyg have the same behavior of the selected A-type systems.

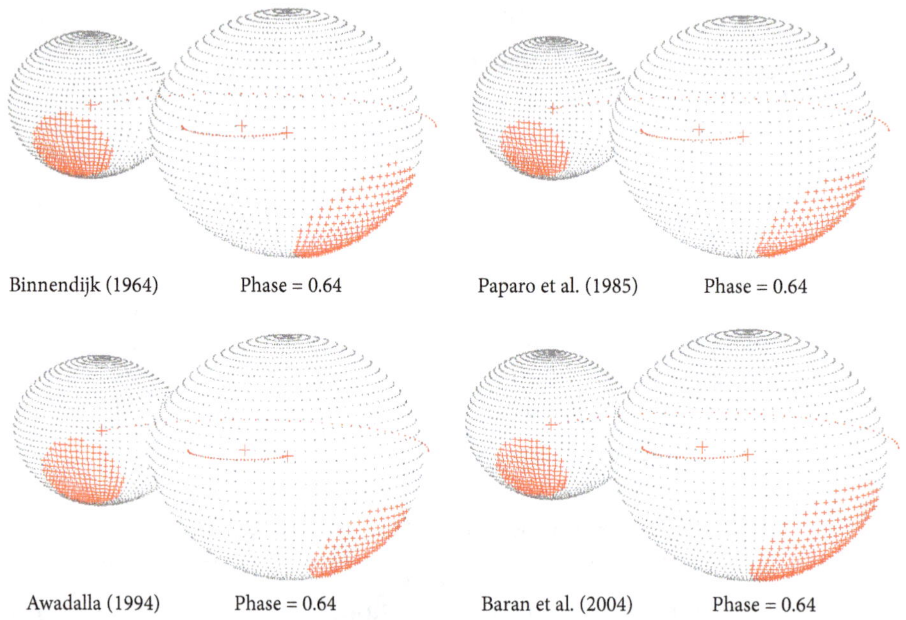

FIGURE 4: Three-dimensional models of the components of DK Cyg.

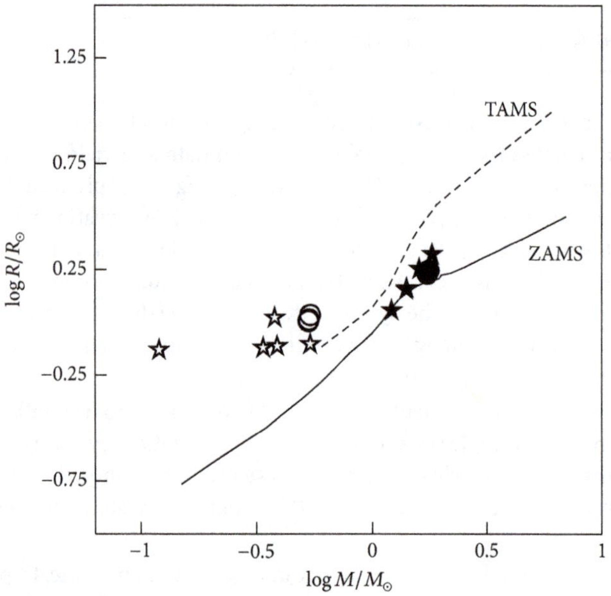

FIGURE 5: The position of the components of DK Cyg on the mass-radius diagram. The filled symbols denote the primary component and the open symbols represent the secondary component. The star symbols denote the sample of the selected A-type systems listed in Table 5.

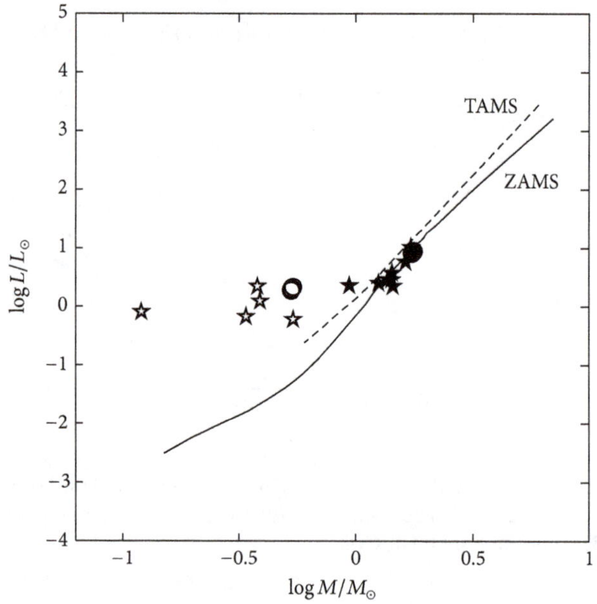

FIGURE 6: The position of the components of DK Cyg on the mass-luminosity diagram. The filled symbols denote the primary component and the open symbols represent the secondary component. The star symbols denote the sample of the selected A-type systems listed in Table 5.

The mass-effective temperature relation (M-T_{eff}) for intermediate and low mass stars [37] is displayed in Figure 7. The location of our mass and radius on the diagram revealed a good fit for the primary and poor fit for the secondary components. This gave the same behavior of the system on the mass-luminosity and mass-radius relations.

Conflict of Interests

The authors declare that they have no conflict of interests regarding the publication of this paper.

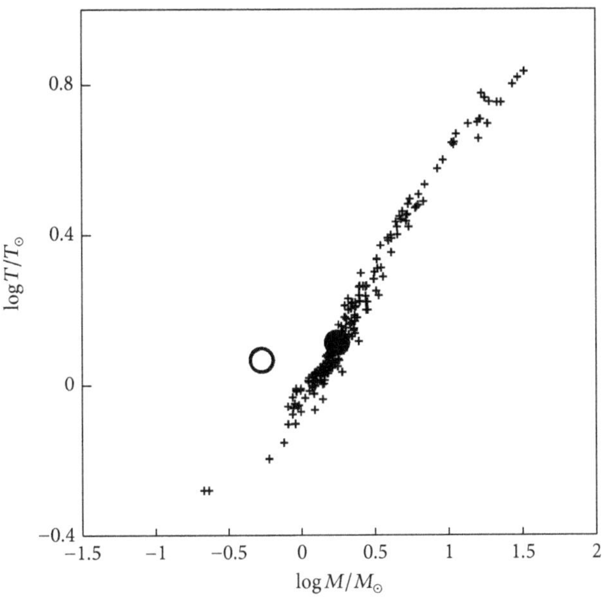

FIGURE 7: Position of the components of DK Cyg on the empirical mass-T_{eff} relation for low-intermediate mass stars by Malkov [37]. The filled symbols denote the primary component and the open symbols represent the secondary component.

Acknowledgments

This research has made use of *NASA's ADS* and the available on-line material of the *IBVS*. The authors sincerely thank Dr. Bob Nelson, who allowed them to use his windows interface code WDwint56a, for his helpful discussions and advice.

References

[1] P. Guthnick and R. Prager, "Benennung von veränderlichen Sternen," *Astronomische Nachrichten*, vol. 10-11, pp. 161–206, 1936.

[2] S. M. Rucinski and W. Lu, "Radial velocity studies of close binary stars. II," *The Astronomical Journal*, vol. 118, no. 5, pp. 2451–2459, 1999.

[3] S. Piotrowski, "Observations photometriques de DK cyg et de BB peg," *Bulletin de L'Observatoire Astronomique de Belgrade*, vol. 1, no. 2, p. 9, 1936.

[4] L. Binnendijk, "The light variation and orbital elements of DK Cygni," *The Astronomical Journal*, vol. 69, pp. 157–164, 1964.

[5] F. Hinderer, "Lichtelektrische untersuchungen an W ursae majoris-sternen," *Journal des Observateurs*, vol. 43, p. 161, 1960.

[6] L. Binnendijk, "The light variation and orbital elements of DK Cygni," *The Astronomical Journal*, vol. 69, pp. 157–164, 1964.

[7] N. Samus, O. Durlevich, V. Goranskij et al., 2014, http://www.sai.msu.su/gcvs/cgi-bin/search.htm.

[8] S. W. Mochnacki and N. A. Doughty, "Models for five W ursae majoris systems," *Monthly Notices of the Royal Astronomical Society*, vol. 156, no. 2, pp. 243–252, 1972.

[9] N. Awadalla, "Photoelectric light curve study of the DK Cygni system," *Astronomy & Astrophysics*, vol. 289, pp. 137–140, 1994.

[10] M. Paparo, M. Hamdy, and I. Jankovics, "Photoelectric observation of DK Cyg," *Information Bulletin on Variable Stars*, no. 2838, 1985.

[11] A. Baran, S. Zola, S. Rucinski, J. Kreiner, and M. Drozdz, "Physical parameters of components in close binary systems: II," *Acta Astronautica*, vol. 54, pp. 195–206, 2004.

[12] T. Borkovits, I. B. Bíró, Sz. Csizmadia et al., "New times of minima of eclipsing binary systems," *Information Bulletin on Variable Stars*, vol. 5579, 2004.

[13] L. Sarounová and M. Wolf, "Precise CCD times of minima of selected eclipsing binaries," *Information Bulletin on Variable Stars*, no. 5594, 2005.

[14] M. Drozdz and W. Ogloza, "Photoelectric minima of eclipsing binaries," *Information Bulletin on Variable Stars*, vol. 5623, p. 1, 2005.

[15] J. Hübscher, A. Paschke, and F. Walter, "Photoelectric minima of selected eclipsing binaries and maxima of pulsating stars," *Information Bulletin on Variable Stars*, vol. 5731, pp. 1–31, 2006.

[16] S. Dogru, A. Dönmez, M. Tüysüz et al., "New times of minima of some eclipsing binary stars," *Information Bulletin on Variable Stars*, vol. 5746, 2007.

[17] J. Hubscher, H.-M. Steinbach, and F. Walter, "Minima and maxima of 292 variables (Hubscher+, 2008)," *Information Bulletin on Variable Stars*, vol. 5830, 2008.

[18] J. Hubscher, P. Lehmann, G. Monninger, H. Steinbach, and F. Walter, "BAV-results of observations—photoelectric minima of selected eclipsing binaries and maxima of pulsating stars," *Information Bulletin on Variable Stars*, vol. 5941, 2010.

[19] N. Erkan, A. Erdem, T. Akin, F. Aliçavus, and F. Soydugan, "New times of minima of some eclipsing binary stars," *Infornation Bulletin on Variable Stars*, vol. 5924, 2010.

[20] R. Diethelm, "Timings of minima of eclipsing binaries," *Information Bulletin on Variable Stars*, vol. 5920, 2010.

[21] S. Dogru, A. Erdem, F. Aliçavus, T. Akin, and Ç. Kanvermez, "CCD times of Minima of some eclipsing variables," *Information Bulletin on Variable Stars*, vol. 5988, 2011.

[22] N. Simmons, 2011, http://www.aavso.org/data-download.

[23] R. Diethelm, "Timings of minima of eclipsing binaries," *Information Bulletin on Variable Stars*, vol. 6011, p. 1, 2012.

[24] R. Diethelm, "Timings of minima of eclipsing binaries," *Information Bulletin on Variable Stars*, vol. 6042, 2013.

[25] L. Kiss, G. Kaszás, G. Fürész, and J. Vinkó, "New times of minima and updated ephemerides of selected contact binaries," *Information Bulletin on Variable Stars*, vol. 4681, p. 1, 1999.

[26] G. Kaszas, J. Vinko, K. Szatmary et al., "Period variation and surface activity of the contact binary VW Cephei," *Astronomy & Astrophysics*, vol. 331, pp. 231–243, 1998.

[27] M. Wolf, P. Molik, K. Hornoch, and L. Sarounova, "Period changes in W UMa-type eclipsing binaries: DK Cygni, V401 Cygni, AD Phoenicis and Y Sextantis," *Astronomy and Astrophysics Supplement Series*, vol. 147, pp. 243–249, 2000.

[28] T. Borkovits, M. M. Elkhateeb, S. Csizmadia et al., "Indirect evidence for short period magnetic cycles in W UMa stars Period analysis of five overcontact systems," *Astronomy & Astrophysics*, vol. 441, no. 3, pp. 1087–1097, 2005.

[29] J. Kreiner, C. Kim, and I. Nha, *An Atlas of O-C Diagrams of Eclipsing Binary Stars*, Wydawnictwo Naukowe Akademii Pedagogicznej, Krakow, Poland, 2001.

[30] R. Szafraniec, "Minima of eclipsing variables observed in 1952," *Acta Astronomica. Series C*, vol. 5, p. 51, 1953.

[31] J. Hubscher, F. Agerer, and E. Wunder, "Beobachtungsergebnisse der Berliner Arbeitsgemeinschaft für Veränderliche Sterne e.V," *Bundesdeutsche Arbeitsgemeinschaft für Veränderliche Sterne Mitteilungen*, vol. 59, 1990.

[32] M. Baldwin and G. Samolyk, AAVSO, No. 8, 2003.

[33] T. Borkovits, I. B. Bíró, T. Hegedüs et al., "New times of Minima of eclipsing binary systems," *Information Bulletin on Variable Stars*, vol. 5313, 2002.

[34] H. Gerner, "Recent minima of 155 eclipsing binary stars," *The Journal of the American Association of Variable Star Observers*, vol. 36, 2008.

[35] G. Samolyk, "Recent minima of 154 eclipsing binary star," *Journal of the American Association of Variable Star Observers*, vol. 37, no. 1, p. 44, 2009.

[36] G. Samolyk, "Recent minima of 154 eclipsing binary stars," *Journal of the American Association of Variable Star Observers*, vol. 37, pp. 44–51, 2009.

[37] O. Y. Malkov, "Mass-luminosity relation of intermediate-mass stars," *Monthly Notices of the Royal Astronomical Society*, vol. 382, no. 3, pp. 1073–1086, 2007.

[38] A. Essam, S. M. Saad, M. I. Nouh, A. Dumitrescu, M. M. El-Khateeb, and A. Haroon, "Photometric and spectroscopic analysis of YY CrB," *New Astronomy*, vol. 15, no. 2, pp. 227–233, 2010.

[39] M. M. Elkhateeb and M. I. Nouh, "Comprehensive photometric study of the eclipsing binary AW UMa," *Astrophysics and Space Science*, vol. 352, no. 2, pp. 673–689, 2014.

[40] M. M. Elkhateeb and M. I. Nouh, "A holistic study of the eclipsing binary EQ Tau," *Journal of Physics and Astronomy Research*, vol. 1, no. 3, p. 15, 2014.

[41] Y. Yang, S. Qian, L. Zhu, J. He, and J. Yuan, "Photometric investigations of three short-period binary systems: GSC 0763-0572, RR centauri, and ϵ Coronae australis," *Publications of the Astronomical Society of Japan*, vol. 57, no. 6, pp. 983–993, 2005.

[42] O. L. Degirmenci, "Photometric analysis of the W UMa type binary V566 ophiuchi," *Information Bulletin on Variable Stars*, vol. 5726, pp. 1–4, 2006.

[43] R. Nelson, 2009, http://members.shaw.ca/bob.nelson/software1.htm.

[44] R. Wilson and E. Devinney, "Realization of accurate close-binary light curves: application to MR cygni," *The Astrophysical Journal*, vol. 166, p. 605, 1971.

[45] R. E. Wilson, "Accuracy and efficiency in the binary star reflection effect," *The Astrophysical Journal*, vol. 356, no. 2, pp. 613–622, 1990.

[46] J. Kallarath and E. Milone, "A study of the O'Connell effect in the light curves of eclipsing binaries," *The Astrophysical Journal Supplement Series*, vol. 55, pp. 571–584, 1984.

[47] L. Lucy, "Gravity-darkening for stars with convective envelopes," *Zeitschrift für Astrophysik*, vol. 65, p. 89, 1967.

[48] R. Ruciński, "The proximity effects in close binary systems. II. The bolometric reflection effect for stars with deep convective envelopes," *Acta Astronomica*, vol. 19, p. 245, 1969.

[49] W. van Hamme, "New limb-darkening coefficients for modeling binary star light curves," *The Astronomical Journal*, vol. 106, no. 5, pp. 2096–2117, 1993.

[50] D. Bradstreet and D. Steelman, "Binary maker 3.0—an interactive graphics-based light curve synthesis program written in java," *Bulletin of the American Astronomical Society*, vol. 34, p. 1224, 2002.

[51] L. Girardi, A. Bressan, G. Bertelli, and C. Chiosi, "Evolutionary tracks and isochrones for low- and intermediate-mass stars: from 0.15 to 7 $M_{\odot}$, and from $Z = 0.0004$ to 0.03," *Astronomy and Astrophysics Supplement Series*, vol. 141, pp. 371–383, 2000.

20

Extension of Cherenkov Light LDF Approximation for Yakutsk EAS Array

A. A. Al-Rubaiee,[1,2] Y. Al-Douri,[2] and U. Hashim[2]

[1] *Department of Physics, College of Science, The University of Mustansiriyah, 10052 Baghdad, Iraq*
[2] *Institute of Nano Electronic Engineering, University of Malaysia Perlis, 01000 Kangar, Malaysia*

Correspondence should be addressed to A. A. Al-Rubaiee; dr.ahmedrubaiee@gmail.com

Academic Editor: Luciano Nicastro

The simulation of the Cherenkov light lateral distribution function (LDF) in extensive air showers (EAS) was performed using CORSIKA code for configuration of Yakutsk EAS array at high energy range for different primary particles (p, Fe, and O_2) and different zenith angles. Depending on Breit-Wigner function a parameterization of Cherenkov light LDF was reconstructed on the basis of this simulation as a function of primary energy. A comparison of the calculated Cherenkov light LDF with that measured on the Yakutsk EAS array gives the possibility of identification of the particle initiating the shower and determination of its energy in the knee region of the cosmic ray spectrum. The extrapolation of approximated Cherenkov light LDF for high energies was obtained for primary proton and iron nuclei.

1. Introduction

Study of the energy spectrum and mass composition of primary cosmic rays (PCRs) in the energy range 10^{13}–10^{17} eV is of a special interest in connection with observed index change of PCR spectrum close to $E = 3$ PeV which is called the "knee" region [1, 2]. The Cherenkov light emitted in the atmosphere by relativistic electrons of cosmic rays (CRs) in EAS carries important information about the shower development and PCR particles. The Cherenkov light LDF depends on energy and type of the primary particle, observation level, height of the first interaction, and direction of shower axis [3]. The Monte Carlo method is one of the necessary tools of numerical simulation for investigation of EAS characteristics and experimental data processing and analysis (determination of the primary particle energy type and direction of shower axis from the characteristics of Cherenkov radiation of secondary charged particles).

Agnetta et al. [4] have discussed the simulation and the experimental setup with detailed information on the detection of Cherenkov light method in EAS. On the other side, Akchurin et al. [5] have presented detailed measurements of high-energy electromagnetic and hadronic shower profiles. The Cherenkov light LDF generated in the shower development process was measured for electrons in the energy range 8–200 GeV. The Cherenkov light profiles are discussed and compared with results of Monte Carlo simulations. Berezhnev et al. [6] have installed the Cherenkov light EAS array (Tunka-133). This array permits a detailed study of cosmic ray energy spectrum and mass composition in the energy range 10^{16}–10^{18} eV with a uniform method. The analysis of LDF and time structure of EAS Cherenkov light allowed estimating the depth of the EAS maximum X_{max}.

In the present work the simulation of Cherenkov light LDF for conditions and configurations of Yakutsk EAS array [7, 8] is performed with the CORSIKA code [9, 10] using two models for simulation of hadronic processes which are QGSJET [11] and GHEISHA [12] models and EGS4 code for the simulation of the EAS electromagnetic component and Cherenkov light radiation. The approximation of the results of numerical simulation of Cherenkov light density was performed on the bases on Breit-Wigner functions [13, 14], an approach to the description of the lateral distribution of EAS Cherenkov light, and analyzes the possibility of its application for the reconstruction of the events registered on the Yakutsk array. The main advantage of this approach is to reconstruct the events of Cherenkov radiation measured with Yakutsk array. The comparison of the approximated Cherenkov light LDF with the reconstructed EAS events registered with Yakutsk EAS Cherenkov array has shown a good opportunity of primary particle identification and definition of its energy around the knee region.

2. Simulation of Cherenkov Light LDF

The simulation of Cherenkov light LDF from EAS was performed using the CORSIKA (COsmic Ray SImulations for KAscade) software package [9, 10] within two models: QGSJET (Quark Gluon String model with JETs) code [11] to model interactions of hadrons with energies exceeding 80 GeV and GHEISHA (Gamma Hadron Electron Interaction SHower) code [12] for energies lower than 80 GeV. The CORSIKA code simulates the interactions and decays of various nuclei, hadrons, muons, electrons, and photons in the atmosphere. The particles are tracked through the atmosphere until they undergo reactions with an air nucleus or, in the case of unstable secondary particles, they decay [9]. The result of the simulations is detailed about the type, energy, momenta, location, and arrival time of the produced secondary particles at a given selected altitude above sea level.

The Yakutsk EAS array consists of 48 Cherenkov light detectors (500 m spacing between detectors); the observation level was assumed to be 100 m above sea level (1020 g/cm^2) and wavelength range from 300 to 600 nm [8].

3. Parameterization of Cherenkov Radiation Density in EAS

The Cherenkov LDF is a function to describe the lateral variation of Cherenkov flux with the core distance that is widely used in event reconstruction, aiming to obtain information about primary particles. Integration over the total range of core distance of LDF results in the shower size, that is, total number of particles. Estimating of core position and age parameter is also made by using the total number of Cherenkov photons (N_γ) radiated by electrons in EAS which is directly proportional to the primary energy E_0 [15]:

$$N_\gamma (E_0) \approx 3.7 \times 10^3 \frac{E_0}{\xi_e} \approx 4.5 \times 10^{10} \frac{E_0}{10^{15}\,\text{eV}}, \quad (1)$$

where ξ_e is the critical energy of electrons which equals 81.4 MeV. The experimental measurement of this magnitude is rather difficult, so one can use the density of Cherenkov radiation, the number of photons (ΔN_γ) per unit detector area (ΔS), which appears as a function of energy and distance from the shower axis [16]:

$$Q(E_0, R) = \frac{\Delta N_\gamma (E_0, R)}{\Delta S}, \quad (2)$$

where R is the distance from the shower axis.

Direct measurements of Cherenkov light showed that the fluctuation of LDF in EAS is essentially less than the total number of photons N_γ [1]. For parameterization of simulated Cherenkov light LDF, we used the proposed function as a function of distance R from shower axis, depth of shower maximum, and energy E_0 of the initial primary particle, which depends on four parameters a, b, s, and r_o:

$$Q(E_0, R) = \frac{Cs \exp [a - A]}{b \left[(R/b)^2 + (R - r_o)^2/b^2 + Rs^2/b \right]}, \quad (3)$$

where A is defined as

$$A = \frac{R}{b} + \frac{(R - r_o)}{b} + \left(\frac{R}{b}\right)^2 + \frac{(R - r_o)^2}{b^2}, \quad (4)$$

where C is the normalization constant [3] and a, b, s, and r_o are parameters of Cherenkov light LDF. The estimation of Cherenkov light density was performed in the energy range 10^{13}–10^{17} eV for different primary particles and different zenith angles. The energy dependence of LDF parameters is approximated as

$$K(E_0) = c_0 + c_1 \lg (E_0) + c_2 (\lg E_0)^2 + c_3 (\lg E_0)^3, \quad (5)$$

where c_0, c_1, c_2, and c_3 are coefficients that depend on the type of primary particles and the zenith angle (see Tables 1 and 2).

The obtained Cherenkov light LDF in EAS due to various cosmic ray particles (p and Fe) below and in the region of the "knee" are presented in Figure 1. It demonstrates the results of the simulated (solid lines) and parameterized (dashed lines) Cherenkov light LDF for vertical showers for primary proton and iron nuclei, respectively, at different primary energies.

The accuracy of the Cherenkov light LDF approximation with that simulated for primary proton is better than 18% at the distances 10–150 m from the shower axis and about 5–15% for the other distances. The accuracy of iron nuclei was found close to 15% at the distances 10–150 m from the core shower and about 5–10% at the other distances.

4. Comparison of the Approximated LDF with Yakutsk Measurements

The Yakutsk EAS array studies cosmic rays of extremely high energies, that is, in the field of cosmic ray astrophysics, an active area at the cutting edge of basic research. The construction of Yakutsk array depends on two main goals; the first is the investigation of cascades of elementary particles in atmosphere initiated by primary particles and the other is the reconstruction of astrophysical properties of the primaries: intensity, energy spectrum, mass composition, and their origin [8]. The main parameters of EAS measurements are zenith and azimuth angles, shower core location, individual LDF, and the density of Cherenkov radiation $Q(R)$. The possibility for reconstruction of the type of EAS primary particles can be demonstrated in Figures 2 and 3.

Figure 2(a) demonstrates the comparison of approximated Cherenkov light LDF (dash lines) with that measured with Yakutsk EAS array (symbols) for three primaries (p, O_2, and Fe) at the distance 100 to 400 m from the shower core.

In Figure 2(b), one may see reasonable agreement between the approximated Cherenkov light LDF (dash lines) and

TABLE 1: Coefficients c_i that determine the energy dependence (5) of the extrapolated parameters a, b, s, and r_o for primary proton, iron nuclei, and primary oxygen for vertical showers of Yakutsk Cherenkov EAS array.

K	c_0	c_1	c_2	c_3
		$\theta = 0^\circ$		
		P		
a	$1.026 \cdot 10^{2}$	$-2.615 \cdot 10^{1}$	$2.063 \cdot 10^{0}$	$-4.914 \cdot 10^{-2}$
b	$-9.709 \cdot 10^{0}$	$2.117 \cdot 10^{0}$	$-1.439 \cdot 10^{-1}$	$-3.190 \cdot 10^{-3}$
σ	$8.126 \cdot 10^{1}$	$-1.669 \cdot 10^{1}$	$1.132 \cdot 10^{0}$	$-2.548 \cdot 10^{-2}$
r_o	$-1.427 \cdot 10^{1}$	$2.918 \cdot 10^{0}$	$-2.007 \cdot 10^{-1}$	$4.430 \cdot 10^{-3}$
		Fe		
a	$-3.851 \cdot 10^{2}$	$7.187 \cdot 10^{1}$	$-4.504 \cdot 10^{0}$	$9.752 \cdot 10^{-2}$
b	$-4.415 \cdot 10^{1}$	$9.802 \cdot 10^{0}$	$-7.120 \cdot 10^{-1}$	$1.711 \cdot 10^{-2}$
σ	$9.887 \cdot 10^{1}$	$-2.102 \cdot 10^{1}$	$1.481 \cdot 10^{0}$	$-3.474 \cdot 10^{-2}$
r_o	$-1.216 \cdot 10^{2}$	$2.505 \cdot 10^{1}$	$-1.714 \cdot 10^{0}$	$3.877 \cdot 10^{-2}$
		O_2		
a	$-9.194 \cdot 10^{1}$	$1.328 \cdot 10^{1}$	$-6.077 \cdot 10^{-1}$	$1.121 \cdot 10^{-2}$
b	$3.847 \cdot 10^{1}$	$-7.855 \cdot 10^{0}$	$5.431 \cdot 10^{-1}$	$-1.254 \cdot 10^{-2}$
σ	$-1.496 \cdot 10^{1}$	$2.781 \cdot 10^{0}$	$-1.784 \cdot 10^{-1}$	$3.830 \cdot 10^{-3}$
r_o	$-2.097 \cdot 10^{2}$	$4.359 \cdot 10^{1}$	$-3.014 \cdot 10^{0}$	$6.910 \cdot 10^{-2}$

TABLE 2: Coefficients c_i that determine the energy dependence (5) of the extrapolated parameters a, b, s, and r_o for primary proton and iron nuclei for inclined showers of Yakutsk Cherenkov EAS array.

K	c_0	c_1	c_2	c_3
		$\theta = 30$∘		
		P		
a	$-7.715 \cdot 10^{1}$	$1.194 \cdot 10^{1}$	$-6.242 \cdot 10^{-1}$	$1.389 \cdot 10^{-2}$
b	$-4.417 \cdot 10^{1}$	$9.292 \cdot 10^{0}$	$-6.400 \cdot 10^{-1}$	$1.458 \cdot 10^{-2}$
σ	$6.133 \cdot 10^{0}$	$-1.353 \cdot 10^{0}$	$9.383 \cdot 10^{-2}$	$-2.180 \cdot 10^{-3}$
r_o	$1.415 \cdot 10^{1}$	$-3.037 \cdot 10^{0}$	$2.137 \cdot 10^{-1}$	$-5.160 \cdot 10^{-3}$
		$\theta = 10^\circ$		
		P		
a	$1.33 \cdot 10^{2}$	$-3.146 \cdot 10^{1}$	$2.365 \cdot 10^{0}$	$-5.467 \cdot 10^{-2}$
b	$-8.468 \cdot 10^{1}$	$1.791 \cdot 10^{1}$	$-1.250 \cdot 10^{0}$	$2.898 \cdot 10^{-2}$
σ	$8.912 \cdot 10^{1}$	$-1.810 \cdot 10^{1}$	$1.218 \cdot 10^{0}$	$-2.733 \cdot 10^{-2}$
r_o	$4.368 \cdot 10^{2}$	$-8.884 \cdot 10^{1}$	$6.011 \cdot 10^{0}$	$-1.356 \cdot 10^{-1}$
		$\theta = 10^\circ$		
		Fe		
a	$-4.151 \cdot 10^{2}$	$8.002 \cdot 10^{1}$	$-5.195 \cdot 10^{0}$	$1.161 \cdot 10^{-1}$
b	$3.609 \cdot 10^{1}$	$-7.492 \cdot 10^{0}$	$5.262 \cdot 10^{-1}$	$-1.232 \cdot 10^{-2}$
σ	$5.043 \cdot 10^{1}$	$-1.043 \cdot 10^{1}$	$7.098 \cdot 10^{-1}$	$-1.604 \cdot 10^{-2}$
r_o	$-4.926 \cdot 10^{1}$	$1.067 \cdot 10^{1}$	$-7.673 \cdot 10^{-1}$	$1.809 \cdot 10^{-2}$

measurements in Yakutsk array (symbols) for primary proton and iron nuclei at the primary energy 5 PeV and zenith angle $\theta = 10^\circ$.

The extrapolation of Cherenkov light LDF parameterization for higher energies (>10^{16} eV) can be seen in Figure 3, where Figure 3(a) displays the comparison of the approximated Cherenkov light LDF that extrapolated to 20 PeV (dash lines) and that LDF measured with Yakutsk EAS array (symbols) for iron nuclei at two zenith angles $\theta = 0^\circ$ and 30°. To illustrate more vividly the errors in approximation (3), the function Q in Figure 3(a) was multiplied by 0.8 for $\theta = 10^\circ$. Figure 3(b) shows the comparison of the approximated Cherenkov light LDF that extrapolated to 50 and 100 PeV (dash lines) and LDF measured with Yakutsk EAS array (symbols) for primary proton at vertical showers. The good agreement between the model parameters of extrapolated Cherenkov light LDF as a function of primary energy of different primaries with that measured with Yakutsk array shows that this model is adequate and is usable for different Cherenkov arrays.

The parameterized Cherenkov light LDF in Figures 2 and 3 slightly differs from the LDF measured with the Yakutsk

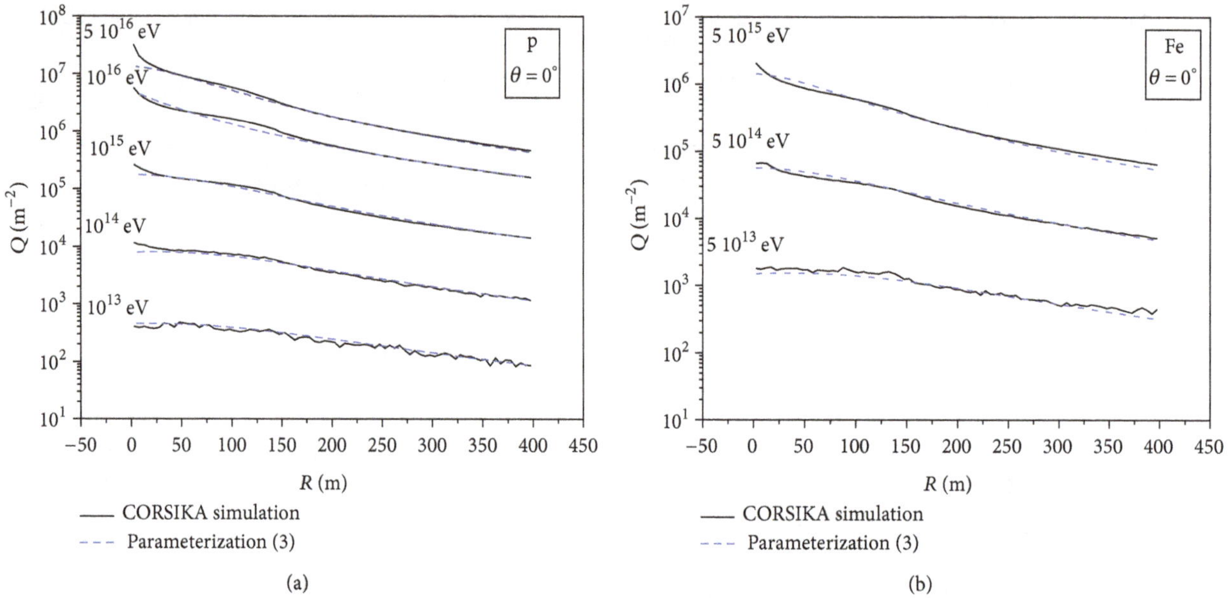

FIGURE 1: Lateral distributions of Cherenkov light that was simulated with CORSIKA code (solid lines) and one approximated (3) (dashed lines) for vertical showers initiated by (a) primary proton at the energy range 10^{13}–$5 \cdot 10^{16}$ eV and (b) iron nuclei at the energies $5 \cdot 10^{13}$, $5 \cdot 10^{14}$, and $5 \cdot 10^{15}$ eV.

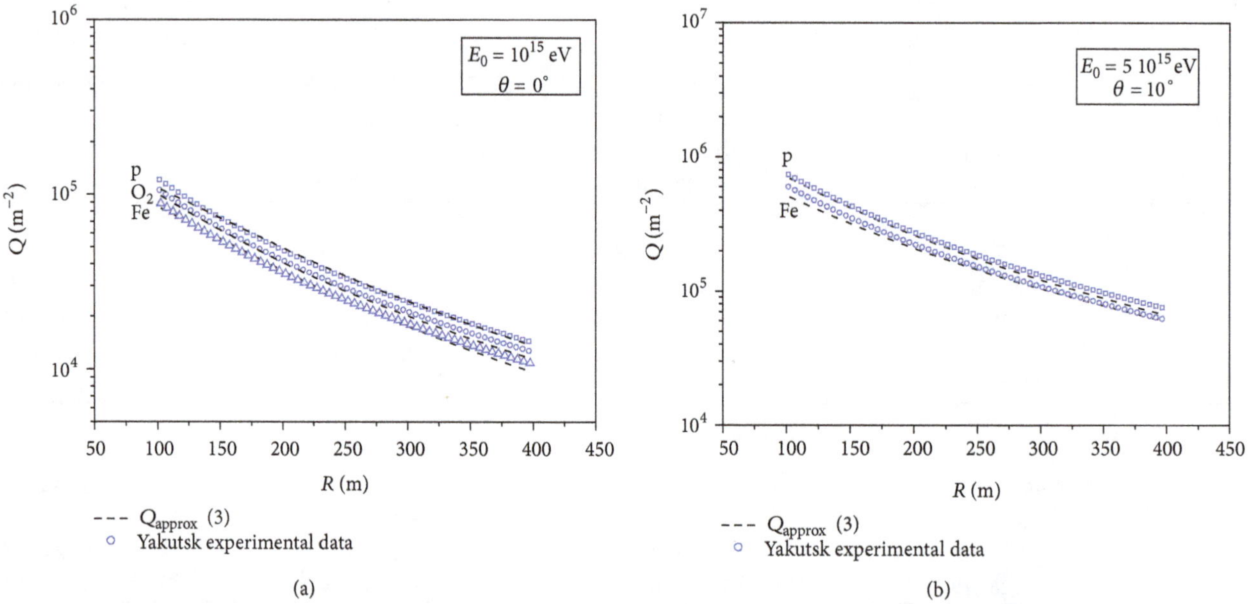

FIGURE 2: Comparison of the parameterized Cherenkov light LDF with the experimental data obtained by Yakutsk EAS array (symbols) for the following: (a) three different primary particles (Fe, O_2, and p) at the energy 10^{15} for vertical showers and (b) two different primary particles (Fe and p) at fixed primary energy $5 \cdot 10^{15}$ eV and fixed zenith angle ($\theta = 10°$).

EAS array; at the distance interval 100–400 m, the distinction is about 5–20% for primary proton, 3–11% for primary oxygen, and 5–13% for iron nuclei for vertical showers. For inclined showers, the distinction at the same distance interval is about 15–20% at $\theta = 30°$ for primary proton and about 8–20% at $\theta = 10°$ for primary proton and iron nuclei.

5. Conclusion

The lateral distribution function of Cherenkov radiation from particles of extensive air showers initiated by primary proton, iron nuclei, and oxygen has been simulated in the energy range 10^{13}–10^{17} eV using CORSIKA code. On the

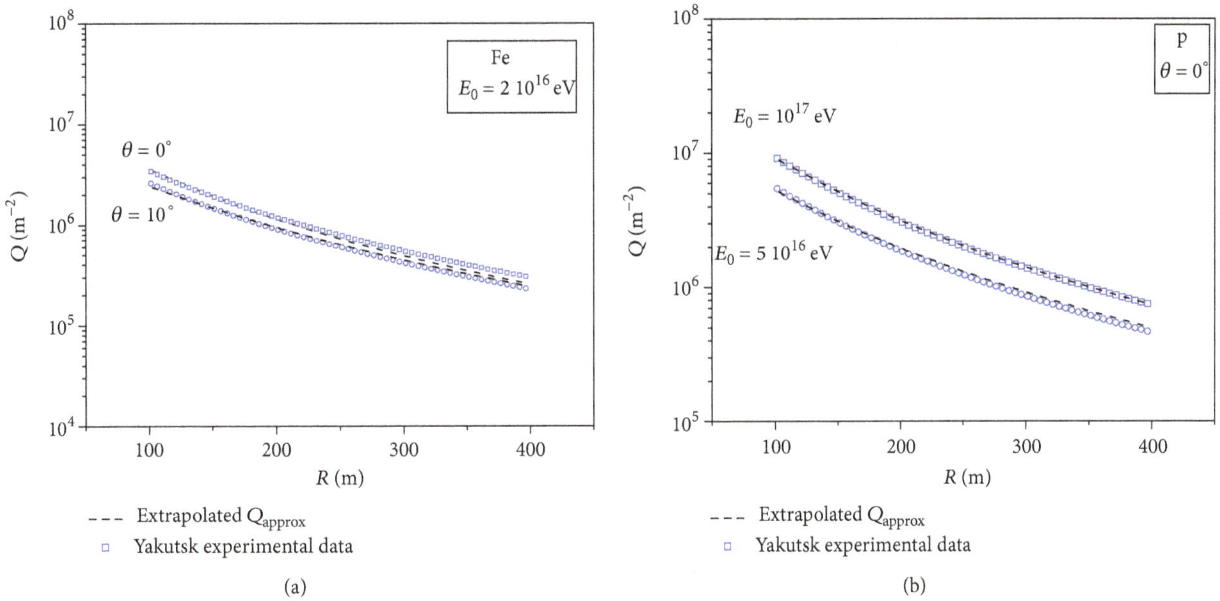

FIGURE 3: Extrapolation of parameterized Cherenkov light LDF (3) (dash lines) in comparison with experimental data obtained by Yakutsk EAS array (symbols) for the following: (a) iron nuclei at the energy $2 \cdot 10^{16}$ eV for different zenith angles ($\theta = 0°$ and $10°$); (b) vertical showers initiated primary proton at two different energies $5 \cdot 10^{16}$ and 10^{17} eV.

basis of this simulation with depending on Breit-Wigner function; sets of approximating functions were constructed for different primary particles and different zenith angles. The comparison of the approximated Cherenkov light lateral distribution functions with that measured with Yakutsk EAS array has demonstrated the ability of identification of the particle initiating EAS showers and determination of its energy around and above the knee region of the cosmic ray spectrum. The extrapolation of the Cherenkov light lateral distribution function parameterization of the obtained data with CORSIKA program for the energies ($E_0 > 10^{16}$ eV) is obtained. The main advantage of the given approach is the opportunity of creation of a representative library of lateral distribution function patterns for a short time which could be utilized for analysis of real events detected with EAS arrays and reconstruction of primary cosmic ray energy spectrum and mass composition.

Conflict of Interests

The authors declare that they have no conflict of interests regarding the publication of this paper.

Acknowledgments

Y. Al-Douri would like to acknowledge the University of Malaysia Perlis for Grant no. 9007-00111 and TWAS-Italy for the full support of his visit to JUST-Jordan under TWAS-UNESCO Associateship.

References

[1] G. B. Khristiansen, Y. A. Fomin, N. N. Kalmykov et al., "The primary cosmic ray mass composition around the knee of the energy spectrum," *Nuclear Physics B: Proceedings Supplements*, vol. 39, no. 1, pp. 235–241, 1995.

[2] M. Amenomori, Z. Cao, B. Z. Dai et al., "The cosmic-ray energy spectrum between $10^{14.5}$ and $10^{16.3}$ eV covering the knee region," *The Astrophysical Journa*, vol. 461, pp. 453–460, 2008.

[3] A. A. Al-Rubaiee, O. A. Gress, K. S. Lokhtin, Y. V. Parfenov, and S. I. Sinegovskii, "Modeling and parameterization of the spatial distribution of Čerenkov light from extensive air showers," *Russian Physics Journal*, vol. 48, no. 10, pp. 1004-1011, 2005.

[4] G. Agnetta, P. Assis, B. Biondo et al., "Extensive air showers and diffused Cherenkov light detection: The ULTRA experiment," *Nuclear Instruments and Methods in Physics Research Section A: Accelerators, Spectrometers, Detectors and Associated Equipment*, vol. 570, pp. 22–35, 2007.

[5] N. Akchurin, K. Carrell, J. Hauptman et al., "Comparison of high-energy electromagnetic shower profiles measured with scintillation and Cherenkov light," *Nuclear Instruments and Methods in Physics Research A: Accelerators, Spectrometers, Detectors and Associated Equipment*, vol. 548, no. 3, pp. 336-354, 2005.

[6] S. F. Berezhnev, D. Besson, N. M. Budnev et al., "The Tunka-133 EAS Cherenkov light array: status of 2011," *Nuclear Instruments and Methods in Physics Research A: Accelerators, Spectrometers, Detectors and Associated Equipment*, vol. 692, pp. 98–105, 2012.

[7] S. Knurenko, V. Kolosov, Z. Petrov, I. Sleptsov, and S. Starostin, in *Proceedings of the 28th International Cosmic Ray Conference (ICRC '03)*, pp. 177–179, Tsukuba, Japan, 2003.

[8] A. A. Ivanov, S. P. Knurenko, and I. Y. Sleptsov, "Measuring extensive air showers with Cherenkov light detectors of the Yakutsk array: the energy spectrum of cosmic rays," *New Journal of Physics*, vol. 11, Article ID 065008, 30 pages, 2009.

[9] D. Heck and T. Peirog, *Extensive Air Shower Simulations at the Highest Energies—A User's Guide*, Institut fur Kernphysik, Heidelberg, Germany, 2013.

[10] J. Knapp, D. Heck, S. J. Sciutto, M. T. Dova, and M. Risse, "Extensive air shower simulations at the highest energies," *Astroparticle Physics*, vol. 19, no. 1, pp. 77–99, 2003.

[11] S. Ostapchenko, "QGSJET-II: towards reliable description of very high energy hadronic interactions," *Nuclear Physics B: Proceedings Supplements*, vol. 151, pp. 143–146, 2006.

[12] D. Heck and R. Engel, in *Proceedings of the 28th International Cosmic Ray Conference*, pp. 279–282, Tsukuba, Japan, 2003.

[13] S. C. Mavrodiev, A. L. Mishev, and J. N. Stamenov, "A method for energy estimation and mass composition determination of primary cosmic rays at the Chacaltaya observation level based on the atmospheric Cherenkov light technique," *Nuclear Instruments and Methods in Physics Research A: Accelerators, Spectrometers, Detectors and Associated Equipment*, vol. 530, no. 3, pp. 359–366, 2004.

[14] A. Mishev, "Analysis of lateral distribution of atmospheric cherenkov light at high mountain altitude towards event reconstruction," *ISRN High Energy Physics*, vol. 2012, Article ID 906358, 12 pages, 2012.

[15] N. Aliev, T. Alimov, M. Kakhkharov et al., in *Proceedings of the 18th International Cosmic Ray Conference (ICRC '83)*, vol. 2, pp. 383–386, Bangalore, India, 1983.

[16] A. Mishev, I. Angelov, E. Duverger, R. Gschwind, L. Makovicka, and J. Stamenov, "Experimental study and Monte Carlo modeling of the Cherenkov effect," *Nuclear Instruments and Methods in Physics Research A: Accelerators, Spectrometers, Detectors and Associated Equipment*, vol. 474, no. 2, pp. 101–107, 2001.

21

Chaos and Intermittency in the DNLS Equation Describing the Parallel Alfvén Wave Propagation

Gustavo Krause,[1,2] Sergio Elaskar,[1,2] and Andrea Costa[2,3]

[1] *Department of Aeronautics, Faculty of Exact, Physical and Natural Sciences, National University of Córdoba, Vélez Sarsfield 1611, X5016GCA Córdoba, Argentina*
[2] *National Council of Scientific and Technical Research (CONICET), Avenue Rivadavia 1917, C1033AAJ Buenos Aires, Argentina*
[3] *Institute of Theoretical and Experimental Astronomy (IATE-CONICET), Laprida 854, X5000BGR Córdoba, Argentina*

Correspondence should be addressed to Gustavo Krause; gustavojavierkrause@gmail.com

Academic Editor: Milan S. Dimitrijevic

When the Hall effect is included in the magnetohydrodynamics equations (Hall-MHD model) the wave propagation modes become coupled, but for propagation parallel to the ambient magnetic field the Alfvén mode decouples from the magnetosonic ones, resulting in circularly polarized waves that are described by the derivative nonlinear Schrödinger (DNLS) equation. In this paper, the DNLS equation is numerically solved using spectral methods for the spatial derivatives and a fourth order Runge-Kutta scheme for time integration. Firstly, the nondiffusive DNLS equation is considered to test the validity of the method by verifying the analytical condition of modulational stability. Later, diffusive and excitatory effects are incorporated to compare the numerical results with those obtained by a three-wave truncation model. The results show that different types of attractors can exist depending on the diffusion level: for relatively large damping, there are fixed points for which the truncation model is a good approximation; for low damping, chaotic solutions appear and the three-wave truncation model fails due to the emergence of new nonnegligible modes.

1. Introduction

Alfvén waves are one of the most characteristic features of magnetized laboratory and space plasmas. They are driven by different sources, for example, nonuniform plasma parameters, beams of charged particles, and electrostatic and electromagnetic waves [1]. In space plasmas, large MHD amplitude fluctuations with typical proton cyclotron local frequencies were detected on the Earth magnetosphere [2]. Also, nonlinear Alfvén waves were extensively detected in the solar wind [3] and they are believed to be responsible for the turbulent heating of stellar coronas [4]. The comprehension of nonlinear properties of dispersive Alfvén waves is of crucial importance to interpret the abundant amount of low frequency data provided by space plasma observations.

On the other hand, the interaction of spatial tethers with the Earth ionosphere and the ambient magnetic field leads to the emission of Alfvén waves forming structures called Alfvén wings. This phenomenon could be applied to produce electric power, generate artificial auroras [5], or to moderate spatial trash [6]. Far from the tethers, a linear analysis could be appropriate [7], but near the conductor intense waves with important nonlinear effects are expected.

For the study of the plasma behavior the magnetohydrodynamics (MHD) equations are usually used, but when the frequencies of interest are of the order of the ion-cyclotron frequency or when the characteristic longitudes are comparable with the ion inertial length, these equations have to be extended to include the effect of a finite ion-cyclotron frequency, which is referred to as the "Hall effect". In the resulting Hall-MHD model, although the magnetosonic and the Alfvén modes can still be identified, they are coupled leading to a dispersive evolution. This allows that, for a homogeneous plasma, nonlinear high-frequency Alfvén waves can propagate parallel to a uniform magnetic field since the nonlinearities are balanced by the dispersive term which is constituted by the Hall effect in the induction equation.

Using a two-fluid, quasineutral approximation with electron inertia and current displacement neglected, the MHD-Hall equations normalized by density and magnetic field reference values, B_0 and ρ_0, are [8]

$$\frac{\partial \rho}{\partial t} + \nabla \cdot (\rho \mathbf{u}) = 0,$$
$$\rho\left(\frac{\partial}{\partial t} + \mathbf{u} \cdot \nabla\right)\mathbf{u} = -\nabla p + V_A^2 (\nabla \times \mathbf{B}) \times \mathbf{B}, \quad (1)$$
$$\frac{\partial \mathbf{B}}{\partial t} = \nabla \times (\mathbf{u} \times \mathbf{B}) - \frac{V_A^2}{\Omega_i}\nabla \times \left[\frac{1}{\rho}(\nabla \times \mathbf{B}) \times \mathbf{B}\right],$$

where ρ is the plasma density, $\mathbf{u}$ the velocity, $\mathbf{B}$ the magnetic field, Ω_i the ion-cyclotron frequency, and $V_A = B_0/(\rho_0\mu_0)^{1/2}$ is the Alfvén speed where μ_0 is the vacuum permeability. The equation of state obeying the polytropic law $p \propto \rho^{-\gamma}$, with γ the polytropic coefficient, completes the set of equations.

Weak nonlinearities were studied by perturbation theory [8] leading to a Korteweg-de Vries (KdV) equation for the magnetosonic modes and a modified KdV equation for the Alfvén mode when the propagation angle is large enough [9–11]. When propagation parallel (or quasiparallel) to the ambient magnetic field is considered, the MHD waves are degenerated; the Alfvén mode is decoupled and the waves are circularly polarized, being described by the derivative nonlinear Schrödinger (DNLS) equation (e.g., [12–15]) although the case of arbitrary propagation angles is also valid for a high-β plasma [16].

The DNLS equation was derived by several authors using different techniques: through the Vlasov equation [12] and using the Hall-MHD equation (e.g., [13–15]). The nondimensional DNLS equation in a frame moving with the Alfvén speed along the magnetic field direction (z) with $V_A \gg c_s$ ($\beta = c_s/V_A \approx 0$, with c_s the sound speed) results in

$$\frac{\partial b}{\partial t} + \frac{\partial}{\partial z}\left(|b|^2 b\right) + \mathrm{i}\frac{\partial^2 b}{\partial z^2} + \hat{\gamma} b = 0, \quad (2)$$

where the positive sign in the dispersive term indicates that left-hand polarized waves are considered, $\hat{\gamma}$ is an appropriate damping/driving linear operator, and the dimensionless variables b, z, and t are defined by

$$b = \frac{B_x + \mathrm{i}B_y}{2B_0}, \qquad 2\Omega_i t \longrightarrow t, \qquad \frac{2\Omega_i}{V_A} z \longrightarrow z. \quad (3)$$

The advantage of using the DNLS is due to the fact that this equation "belongs" to soliton theory; thus much about its solutions is known although it is nonlinear. On the other hand, soliton theory was used to explain many observations of quasiparallel finite-amplitude MHD waves where the competition between nonlinear steepening and dispersion occurs, such as in the foreshock regions of quasiparallel shocks occurring in connection with the bow shocks of planets or comets [17]. Therefore, the solutions of the DNLS equation can be attributed to specific physical problems [18].

Many exact solutions of the DNLS equations are known and stability analysis has been made for some of them [19]. Among these solutions, the circularly polarized wave is of particular interest because the stability of this solution is easily investigated (see Section 3).

On the other hand, different analytical approaches were used, which consist of reducing the DNLS equation to a system with low dimensions, either considering stationary waves or truncating the system by including only a finite number of modes. In the first case, the DNLS equation is reduced to a set of three ordinary differential equations where the free variables are the two components of the transverse magnetic field and the phase wave [20]. In this way, a continuous three-dimensional dynamical system is obtained, which would allow retaining the nonlinear evolution of driven conservative and dissipative Alfvén waves that is registered in more complicated high-dimensional models [21]. This approach was extensively used to study the Alfvén intermittent turbulence in space plasmas [22, 23], showing that the onset of Alfvén turbulence can occur via a crisis-induced intermittency.

The other analytical approach, consisting in the truncation of the system, was also widely studied. This approximation supposes that the solution is the sum of a finite number of modes, by which a set of ordinary differential equations is obtained, where the order of the system depends on the number of modes used for the truncation. Based on numerical studies where the DNLS was fully integrated, three-wave truncation models with a resonance relation $2k_0 = k_1 + k_2$ (k_0 is the mother wave) have been proposed. Although the results of the first analysis of this model did not exhibit a chaotic dynamics [21], subsequent works showed that the three-wave truncation model does present chaos through different routes [24–27]. Similar studies were performed in the context of the nonlinear Schrödinger equation (NLS) [28].

Numerical studies of the DNLS have been carried out by numerous authors exploring the integrability properties of this equation [21, 27, 29–31]. Most of these works are based on spectral methods assuming periodic boundary conditions, but different integration schemes have been used to solve stability problems.

In this paper, a numerical analysis of the driven dissipative DNLS equation for the particular case of three resonant Alfvén waves is presented. The numerical results are compared with those obtained by a three-wave truncation model which was carried out to represent Alfvén wave fronts generated by orbiting conductive tethers interacting with the ambient magnetic field in the ionosphere [5, 26, 27]. This analysis allows establishing the application range of the truncation method by inspecting the power spectrum, in a similar way as the work by Ghosh and Papadopoulos [21]. However, in this case a simpler numerical scheme is used with an aliasing filter that only cancels 1/3 of the wavenumber domain, allowing retaining a greater number of effective modes [32]. In addition, new features of the dynamic solutions are obtained, showing that there are more configurations which present bifurcation diagrams exhibiting a Hopf bifurcation that separates the more diffusive region, with stationary solutions, from the less diffusive one, with complex dynamic solutions, where chaotic attractors are developed by different routes (Feigenbaum cascades, intermittency, and crises).

Also, new coherence relations are found for the stationary solutions, where the existence of five resonant modes is detected. In this way, the analysis of the present paper extends the results of previous works [21, 27]. On the other hand, another new aspect of this research is the detailed analysis of the stability of the numerical scheme presented here, which not only allows verifying the code, but also permits to study the evolution of the conservative DNLS equation under unstable conditions. These results show that the presence of periodic or chaotic evolutions depends on the parameters of the initial wave.

The organization of the paper is as follows. In Section 2, a brief description of the numerical method is presented. In Section 3, the numerical method is tested by verifying the analytical conditions of modulational stability in the non-diffusive case. In Section 4, numerical results for the driven diffusive DNLS equation under a three-wave resonant initial condition are shown, comparing the solutions with those of the three-wave truncation model. Finally, in Section 5 the main conclusions of the work are presented.

2. The Method

The DNLS equation, as type-KdV equations, admits that periodic boundary conditions are used when the initial conditions are periodic [33]. Therefore, boundary conditions of the form

$$b(z,t) = b(z+L,t) \tag{4}$$

are appropriate to implement numerical simulations [19], where L is the domain size.

Spectral methods based on trigonometric functions automatically and individually satisfy the periodic boundary conditions. The discrete Fourier expansion allows computing the Fourier coefficients of a function $b(z)$ defined in the domain $(-L/2, L/2]$ and known at points

$$z_j = -\frac{L}{2} + \frac{L}{N}j, \quad j = 1,\dots,N, \tag{5}$$

where N is the number of grid points. The discrete Fourier transform and its inverse are

$$\begin{gathered}\mathcal{F}\left[b_j\right] = \overline{b}_k = \frac{1}{N}\sum_{j=1}^{N} b_j \exp\left(-\mathrm{i}\frac{2\pi n}{L}z_j\right),\\ \mathcal{F}^{-1}\left[\overline{b}_k\right] = b_j = \sum_{n=-(N/2)+1}^{N/2} \overline{b}_k \exp\left(-\mathrm{i}\frac{2\pi n}{L}z_j\right),\end{gathered} \tag{6}$$

where the notations $b_j = b(z_j)$ and $\overline{b}_k = \overline{b}(k)$ are used. The wavenumbers k are given by

$$k = \frac{2\pi n}{L}, \qquad n = -\frac{N}{2} + 1,\dots,\frac{N}{2}. \tag{7}$$

Equations (6) define an exact transform pair between the N grid values b_j and the N discrete Fourier coefficients $\overline{b}_k$. The transformation can be performed from one representation to the other without loss of information [34]. Note that from (6) the derivatives at the grid points are given by

$$\frac{\mathrm{d}^n b_j}{\mathrm{d}z^n} = \mathcal{F}^{-1}\left[(\mathrm{i}k)^n \overline{b}_k\right], \tag{8}$$

and then we can write the DNLS equation as

$$\frac{\partial b}{\partial t} = \mathcal{F}^{-1}\left[-\mathrm{i}k\mathcal{F}\left[|b|^2 b\right] + \mathrm{i}k^2\mathcal{F}\left[b\right]\right] - \widehat{\gamma}b; \tag{9}$$

thus, the time derivative can be computed for a defined damping/driving operator $\widehat{\gamma}$. In this way, the time integration can be explicitly performed using a fourth order Runge-Kutta scheme.

It should be noted that the N truncation of the series introduces the "aliasing" error, which is associated with the difference between the discrete and the continuous Fourier expansion. This error can cause numerical instabilities in the time integration of nonlinear equations [35]. The usual strategy employed to avoid these effects is to remove the "aliased" modes by applying an "all-or-nothing" filter where the Fourier coefficients external to a certain portion of the central domain are canceled. In this paper, contrary to previous works where half of the modes are preserved [21, 27], the 2/3 of the central domain is used, allowing to extend the amount of effective modes used in the simulations [32]. The subsequent results show that this is a good implementation.

2.1. Modeling of the Diffusive Term. In (2), $\widehat{\gamma}$ is used to represent an appropriate damping/driving linear operator, which is assumed to have the characteristic diffusive form

$$\widehat{\gamma} = -\eta\frac{\partial^2}{\partial z^2}, \tag{10}$$

where η is the damping coefficient. Taking into account this definition and (8), it is obtained for the generic sample point z_j that

$$\widehat{\gamma}_j = -\eta k_j^2. \tag{11}$$

This operator is equivalent to consider a resistive damping model, where the dissipation is proportional to k_j^2 [24]:

$$\left|\widehat{\gamma}_j\right| \approx \frac{V_A^2 k_j^2}{2\Omega_e\Omega_i\tau_c} = \eta k_j^2, \tag{12}$$

where Ω_e is the electron-cyclotron frequency and τ_c is the characteristic Braginskii collision time.

To represent the excitation of the mother wave (identified with k_0), the Dirac function δ is used:

$$\widehat{\gamma}_j = \gamma_g\delta\left(k_j - k_0\right) - \eta k_j^2, \tag{13}$$

where γ_g gives the excitation level of the wave. For k_j other than k_0, the Dirac function is zero and (13) reduces to (10).

With the latest definitions, the real and the imaginary parts of (9) are

$$\begin{aligned}\operatorname{Re}\left[\frac{\partial b_j}{\partial t}\right] &= \mathcal{F}^{-1}\left[k_j \operatorname{Im}\left[\mathcal{F}\left[\left|b_j\right|^2 b_j\right]\right] - k_j^2 \operatorname{Im}\left[\overline{b}_{kj}\right]\right] \\ &\quad + \mathcal{F}^{-1}\left[\left(\gamma_g \delta\left(k_j - k_0\right) - \eta k_j^2\right) \operatorname{Re}\left[\overline{b}_{kj}\right]\right], \\ \operatorname{Im}\left[\frac{\partial b_j}{\partial t}\right] &= \mathcal{F}^{-1}\left[-k_j \operatorname{Re}\left[\mathcal{F}\left[\left|b_j\right|^2 b_j\right]\right] + k_j^2 \operatorname{Re}\left[\overline{b}_{kj}\right]\right] \\ &\quad + \mathcal{F}^{-1}\left[\left(\gamma_g \delta\left(k_j - k_0\right) - \eta k_j^2\right) \operatorname{Im}\left[\overline{b}_{kj}\right]\right],\end{aligned} \tag{14}$$

and the time derivative of the DNLS equation is finally obtained.

3. Verification of the Numerical Method

In order to test the numerical scheme, the nondiffusive DNLS equation ($\widehat{\gamma} = 0$) is analyzed to verify the accomplishment of the analytical conditions of modulational stability, which can be established for an initial magnetic field condition

$$b(z, 0) = b_0 = A_0 \exp\left(ik_0 z\right), \tag{15}$$

where the initial wavenumber is

$$k_0 = \frac{2\pi n_0}{L}, \tag{16}$$

n_0 being an integer, L the domain size, and A_0 the initial amplitude.

The DNLS modulational instability analysis is performed using the modulation of the constant amplitude wave train expressed by

$$b(z,t) = A_0 \exp\left[i\left(k_0 z - \omega t\right)\right]\left[1 + \varepsilon(z,t)\right], \tag{17}$$

where b and ε are complex periodic functions in the range $[-L/2, L/2]$, $|\varepsilon|^2 \ll 1$ and $\omega = A_0^2 k_0 - k_0^2$. According to the positive sign in the dispersive term of (2), only the left-hand polarized waves are considered ($k_0 > 0$). On the other hand, the right-hand polarized waves ($k_0 < 0$) are unconditionally stable for parallel modulation [30]. These conditions are due to the left-hand polarized waves leading to the resonance of the ion cyclotron mode at higher values of k, while the right-hand polarized branch connects to the whistler mode without resonance [18].

Expanding $\varepsilon(z,t)$ in a Fourier base gives

$$\begin{aligned}\varepsilon(z,t) &= \sum_{j=-\infty}^{\infty} \widehat{\varepsilon}_j(t) \exp\left(i\ell_j z\right), \\ \widehat{\varepsilon}_j(t) &= \widehat{\varepsilon}_j(0) \exp\left(i\lambda_j t\right),\end{aligned} \tag{18}$$

where $\ell_j = 2\pi j/L$ is the modulational wavenumber. Replacing this expression in the DNLS equation reads [36]

$$\lambda_j = i2\left(A_0^2 - k_0\right)\ell_j \pm \left|\ell_j\right| \sqrt{\left(2k_0 - A_0^2\right) A_0^2 - \ell_j^2}. \tag{19}$$

From (19), the instability conditions for the left-hand polarized waves are obtained:

$$\begin{aligned}2k_0 < A_0^2 &\longrightarrow \text{Marginally stable}, \\ 2k_0 > A_0^2 &\longrightarrow \text{Unstable}.\end{aligned} \tag{20}$$

To numerically verify this condition the parameter E_{k_0} is defined, which represents the ratio between the energy carried by the initial wave and the total energy of the system in the Fourier space:

$$E_{k_0} = \frac{\left|\overline{b}_{k_0}\right|}{\sum_{j=1}^{N}\left|\overline{b}_{kj}\right|}, \tag{21}$$

where $\overline{b}_{k_0}$ is the discrete Fourier transform of the initial wave, that is, the amplitude of wave k_0, N is the total number of modes used in the simulation, and k_j indicates the wavenumber. $E_{k_0} = 1$ implies that the energy is concentrated in the initial wave. When the instability is triggered, the energy initially concentrated in k_0 is transferred to other modes and $E_{k_0} < 1$. Figure 1 shows, for different initial wavenumbers, the triggering time of the instability as a function of the initial amplitude A_0.

Figure 1 indicates that the instability condition $A_0 < \sqrt{2k_0}$ is always verified. For larger amplitudes, the stability was confirmed for initial amplitude values twice the limit amplitude and a simulation time of 50000 units. The time step used in the simulations is $\tau = 10^{-3}$, with a grid of $N = 256$ points. These values were established searching for the minimum computational time for which conditions (20) and the stability of the right-hand polarized waves were verified.

This analysis not only allows verifying the stability of the numerical scheme but also permits to study the evolution of the unstable solutions, which is not possible by the theoretical analysis. Figure 2 shows the parameter E_{k_0} as a function of time for different initial wavenumbers k_0 and amplitudes A_0. In addition, the energy distribution is also shown in order to know how the initial energy (curve $|\overline{b}_k|_0$) is transferred when E_{k_0} reaches its minimum value (curve $|\overline{b}_k|_{min}$).

It can be seen that the evolution of E_{k_0} is similar in all cases independent of the initial wavenumber k_0. This evolution can be divided in two different behaviors depending on the initial amplitude A_0. For small initial amplitudes, quasiperiodic solutions are registered in which the initial wave recovers its energy and E_{k_0} has a cyclic evolution between a maximum ($E_{k_0} \approx 1$) and a minimum value. This minimum value is associated with the amplitude A_0, being smaller with shorter periods for larger amplitudes. Concerning the energy transfer, it is observed that it occurs mainly among few modes, the initial wave (mother wave k_0) and the two adjacent modes (daughter waves) with wavenumbers k_1 and k_2 that satisfy the resonance relation $2k_0 = k_1 + k_2$.

For large amplitudes, the regular solution is lost and a spatio-temporal chaotic evolution is produced, which is independent of the initial wavenumber. Unlike the previous case, the initial wave does not recover its initial energy and a random distribution among the different modes occurs.

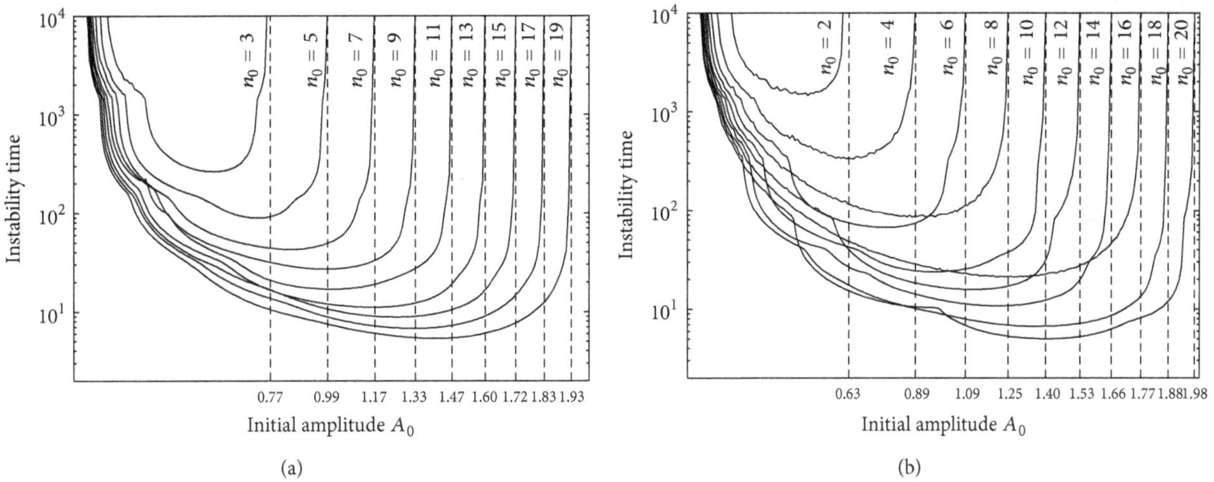

FIGURE 1: Instability time as a function of the initial amplitude A_0 for different initial wavenumbers $k_0 = 2\pi n_0/L$. Dashed lines indicate the stability limits for each configuration, (20). The stable region is bounded by the dashed line and $A_0 \to \infty$ (for clarity, the figure is divided into even and odd n_0 cases).

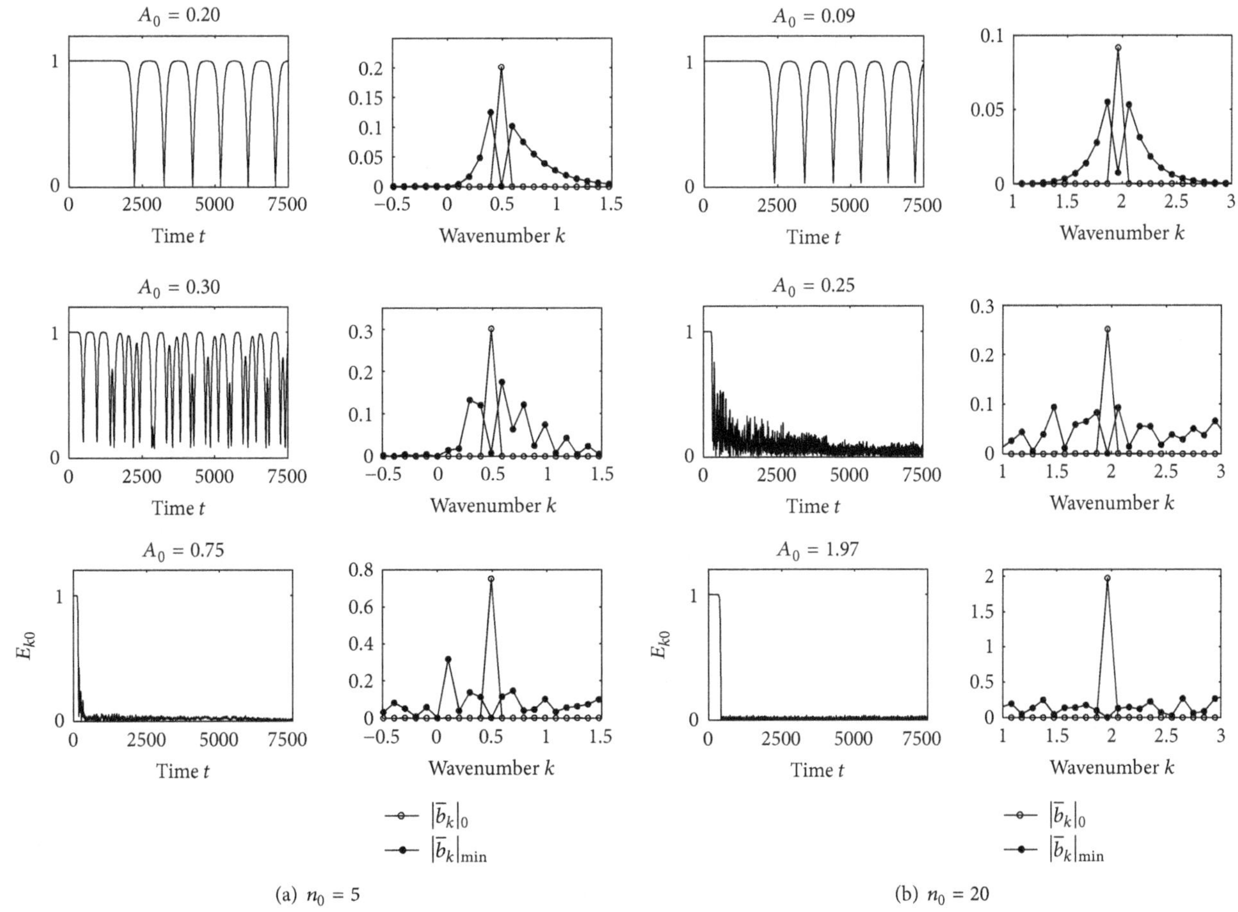

FIGURE 2: Left panels in (a) and (b): E_{k_0} evolution as a function of time for different initial wave amplitudes A_0. Right panels in (a) and (b): energy distribution among the different modes for the initial condition $|\bar{b}_k|_0$ and the corresponding to the minimum E_{k_0} value ($|\bar{b}_k|_{min}$).

The transition between the quasiperiodic and the chaotic behavior occurs for A_0 values where the results are a combination of both solutions. In the transition range, the initial wave partially recovers the initial energy with a cyclic performance as in the first case, but a descendant behavior of the maximum energy occurs until the irregular trend is established.

The initial amplitude values associated with the different behaviors depend on the initial wavenumber. From the figures, it is noted that higher initial modes have periodic behavior ranges smaller with bounded transition intervals. On the contrary, low initial modes have periodic behavior at higher amplitudes and the transition is smoother.

4. Results of the DNLS Equation with Diffusive and Excitatory Effects

In this section the results for the driven diffusive DNLS equation are presented, where a three-wave initial condition under a resonant relation is used. The wave k_0 (mother wave) is linearly unstable and the others are damped. Such simplified system was extensively studied since it can be useful in making a preliminary analysis of Alfvén waves generated by tethers [24, 26, 27]. On the other hand, the system driven by a one unstable mode in the Fourier spectrum was also widely used as a physical analogy of a warm-low-density field-aligned ion beam creating growth along a finite bandwidth of modes through the electromagnetic ion-beam cyclotron instability [21].

The three-wave initial condition is

$$b(z,0) = \sum_{j=0}^{2} a_j \exp\left(ik_j z\right), \tag{22}$$

with wavenumbers $k_1 = (1-\Delta)k_0$ and $k_2 = (1+\Delta)k_0$ that satisfy the resonance relation $2k_0 = k_1 + k_2$ and the periodic boundary conditions, (4).

The numerical results are compared with a three-wave truncation model under resonant interaction. The truncation model is obtained looking for solutions of the DNLS in the form

$$b(z,t) = \sum_{j=0}^{2} a_j \exp\left[i\left(k_j z - \omega_j t + \phi_j\right)\right], \tag{23}$$

involving three modes satisfying the resonance relation and the dispersion relation $\omega_j = \mp k_j^2$. Substituting (23) in the DNLS and only considering terms involving k_0, k_1, and k_2, the truncation equations can be derived to obtain the amplitudes a_0, a_1, and a_2 and the relative phase $\theta = \nu t + \phi_1 + \phi_2 - 2\phi_0$ with $\nu = 2\omega_0 - \omega_1 - \omega_2 = \pm 2\Delta^2$, a frequency mismatch. The upper (lower) sign corresponds to the left-hand (right-hand) polarized waves [26, 27].

The comparison between both methods is done considering the amplitudes of the resonant waves a_0, a_1, and a_2 and the energy in the whole domain

$$E_m = \int_{-L/2}^{L/2} |b|^2 dz. \tag{24}$$

TABLE 1: Range of existence of fixed points as a function of the damping coefficient with $\hat{\gamma}_0 = 0.02$.

Attractor	Damping range	n_1	n_0	n_2
A_8	$\eta \geq 0.04$	1	9	17
A_7	$\eta \geq 0.08$	2	9	16
A_6	$\eta \geq 0.05$	3	9	15
A_5	$\eta \geq 0.04$	4	9	14
A_4	$0.05 \leq \eta \leq 0.15$	5	9	13
A_3	$0.03 \leq \eta \leq 0.13$	6	9	12

In the case of the truncation model, where the only non-zero amplitudes are a_0, a_1, and a_2, the last equation results in:

$$E_m = \int_{-L/2}^{L/2} |b|^2 dz = \left(a_0^2 + a_1^2 + a_2^2\right) L. \tag{25}$$

The analysis is carried out considering attractors with a mother wave with wavenumber given by $n_0 = 9$ and an excitation $\hat{\gamma}_0 = 0.02$. The attractors are numerically obtained by the simulation for certain damping coefficient values depending on the wavenumbers of the daughter waves. It should be noted that in this analysis only long term dynamic solutions are considered; therefore, transient solutions as those due to chaotic saddles are not presented. However, it is mentioned that transient chaos appears in almost all the studied configurations, either in the case of stationary solutions or before the convergence to limit cycles. A very detailed study of this type of nonattracting chaotic sets can be found at [22, 23, 28].

To clarify the analysis, the results are separated into stationary solutions and dynamic solutions as follows.

4.1. Stationary Solutions. When intermediate and strong diffusion levels ($\eta \gtrsim 0.05$) are considered, a set of attractors is obtained which consist of stationary traveling waves with most of the energy concentrated in the three initial modes, such that its amplitudes and the energy can be considered as stable fixed points. Table 1 shows the damping range where each attractor is reached.

To obtain the attractors, The attractors are obtained considering the convergence of the energy E_m and the amplitudes of the resonant modes, a_0, a_1, and a_2. For coefficients η less than the indicated values, in some cases the solutions diverge ($E_m \to \infty$) and the attractor disappears. In other cases, the lower limit is a supercritical Hopf bifurcation that gives rise to periodic solutions which lead to complex attractor, as explained in the next section.

The attractors of Table 1 are classified as stable fixed points; however, this feature depends on the integration time that is considered. For a maximum time less than 5000 units, the attractors can be still assumed as stable fixed points, but for longer times it is found that, independent of the daughter wavenumbers n_1 and n_2, all configurations converge to the attractor A_8 via a jump. This is, the energy of the daughter waves is transferred to the modes given by $n_1 = 1$ and $n_2 = 17$, and the system evolves towards the mentioned attractor. Figure 3 shows the energy evolution for

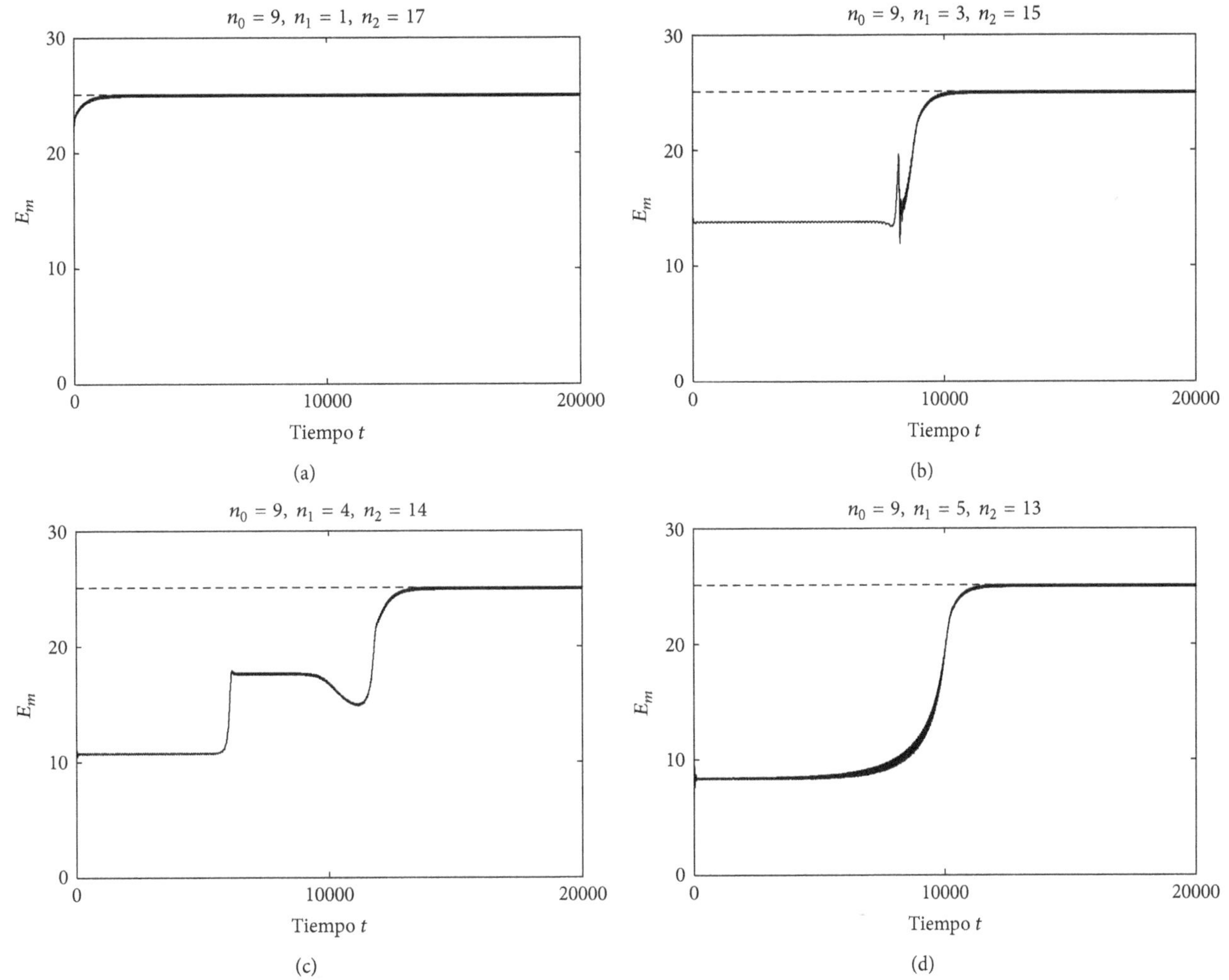

FIGURE 3: Energy evolution E_m for different configurations of resonant initial waves with $n_0 = 9$, $\eta = 0.12$, and $\hat{\gamma}_0 = 0.02$ (in configurations with $n_1 > 4$ and $n_2 < 14$, the convergence to the attractor A_8 occurs with small transitory stability intervals and thus they are not considered in the analysis).

different initial configurations. Note that the convergence to the attractor A_8 occurs from a pseudo-stable position. Similar behaviors can be observed for the evolution the of mother wave amplitude a_0, while for the daughter waves the amplitudes a_1 and a_2 rapidly converge to zero as the energy jumps to the waves of A_8. The convergence of the solutions to this attractor is due to the strong dissipation, for which the system is obligated to put the energy in the unstable mode and in the mode $n_1 = 1$ that has the lower diffusion, according to the resistive damping model. Consequently, the solution can be affected by the boundary conditions [27].

In Figures 4 and 5, the results for attractors A_8 and A_5 are shown. The results of the others attractors are very similar; therefore, the next comments are valid for all attractors presented in Table 1. In the figures, the energy E_m and the amplitudes a_0, a_1, and a_2 are shown as a function of the damping coefficient η, where circles indicate the numerical simulations and the full lines the results of the truncation model. In addition, it also shows the Fourier energy spectrum of the solution, that is, the amplitude of each mode as a function of the wavenumber k, highlighting with dashed lines the resonant waves. In this way, it can be known how the energy is distributed among the different modes when the attractor is formed. It should be noted that, for attractors other than A_8, the comparison between the numerical and the truncation results is only valid at times before the jump to A_8, since the truncation model only considers the three initial waves.

Observing Figures 4 and 5, it can be seen that both methods are in good accordance with most of the damping domain, excepting the region $\eta \lesssim 0.15$, where the results are different for each method. The energy spectrum shows that, in addition to the discrepancy in the evaluation of the amplitude of the daughter waves, the differences between the results of E_m are also due to the emergence of two new modes with nonnegligible energy values. These new modes identified as k_3 and k_4, being $k_3 < k_1 < k_0 < k_2 < k_4$, satisfy the resonance relation $2k_0 = k_1 + k_2 = k_3 + k_4$ and an additional relation $k_4 - k_3 = 2(k_2 - k_1)$. This behavior is registered in all the attractors of Table 1 when intermediate

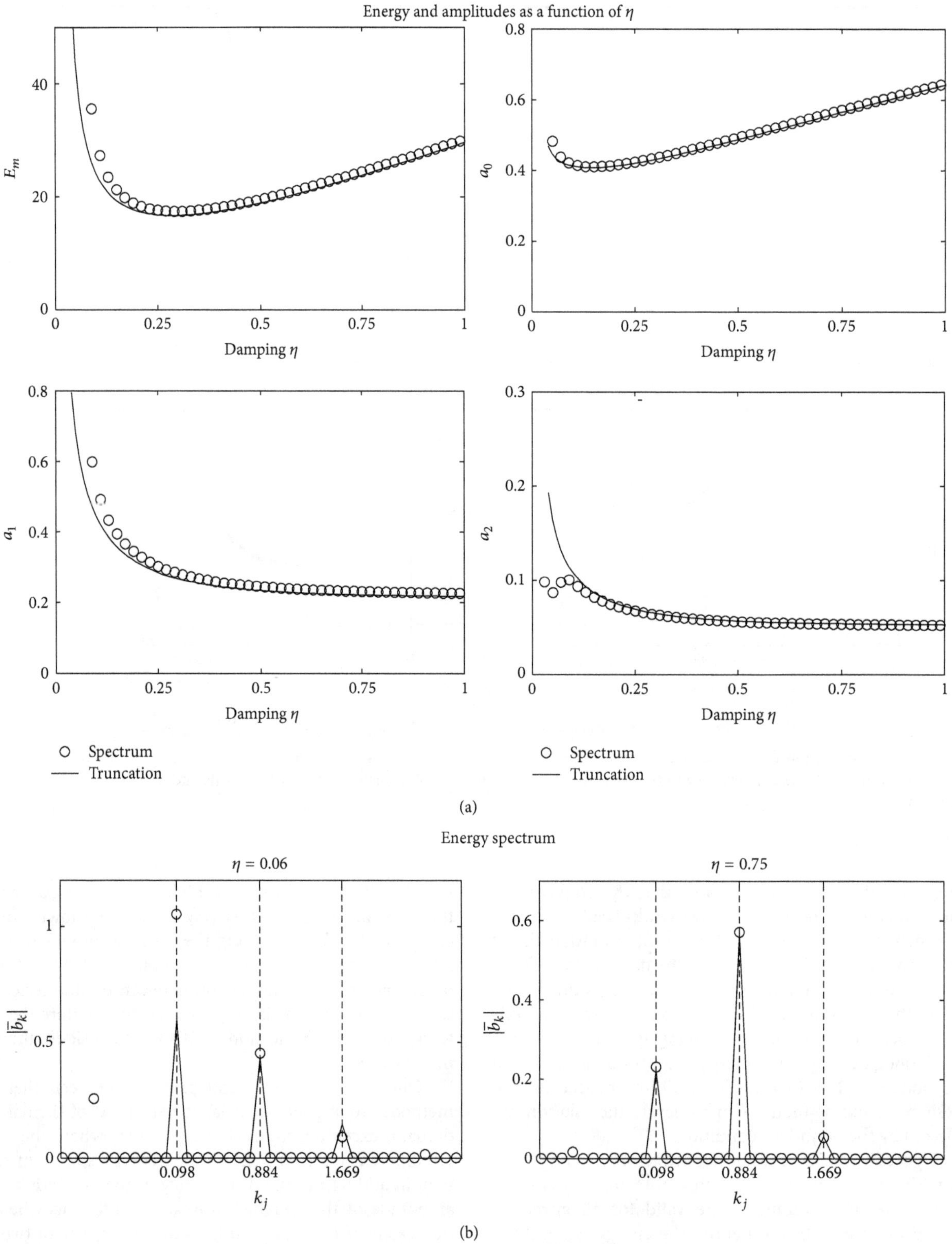

FIGURE 4: Attractor A_8 ($n_0 = 9$, $n_1 = 1$, and $n_2 = 17$ with $\hat{\gamma}_0 = 0.02$).

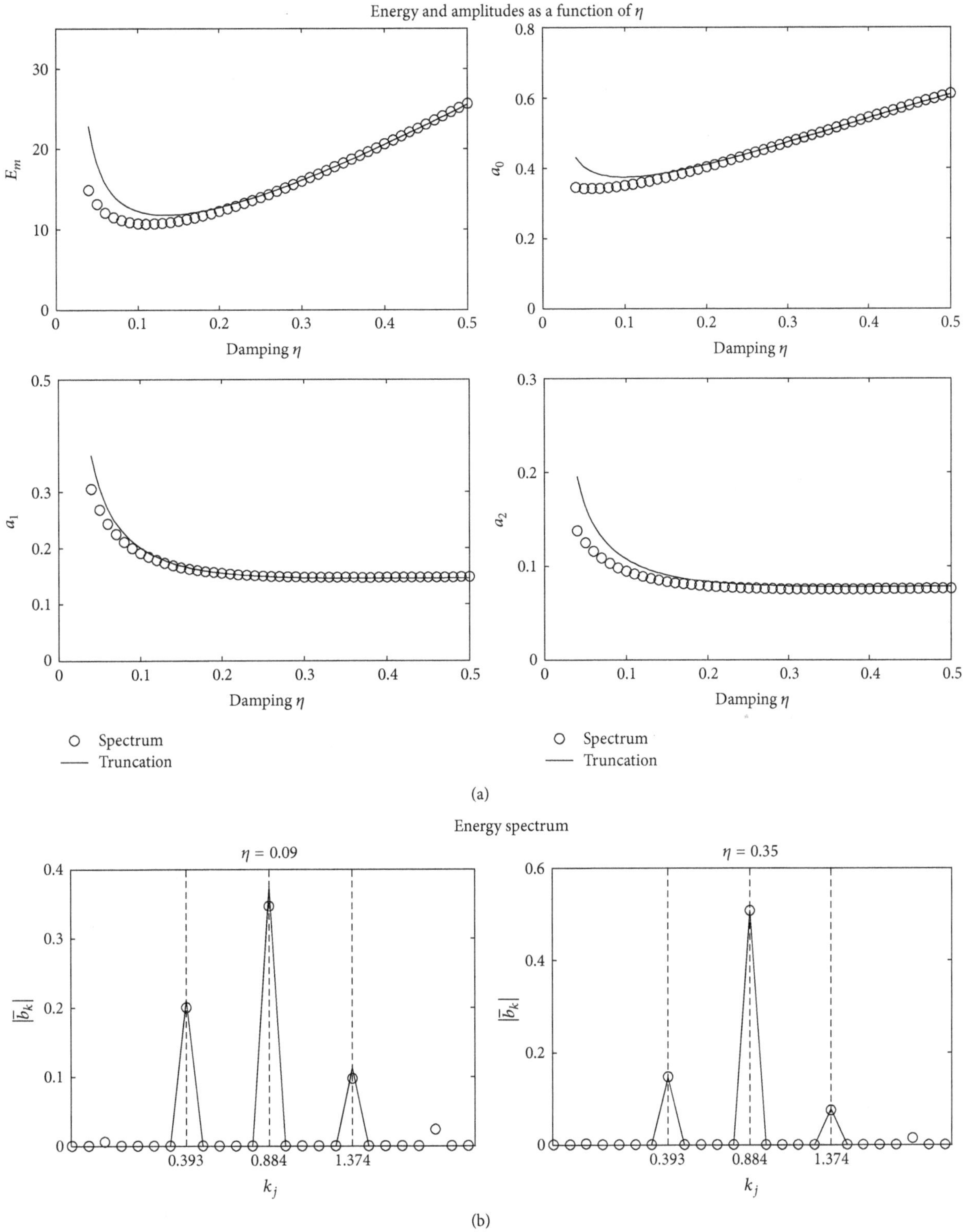

FIGURE 5: Attractor A_5 ($n_0 = 9$, $n_1 = 4$, and $n_2 = 14$ with $\widehat{\gamma}_0 = 0.02$).

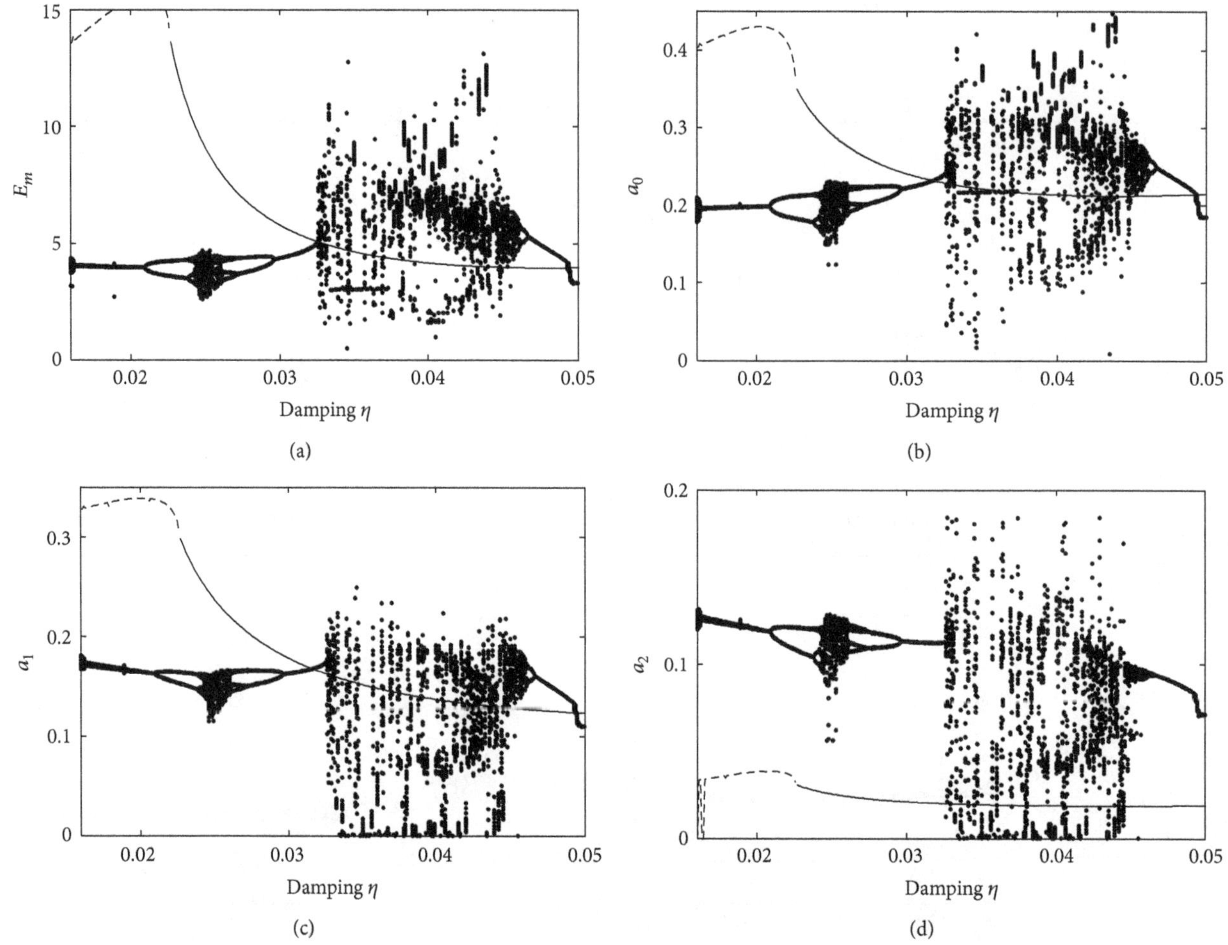

FIGURE 6: Bifurcation diagram of attractor A_2 ($n_0 = 9$, $n_1 = 7$, and $n_2 = 11$ with $\widehat{\gamma}_0 = 0.02$). Points indicate the numerical simulations. The truncation solutions are indicated by the full line (branch of fixed points) and the dashed line (maximum of the periodic solution).

TABLE 2: Range of existence of dynamic solutions as a function of the damping coefficient with $\widehat{\gamma}_0 = 0.02$.

Attractor	Damping range	n_1	n_0	n_2
A_6	$0.0256 \leq \eta \leq 0.05$	3	9	15
A_4	$0.0162 \leq \eta \leq 0.05$	5	9	13
A_2	$0.0163 \leq \eta \leq 0.05$	7	9	11

values of the damping coefficient are used. On the other hand, when larger values of η are considered, there are no new modes with relevant energy levels and the results are similar for both methods.

4.2. Dynamic Solutions. In this section, attractors corresponding to small damping coefficients are presented. In these cases a series of complex solutions is obtained, which are formed by limit cycles, chaotic attractors, and processes of intermittency and crises.

Table 2 shows the damping range for existence of dynamic solutions for the indicated initial wave configurations. These solutions are only captured by the numerical simulations, while the truncation model mainly converges to spurious fixed points. Wave configurations of attractors A_8, A_7, A_5, and A_3 not included in Table 2 have no dynamic solutions, but they diverge for small damping values ($E_m \rightarrow \infty$).

In Figure 6, bifurcation diagrams of the maximums of the energy and the amplitudes are plotted versus the damping coefficient η for attractor A_2. In this way, a one-period limit cycle for a given damping coefficient is identified as a single point in the bifurcation diagram, a two-period limit cycle as two points, and so on, while for a chaotic attractor the solution is broad.

According to the numerical results, the attractor A_2 starts at $\eta_c \approx 0.05$ with a supercritical Hopf bifurcation where the fixed points solutions that rapidly converge to attractor A_8 become stable periodic solutions modulated in time for $\eta \lesssim \eta_c$. In $\eta \approx 0.0467$, there is a period doubling bifurcation followed by a Feigenbaum sequence ending in a chaotic attractor. The periodic solution emerges again at $\eta \approx 0.0325$ through what appears to be an intermittent process. Then a new Feigenbaum sequence continues, which starts at $\eta \approx 0.029$ ending in a new chaotic attractor. By means of another similar sequence, the periodic solution returns to finally culminate in a chaotic attractor that ends the periodic solutions at $\eta \approx 0.016$. Finally, the chaotic attractor is

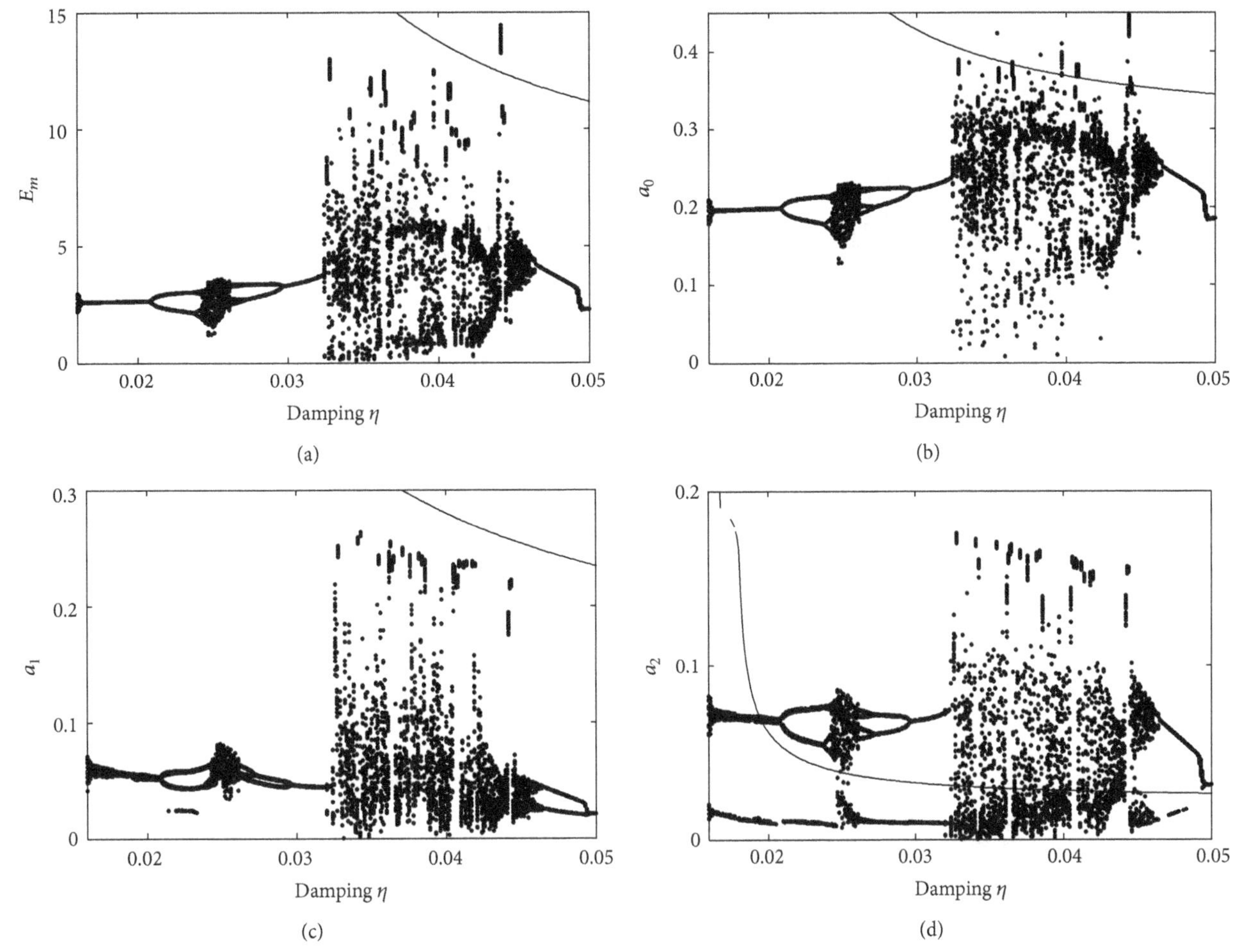

FIGURE 7: Bifurcation diagram of attractor A_4 ($n_0 = 9$, $n_1 = 5$, and $n_2 = 13$ with $\hat{\gamma}_0 = 0.02$).

suddenly destroyed at $\eta \approx 0.0163$ where a boundary crisis appears to take place.

With respect to the solutions of the truncation model, it can be seen that the method does not capture the chaotic attractors, but it predicts the existence of a branch of stable fixed points that ends in a Hopf bifurcation at $\eta \approx 0.023$, where periodic solutions take place.

The discrepancies between both methods are due to the energy transfer, since the energy initially contained in the three resonant waves is transferred to other modes which are not considered in the three-wave truncation model. This feature confirms the results obtained in the analysis of stationary solutions, where it was shown that for small damping new additional modes emerge; although in this case the energy spectrum is broader than only five modes.

Figures 7 and 8 show the bifurcation diagrams for attractors A_4 and A_6, respectively. For A_4 similar results to attractor A_2 can be observed, where the same Feigenbaum sequences and chaotic attractors at similar damping ranges are registered; although in this case the results of the truncation model are rather worse than those obtained for A_2. On the other hand, in the bifurcation diagram of the daughter waves can be observed what appears to be an attractor merging crisis. With respect to attractor A_6, the above comments are valid for damping coefficients $\eta \gtrsim 0.0256$, for which the attractor behaves similarly to attractor A_4. But in this case the boundary crisis take place for $\eta \approx 0.0256$ when the attractor is suddenly destroyed.

To complete the analysis, Figure 9 is presented in order to visualize the intermittent process by which the periodic solution is restored at $\eta_c \approx 0.0325$. In the figure the time evolution of the mother wave is shown, where it can be seen that for a minimum increment of the damping coefficient the periodic solution is replaced by pseudoperiodic evolutions that are interrupted by chaotic explosions, responding to the characteristic shape of the intermittent phenomenon. According to the bifurcation diagrams, a tangent bifurcation appears to take place at η_c, which indicates the presence of type-I Pomeau-Manneville intermittency [37]. In such a phenomenon, the emergence of chaotic solutions is related to the coalescence of two fixed points in a tangent bifurcation that disappear when the control parameter is changed, generating a narrow channel through which the orbits slowly evolve forming the characteristic laminar phases of the intermittency phenomenon. To visualize this situation, Figure 10 shows a one-dimensional Poincaré map built with the values of the peaks of a_0 for $\eta = 0.032507$. This figure shows the presence of a narrow channel between the map and the

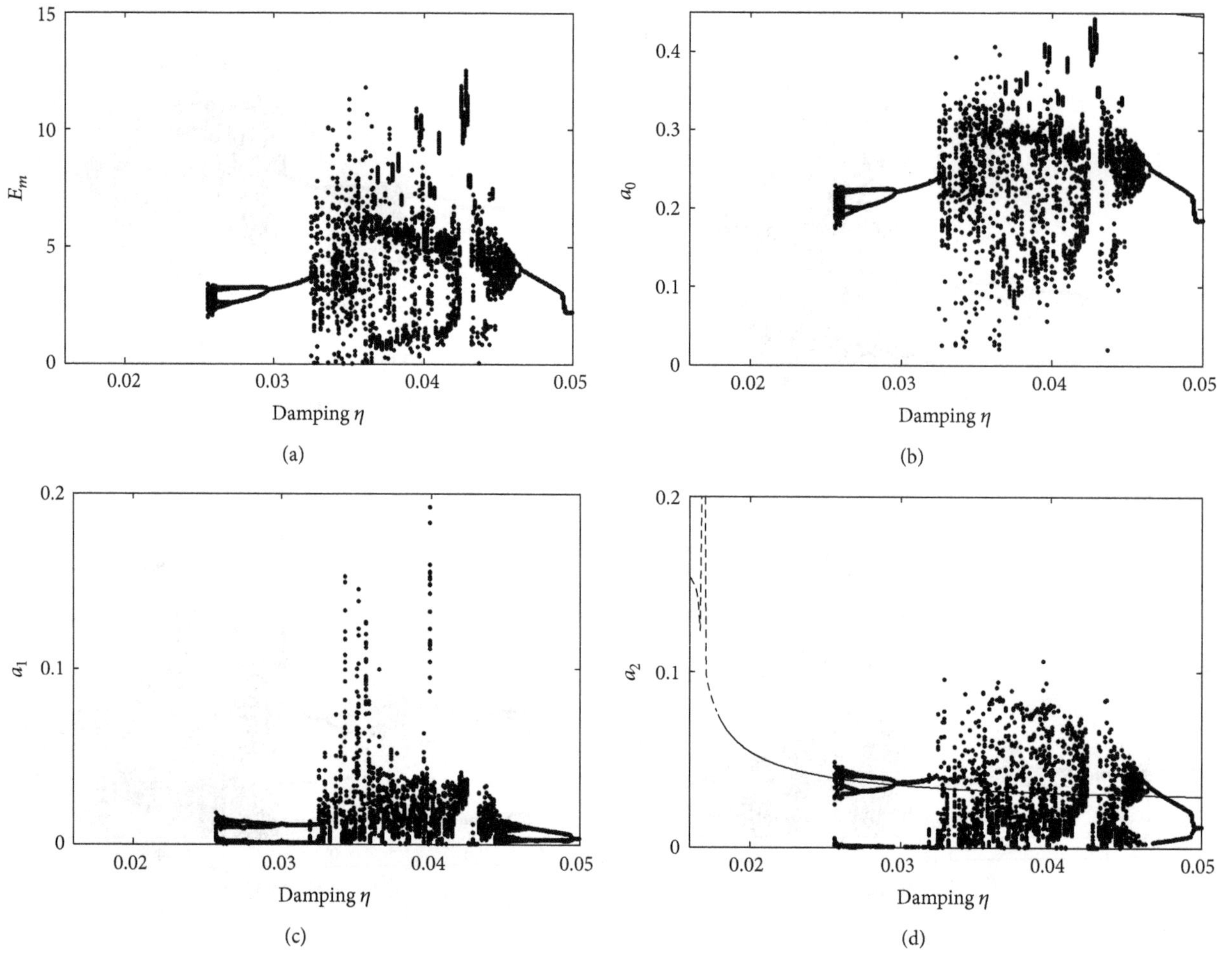

FIGURE 8: Bifurcation diagram of attractor A_6 ($n_0 = 9, n_1 = 3$, and $n_2 = 15$ with $\hat{\gamma}_0 = 0.02$).

bisectrix, evidencing the process of type-I intermittency. A detailed analysis of this phenomenon was recently carried out, where a new methodology to obtain the statistical properties of type-I intermittency with discontinuous reinjection probability density is presented for the solutions of the DNLS equation [38].

5. Conclusions

In this work, the derivative nonlinear Schrödinger (DNLS) equation for periodic boundary conditions has been numerically solved using spectral methods based on Fourier expansions for the spatial derivatives and a fourth order Runge-Kutta scheme for time integration. The left-hand polarized Alfvén waves in a homogeneous uniform plasma with $\beta \approx 0$ were considered in the analysis.

The DNLS equation is extensively used to describe Alfvén waves propagating parallel (or quasiparallel) to the magnetic field direction. Alfvén waves are ubiquitous in space and play an important role in the solar wind, interstellar medium, turbulent heating in magnetosphere, and among others [39, 40]. In these astrophysical processes the Hall effect becomes relevant; thus the magnetohydrodynamics (MHD) equations have to be modified, resulting in a computationally demanding problem. However, for the description of the nonlinear dynamics of large amplitude Alfvén waves traveling along to the background magnetic field, the solution of the full Hall-MHD equations can be replaced by the analysis of the DNLS equation, which represents a large reduction in the computational cost when numerical methods are used. The validity of this idealized model will always depend on the validity of the assumptions carried out during its derivation, in addition to the appropriateness of the use of a fluid-like damping model and a one-wave driver term [8, 18, 20]. Having into account these circumstances, the DNLS equation is widely used to explain many phenomena of astrophysics, such as the Alfvén turbulence [22, 23], the solar wind [41], and the wave-wave interactions [25].

The first analysis of this work was carried out to validate the numerical scheme by testing the analytical conditions of modulational stability in the nondiffusive case with a one-wave initial condition. The stability conditions were adequately reproduced by the code. In addition, the instability time could be calculated and a study of the evolution of the unstable configurations was carried out, features not predicted by previous theoretical analysis. In this way the

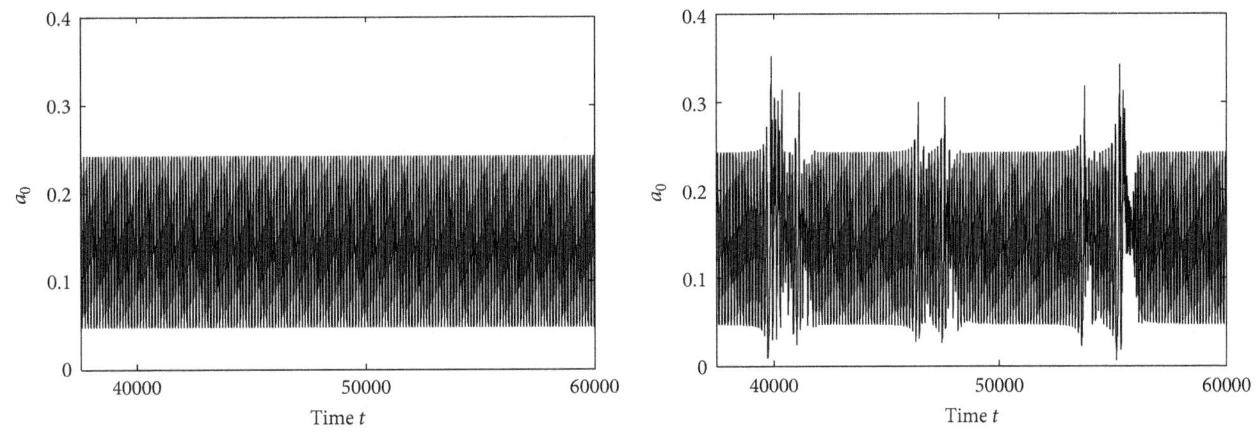

(a) $\eta = 0.032506$ (b) $\eta = 0.032507$

FIGURE 9: Time evolution of the mother wave for attractor A_2. (a) Periodic solution. (b) Chaos.

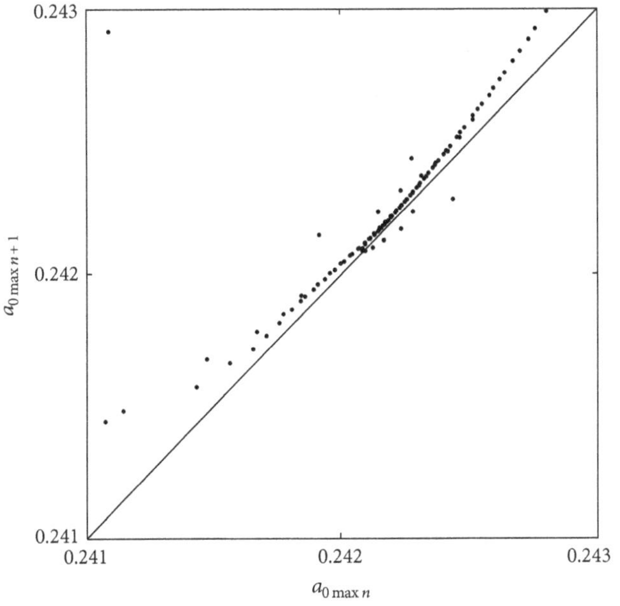

FIGURE 10: Poincaré map of the peaks of a_0 for $\eta = 0.032507$ in the region of the tangent point $a_0 \approx 0.242$ where the phenomenon of type-I intermittency is observed.

numerical results showed that the wavenumber of the initial wave defines the instability time of the system (higher modes destabilize faster), while the amplitude governs the shape of the evolution: small amplitudes evolve as periodic solutions where the initial wave recovers its energy in each cycle, while greater amplitudes produce spatio-temporal chaos where the energy is randomly distributed among all the modes with an irregular evolution.

In the study of the driven dissipative DNLS equation a three-wave initial condition was considered, with the mother wave (k_0) linearly unstable and the remaining waves damped, with the daughter waves (k_1 and k_2) satisfying the resonant condition ($2k_0 = k_1 + k_2$). This analysis showed that the diffusion level of the system defines the kind of solutions. For relatively large damping a set of attractors consisting in pseudo-stable fixed points is found, which converge to the less diffusive wave configuration via a jump for sufficiently long times. Before the jump, each attractor behaves as a conventional fixed point. In these cases, for larger damping levels, the energy remains contained in the three initial modes, while for smaller diffusion it is distributed to two new modes k_3 and k_4 (being $k_3 < k_1 < k_0 < k_2 < k_4$) that always satisfy the resonant condition $2k_0 = k_1 + k_2 = k_3 + k_4$ and are related to the daughter waves by $k_4 - k_3 = 2(k_2 - k_1)$. The emergence of these new modes produces the errors in the results of the three-wave truncation model, which can be considered a good approximation for large damping but fails when the new modes become relevant.

When very small damping coefficients are used, the numerical solutions drastically change resulting in a complex dynamics. The bifurcation diagrams of these solutions exhibit limit cycles, chaotic attractors, and processes of intermittency and crisis with the change of the damping coefficient. On the other hand, the three-wave truncation model cannot capture the dynamic solution and predicts a spurious branch of fixed points instead. This is because the power spectrum for $\eta \rightarrow 0$ is broad and consequently the truncation model is unable to represent the dynamics of the system. As a result, this simplified model cannot be used for small diffusion levels.

These results allow concluding that the numerical scheme proposed in this paper is a good option to solve the driven dissipative DNLS equation, while the three-wave truncation model is a good approach when relatively large damping levels are considered, but for small diffusion the energy transfer does not allow the use of the simplified model.

Finally, according to the results, it is concluded that the numerical method presented in this work is a good option to solve the driven dissipative DNLS equation, which has been properly validated and by which new results were obtained.

Conflict of Interests

The authors declare that there is no conflict of interests regarding the publication of this paper.

Acknowledgments

This work has been supported by the following institutions: CONICET, Universidad Nacional de Córdoba, and Ministerio de Ciencia y Tecnología-Córdoba.

References

[1] P. Shukla and L. Stenflo, "Nonlinear phenomena involving dispersive Alfvén waves," in *Nonllnear MHD Waves and Turbulence*, T. Passot and T. Sulem, Eds., chapter 1, pp. 1–30, Springer, Heidelberg, Germany, 1999.

[2] B. Tsurutani, I. Richardson, R. Thorne et al., "Observation of the right-hand resonant ion beam instability in the distant plasma sheet boundary layer," *Journal of Geophysical Research*, vol. 90, pp. 12159–12172, 1985.

[3] E. Smith, A. Balogh, M. Neugebauer, and D. McComas, "Ulysses observations of Alfvén waves in the southern and northern solar hemispheres," *Journal of Geophysical Research*, vol. 22, no. 23, pp. 3381–3384, 1995.

[4] M. Pettini, L. Nocera, and A. Viulpiani, "Compressible MHD turbulence: an efficient mechanism to heat stellar coronae," in *Chaos in Astrophysics*, J. Buchler, J. Perdang, and E. Spiegel, Eds., pp. 305–316, D. Reidel, Dordrecht, The Netherlands, 1985.

[5] J. Sanmartin, M. Charro, J. Pelaez et al., "Floating bare tether as upper atmosphere probe," *Journal of Geophysical Research A: Space Physics*, vol. 111, no. 11, Article ID A11310, 2006.

[6] E. Ahedo and J. R. Sanmartín, "Analysis of bare-tether systems for deorbiting low-earth-orbit satellites," *Journal of Spacecraft and Rockets*, vol. 39, no. 2, pp. 198–205, 2002.

[7] J. R. Sanmartín, "Alfvén wave far field from steady-current tethers," *Journal of Geophysical Research A: Space Physics*, vol. 102, no. 7, Article ID 97JA00346, pp. 14625–14630, 1997.

[8] N. Cramer, *The Physics of Alfvén Waves*, Wiley- Verlag VCH, Berlin, Germany, 1st edition, 2001.

[9] K. Morton, "Finite amplitude compression waves in collision-free plasmas," *Physics of Fluids*, vol. 7, pp. 1800–1815, 1964.

[10] H. Kever and G. K. Morikawa, "Korteweg-de Vries equation for nonlinear hydromagnetic waves in a warm collision-free plasma," *Physics of Fluids*, vol. 12, no. 10, pp. 2090–2093, 1969.

[11] T. Kakutani, H. Ono, T. Taniuti, and C. Wei, "Reductive perturbation method in nonlinear wave propagation II. Application to hydromagnetic waves in cold plasma," *Journal of the Physical Society of Japan*, vol. 24, no. 5, pp. 1159–1166, 1968.

[12] A. Rogister, "Parallel propagation of nonlinear low-frequency waves in high-β plasma," *Physics of Fluids*, vol. 14, no. 12, pp. 2733–2739, 1971.

[13] E. Mjolhus, "On the modulation instability of hydromagnetic waves and the DNLS equation," *Journal of Plasma Physics*, vol. 16, no. 3, pp. 321–334, 1976.

[14] K. Mio, T. Ogino, K. Minami, and S. Takeda, "Modified nonlinear Schrödinger equation for Alfvén waves propagating along the magnetic field in cold plasmas," *Journal of the Physical Society of Japan*, vol. 41, no. 1, pp. 265–271, 1976.

[15] S. R. Spangler and J. P. Sheerin, "Properties of Alfvén solitons in a finite-beta plasma," *Journal of Plasma Physics*, vol. 27, no. 2, pp. 193–198, 1982.

[16] M. S. Ruderman, "DNLS equation for large-amplitude solitons propagating in an arbitrary direction in a high-β Hall plasma," *Journal of Plasma Physics*, vol. 67, no. 4, pp. 271–276, 2002.

[17] B. T. Tsurutani, E. J. Smith, A. L. Brinca, R. M. Thorne, and H. Matsumoto, "Properties of whistler mode wave packets at the leading edge of steepened magnetosonic waves: comet Giacobini-Zinner," *Planetary and Space Science*, vol. 37, no. 2, pp. 167–182, 1989.

[18] E. Mjolhus and T. Hada, "Nonlinear phenomena involving dispersive Alfvén waves," in *Nonllnear Waves and Chaos In Space Plasmas*, T. Hada and H. Matsumoto, Eds., chapter 4, pp. 121–169, Terra Scientific, Tokio, Japan, 1997.

[19] V. Belashov and S. Vladimirov, *Solitary Waves in Dispersive Complex Media*, Springer, Berlin, Germany, 2005.

[20] T. Hada, C. F. Kennel, B. Buti, and E. Mjølhus, "Chaos in driven Alfvén systems," *Physics of Fluids B*, vol. 2, no. 11, pp. 2581–2590, 1990.

[21] S. Ghosh and K. Papadopoulos, "The onset of Alfvén turbulence," *Physics of Fluids*, vol. 30, no. 5, pp. 1371–1387, 1987, http://scitation.aip.org/content/aip/journal/pofl/30/5/10.1063/1.866252.

[22] A. Chian, F. A. Borotto, and W. D. Gonzalez, "Alfvén intermittent turbulence driven by temporal chaos," *Astrophysical Journal Letters*, vol. 505, no. 2, pp. 993–998, 1998.

[23] E. Rempel and A. Chian, "Space plasma dynamics: Alfvén intermittent chaos," *Advances in Space Research*, vol. 35, no. 5, pp. 951–960, 2005.

[24] J. Sanmartin, O. Lopez-Rebollal, E. Del Rio, and S. Elaskar, "Hard transition to chaotic dynamics in Alfvén wave fronts," *Physics of Plasmas*, vol. 11, no. 5, pp. 2026–2035, 2004.

[25] R. Miranda, E. Rempel, A. Chian, and F. A. Borotto, "Intermittent chaos in nonlinear wave-wave interactions in space plasmas," *Journal of Atmospheric and Solar-Terrestrial Physics*, vol. 67, no. 17-18, pp. 1852–1858, 2005.

[26] G. Sánchez-Arriaga, J. R. Sanmartin, and S. A. Elaskar, "Damping models in the truncated derivative nonlinear Schrödinger equation," *Physics of Plasmas*, vol. 14, no. 8, Article ID 082108, 2007.

[27] G. Sánchez-Arriaga, T. Hada, and Y. Nariyuki, "The truncation model of the derivative nonlinear Schrödinger equation," *Physics of Plasmas*, vol. 16, no. 4, Article ID 042302, 2009.

[28] R. Miranda, E. Rempel, and A. Chian, "Chaotic saddles in nonlinear modulational interactions in a plasma," *Physics of Plasmas*, vol. 19, Article ID 112303, 2012.

[29] S. Spangler, J. Sheerin, and G. Payne, "A numerical study of nonlinear Alfvén waves and solitons," *Physics of Fluids*, vol. 28, no. 1, pp. 104–109, 1985.

[30] B. Buti, M. Velli, P. C. Liewer, B. E. Goldstein, and T. Hada, "Hybrid simulations of collapse of Alfvénic wave packets," *Physics of Plasmas*, vol. 7, no. 10, pp. 3998–4003, 2000.

[31] Y. Nariyuki and T. Hada, "Self-generation of phase coherence in parallel Alfvén turbulence," *Earth, Planets and Space*, vol. 57, no. 12, pp. e9–e12, 2005.

[32] S. Orzag, "Numerical simulation of incompressible ows within simple boundaries. Galerkin spectral representation," *Studies in Applied Mathematics*, vol. 50, pp. 293–327, 1971.

[33] G. Drazin and R. Johnson, *Solitons: An Introduction*, Cambridge University Press, New York, NY, USA, 1989.

[34] S. Jardin, *Computational Methods in Plasma Physics*, Chapman & Hall, CRC Press, New York, NY, USA, 2010.

[35] J. Boyd, *Chebishev and Fourier Spectral Methods*, Dover Publications, New York, NY, USA, 2001.

[36] T. Flå, "A numerical energy conserving method for the DNLS equation," *Journal of Computational Physics*, vol. 101, no. 1, pp. 71–79, 1992.

[37] Y. Pomeau and P. Manneville, "Intermittent transition to turbulence in dissipative dynamical systems," *Communications in Mathematical Physics*, vol. 74, no. 2, pp. 189–197, 1980.

[38] G. Krause, S. Elaskar, and E. del Rio, "Type-I intermittency with discontinuous reinjection probability density in a truncation model of the derivative nonlinear Schrodinger equation," *Nonlinear Dynamics*, 2014.

[39] T. K. Suzuki and S. Inutsuka, "Solar winds driven by nonlinear low-frequency Alfvén waves from the photosphere: parametric study for fast/slow winds and disappearance of solar winds," *Journal of Geophysical Research A: Space Physics*, vol. 111, no. 6, Article ID A06101, 2006.

[40] X. Blanco-Cano, N. Omidi, and C. T. Russell, "Macrostructure of collisionless bow shocks: 2. ULF waves in the foreshock and magnetosheath," *Journal of Geophysical Research A: Space Physics*, vol. 111, no. 10, Article ID A10205, 2006.

[41] Y. Nariyuki and T. Hada, "Density uctuations induced by modulational instability of parallel propagating Alfvén waves with $\beta \sim 1$," *Journal of the Physical Society of Japan*, vol. 76, Article ID 074901, 2007.

Permissions

All chapters in this book were first published in JAS, by Hindawi Publishing Corporation; hereby published with permission under the Creative Commons Attribution License or equivalent. Every chapter published in this book has been scrutinized by our experts. Their significance has been extensively debated. The topics covered herein carry significant findings which will fuel the growth of the discipline. They may even be implemented as practical applications or may be referred to as a beginning point for another development.

The contributors of this book come from diverse backgrounds, making this book a truly international effort. This book will bring forth new frontiers with its revolutionizing research information and detailed analysis of the nascent developments around the world.

We would like to thank all the contributing authors for lending their expertise to make the book truly unique. They have played a crucial role in the development of this book. Without their invaluable contributions this book wouldn't have been possible. They have made vital efforts to compile up to date information on the varied aspects of this subject to make this book a valuable addition to the collection of many professionals and students.

This book was conceptualized with the vision of imparting up-to-date information and advanced data in this field. To ensure the same, a matchless editorial board was set up. Every individual on the board went through rigorous rounds of assessment to prove their worth. After which they invested a large part of their time researching and compiling the most relevant data for our readers.

The editorial board has been involved in producing this book since its inception. They have spent rigorous hours researching and exploring the diverse topics which have resulted in the successful publishing of this book. They have passed on their knowledge of decades through this book. To expedite this challenging task, the publisher supported the team at every step. A small team of assistant editors was also appointed to further simplify the editing procedure and attain best results for the readers.

Apart from the editorial board, the designing team has also invested a significant amount of their time in understanding the subject and creating the most relevant covers. They scrutinized every image to scout for the most suitable representation of the subject and create an appropriate cover for the book.

The publishing team has been an ardent support to the editorial, designing and production team. Their endless efforts to recruit the best for this project, has resulted in the accomplishment of this book. They are a veteran in the field of academics and their pool of knowledge is as vast as their experience in printing. Their expertise and guidance has proved useful at every step. Their uncompromising quality standards have made this book an exceptional effort. Their encouragement from time to time has been an inspiration for everyone.

The publisher and the editorial board hope that this book will prove to be a valuable piece of knowledge for researchers, students, practitioners and scholars across the globe.

List of Contributors

V. Vasanth and S. Umapathy
School of Physics, Madurai Kamaraj University, Madurai 625 021, India

E. E. Benevolenskaya
Pulkovo Astronomical Observatory, Pulkovskoe sh. 65, Saint Petersburg 196140, Russia
Saint Petersburg State University, Saint Petersburg 198504, Russia

I. G. Kostuchenko
Karpov Institute of Physical Chemistry, Ul. Vorontsovo Pole 10, Moscow 105064, Russia

A. A. Kirillov and E. P. Savelova
Dubna International University of Nature, Society and Man, Universitetskaya Street 19,Dubna 141980, Russia

Gagik Ter-Kazarian
Division of Theoretical Astrophysics, Ambartsumian Byurakan Astrophysical Observatory, Byurakan, 378433 Aragatsotn, Armenia

Alla V. Suvorova and Alexei V. Dmitriev
Institute of Space Science, National Central University, No 300 Jungda Road, Jhongli, Taoyuan 32001, Taiwan
Skobeltsyn Institute of Nuclear Physics, Lomonosov Moscow State University, Moscow 119234, Russia

Chien-Ming Huang
Institute of Space Science, National Central University, No 300 Jungda Road, Jhongli, Taoyuan 32001, Taiwan

M. R. Hossen, L. Nahar and A. A.Mamun
Department of Physics, Jahangirnagar University, Savar, Dhaka 1342, Bangladesh

Nayem Sk
Department of Physics, University of Kalyani, Nadia 741235, India

Abhik Kumar Sanyal
Department of Physics, Jangipur College, Murshidabad 742213, India

Riou Nakamura and Masa-aki Hashimoto
Department of Physics, Graduate School of Sciences, Kyushu University, 6-10-1 Hakozaki, Higashi-ku, Fukuoka 812-8581, Japan

Shin-ichiro Fujimoto
Department of Control and Information Systems Engineering, Kumamoto National College of Technology, 2659-2 Suya, Koshi, Kumamoto 861-1102, Japan

Katsuhiko Sato
Institute for the Physics and Mathematics of the Universe, University of Tokyo, Kashiwa, Chiba 277-8568, Japan
National Institutes of Natural Sciences, Kamiyacho Central Place 2F, 4-3-13 Toranomon, Minato-ku, Tokyo 104-0001, Japan

Basudev Ghosh
Department of Physics, Jadavpur University, Kolkata 700 032, India

Sreyasi Banerjee
Department of Electronics, Vidyasagar College, Kolkata 700 006, India

Hans J. Haubold
Office for Outer Space Affairs, United Nations, Vienna International Centre, P.O. Box 500, 1400 Vienna, Austria
Centre for Mathematical Sciences Pala Campus, Arunapuram P.O., Palai, Kerala 686 574, India

Dilip Kumar
Centre for Mathematical Sciences Pala Campus, Arunapuram P.O., Palai, Kerala 686 574, India

S. Thirukkanesh
Department of Mathematics, Eastern University, 30350 Chenkalady, Sri Lanka

Kingsley Chukwudi Okpala
University of Nigeria, Nsukka 410002, Nigeria

Noble P. Abraham, Sijo Sebastian, G. Sreekala, R. Jayapal and Venugopal Chandu
School of Pure & Applied Physics, Mahatma Gandhi University, Priyadarshini Hills, Kottayam, Kerala 686 560, India

C. P. Anilkumar
Equatorial Geophysical Research Laboratory, Indian Institute of Geomagnetism, Krishnapuram, Tirunelveli, Tamil Nadu 627 011, India

Masa-aki Hashimoto, Reiko Kuromizu and Masaomi Ono
Department of Physics, Kyushu University, Fukuoka 810-8560, Japan

Tsuneo Noda
Kurume Institute of Technology, Fukuoka 830-0052, Japan

Masayuki Y. Fujimoto
Department of Physics, Hokkaido University, Sapporo 060-8810, Japan

Jagadish Singh
Department of Mathematics, Faculty of Science, Ahmadu Bello University Zaria, PMB 2222, Samaru-Zaria, Kaduna, Nigeria

Oni Leke
Department of Mathematics, College of Science, University of Agriculture, PMB 2373, North-Bank, Makurdi, Nigeria

Arthur G. Nikoghossian
Ambartsumian Byurakan Astrophysical Observatory, Byurakan, 378433 Aragatsotn, Armenia

M. Khalid, Mariam Sultana and Faheem Zaidi
Department of Mathematical Sciences, Federal Urdu University of Arts, Sciences & Technology, Karachi 75300, Pakistan

Motoaki Saruwatari, Masa-aki Hashimoto and Ryohei Fukuda
Department of Physics, Kyushu University, Hakozaki, Fukuoka 812-8581, Japan

Shin-ichiro Fujimoto
Department of Electronic Control, Kumamoto National College of Technology, Kumamoto 861-1102, Japan

M. M. Elkhateeb, M. I. Nouh and E. Elkholy
Astronomy Department, National Research Institute of Astronomy and Geophysics, Helwan, Cairo 11421, Egypt
Physics Department, College of Science, Northern Border University, Arar 1321, Saudi Arabia

B. Korany
Astronomy Department, National Research Institute of Astronomy and Geophysics, Helwan, Cairo 11421, Egypt
Physics Department, Faculty of Applied Science, Umm Al-Qura University, Makkah 715, Saudi Arabia

A. A. Al-Rubaiee
Department of Physics, College of Science, The University of Mustansiriyah, 10052 Baghdad, Iraq
Institute of Nano Electronic Engineering, University of Malaysia Perlis, 01000 Kangar, Malaysia

Y. Al-Douri and U. Hashim
Institute of Nano Electronic Engineering, University of Malaysia Perlis, 01000 Kangar, Malaysia

Gustavo Krause and Sergio Elaskar
Department of Aeronautics, Faculty of Exact, Physical and Natural Sciences, National University of Córdoba, Vélez Sarsfield 1611, X5016GCA Córdoba, Argentina
National Council of Scientific and Technical Research (CONICET), Avenue Rivadavia 1917, C1033AAJ Buenos Aires, Argentina

Andrea Costa
National Council of Scientific and Technical Research (CONICET), Avenue Rivadavia 1917, C1033AAJ Buenos Aires, Argentina
Institute of Theoretical and Experimental Astronomy (IATE-CONICET), Laprida 854, X5000BGR Córdoba, Argentina

www.ingramcontent.com/pod-product-compliance
Lightning Source LLC
Chambersburg PA
CBHW080625200326
41458CB00013B/4506